T0338452

For Profit and Prosperity

For Profit and Prosperity

The Contribution made by Dutch Engineers to Public Works in Indonesia

1800 - 2000

Wim Ravesteijn and Jan Kop
editors

APRILIS - ZALTBOMMEL

KITLV Press - LEIDEN

The Dutch Prince Bernhard Culture Fund and the Dutch Ministry of Foreign Affairs made the realization of this publication possible through their financial support.

Other supporting organizations include: DHV, Dutch Foundation for the History of Technology, Nederlandse Bruggen Stichting, Netherlands Water Partnership, PWN Waterleidingbedrijf Noord-Holland, Stichting Het Lamminga Fonds, Water Research Centre Delft, Waterleidingmaatschappij Drenthe, Witteveen+Bos.

Cover and text layout: Foxy Design – Zaltbommel, the Netherlands

APRILIS ISBN: 978 90 5994 221 9
KITLV Press ISBN: 978 90 6718 323 9

NUR 688

APRILIS
P.O. Box 141
5300 AC Zaltbommel
The Netherlands
T 00.31.418.512088
F 00.31.418.684908
info@aprilis.nl
www.aprilis.nl

KITLV Press
Reuvensplaats 2
P.O. Box 9515
2300 RA Leiden
The Netherlands
T 00.31.71.527 24 65
F 00.31.71.527 26 38
info@kitlv.nl
www.kitlv.nl

CONTENTS

FOREWORD

Lasting testimony

Is the colonial past of the Netherlands flawed? This is a question that continues to be raised in different ways. Clearly the Netherlands has not yet finished dealing with its colonies. Can the police campaigns be justified? To what extent were the native inhabitants exploited? Did the colonial policy adopted by the Netherlands contribute to the big famines? To what extent did the Netherlands line its own pockets with colonial revenue? When delving into the colonial past these are the kinds of moral questions that often arise. Sometimes they are the main themes underlying research whilst at other times they are themes that are foisted upon the researcher. The reality is that historical research never will provide all the answers for the simple reason that moral questions are not historical issues. However, before moral judgement can be cast, historical analyses need to be made.

Moral motives also played a part in the realisation of this book on public works in the Dutch East Indies and Indonesia. Ironically, though, it was not any kind of consciousness of guilt that provided the incentive but rather an awareness of the true greatness of all the works created by Dutch engineers. They designed irrigation systems, drew up plans for harbours, constructed railway lines, built bridges, provided public drinking water supplies and developed cities. The legacy of all their efforts can still be seen in present-day Indonesia. Such works almost automatically give rise to feelings of pride and enthusiasm but one must remain level-headed. The editors of this book correctly surmise that this publication is not so much about casting moral judgement as about providing a lasting testimony of all the work that was carried out.

In this book that 'lasting testimony' is expressed in a variety of ways. In some chapters the emphasis is on the actual technological projects: the design side, the construction methods, the use of materials, the practical problems faced and the creative solutions found. Yet other chapters provide the more administrative and political frameworks by concentrating on planning, decision-making, legislation, organisations and institutes. In that way, the various projects come to be framed in the political and nation-building process. Finally, there are also chapters that set out to expound on the technological developments. In so doing, the broader socio-economic, political, cultural and technical colonial context is analysed. That context is then linked to prevailing technological practice, to the initiatives of actors, to the technological alternative options that arose and to the way in which those facets influenced the technological development selection process.

This whole range of approaches shows us just what Dutch colonial technology historians have to offer. It also defines the present situation as far as colonial technological history is concerned. In the Netherlands, the history of technology is a young field. It is a mere three decades old. Within that field the attention paid to the colonies is even more recent, the first doctoral studies in that direction both having emerged in 1997 with the completion of the respective research of Wim Ravesteijn and Margaret Leidelmeijer into Javanese irrigation and the Javanese sugar industry. In some respects it is therefore courageous to produce such an overview of public works in the archipelago by compiling this book *For Profit and Prosperity* so soon. The editors are, however, aware of the inherent limitations and recognise that this book will certainly not constitute the final word on the subject. Their 'lasting testimony' can thus be taken to mean that this book also aims to inspire historians for many years to come to continue to subject the works and ideals of Dutch engineers of the past to further research.

HARRY LINTSEN

Harry Lintsen (1949) gained his doctorate in 1980 after having studied technological physics at Eindhoven University of Technology. His dissertation was on the engineering profession in the Netherlands in the nineteenth century. As a follow-up to his dissertation he wrote the book *Ingenieur van beroep* [Engineer by Profession] (1985). In 1990 he was made history of technology professor both at Delft University of Technology and at Eindhoven University of Technology. Between 1988 and 1995 he was chief editor of a project that culminated in a series of six books entitled *Geschiedenis van de techniek in Nederland. De wording van een moderne samenleving 1800-1890* [The history of technology in the Netherlands. The creation of a modern society 1800-1890] (1992-1995). Between 1994 and 2003 he chaired the editorial team of the seven-part series entitled *Techniek in Nederland in de twintigste eeuw* [Technology in the Netherlands in the twentieth century] (1998-2003). In 2005 Lintsen published the book *Made in Holland. Een techniekgeschiedenis van Nederland, 1800-*2000 [Made in Holland. A history of technology in the Netherlands, 1800-2000]. In 2006 he was one of the editors of the book *Gedreven door nieuwsgierigheid. Een selectie uit vijftig jaar TU/e onderzoek* [Driven by curiosity. A selection from fifty years of TU/e research]. He is currently participating in a research project into 'the Netherlands as a knowledge society'.

PREFACE

A development project avant la lettre

The idea for this book started with a letter dated 15th October 1998 from Professor J.H. Kop that was addressed to Professor H.W. Lintsen, instigator of the academic history of technology studies in the Netherlands. In that letter Jan Kop requested that some attention be paid to the history of public works in the Dutch East Indies. It was a field that had always interested him as I was later to learn from a student who informed me in an e-mail that his former professor, Jan Kop, had always done much to promote that side of historical matters. One of the reasons for Jan Kop's intense interest in the field is the fact that he spent part of his youth in Indonesia at a time when the country was still in Dutch hands.

The son of an agricultural-chemical engineer, Jan Kop was born in Djatiroto (East Java) in 1930. He was educated in Sumatra and in the Netherlands where, in 1957, he graduated from the Technical College in Delft as a civil engineer specialised in irrigation and hydropower. His first job was as an irrigation and drainage engineer on the Ganges-Kobadak project in Bangladesh under Professor W.J. van Blommestein. He pursued his irrigation work in Tunisia on the Lower Medjerda Development Project. After that he was made head of the Architecture, Construction, Hydraulic Engineering and Sanitary Engineering department of the Grontmij NV consulting firm in the Netherlands. In that final capacity he was responsible for the creation (and also in part for the execution) of a master plan for drainage and flood control in Jakarta. He was also engaged in various international projects, especially in the field of urban drainage and flood control, including in various New Parisian Towns (St. Quentin en Yvelines, Marne la Vallée), Abidjan (Ile de Petit Bassam), Manila and Bangkok. He then went on to become head of the Planning Bureau of the Dutch Water Company Association (VEWIN) and after that a full professor in the field of sanitary engineering – in particular in relation to public drinking water supplies – at Delft University of Technology until his retirement in 1994. However, Jan Kop remained interested and involved in Indonesian civil engineering, also from the historical angle. In 1998 he gave a presentation on public works in the Dutch East Indies and in Indonesia on the occasion of a special Hydraulic Engineering day. A copy of that speech was enclosed in the above-mentioned letter and it was to form the basis for this book.

That Hydraulic Engineering day had been arranged in conjunction with the two-hundred-year anniversary of the Dutch Public Works agency (i.e. Rijkswaterstaat). It was an anniversary that drew much attention and which was even to become the immediate inspiration for a major research project led by Harry Lintsen within the History of Technology Department in the Faculty of Technology, Policy and Management at Delft University of Technology. One of the results of that project was the commemoration publication entitled *Twee eeuwen Rijkswaterstaat* (Two centuries of Public Works agency) by A. Bosch and W. van der Ham (Zaltbommel 1998). At the time, Harry Lintsen was professor in both Delft and Eindhoven (since 1st January 2003 he has worked full-time in Eindhoven). It was the above-mentioned book that was really to definitely establish public works history as the specialist research field for his department in Delft.

In Delft the history of public works was also being studied in relation to the former Dutch colonies. When Jan Kop's letter arrived in our department I had just completed my doctoral research into the development of modern irrigation in Java and was busy setting up a sequel study. In short, Jan Kop's letter fell on fertile ground which meant that after several meetings we were ready to go ahead and give shape to the relevant research project under the supervision of Harry Lintsen. The research proposal did not take very long to write but finding the required financial backing proved to be a harder task. In the end, Delft University proved willing to finance a substantial part of the project and further subsidies were obtained from the Dutch Ministry of Foreign Affairs and the Dutch Ministry of Education, Culture and Science, from the Dutch Public Works agency, from the Bandung Technical College Fund and from the Dutch Lamminga Fund (via a subsidy provided for the doctoral research being conducted by Maurits Ertsen). We are most grateful to all these different organisations for their support.

The whole research project soon became a central activity within the History of Technology Department. Alongside of Harry Lintsen (supervision and foreword) there were other departmental members who also got involved such as Dr. Adrienne van den Bogaard (who assisted with Chapter 7), the engineer Dr. Maurits Ertsen (Chapters 6 and 10), Louise van Gemert, M.A. (illustrations editor and editor of the epilogue), Dr. Marie-Louise ten Horn (Chapter 2) and Frida de Jong, M.A. (Chapter 1 and annotator). Other researchers and colleagues also showed interest in the project which, in some cases, came to form a part of their respective doctoral research projects. Professor Charles Vos who sadly died suddenly in 2001 played an important intervening role. It was thanks to him that we were able to obtain the cooperation of the Dutch Bridges Association. Both the Institut Teknologi in Bandung and the Petra Christian University in Surabaya were prepared to contribute to the book. In that way we ended up with an international research team involving some 30 different researchers, the Dutch branch of which met up a number of times and was coordinated by Delft University. Ultimately all the different research projects culminated in the written contributions to the present book. Some of these contributions constitute rounded off studies while yet others represent phases in larger research projects that are still in progress. Whilst all the various chapters have been attuned to each other and made to fit into a clear overall structure, we have taken care to ensure that they can also be read as autonomous texts.

This book is essentially a translation of its preceding Dutch version, entitled *Bouwen in de archipel. Burgerlijke openbare werken in Nederlands-Indië en Indonesië 1800-2000* (Building in the archipelago. Civil public works in the Dutch East Indies and Indonesia 1800-2000 – Zutphen 2004). We have added an extra appendix containing the comments of the civil engineer H.J. Schoemaker, the last professor of irrigation at Delft University of Technology, who was educated in the true tradition of irrigation and hydraulic engineering in the Dutch East Indies and who gained there experience in the field. Some minor improvements and updates concerning references have also been made. The publication of the Dutch edition was financed by the Prince Bernhard Culture Fund, the Unger van Brero Association Fund and the J.E. Jurriaanse Association. For the present book, we are grateful, again, to the Prince

Bernard Culture Fund and to the Ministry of Foreign Affairs, both of which supplied us with the necessary means. In particular, we thank Diane Butterman who did all the translation work and spent many hours making our study available to an international public. We are very pleased with the result and take full responsibility for any irregularities that escaped our notice. A special word of thanks goes to Dick van der Spek for his excellent maps.

The Dutch East Indian Public Works department played just as great a part in the formation of the Dutch East Indies as *Rijkswaterstaat* in the formation of the Netherlands. More so than in the Netherlands at the same period, engineers played a vital part in the development of the East Indies. Following in the footsteps of the publications of Professor J.A.A. van Doorn on this subject we can, in retrospect, definitely assert that between 1800 and 1950 the East Indies was one huge technological development project conducted by the Netherlands. The Dutch did indeed achieve great things in the East Indian colony, even if sometimes big things also went wrong and even though it was Dutch interests that always remained central. This book gives an overview of the results of that project, at least as far as it relates to public works. The various chapters remind one of the endeavours of Dutch East Indian engineers in the different fields to which they applied themselves whilst at the same time focusing on the contribution made by European civil engineering to the formation of present-day Indonesia, where many existing works still date from the colonial era. Relevant examples of public works that have been created in Indonesia since the time of independence which either radiate a distinctly Dutch style or have actually come about through Dutch intervention, are also considered in this book which is why the period covered encompasses two centuries.

In the words of Jan Kop, we are extremely conscious of the fact that as an overview this publication remains 'incomplete, imperfect and insufficient'. Much has remained unresearched and a considerable amount of the available material could not be included. We feel privileged that three of the authors were in the process of gaining their doctorates which means that at least in their particular areas (harbours, irrigation and urban development) the research has been and will be continued. We sincerely hope that in certain of the other fields follow-up research projects will be undertaken (especially where public health works and transport are concerned). We shall not hesitate both to initiate research in these fields and to support it whenever it is undertaken.

WIM RAVESTEIJN

The road leading to Garoet on the south coast of West Java, approximately 1925.

INTRODUCTION **COVETOUSNESS AND VOCATION**

The infrastructure of Dutch colonialism in Indonesia

JAN KOP | WIM RAVESTEIJN

Indonesia's 'official' gaining of independence in 1949 really did amount to a dream come true. In the twentieth century the nationalist movement had come to treasure that dream. For activists like Soekarno, Shahrir and Hatta independence certainly constituted part of 'the cause' but centuries before them there had been Javanese and Sumatran rulers who would have liked to turn the archipelago – or at least a large part of it – into a kingdom, even though their imperialistic efforts remained limited. Paradoxically one of the things that accelerated Indonesian independence was European expansionism, notably in the form of Dutch colonialism and imperialism.[1] As the 'Dutch East Indies' (in short, the East Indies[2]), Indonesia had been a Dutch colony where first the United East India Company (Verenigde Oost-Indische Compagnie, VOC) and then the Dutch state had held sway. One important factor in all of this had been technology. Indeed it was technology that made it possible for the Dutch to travel to the Indonesian archipelago, to repress the population and, as of 1800, to create a colonial state which, after the Second World War, could easily be taken over by the Indonesians. What is central to this study is the technologies that formed the East Indies and Indonesia as a state, notably the road building and hydraulic works facets but also the harbours and railway links that were created and all the urban developments that took place.

The technological foundations of modern Indonesia were laid by Dutch or Dutch East Indian engineers. It all happened at the time when the Indonesian archipelago was ruled by the Dutch, officially between 1798 when the possessions of the bankrupt United East India Company fell to the Dutch state and 1949 when the Netherlands finally recognised the country's earlier proclaimed independence and Indonesia became a sovereign state. In retrospect it might be concluded that for a period of 150 years Indonesia was a development project of the Dutch state, a project which – over the course of time – would gradually turn into a technological or engineering project in which the year 1885 could be seen as the turning point.[3] That was the year when Dutch East Indian Public Works rulings, giving engineers extensive powers, were adopted. The results, less impressive than colonial historiographers would have had us believe[4], did remain achievements in their own rights. When, in 1942, the Japanese invaded the East Indies, which incidentally is an area at least as big as Europe from the point of view of its geographical extensiveness, the following had been achieved in terms of civil public works:

- 1.5 million hectares of agricultural land had been devoted to 'technical irrigation'; in Java and Madoera 1.3 million hectares of the 3.3 million hectares of paddy fields had been provided with modern irrigation (two-fifths of the area of the Netherlands);
- there was about a 5,500 km long network of railway lines in Java and at least 2,000 km of tracks in Sumatra (the Netherlands had roughly 3,300 km of rails in 1940);
- a road network comprising 12,000 km of asphalted surface, 41,000 km of metalled road area and 16,000 km of unimproved surfaces (the total of 69,000 km is equivalent to 1.7 times the earth's circumference);
- harbours for international shipping in places such as Medan, Batavia and Soerabaja;
- 140 public – particularly urban – drinking water facilities, largely on the island of Java.

These public works, many of which either in a 'rehabilitated' state or not are still widely in use, formed and still form the material substrate of the Dutch East Indian and Indonesian state. This list is merely a preview: the present book provides an extensive impression (though not without various constraints that will be elucidated) of the results of Dutch overseas intervention in technological administrative affairs, how these results were achieved and why, in terms of the effort involved, people went to such lengths.

Obviously there were other factors that made the united Indonesian state possible. It was not only technologically but also socio-politically that the republic was formed with the help of Dutch colonial efforts. The Dutch stimulated the use of Malaysian as the common language.[5] In present-day Indonesia it is indeed Malaysian that has turned into the national language 'Bahasa Indonesia' and has further evolved. Under colonial rule, Indonesia had also become a cohesive state in other ways, for instance as regards its legislation, economic activities and education. In addition to this, the Indonesian nationalistic movement was obviously of pre-eminent importance, even though it was indirectly stimulated by the Dutch Ethical Policy that had been directed at protecting the interests of the population and at giving education an impetus.[6] We can safely assert that without all these various infrastructural networks Indonesia as we know it today would not have come into being. This bold hypothesis that will be supported at length in this book confirms the broader hypothesis of present-day academic history of technology studies to the effect that the development of modern technology, particularly in the form of 'large technological systems', including 'infrasystems', has superseded the influential roles of geography and politics when it comes to the matter of the structuring of society.[7]

This does not, however, mean to say that all the civil engineering developments in colonial Indonesia were responsible for defining Indonesian society, colonial or otherwise. In modern history of technology studies, technology and society are seen as one seamless fabric and technological development as a social process. For Indonesia it is, for instance, difficult to separate the process of civil technological development from the process of colonial state formation. In this introduction we shall examine our approach to technological development (in the light of two fields: history of technology and technology dynamics) and the themes and objectives of this book but first we shall give an overview of the amazing connectedness of the Netherlands and Indonesia.

The Netherlands and Indonesia: a world of difference

Today Indonesia has a population of more than 220 million, thus making it the fifth biggest country in the world in terms of population size. More than half of the population lives on the island of Java which is no more than four times as big as the Netherlands. With these calculations it must be borne in mind that because of the large number of volcanoes on that island only three-quarters of the area is habitable and in that habitable area the average population density is around 1170 inhabitants to the km^2; the densely populated Dutch state has around 470 inhabitants to the km^2. The other half of the population is dispersed over the other thousands of relatively thinly populated smaller and larger islands (Sumatera/Sumatra, Kalimantan/Borneo, Sulawesi/Celebes, Papua/Irian Jaya/New Guinea), with a total land area of 1,794,443 km^2 where there are numerous huge unbroken tracts

Colonial technology in Indonesia: sawa landscape with a railway bridge; the Tasikmalaja-Bandjar, Preanger area (West Java) railway line, c.1930.

of virgin forestland. Java is becoming more and more industrialised whilst the number of large metropolises with fast growing populations is rapidly increasing. The Outer Regions or Outer Islands, that is to say, the areas or islands outside Java and Madura/Madoera still reflect, by contrast, a rural scene with vast expanses of agricultural land and cultivated areas (like in Deli) broken only by 'pockets of urbanisation', cities like, for instance, Medan with its millions of inhabitants and massive industrial complexes in the form of its oil refineries (Palembang, Lhokseumaweh) and the aluminium forge situated in Kuala Tanjung.

If one studies and compares the present geographic and demographic proportions between the two countries, the one-time colonisation of Indonesia by the Dutch appears remarkable, to say the very least. Indonesia has a total land area of 1,919,943 km², whilst the Netherlands is less than 2 percent of that size with its 33,811 km². The Netherlands has 16 million inhabitants, which is roughly 7 percent of the more than 220 million people inhabiting the former colony.[8]

However, the picture becomes less sharply contrastive if we go back in time and compare the population situation in the period when, as a colony, Indonesia was part of the Kingdom of the Netherlands, that is to say, in the period between 1798 and 1945/49. In 1815 the Dutch East Indies had around 6 million inhabitants, 4,5 million of whom lived in Java and Madoera. At that time there were 2,5 million Dutch people. By the end of the colonial era the difference in population numbers was greater. The figures for 1930 show that the East Indies had roughly 60 million as opposed to 8 million inhabitants in the Netherlands itself. Even then, with their approximately 41 million people, the Javanese formed the majority by a long way; in time that would change. In time the population difference between the Netherlands and the East Indies diverged and that trend only continued after independence (see Table 1).

There is also another developmental trend to illustrate how the Netherlands and Indonesia continued to diverge after independence. Around 1940 Indonesia was predominantly rural and at that time, some 60 to 70 years ago, there were no huge metropolises in the country, not even on the densely populated island of Java. Jakarta, now a world city with approximately 12 million inhabitants, had a population of less than half a million in 1930, some 31,000 of which comprised Europeans. Similarly, all the other Javan cities which today have population figures that run into the millions, were much more modestly populated throughout the colonial era (see Table 2).

Table 1. **Population totals in the Dutch East Indies/Indonesia and the Netherlands***

Year	Java and Madoera	Dutch East Indies/ Indonesia	The Netherlands
1815	4,5 million	± 6 million	2,5 million
1845	9,4 million	± 14 million	3,0 million
1880	12,5 million	± 18 million	4,0 million
1900	28,4 million	± 42 million	5,1 million
1920	34,4 million	± 50 million	6,9 million
1930**	40,9 million	59,1 million	7,9 million
2000	> 100 million	> 200 million	16,0 million

* The reported population totals for Java and Madoera and for the Dutch East Indies/Indonesia only apply to the native population/Indonesians. In 1930 the number of Europeans (or people of similar status) totalled 193,618 for Java and Madoera and 242,372 for the entire Dutch East Indies; the respective numbers of Foreign Easterners was 635,662 and 1.344,878. See Lekkerkerker, *Land en volk van Java* and PERPAMSI, Direktori 2000.

**The year of the last official population census in the Dutch East Indies. See Gonggryp's, *Geïllustreerde encyclopaedie van Nederlandsch-Indië*.

Sources: The Dutch Central Statistical Office; Lekkerkerker, *Land en volk van Java; De Ingenieur in Nederlands-Indië; PERPAMSI, Direktori 2000;* Boomgaard and Gooszen, *Population trends 1795 – 1942.*

Table 2. **The larger cities of Java**

City	2000 Totaal number of inhabitants	1930 Total number of inhabitants	Europeans
Jakarta/Batavia city*	c. 12 million	437,433	31,340 (7.2%)
Surabaya/Soerabaja	c. 4 million	342,000	26,000 (7.6%)
Bandung/Bandoeng	c. 3 million	166,722	19,664 (11.8%)
Semarang	c. 2 million	218,000	12,600 (5.8%)
Cirebon/Cheribon	c. 1 million	52,000	1,500 (2.9%)

* Around the year 2000 the total population of Greater Jakarta (including Tangerang, Bogor and Bekasi) amounted to over 20 million.

Sources: Lekkerkerker, *Land en volk van Java*; PERPAMSI, *Direktori 2000*.

Still, it does indeed remain a fascinating fact that such a small country as the Netherlands was able to turn the Indonesian archipelago with its more than 13,000 islands into colonial 'property' and more than that. The VOC was chartered and set up in 1602 as a semi-private trade-oriented concern. What Jan Pieterszoon Coen (1587-1629, governor-general of the Dutch East Indies 1619-1623 and 1627-1629) required for that purpose was a bridgehead which was why he founded the settlement of Batavia on the spot where 'Jakatra', seized and ravaged by him, had once lain. The capitulation of the Dutch East Indies for the Japanese occupying forces in 1942 thus in effect brought to an end nothing less than a colonial state or perhaps, more precisely, a colonial empire.[9] In the space of 350 years a single trading post had evolved into an immense group of islands 'Insulinde' which, in very many places, was administered by the Netherlands in every single detail.

In the years that the Netherlands and Indonesia were so inextricably interconnected, the Netherlands was slowly emerging as a modern and wealthy trading and industrial nation and, in the process, it was very much profiting from the East Indies. The Dutch Golden Age (of the seventeenth century) was largely founded on Baltic-based grain trading but the VOC played an important part as well. In the nineteenth century, the Netherlands derived much income from the Cultivation System (1830-1870) period in which Javanese farmers were compelled to grow cash crops. A portion of the profit was pumped into the funding of canal and railway construction in the 'mother country'. Yet later, in the age of 'free trade imperialism', the East Indies especially became such an important niche for the marketing of industrial products that it came to be seen as the cork upon which the Netherlands was being kept afloat. Not surprisingly therefore many feared, around the time of independence, that the loss of Indonesia might well equate with disaster. Fortunately for the Netherlands, things turned out better than some had feared.[10]

What were the consequences of these colonial links for Indonesia? This question requires a specific deconstruction. On the one hand it must be observed that the links between the Netherlands and Indonesia did not constitute a win-win situation: while the Netherlands became relatively wealthy Indonesia became (or remained) relatively poor. After independence Indonesia did, it is true, witness rapid economic development partly from the oil reserves already extracted within the archipelago in the colonial era and partly thanks to all the infrastructures, as detailed in this book, constructed in those times. Nevertheless it has not been able to rid itself of the 'developing country' label. On the other hand, as has been sketched, Indonesia largely owed its existence to the Netherlands. The country had not only been infrastructurally developed by the Netherlands but also, even if to a limited extent, industrially developed as well. Whilst in the past things had clearly been different there were, in the Ethical Policy period, net private and government investment funds flowing into Indonesia.

Precisely how the Netherlands was able to impose rule in the Dutch East Indian archipelago, how generally 'the West' was able to dominate 'the rest' of the world and how it was possible for the West to thrive whilst the South became or remained poor are interesting questions pertaining to world history. An interesting and authoritative analysis of the situation is to be found in the *The Wealth and Poverty of Nations: Why Some are so Rich and Some are so Poor* by the celebrated American economic and technology historian, David Landes. This book deals with the whys and wherefores of the European shift towards world hegemony. Landes points to a range of factors, geographical included (and even climatic variables), but places the emphasis upon economic, political, technological and scientific, and ultimately, cultural aspects: the free enterprise style production machine for the capitalist market, the open society of liberalism, political pluriformity and competition at super-national level, the 'invention of invention', the hard work ethic of Calvinism and the Judaic-Christian tradition with its respect for handiwork, the conviction that nature is there to serve man and, finally, the linear concept of time. Under the revealing title 'For love of gain' there is also a chapter that is devoted to the Dutch East Indies. For the time being we shall leave these and other all-embracing considerations for what they are[11] and confine ourselves to the sketching of several high points in the shared history of the Netherlands and Indonesia. In order to gain a better understanding of Dutch colonialism in Indonesia it is essential to realise that it was not only a colonial relationship that was developing between the two countries but that the two countries were also busy individually creating themselves over the course of time.

How the Netherlands became a coloniser

It all started in 1596 when the seaman Cornelis de Houtman sailed into the Indonesian archipelago in supreme command of his four ships that had been stocked by Amsterdam merchants and then proceeded to draw up a treaty with a Javanese ruler. It was a time when the Mediterranean Sea was no longer the focal point of world trade and sailors and merchants – following in the footsteps of men like Columbus who had found his way to America in 1492 and Vasco da Gama who had navigated his way to Asia in 1498 – were busy 'discovering' the world and making money through trade. At the time, the Netherlands was merely a loose confederation of regions or provinces dominated by Holland: the Republic of

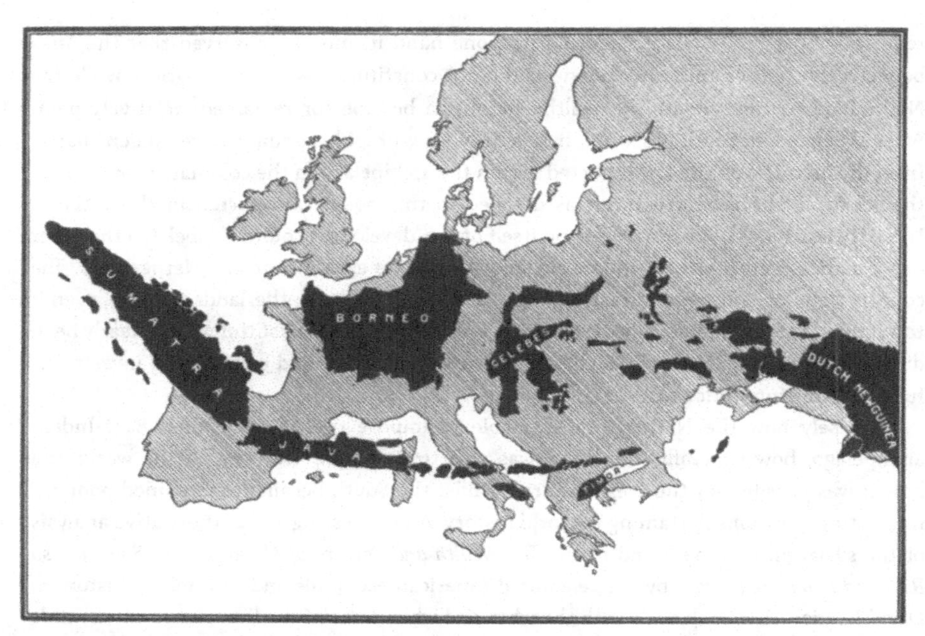

In the East Indies people thought 'big'. This superimposing of Indonesia onto a map of Europe illustrates such thought patterns in a very geographical way.

the United Provinces. The States General was the general administrative body but it only had control over foreign political affairs and defence.

In 1602, the United East India Company (VOC) was established at the instigation of Johan van Oldenbarnevelt, Advocate of Holland and the Stadholder, Prince Maurits and at the behest of the States General. In that way a stop was finally put to all the random companies, the forerunners which had up until then existed. The idea was that what was required in the interests of the Netherlands was cooperation and not competition and fragmentation. It was also a time when the Netherlands was caught up in its Eighty Years' War and the struggle to wrest itself from Spanish dominion. The VOC, the first public limited company in the world, did business that was beneficial to the Netherlands, not only with the East Indies but also with – and between – all other kinds of regions in South and Southeast Asia. Empowered by the States General, the VOC also administered the areas which, in the interests of trade, were being occupied and influenced and it was Batavia, founded by Jan Pieterszoon Coen that was to serve as the bridgehead. The widespread success of the VOC rested in its striving for monopolies and, in the case of: nutmeg, cloves and mace from the Spice Islands (the Banda Islands and the Moluccas) everything worked quite well, at first. Coupled with the political aspirations of the larger and smaller European countries this form of colonialism led to wars. After things started going from bad to worse for the VOC in the eighteenth century, the enterprise finally went bankrupt in 1796 as a direct result of a war between Britain and the Netherlands (the Fourth Anglo-Dutch War, 1780-1784) and the protecting of personal interests including corruption within the company's own ranks. When the VOC was disbanded a couple of years later the Dutch state absorbed all its debts and possessions and established a satellite state in the East Indies.[12]

Meanwhile the French Revolution of 1789 had just taken place and Napoleon was caught up in desperate attempts to export the product of that same revolution: a 'democratic' central state. The country was thus granted its Batavian Republic in 1795. In 1806 and under Louis Bonaparte it was declared the Kingdom of Holland only to be later annexed, in 1810, by France. After the defeat of Napoleon, Europe slowly gave way to a wave of restoration and in 1815 the Netherlands (temporarily fused with Belgium) emerged as a kingdom in its own right under the House of Orange. The independence of the Dutch nation was made possible and safeguarded by the bigger European powers of France, Germany and Britain. Much the same went for the Dutch colonial realm. The East Indies which the British had occupied in 1811 was returned to the Dutch in 1816.

In the nineteenth century the national government decided to organise its colonial affairs. From 1808 until 1842 there was a Ministry for the Colonies which operated in association with other ministries. As of 1845 there was a separate Ministry for the Colonies that was to last just over 100 years. Until 1848 supreme administrative power over the East Indies was 'the exclusive prerogative of the King'. After 1848 the cooperation of the Dutch legislator (parliament) was made possible and an annual *Koloniaal Verslag* (Colonial Report) was produced for the benefit of parliament. As of the seventies (in the nineteenth century) that report formed a permanent appendix to the *Handelingen der Staten-Generaal* (Proceedings of the States General). At the end of the First World War, the annual publication of this report was abolished.

State formation in the Dutch East Indies

By around 1600 there were a number of large kingdoms in the Indonesian archipelago, the most successful of which extended over – or embraced – several islands. The VOC trade and its intensification led, in the Dutch East Indies, to an administration that started to take care of order, lawfulness and peace in ever more areas and which paved the way for the public Dutch 'national' government which, as of 1798, laid down the law in the archipelago. Though the takeover was abrupt, the transition from a trading company to a colonial state was gradual. The state that had been established in the East Indies[13] underwent a slow transformation in which the early colonial state made way for a modern colonial state. In this connection three distinct phases may be identified (the dates are just used to give rough guidelines)[14]:

1. The early colonial state (1800-1870). During that period the institution of colonial state was established and the Netherlands expanded to become a proper colonial empire. The relevant authorities maintained a power political grip aimed at conquering and demarcating the area that had been claimed and then established there an effective administration, both in the interests of contacts with other colonial powers and contacts with the local population. Within the area brought under colonial rule, the striving was for 'peace and order' under the local population and the exploitation of the land and its

people (the Cultivation System 1830-1870 and the Credit-Surplus Policy until 1901: the colony had to yield profit). A simple administrative body was developed that was headed by the Dutch East Indian government (the 'gouvernement') and given the backbone of the domestic administration (the Inland Administration, the Dutch Civil Service).

2 The modernising colonial state (1870-1920). The early colonial state came to an end when private initiative was given the freedom to expand and thus also 'free trade imperialism' (i.e. free trade especially within the framework of a colonial empire). Those in power supported private capital by creating infrastructural works and developing a welfare policy (or Ethical Policy after 1901) for the mixed colonial population. The Dutch state consolidated its sphere of influence and the colonial administrative machinery was expanded. Other processes that became apparent at that time were: the increasing relative independence of the East Indies in relation to the Netherlands, the decentralisation of the colonial administration and the 'emancipation' of the Dutch East Indian population.

3 The modern colonial state (1920-1945/49). The state, persisting with its Ethical Policy strategy, achieved a certain degree of autonomy in relation to the mother country and the East Indian population became involved in the administration side of matters.

The modern colonial state was superseded by the independent Indonesian state. This could therefore be termed the fourth and final phase of the process of Dutch East Indian/ Indonesian state formation.

This process of state formation can be elucidated in three ways, starting with the pure power aspect.[15] Round about the year 1800 the Dutch did not have total control of the Indonesian archipelago by a long way; the Dutch East Indian state might have been established but from the point of view of its scope and the managing of the land and the population, all of that still had to be realised. Dutch administration had only penetrated into part of the area and it existed alongside the separate parallel authority of sultans and tribal chiefs. The sphere in influence was thus limited and partial. During the course of the nineteenth century, Dutch presence and power in the archipelago gradually increased, though that still differed greatly from place to place. In this respect, too, the year 1870 may be seen as something of a turning point.

Before 1870 the Dutch colonisation of the East Indies was really confined to Java and Madoera and to several islands in the Moluccas. That was a total area not much more than five to six times as big as the Netherlands that had a population of approximately five million, which was not much more than double the population of the mother country. These, then, were the proportions underlying the management, control and domination of the East Indies by the Netherlands. In the Outer Regions where, unlike in Java, much of the population was illiterate, there was hardly any evidence of Dutch dominion. At that time – pre 1860 – there was no tobacco cultivation in Sumatra, Deli was 'empty' and the island was virtually completely covered in natural forest growth. Head hunting was still practiced in Borneo and there was cannibalism in the Batak regions of North Sumatra and in New Guinea.

When it came to the matter of Dutch penetration and the exercising of administration, chiefly in the inland areas of Java, the Cultivation System (1830-1870) was of vital importance. The Cultivation System forced the civil servants engaged in Inland Administration to intervene directly with the population, partly through the direct taxation (i.e. land rents) that superseded the taxes enforced by the indigenous rulers. The opening up (and simultaneous establishing) of rule progressed at a slow pace. When, for instance, around 1835 the famous natural scientist, F.W. Junghuhn, explored the volcanoes of Java, much of the island was still forested. He made use of bearers and lit fires at night to ward off wild animals (tigers, rhinoceroses).[16] Even after 1870 Java still had its areas of wilderness: according to accounts of the day, whilst the railway line between Batavia and Bandoeng was under construction four tigers had to be put down.

In the period leading up to 1870 Dutch hegemony in the archipelago – apart from in Java, Madoera and the Moluccas – was certainly not definite. In the Treaty of London (1814) Great Britain resolved to relinquish its possessions in Sumatra – Benkoelen, Tapanoeli and Natal (South Tapanoeli) – in exchange for Dutch recognition of British control of Malacca, Singapore included, and the Dutch guarantee that Atjeh would remain independent. In 1839-1840 the English explorer, James Brooke, helped the sultan of Sarawak in his battle against the Dajaks who were assisted by the sultan of Brunei. He claimed an area of Borneo as big as the Netherlands, in 1841, and then promptly handed over sovereignty of the region to Britain. He was rewarded with the title 'Radjah (king) of Sarawak'. In 1846 he obtained the 'pit coal island' Laboean. In 1847 the sultanate of Brunei also fell under British rule.

The Netherlands did not have its hands free until 1871 when, after the signing of a new treaty with Great Britain, a start could be made on combating the piracy in the Straits of Malacca by the people of Atjeh who operated under the leadership of the sultan of Kota Radja (Banda Aceh). Even though the signing of an agreement with the sultans of Langkat, Deli, Serdang and Ashan in 1858 had meant that the area between Atjeh and Riouw fell de jure under Dutch rule, the effects of this agreement were minimal until 1884. Partly as a result of the introduction of tobacco cultivation to Deli by the pioneer Jacobus Nienhuys, the residence (or region) of the East Coast of Sumatra, could be established in 1873. Similarly, it was not until 1887 that Medan became the dwelling-place of the resident (the head of the region) and the de facto headquarters of Sumatra's East Coast. In reality the East Indies did not actually become unified, in the sense of becoming one nation – albeit a colonial nation – until the time of J.B. van Heutsz (1851-1924), governor-general from October 1904 until December 1909. During his time in office every local monarch in the entire archipelago was forced to bow to Dutch supremacy, often in bloody conflicts (Atjeh, Bali).

A second facet of the whole state formation process that should be given some attention here was the fact that the colonial administration started to involve itself more and more with the local population. That brought with it a degree of concern and care for the population. By means of the Governmental Regulations of 1854 which prescribed that it was the duty of the governor-general to protect the population from arbitrary action, such protection was enforced but after the royal parliamentary speech of 1901 for which A.W.F. Idenburg, Minister for the Colonies was responsible, this was elevated to the status of Ethical Policy. The way in which the East Indian population was approached displayed 'moral revival' parallels with the way in which, in the Netherlands, socialism had tackled working class problems. In both cases the focus was upon improving the social circumstances of

Modern reinforced concrete in the jungle: a rod stiffened arch bridge with a raised driving surface. Located in Tapanoeli (Sumatra), 1934.

the relevant population groups as well as working on the individual's mental elevation. The approach in the East Indies was just as strongly paternalistic as it had been in the Netherlands. The poor brown brother's intentions might have been good but he needed to be steered towards an adult state by means of basic education: that was the attitude. Schools were built and health centres providing all the necessary advisory services were opened. The Government Savings Bank[17] was also initiated, partly in order to protect the 'Inlander' from the extortionate interest rates of the Chinese and Arabs. What also fitted into the same picture was the introduction of the Agricultural Credit System (which gave rise to the Service for the People's Credit System) in 1905, the Cooperative Service in 1930 and the General People's Credit Bank of 1934.

A third important administrative development had to do with increasing the degree of involvement from those lower down in the hierarchy. Just as in the VOC era, the Dutch East Indian state recognised pronounced centralised authority in the form of the governor-general who headed the colonial government which further consisted of members of the Council of the East Indies. The administration was characterised by strict divisions into the European Administration, the Inland Administration and the Foreign Easterners Administration. Under Ethical Policy all of this was to change. In 1903 inhabitants were given a voice through the various local councils. The *desas* retained their traditional autonomy. Administrative decentralisation also involved the recognising of various powers for the different regions so that they could 'provide for their own needs' (1903), the establishing of communes (1905) and the creation of the provinces of: West Java (1925), East Java and Madoera (1929) and Central Java (1930) which excluded the Royal Domains of Djokjakarta and Soerakarta.

A further significant development in the whole integration of indigenous and non-indigenous bodies when it came to the ruling of the entire country was marked by the establishment, in 1918, of the People's Council comprising a chairman appointed by the King, 30 local population members and 30 non-local members. At first, the People's Council merely had an advisory status but after the reforms of 1925 (i.e. the East Indian State Ruling of that year that built on the Dutch constitutional revisions of 1922) the People's Council obtained further joint legal powers for 'internal affairs'.

The process of colonial state formation finally gained its definitive form within the framework of a different kind of long-term process, namely that of the integration of the East Indies into the emerging world economy. That also exposed the country to the various crises that periodically arose in that connection, not only to falling export levels and all other kinds of attendant problems but also to the consequent loss of revenue which meant that the purchasing power of the East Indian state would always decrease at such times. It was especially the Wall Street Crash of 1929 and the subsequent years of depression in the thirties that was to prove disastrous for production and employment in the East Indies and thus also for the colonial government's reserves and the measures aimed at stimulating the welfare of a population that had already been badly hit. The Javanese sugar production level which, in 1929, had amounted to 2,900,000 tons fell to just 1,400,000 tons in 1933. Six sugar refineries had been forced to shut down. Some 130,000 of the 320,000 contracted manual labourers employed on Sumatra's eastern coast in 1929 were sent back to Java in the 1930-1935 period. It was in 1934 that things reached rock bottom and by 1941 the numbers had crept up again to 210,000. Whilst East Indian exports for the year 1929 had brought in 1,446 million guilders, by 1935 that sum had dropped to 472 million guilders. The East Indian state revenue fell from 525 million guilders in 1929 to 246 million guilders in 1933.

Clearly the moral revival that had manifested itself after 1900 had speeded up the education process and thus also the raising of self-awareness and nationalism among Indonesians. These feelings of nationalism were only consolidated during the crisis years of the thirties. The Japanese occupation that commenced in 1942 created the opportunity for Indonesians to seize their independence. Dutch dominion over Indonesia ended in 1945 when Soekarno declared his country an independent republic and when, in 1949, the Netherlands finally agreed to independence.

200 Years of civil engineering

Around about the year 1800 private economic and administrative activities in Indonesia made way for similar kinds of efforts undertaken by the state. What was crucial in all of this was the construction of public works which was something that would not only pave the way for intensified trading activities but which would also simultaneously lay the foundations for well-organised administration. The expansion of the administrative services, the branching out and consolidating of the infrastructure and the introduction of the various accompanying technologies only served to ensure that the imposed pacification filtered ever-deeper into

The weir in the Tjitaroem river, complete with its main canal and inlet sluice (West Java), 1938. Weirs were 'dry built'. Here one can clearly see the original bend in the river.

the whole social fabric. Due to the public works that made all these developments possible, engineers were starting to play a more and more important part.

Prior to 1800 it had been the VOC that had given the orders for all kinds of civil engineering projects to be undertaken, such as the building of roads, waterways and drainage canals together with all the necessary accompanying structural works and several health-technical provisions, predominantly in Batavia. As trading increased so the need for good infrastructural links became more acute. Later road and hydraulic works were constructed at the instigation of the Dutch East Indian state and slowly this was done by civil engineers employed by the East Indian Public Works department. When roads began to be asphalted, access to the country's interior started to improve and gradually everything was expanded and modernised which, after 1870, was also augmented by the laying of railway lines. Initially these railway connections together with, for instance, irrigation and drainage works, served the purposes of colonial exploitation. The increasing concern for the population clearly manifested itself in the stepping up of the creation of public works, especially in the area of irrigation and drainage, public water supplies, sanitation and measures to protect people in urbanised areas from *bandjirs* (sudden seasonal flooding of rivers).

One important factor in the realisation of these civil public works was the introduction of the 1854 Governmental Regulations which stipulated that five directors be made responsible for the managing of the various aspects of civil administration. This included the foundation of the Public Works office. After the establishment of general administrative departments in 1866 the office continued under the name of the Department of Civil Public Works (Departement voor Burgerlijke Openbare Werken or BOW). Other departments that were later involved in the construction and maintenance of public works were: the Department of Agriculture, Industry and Trade (established in 1905; the agricultural experts were involved in the creation of 'tertiary divisions' for irrigation systems in which the population's *sawas* or rice fields were to be found, and with water distribution; see Chapter 6) and the Department of Government Companies (1909-1934; at that time the transport sector, which included responsibility for the constructing and maintaining of the railways, fell under this department). On the whole, the initiation and realisation of public works enjoyed renewed incentive under the Ethical Policy ruling, partly also through the introduction of the decentralised system of administration. In Chapter 1 we shall examine the development of the Dutch East Indian Public Works department in further detail.

Alongside civil public works there were also a number of extensive private civil engineering projects that were completed, such as the various roads and railway lines with all the accompanying bridges, landing stages, drainage and irrigation works, hydropower works and provisions pertaining to public health – including drinking water provisions, sanitation plants et cetera – both in relation to companies (such as the Batavian Petroleum Company) and big crop cultivation industries such as the sugar enterprises in East Java and the tobacco, oil palm and rubber enterprises situated in Deli.

The colonial era which ended in 1942 was technologically characterised as one in which public works were created, managed and maintained. After that period the Japanese occupation, the Indonesian struggle for independence and the final transference of sovereignty were to bring many changes. At that time the Dutch contribution to infrastructural development in Indonesia stagnated whilst the intervention of other countries and international organisations increased. The Japanese period of occupation (1942-1945) brought destruction and decline. As far as Dutch technological contributions to the progress of the East Indies and Indonesia in the 1942 to 2000 era go there are three distinct periods which may be distinguished: 1945-1949, 1949-1958 and 1965-2000. The turbulent years leading up to 1949 were characterised by recovery[18], new town development (e.g. of Kebayoran, the satellite city in the southwest of Jakarta) and planning (e.g. the welfare plan for West Java that will be discussed in Chapter 6). The 1949-1958 period was devoted to helping in the area of recovery and reconstruction. Between 1958 and 1965 there was a total lack of Dutch influence for various political reasons. The final period, leading up to 2000, was characterised by intensive cooperation surrounding the rehabilitation of irrigation provision or, in other words, the repairing of structural works and water systems which, due to lack of maintenance, had fallen into disrepair and were either not operational or hardly usable, and land reclamation (e.g. the Pasang-Surut works constructed for the purposes of transmigration). It was characterised also by master planning (e.g. the Master Plan for the Drainage and Flood Control of Jakarta), by the reconstructing and improving of traffic bridges, rail bridges and harbour constructions and by the reconstruction, improvement and expansion of public drinking water provisions (some of these subjects will be dealt with in Chapter 9).

Whilst Dutch technological input was very centrally directed, bureaucratic, 'top-down' and paternalistic in the colonial era, especially in the Ethical Policy period, the external contribution became more diversified in the post-colonial era with organisations such as the United Nations and the Dutch Ministry for Development Aid becoming involved and attention turning more to locally identified and lobbied for needs, to offering support and to taking care of knowledge transference, in short, to development cooperation.

Historiography

The transfer of sovereignty in 1949 put an abrupt end to all Dutch dreams of further developing the East Indies. In fact, how far did the Dutch actually get in realising all those dreams? As the Dutch period of administration drew to an end in the Indonesian archipelago, all kinds of triumphant stories started being told about what had been achieved in the Indonesian archipelago during the colonial period. The compilation entitled *Daar wèrd wat groots verricht* (Great things were achieved there) that had appeared in 1941, just before the Japanese invasion, constituted the most important of all these publications.[19] Led by the lawyer, W.H. van Helsdingen, former chairman of the Dutch East Indian Peoples' Council, who was supported by the lawyer, Dr. H. Hoogenberk, some 40 experts reviewed what had been achieved during the period of 'Holland's great striving to bring peace and prosperity to its overseas State'.[20] As indicated by the title, the conclusions were generally positive. According to the book's editors, the Netherlands had succeeded in completely realising all the expectations expressed by Jan Pieterszoon Coen in the early days of Dutch hegemony in the archipelago to the effect that 'Great things could be achieved in the East Indies' and so there was every reason to feel proud.

The book devoted considerable attention to 'Western technology'. The civil engineer H.C.P. de Vos, the first professor of hydraulic engineering at the Technical College in

Rice harvesting (East Java), 1993.

Bandoeng, wrote a section on 'The battle for and against water'.[21] He concluded that the land area provided with 'fully technical irrigation works' was equivalent to an area two-fifths as big as the Netherlands (1.3 million hectares). What was important in all of this was the different types of fixed and mobile weirs that were used in rivers and canals. Large reservoirs had also been built with a collective capacity of some 400 millions m³ of water. De Vos also determined that all the work carried out by engineers had led to a situation in which 'all the water available in Java had become technologically controlled'. He also reported that between 1900 and 1940 the total expenditure on irrigation activities had amounted to 270 million guilders.

In the light of further information given in the same publication this picture can be completed. During the colonial period, the population saw rapid expansion as evidenced in the section entitled 'From four to forty-four million souls in Java'.[22] Modern irrigation methods helped the growing population, which was totally dependent upon rice crops, to support itself. It was certainly a successful operation. Another contribution reports that because of such rural agricultural facilities in Java the population had become virtually totally self-supporting.[23] In the book in question the attention devoted to Western technology does not remain limited to irrigation works. The railways and other transport and communication means in the East Indies (inter-island shipping, road transport, air traffic, radio telegraphy, telephony and broadcasting systems) were all considered under the heading, 'The conquering of distance'.[24]

Since its independence, the colonial history of Indonesia has been examined in some breadth and depth so that a more balanced picture has emerged giving a somewhat more critical view of Dutch overseas efforts than the one presented in the publication referred to above.[25] One facet which, up until now, has been given relatively little attention is that of the role played by the influence of modern Western technology. In view of the tremendous significance of technology in the development of East Indian society (so much so that it was termed a technocracy[26]) this is, in itself, rather strange. Studies have been made of the earliest Javanese railway link[27], of the divergent types of Dutch East Indian engineers[28], of the Royal Package Shipping Company[29] and of the technological developments supporting the Javanese sugar industry.[30] The development of modern irrigation has further become the subject of two doctoral theses.[31] In addition, attention has been devoted to civil and other engineering branches in the East Indies in relation to Indonesian nationalism.[32] Except for in the area of irrigation, the subject of public works in the East Indies has not been systematically examined in any depth and much the same goes for the history of the East Indian Public Works department.[33] This book does, to an extent at least, manage to fill that historiographic gap.

Themes

The road, hydraulic and other civil engineering activities of Dutch engineers – Dutch East Indian engineers[34] – in the Dutch East Indies and in Indonesia as well as the development of the main relevant service, the Dutch East Indian Civil Public Works department, are the themes that are central to this book. From time to time the link with the Netherlands will be emphasised, particularly where this pertains to the Dutch Public Works agency, its structural works and Dutch as opposed to East Indian civil engineering. It has, obviously, been necessary to draw a number of lines and to give ourselves a number of editorial restrictions.

Our focus will be particularly on civil public works in the East Indies roughly between 1800 and 1950 and – in Chapters 9 and 10 and also in the Epilogue – upon similar types of works with a 'Dutch stamp' constructed in Indonesia in the 1950 to 2000 period. The decision not to highlight the period before that (with a few exceptions) evidently has to do with the fact that in the seventeenth and eighteenth centuries there were no civil public works in the East Indies in the true sense of the word. In fact such works depend, by definition, on the existence of a government and form part of the governmental task. From the point of view of relevance (bearing in mind the necessary constraints) the spotlight will be notably on the colonial period up until 1942 and the period after 1965 when relations between the Netherlands and Indonesia were intense.[35]

In this connection 'civil public works' should be understood to mean public works of a civil engineering nature, railways included. We shall just permit ourselves one deviation in order to contemplate the urban development that was important in conjunction with all kinds of civil engineering projects linked to water, in particular health-related works, and works designed to protect areas against flooding.[36]

Technology involves manpower: the building of a temporary bridge over the Djati river between Malang and Blitar (East Java), c.1922.

The appendage 'bearing a Dutch stamp' means that in examining the post-1950 Indonesian civil public works we shall, of course, only be looking at those characterised by a Dutch share in the design. In the Epilogue which explores developments following independence in Surabaya and neighbouring Malang (East Java), works that clearly do not have a Dutch identity are also discussed. The research underlying this part of the book was more directed towards developments since 1945/49 in general than towards strictly Dutch repercussions and we did not feel that it would be a problem to retain that character (see below).

All the various private civil works, regardless of how important they were to the country and its people, do not fall under civil public works and so will not be specifically discussed in this book except in the chapter on the railways, that is to say, as far as railways for public transport (passengers, goods et cetera) are concerned; private industrial railways on plantations, industrial areas and such like will not be considered. In that chapter private and state enterprises will thus only be discussed in that particular light. For quite some time the matter of whether railways should be constructed and operated by the state or by private companies remained a moot point leading to a number of fiery debates. Ultimately the railways fell into private and state hands at various different times and so, accordingly, the emphasis shifted.[37]

In view of the 'mer à boire' of civil public works that emerged in the East Indies and Indonesia after 1800 we shall first limit ourselves to a brief overview of those works before then, when it comes to an in-depth exploration of those works, examining just a selection. The stress lies on the works that have somehow contributed to the cultivating and developing of the land and to the augmenting of the population's prosperity, as indeed should be the case with civil public works. We shall be dealing here with works that may be viewed as representative in the fields of irrigation[38], hydropower, railways, roads, harbours, public health (domestic water provision, sanitation[39], malaria prevention et cetera.) and protection against floods, notably in urbanised areas.[40] We hope that this research will challenge other writers such as historians and technologists to further investigate the subjects touched on here.

There will be no summing up and discussing of all the civil public works with a purely technological character, instead we shall endeavour to illustrate the spirit of the times and all the ideological clashes, ideas and interests within the various smaller and bigger power structures that formed the very foundations of these works. In so doing we shall draw on images and concepts taken from modern history of technology studies.

Actors, systems and regimes

Traditional discourse on the subject of technological history tends to dwell especially on the describing of inventions, discoveries and successful technological scientific applications viewed from the angle of human and social progress.[41] The writers of such tracts, who are invariably engineers or others directly involved in technology are thus partly responsible for the image that has emerged of the technological epos as being one in which nature is subjected to human ingenuity where inventors and technologists in general play the parts of the heroes. The tales told in the collection 'Great things were achieved there' fits perfectly into this genre.

It is a view that returns in the 'technology push model', the notion that technology literally forces society to forge ahead. The 'technology as applied science model' variant, places science in a central position. The opposite outlook, especially popular in economist and social scientist circles, is that of the 'market pull' where it is society that tends to be seen as the dictator of technological progress. This is also the view that sometimes resounds in colonial literature.[42] Both approaches tend to be rather blinkered and one-sided.[43] Those who

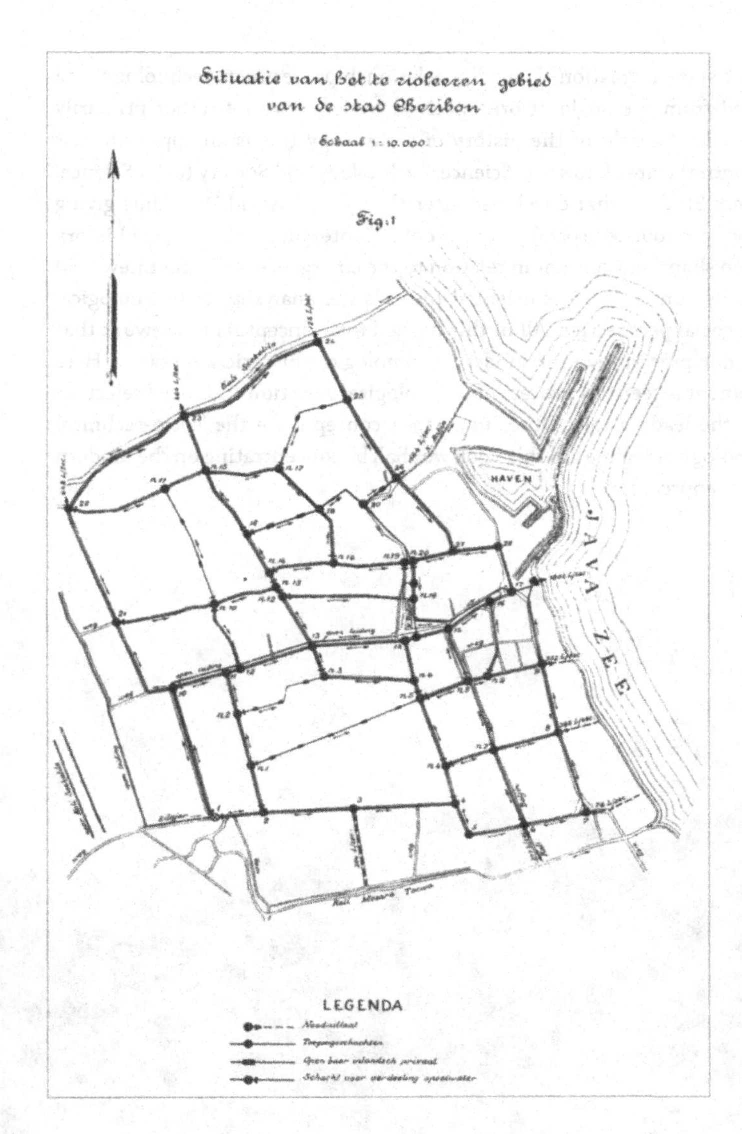

Sewerage according to the mixed system method in the centre of Cheribon (West Java), 1920.

maintain that the engineers in the Dutch East Indies were autonomous white people who, on the strength of their expertise and supported by an independently operating organisation were easily able to make their mark on East Indian society or were simply able to meet the needs of the day have absolutely no idea of the – sometimes extremely subtle – processes that went hand in hand with emerging technology in the developing East Indies!

The above-mentioned one-sided technological development models paved the way for what came to be known as 'contextual history of technology' where, on the one hand, attention is paid to technological development at concrete artefact level and to the actors involved while, on the other hand, there is thought for social-historical processes, such as modernisation and state formation. Here the relationships between technology and society are not viewed from the angle of preconceived notions but are rather primarily empirically researched. In the field of the history of technology this is an approach that arose in conjunction with the area known as Science, Technology and Society (or as Science, Technology and Society Studies) that developed after the Second World War, thus giving science and technology pronounced socially-based roots. Contextual technological history was subsequently given shape and content in relation to the emergence of another new field known as technology dynamics which is oriented towards the analysing of technological developments from a social perspective. All of this has led to a conceptual framework that is direction-giving if not prescriptive for modern technological historical research. Here technological development is seen as a process of technological variation and social selection in which actors fulfil the leading role. Other important concepts are the 'socio-technical system' and the 'technological regime'. In this book we shall be concentrating on the modern technological historical approach.[44]

The harvesting of sugar cane (Central Java), 1993.

Much attention will be devoted to actors in the sense of people, groups and organisations and to their positions of power, interests and conceptions. One important collection of actors in Dutch East Indian society was linked to the state formation process described here above. In the first place this involved East Indian Public Works department and its engineers but also, for instance, inland administrative employees (united in the Inland Administration), agricultural experts (from the Agricultural department), the Navy and the colonial government. Other actors were private enterprise and groups drawn from the colonial population. As East Indian society became more complex so the number of actors involved in technology increased which gave technological development in general the character of a dynamic socio-technical system steered by technological regimes.[45]

A socio-technical system is really a conglomeration of technological and social elements. This is a concept that is especially applicable to the kinds of infrastructure systems or infrasystems that remain central to this book. The system-based character of technological development manifests itself in the way in which system mechanisms spring into action like, for instance, the mechanism of momentum that lends any development a certain degree of autonomy.[46] A technological regime may be viewed as a collection of technological and other rules serving to influence the design process of engineers and the process of technological development in general.[47] It might be something of a simplification but we could assert that socio-technical systems have to do with the hardware of technology (e.g. an infrastructural network with its accompanying organisations) and that essentially technological regimes create the relevant software (like, for instance, that amassed when a person is educated as an engineer) and furthermore that a socio-technical system is especially founded on the natural environment while a technological regime tends to reflect the social context in which that environment is controlled. In this connection we shall not treat civil public works like a random cluster of artefacts but rather as something to be viewed from a socio-technical system perspective.[48] In addition, we shall adhere to the regime concept, albeit in a fairly global sense. As will be revealed below we shall see that with East Indian technological development there were, in very broad outline, two different technological regimes that were involved and which were rooted in different colonial interests and aspirations.

Imperialism and modernisation

In the case of technological advancement the importance of long-term processes cannot really be overestimated[49] and that is certainly something that applied to civil technological developments in the Dutch East Indies. We shall examine here two essential long-term processes that constitute part of the historical-dynamic framework of East Indian civil engineering: the earlier discussed establishing of a satellite state and the way in which it was transformed into a modern colonial state (colonial state formation) and the process of being absorbed into the world trade network (globalisation), also mentioned above.[50] The thread linking these two factors is imperialism and what we mean in this case is the establishing of colonial states by European powers for political and economic reasons.[51] The colonial state in the East Indies was, in effect, a local agent of Dutch imperialism.[52] Science and technology not only played a prominent part in the rise of Europe but also, as the American colonial technology historian, Daniel Headrick perfectly illustrates, in colonial relations.[53] In each developmental phase of European expansion, technology played an important part though the ways in which and degree to which varied.

When it came to the matter of exploring areas outside of Europe, steam ships and the malaria fever prevention drug, quinine, were important parts of the picture. Furthermore, without the availability of rapid firing firearms European dominion over these areas would also have been unthinkable. In order to consolidate colonial administration, communication and transport networks had to be developed: shipping lines, telegraph cables and rail links. These kinds of infrastructure networks were also constructed for the purposes of colonial exploitation. Headrick demonstrates how modern technology helped to alter the nature of colonialism. In conjunction with industrialisation and the rise of modern science and technology, colonialism gained the allure of 'free trade imperialism': colonial administration expanded but the actual colonial exploitation fell into the hands of private enterprise. It was notably during this phase of European hegemony that the 'technology transfer' process became apparent in which flows proceeded mainly, though not exclusively, in the direction of the colonies. This pattern was also visible in the Dutch East Indies, both throughout the period of colonial history and during the process of colonial state formation when technology was used to discover, conquer, open up and exploit. In the East Indies (and in other colonies as well) it did not stop at exploitation; the interests of the population were also considered.

This brings us to another long-term process, namely the process of modernisation. Just like the Netherlands, the Dutch East Indies experienced a process of modernisation during the nineteenth century. [54] From the political point of view the modernisation of the East Indies reflected two components:

- the liberalisation of the economy by means of state-supported free enterprise exploitation rather than straightforward state exploitation;
- an increasing need to bear in mind the needs of the colonial population while striving to achieve public prosperity and to develop.

It was after about 1870 that the Dutch East Indian political modernisation process began to take shape. The civil public works that had been constructed during the early colonial state period had exclusively served the imperialist goals of establishing power-based monopoly and state exploitation in which exploitation was the chief aim and pacification a subsidiary goal. Instances of this are the Great Post Route in Java (see Chapter 2) created by the governor-general, H.W. Daendels, for military-strategic reasons and the first irrigation works, primarily created for sugarcane cultivation and the sugarcane industry (see Chapter 6). Once colonial state modernisation gained a hold, the objectives that became important within the imperialistic policy framework were the supporting of private enterprise in the plantation sector and the improving of the prosperity of all population categories (especially Europeans, Chinese and the native population). Certain works were undertaken with a view to the interests of the colonial population, like for example works for health care purposes (see Chapters 7 and 8). Yet others – such as some of the railway projects (see Chapter 3) – linked up more with interests that had to do with private capital. However, it was not just the underlying reasons for undertaking works projects that differed. To a certain extent it was also objectives that determined the categories into which works fell and the form they took: health-related works were, for instance, especially undertaken in the interests of the population whilst, for example, the form taken by road and rail works differed depending on the anticipated use to which such provisions were to be put.

The construction of public works not only varied over the course of time but also depended very much on the location in question. When, in the nineteenth century, the exploitation of Java was in full swing the Outer Regions still had to be pacified. Java was also the first island where governmental care for the population manifested itself. Admittedly the process of colonial state formation passed through the same phases in all the different areas but the precise timing and duration varied.

Imperialism and state modernisation were two processes that were really felt in the East Indies. Whilst imperialism brought with it the domination and exploitation of the country and its people, the rise of the modern colonial state proved to be a significant factor in which the population became important actors.[55] These two processes had consequences for technological development, both in different ways. The discrepancy between imperialistic and 'modernistic' works can be further distinguished on the basis of the technological regime concept.

Within the framework of the colonial state there are, roughly speaking, two technological regimes to be envisaged that steered civil engineering developments in the East Indies: that of the exploitation regime and that of the developmental regime. The first regime constituted an expression of imperialist interests and falls into the category described by Landes as love of gain (or covetousness).[56] The second regime represents modernity and links up very much with the views of Van Doorn who saw the East Indies as a technological development project.[57] Both these technological regimes may be said to be period bound, or at least phase bound, and to display a progression from exploitation to development regimes but to a certain extent they also took place simultaneously. The exploitation regime presented the true face of the early colonial state in which the colonial government operated as an instrument of Dutch imperialism. It was via the exploitation regime and within the framework of colonial state modernisation that the developmental regime reached fruition. Ultimately all of this gave the state a new role: it was forced to stimulate and coordinate or manage different interests. In that way political modernisation led to – the endeavour at least – to realise social modernisation. Ultimately these general technological regimes underlie all the works discussed in this book, though it is perpetually necessary to empirically affirm which regime matched and steered which works or series of works and to what extent that was really the case.[58]

The structure of this book

The differentiation between the above-mentioned technological regimes will also be reflected in the chapter divisions. The subjects will be examined chronologically depending on the initiation point of the relevant works or the period in which the centre of gravity of the effort involved lies. This means that the first works to be dealt with will be those that had their origins and found their main inspiration in the exploitation regime: the roads, the railways, the bridges and the harbours. After that the attention will switch more to works that were allied to the developmental regime: the irrigation and hydropower works, urban development and health-linked technological works.

The technological chapters (2 to 8) will be preceded by a chapter on the rise of the Dutch East Indian Public Works department (Chapter 1) and will be followed by four chapters that dwell on the developments that took place after independence and contemplate the various conclusions that can be drawn. Chapter 9 deals with technological developmental cooperation between the Netherlands and independent Indonesia and concentrates on irrigation, health technology and bridge building. Chapter 10 gives a view of East Indian civil engineering and the international origins and distribution of the accompanying know-how. There again, irrigation will be a main topic of discussion. This chapter will provide a number of conclusions that can be drawn from all the preceding chapters whilst also considering the changes that took place after independence in a more extensive way than in any of the preceding chapters but not as comprehensively as in Chapter 9 and the Epilogue. The Epilogue stops to consider post-independence developments from the Indonesian angle by focusing on the East Javan cities of Surabaya and Malang. As a sequel to Chapter 10, which reveals what is specific about Dutch East Indian civil engineering as a science, the Concluding Considerations chapter will sketch what is special about East Indian technology as such and several theoretical conclusions will be drawn. In the process, certain main points from this introductory chapter will once again be covered.

Sources

Whilst Headrick provides several points of reference for an analysis of technological developments in overseas areas his fellow American, the technological historian Michael Adas, warns us not to develop a Eurocentric view of matters. In his study entitled *Machines as the Measure of Man*[59] he demonstrates how it is that as modern technology developed in Europe this increasingly became the criterion for judging non-Western peoples, where technology substituted religion, notably Christianity. The turning point came round about 1800 at a time when in Europe, in Britain especially, the Industrial Revolution was just getting underway. Though most of Adas' sources are European his study turns, in effect, into an indirect plea for non-Western source research. This book partly satisfies that plea, on the one hand because some of the authors carried out much of their research in Indonesia or are able to draw on extensive work experience there and, on the other hand, thanks to cooperation received from the Indonesian side.[60]

Regarding the data sources consulted for the purposes of compiling the descriptions and considerations included in this book, the following may be said. Tremendous volumes of information have been recorded in archive material kept in the Netherlands and also in various publications such as articles, books and published reports. Here below is a selection of some of the more important sources.

a. The editions of the *Verslag over de burgerlijke openbare werken in Nederlandsch-Indië* (Civil public works in the Dutch East Indies report), which at one point became divided into seven different sections.[61] Other official publications are: the *Koloniaal Verslag* (Colonial Report), appendix to the *Handelingen der Staten-Generaal, Eerste en Tweede Kamer* (Proceedings of the States General, First and Second Chamber) and the *Staatsblad van Nederlandsch-Indië* (Law Gazette for the Dutch East Indies).[62]
b. Various professional journals including the Dutch publication: *De Waterstaats-Ingenieur*

(The Public Works Engineer, the journal of the Association of Public Works Engineers in the Dutch East Indies).

c. Archive material kept in the National Archive, The Hague:
   - the Ministry for the Colonies collection including what were known as the mail reports;
   - the J. Haringhuizen/H.J. Schoemaker collection;
   - the J. Groothoff collection;
   - the W. Cool collection.

d. Material kept in the archives of the former Department of Civil Public Works (later: Transport and Public Works) in Citeurup, Indonesia.[63]

There are also, however, a number of hiatuses. The BOW archive in Citeurup is only partly accessible as the material has not been properly catalogued but only filed according to year. In general a great deal of archive material in Indonesia has been lost because of the ravages of time and because of various wars and conflicts (1942, 1945-1949, 1950-1952 and 1965). In addition, historic information has been lost due to the fact that representative civil engineering structures have disappeared or have been replaced, due to inadequate upkeep and, once again, because of wars and certain projects having been insufficiently described and catalogued in the first place.

Attempts have been made to fill up several of the data gaps with two questionnaires. One of the questionnaires was conducted among water works companies in Indonesia and undertaken in conjunction with research activities conducted for this book. The questionnaire was set up in connection with the Indonesian Association of Water Supply Companies (PERPAMSI[64]) and the Institute of Technology in Bandung (ITB). The other questionnaire we did was directed particularly towards several of the Dutch engineering bureaux once involved in establishing cooperation between the Netherlands and Indonesia in the field of domestic water supplies, sanitation and the protecting of urbanised areas from *bandjirs* (floods) in the post-colonial era, notably since 1965.

We have done everything in our power to make sure that the subject-matter of the present book has not been approached in a narrow-minded way and just from the Dutch perspective. In an attempt to prevent this from happening use was made of data obtained from staff at the Bandung Institute of Technology and at PERPAMSI. There was also the Petra Christian University in Surabaya which, when approached by us, came up with an informative contribution relating to developments in various technological areas in the new country of Indonesia where the main emphasis was on the own particular region. The latter contribution is included in the Epilogue in an abridged form.

Lasting testimony

All historiography remains a reconstruction of the past from the particular present in which it is written and so is thus never impartial. Ultimately the words of the historian, R.F. Beerling, may be said to hold: 'Every present has the past which it asks for'.[65] Despite all of this we have endeavoured to paint a picture that is as unprejudiced as possible. In view of the time that has passed with respect to the colonial Dutch East Indian past that is something that is easier to achieve nowadays than it would have been some years ago. We are also guided by the engineer J.H. Thal Larsen who, in 1932, in his capacity as vice chancellor of the

Urbanisation: illegal slums on the levees and verges of the Jakarta Bandjir Canal, around 1970. Note the 'accommodation' beneath the drinking water mains.

Agricultural College in Wageningen, gave a commemorative speech in which he alluded to the situation in Demak – a region in Central Java that saw famine in 1848-1850, 1872 and 1902 – where 'thanks to the safeguarding of agriculture against drought and flooding' there was no more hunger, a situation to which he attached the following conclusion:

> *Together with many other good works that which has been achieved in this field will amount to a lasting and favourable testimony of Dutch domination over these areas in spite of the opinion of all those who do not wish to judge the deeds of their forefathers by the mentality of the times but who would rather measure them against the elevated theories of their own times.* [66]

Despite the fact that this claim was made against the backdrop of the triumphant spirit also underscoring 'Great things were achieved there' it does, nevertheless, embrace the valuable plea to view matters in their rightful time framework. By placing oneself in the position of Indonesia as it was in the time of the Dutch East Indies and the Netherlands as it then was and by assessing and sensing the developments of that time in their true historical perspective, it becomes possible to see the achievements of the Dutch East Indian Public Work engineers for what they really were. Rather than making favourable or unfavourable statements about the past, however, we hope that this book will stand as a lasting testimony to the work done in this field in Indonesia.

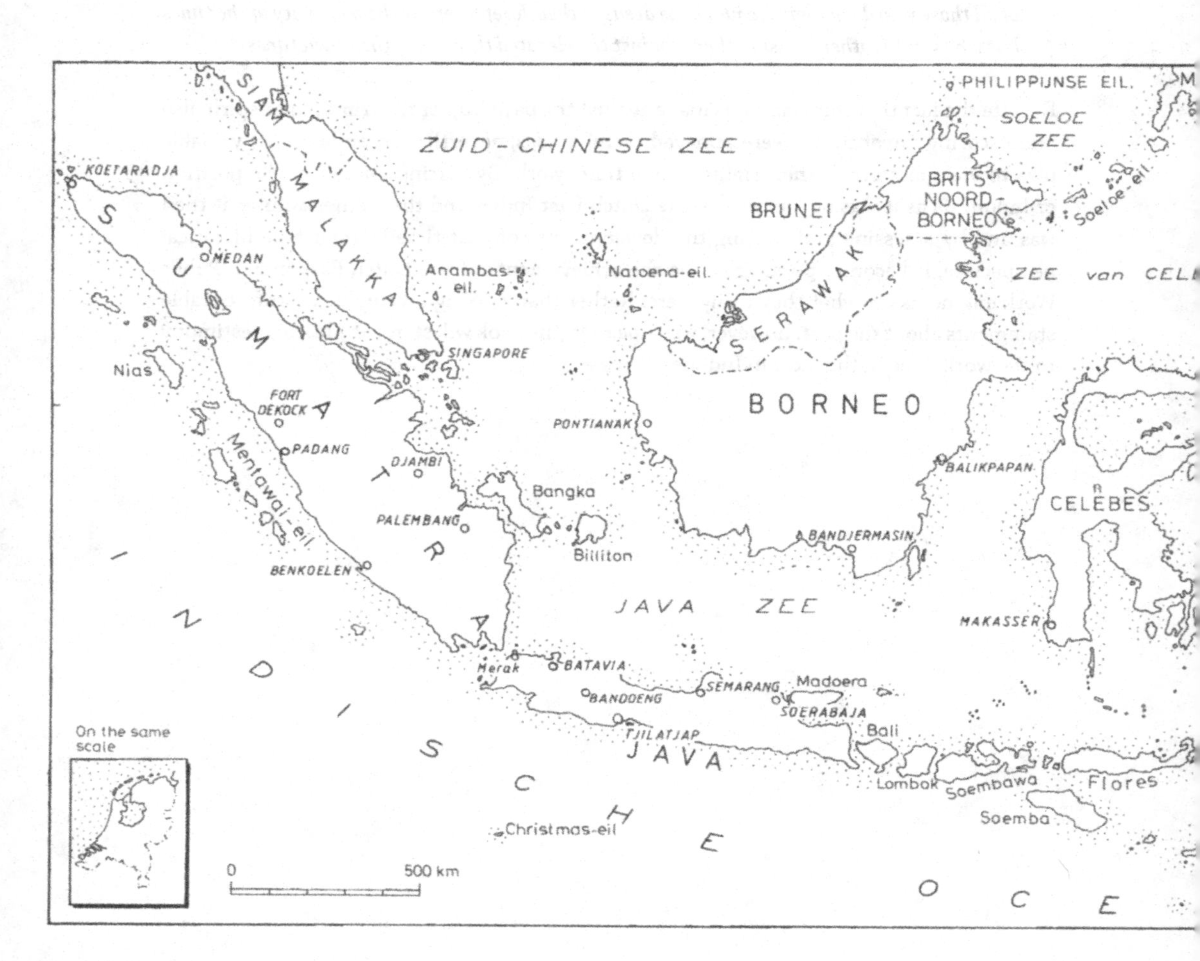

Map of the Dutch East Indies.

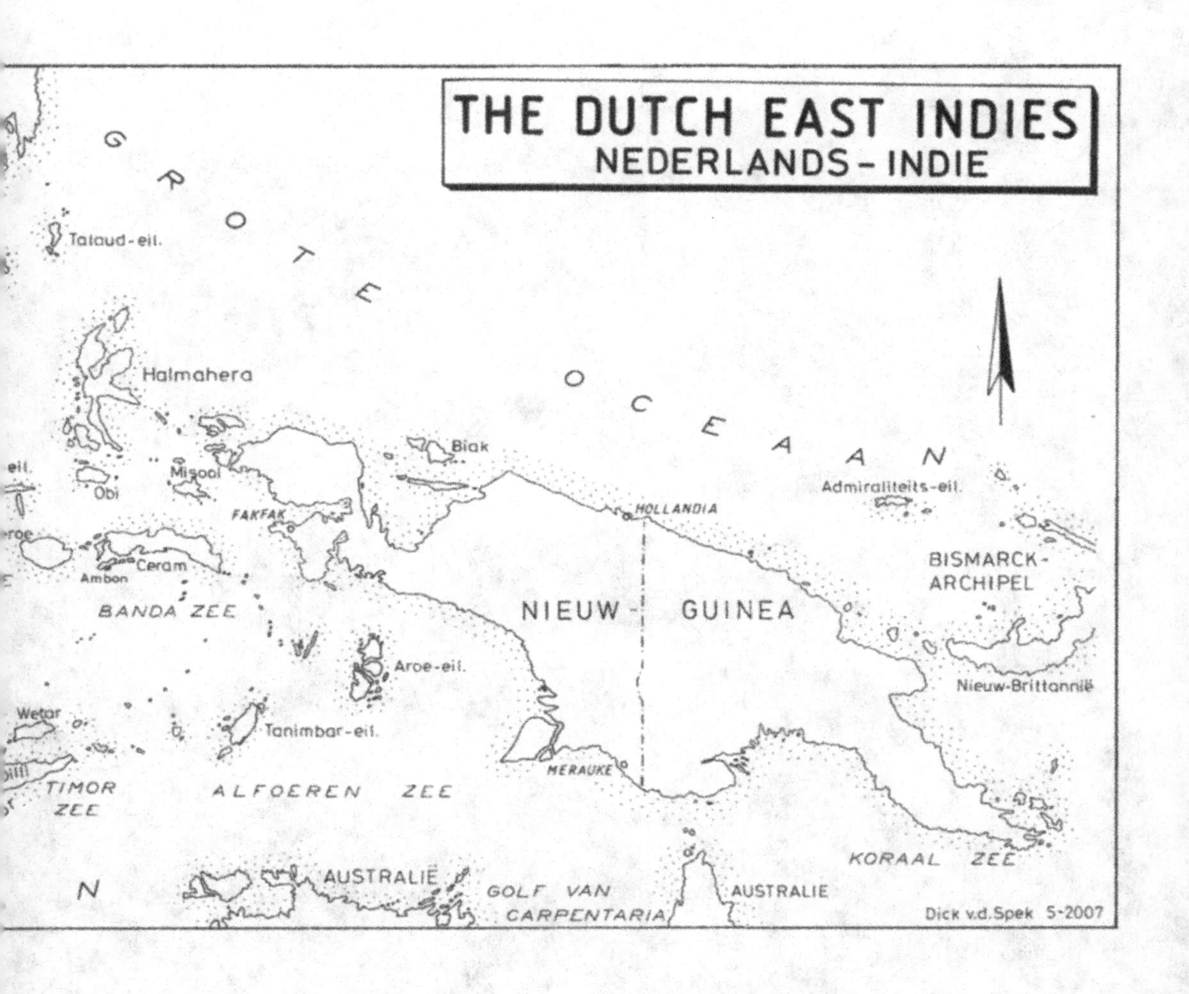
THE DUTCH EAST INDIES
NEDERLANDS - INDIE
GROTE OCEAAN
Talaud-eil.
Halmahera
Biak
Misool
Obi
Ceram
Ambon
FAKFAK
HOLLANDIA
Admiraliteits-eil.
BISMARCK-ARCHIPEL
BANDA ZEE
NIEUW - GUINEA
Aroe-eil.
Nieuw-Brittannië
Wetar
Tanimbar-eil.
MERAUKE
TIMOR ZEE
ALFOEREN ZEE
KORAAL ZEE
AUSTRALIË
GOLF VAN CARPENTARIA
AUSTRALIE
Dick v.d.Spek 5-2007

Bridge over the Air Tetap on the Manna-Bintoehan road in Benkoelen (Sumatra), c. 1918. This bridge is identical to the bridges (described in chapter 4) that were to be found along the Loemadjang-Pasirian road (East Java).

CHAPTER 1 **TECHNOLOGY AND ADMINISTRATION**

The rise and development of Public Works in the East Indies

FRIDA DE JONG | WIM RAVESTEIJN

H. de Bruyn (1815-1885).
Twice BOW (i.e. Civil Public Works) director.

J.A. de Gelder (1841-1912). Constructor of the Tandjong Priok (Batavia) harbour and compiler of the 1885 public works ruling that was to give BOW great independence.

The consequences of the1883 Krakatau eruption were catastrophic. After a 200-year period of inactivity following an insignificant eruption in the seventeenth century the volcano, situated between Java and Sumatra on the island of the same name, started to once again show signs of life in May of that year. There were earth tremors, emissions of ash and steam and explosions. The explosions could be heard up to 350 km away but the showering of ash remained confined to the immediate vicinity and caused no real damage. Though the phenomena persisted, the big eruption of August came as an unexpected surprise. The volcano produced some 18 km^3 of material. The smaller particles of ash formed a cloud which then proceeded to circle the earth a number of times, originally at a speed of 72 miles per hour whilst, in the process, colouring the sun blue, green or red, depending on the viewer's angle of observation. The eruption and the ensuing rain of pumice stone and ash covering an area 23 times as big as the Netherlands caused little damage; the island in question and the neighbouring islands (Long Island, Deserted Island and the Polish Hat) all constituted old crater side parts protruding above sea level and were uninhabited. What was disastrous was the high tidal waves accompanying the eruption which then went on to ravage the Javan and Sumatran coastlines. In the process, 37 Europeans and 36,380 native inhabitants lost their lives while 297 *kampongs* (local villages) were wiped from the face of the earth or seriously damaged.[1]

The success of the engineer De Gelder

The disaster proved to be the ultimate test for the new harbour of Tandjong Priok, in the Batavia vicinity. The harbour had been created when it was found that the old harbour situated

M.J. van Bosse (1843-1917). BOW director, creator of the General Irrigation Plan for Java.

A.G. Lamminga (1855-1920). Founding father of modern irrigation technology in the East Indies.

in the mouth of the river Tjiliwoeng no longer sufficed. The opening of the Suez Canal in 1869 precipitated intensified trade relations between the Netherlands and the Dutch East Indies which, in combination with the dawn of steam power, led to the introduction of larger ships. It was not only the case that the harbour was too small but there was also the problem of the river's large sediment deposits, a factor that placed constraints upon the draughts of the ships. Generally speaking it was a time when colonialism was evolving in the direction of free trade imperialism. European powers competed with each other to seize all of Africa and Asia. With their colonial possessions they banded together to form huge empires so that the free trade of private enterprise was able to flourish. The Netherlands formed a part of all of this and so it, too, took appropriate infrastructural measures and it was the harbour of Batavia that had priority. Many designs were presented and many meetings were convened before finally a plan was chosen that was implemented by the East Indian Public Works department in 1877. The way in which the head of works, the engineer J.A. de Gelder, tackled the whole project gained respect from far and wide though it was also recognized that he controlled the department with an 'iron grip'. By May 1883, when the Krakatau was beginning to rumble, the works had reached such an advanced stage that the chief engineer was able to bid farewell to the project's personnel. On that particular occasion the engineer M.J. van Bosse, who was later to become director of East Indian Public Works, gave a speech which ended with the following quote from Goethe: 'Das Werk soll seinen Meister loben'.

His words proved to be prophetic. Despite the fact that the bridgeheads, the quay wall and the other structural works had certainly not been constructed to withstand large tidal waves they weathered the consequences of the Krakatau eruption with glowing colours: 'The strong currents and surging vortexes [...] have not damaged the works in any significant way'.[2] More to the point, the efficient organisation surrounding the works had paid off:

> *One of the consequences of the disaster was that every sea-worthy vessel was immediately requisitioned to assist in all kinds of operations. As the lighthouses in the Straits of Sunda [between Java and Sumatra] had been destroyed, two hopper barges were instantly deployed to alert passing ships of any danger. Another two barges were used to take provisions to the distressed population. Contact with Kali Anda in the Lampong Bay [Sumatra], which was lost when the telegraph cable in Anjer broke, was frequently maintained by the harbour works hoppers. In short, there were numerous other services that came to depend on the assistance of equipment and personnel borrowed from the harbour works.*[3]

The personnel, none of whom had had an easy time under De Gelder, were thus praised just like their boss himself:

> *Much was demanded of the personnel at that time and I must admit that had they not been so well trained by De Gelder and had the good spirit cultivated by him among his subordinates not been so powerfully in evidence, then the services provided in those days could never otherwise have been achieved.*[4]

The success of De Gelder, both as a technical man and as a manager, was complete and his further career ensured. In 1884 De Gelder, who had previously proved himself through his 'heroic deeds in combat against pirates'[5], was made a member of the Council of the East Indies after having turned down the offer to be made Minister for the Colonies. In council he was able to devote himself to the Dutch East Indian Public Works department, a thing that was certainly necessary. The Krakatau had erupted at what amounted to a crucial moment in the whole struggle for emancipation on the part of East Indian Public Works engineers! This process, and what preceded it, will be discussed in this chapter as well as how, partly thanks to the efforts of De Gelder, things then continued.[6]

Colonial control as a moral duty

The end of the eighteenth century was a restless time in the Netherlands and elsewhere as well.[7] The United Provinces Republic had been dissolved and in the year 1795 a central state had been created. With the help of the French who were determined to export their revolution of 1789, the patriots had managed to establish the Batavian Republic. As the Dutch United East India Company (the VOC) had just gone into liquidation and as its possessions had been requisitioned by the Dutch state, the republic was rewarded with the East Indies. The next question that arose was that of how to deal with colonial control. The former company partner, Dirk van Hogendorp, had an idea and so he published a colonial plan. He recognised the right to a sovereign state for the East Indian population but recommended that the Netherlands should not relinquish its colony. One of the arguments advanced in this connection derived from the 'imperfect state' view of East Indian civilisation. It was simply the case that one could not afford to burden the indigenous population with independence.

The contrast between 'a right to' and 'the inability to' was something that he resolved in a way that would come to permanently characterise Dutch colonial rule and which, even up until the present day, would give historians grounds to view the Netherlands as a positive exception when compared to all the other colonial powers of the time.[8] The Netherlands did indeed have a moral duty to fulfil and that was:

> *the obligation to protect these territories and all their inhabitants from any external and internal violence and hostilities and to, by providing for them good management, give and ensure civil freedom, ownership of property, and protection against all kinds of repressive actions while practising and enforcing totally fair and unbiased legislation.*[9]

Van Hogendorp maintained that ultimately the Netherlands would also profit from this in view of the fact that the enlightened colonial administration would want to uphold peace.

In the debate that persisted around 1800 on the matter of the form and content of colonial administration two opposing models emerged: the liberal model involving a rational-bureaucratic administration that was directly imposed upon the population and the traditional, VOC type of model, which provided a feudal and indirect kind of administrative structure operated via the residents who, as heads of regional management, functioned like viceroys via the Javanese rulers.[10] Ultimately it was the traditional model that was chosen, though the line followed by the governor-general, H.W. Daendels (1808-1811), was a liberal one just like that of his British counterpart, T.S. Raffles, at the time of British Interim Administration (1811-1816). When the British gave the East Indies back to the Dutch it was once again a feudal approach that prevailed. The Van Hogendorp doctrine which was, in effect, the one that fitted in best with liberal thinking would not define colonial policy until Ethical Policy was introduced in 1901, precisely at the time when the national movement started to contest that view! Prior to that, steps had perpetually been taken to approximate Van Hogendorp's ideology. As will emerge later on in this chapter, the rise of East Indian Public Works department would come to play a key part in the whole process.[11]

Public Works up until 1854

In the early days the VOC confined its activities to Batavia and the surrounding parts. Ever since the time that the VOC was involved in the constructing of a quay, the creating of access roads or the digging of drainage systems there had, in effect, been evidence of public works avant la lettre. In 1627 a man by the name of Berckenroode was made surveyor and clerk of works in conjunction with building stipulations. One might say that he was the first public works official. His activities pertained to the construction of civil works such as a church, a school and a harbour. Later on the deepening of the city canals could also be added to that list. After a number of years the Batavian administration was reinforced with a bailiff and sheriffs and an administrator or 'landdrost' was appointed for the outlying regions. In 1664 the surrounding regions started being administered by a board of dike reeves which did not in fact receive any official regulations until 1680 when a dike reeve was simultaneously appointed to do the surveying. In 1685 rulings were laid down in which local inhabitants were given responsibility for the maintenance of roads, bridges and canals under the supervision of the above-mentioned board of dike reeves. During the course of the eighteenth century, Dutch tradesmen started to establish themselves in the Batavian Trade District and the official put in charge of them was known as 'the fabricq'. In 1795 that incumbent was replaced by a Fortifications, Buildings and Waterworks director. There were official instructions for this function as well as support in the form of an engineering captain who was the supervisor and two ordinary engineers. For the duration of certain projects an ad hoc military engineer would sometimes be engaged.

French dominion in the Netherlands was to herald the end of the VOC. It was at the command of the French government that Daendels journeyed to Batavia where he proceeded to make rigorous changes. With real French élan the city was improved and reorganised at the expense of local inhabitants who were forced to supply the necessary materials. In true Napoleonic style, public works facilities were centralised and expanded. Daendels' greatest achievement was the construction of the Great Post Road.[12] After Napoleon had been defeated the British briefly appeared on the scene but, as it was clear from the start that the colony would inevitably be returned to the Dutch, the methods of the British did not catch on.

It was in August 1816 that Dutch rule over the East Indies was resumed and the Dutch state stepped in. On the one hand the colonial administration provided an East Indian government (or 'gouvernement' as it was often called) headed by a governor-general, the first being G.A.G.P. baron van der Capellen (1816-1826) and on the other hand there was the local administration service (the Inland Administration) divided into regions (or residencies), each overseen by a resident. In the Netherlands, it was the king who was directly responsible for the colony. It was not until 1842 that there an independent Ministry for the Colonies was formed and it was not until after the new constitution had been introduced in 1848 that the Dutch parliament was given control over the colonial budget. In the East Indies it was only after a ruling in 1818 that a chief Public Works inspector (J. Peereboom) was appointed, together with three other inspectors (J. Kortenbout van der Sluijs, J. Tromp and J. Bootsgezel). The control of all the hydraulic works did, however, remain in the hands of local authorities, i.e. the residents. Unlike his predecessors like the director of Fortifications and so on, this particular chief inspector was officially allowed to confine himself to hydraulic works.

At first there was a separate department for civil buildings but because the expensive Javan war (1825-1830) had really eaten into the financial reserves the two departments (hydraulic works and buildings) were merged and shrunk in 1827, for economic reasons. Supervision was transferred to the resident or to other people in high places who corresponded to the Land's Products and Civil Warehouses service (later also to be known as the service for Products and Civil Buildings). The resident of Batavia had his own 1st class engineer, the above-mentioned J. Tromp (1818-1853), and a staff of some ten people; the resident of Soerabaja had a 2nd class engineer (P. van der Hilst) and in Semarang there was a 3rd class engineer (H.A. Benit). All these engineers were in actual fact the first civil engineers in the East Indies. They were no longer affiliated to the army. In all the remaining residencies it was specially delegated supervisors who were responsible for structures and for the various hydraulic works.

When, in 1830, the Cultivation System was introduced: the system involving the compulsory cultivation of cash crops, the demand for roads and water supplies only increased. In many cases it was the local population that first set about tackling such projects but at a later stage they needed assistance. In such cases ambulant engineers were randomly appealed to for assistance (in 1854 that was some seven 3rd class engineers) and they were sometimes assisted by engineers in training who were known as 'élèves'. One of the structural problems was that of the corvée services or forced labour. Much was done in an improvised fashion and on the basis of insufficient knowledge. On top of that there were a number of ambitious plans that were either never realised or which simply ended in disaster. The governor-general J.J. Rochussen (1845-1851) was certainly aware of these shortcomings: 'each resident works, endeavours or strives on his own', all of which resulted in waste. In 1847 he wrote a letter to the Minister for the Colonies, J.C. Baud (1840-1848), in which he pleaded for a more systematic approach to irrigation and transport and one that might be based on thorough research. To his mind this would require 17 fully qualified engineers.

The establishing of Public Works

The ideas put forward by governor-general Rochussen were favourably viewed by the Minister for the Colonies, C.F. Pahud (1849-1856), and resulted in the institutionalisation of East Indian Public Works. By Royal decree an independent service for civil public works was established on 4th November 1854 (i.e. the Bureau of Public Works) and it fell directly under the governor-general. The service was responsible for the supervising of public works activities and for maintenance which, at that time, included bridges, harbour works, canals and major structures. Management was delegated to the various regions. Just like the Directorate General of Public Works and Water Management (Rijkswaterstaat) in the Netherlands, the service comprised centralised and decentralised elements. The centralised component embraced the Public Works bureau established in Batavia that fell under the supervision of the Waterways inspector and the director of Public Works. The decentralised component was divided into some eight regional Public Works divisions (or inspectorates), four of which were situated in Java. Each division had an engineer and a number of supervisors. In the main division in Java, Soerabaja/Brantas, the incumbent was a chief engineer. The other three divisions in Java were headed by 1st class engineers whilst the divisions outside Java had to make do with 2nd or 3rd class engineers.

The bureau was divided into a Technical Office (including a 1st and a 3rd class engineer and two supervisors) and an Administrative Office (headed by a principal clerk supported by various assistant clerks). According to the Royal decree of 1854 the entire organisation was to comprise 82 employees. In the relevant budget a sum of 289,000 guilders had been set aside for the new service. G.H. Uhlenbeck, a former army major, was made the first director. Indeed, in a capacity as advisor, he had already been involved in the preparations.

There were statutory training demands for all Public Works engineers. The position of 'aspiring engineers' (i.e. those in training) facilitated the entry of engineers from the Polytechnic School (the Royal Academy) in Delft into the East Indian engineering corps. Engineers were given impressive official uniforms befitting their respective ranks and they were not allowed to accept any kind of special payment alongside of their normal salary. In short, this constituted an important step forward for this particular professional group as far as status, discipline and identity was concerned. The fact that in daily practice they

Irrigation in the Tjihea area, Preanger Regencies (West Java), 2003. The Bureau of Public Works (1854) was established following the Javanese famines. The creation of irrigation works for rice crops was therefore also one of the service's main tasks. After 1890 that occurred in a structured fashion, also in Tjihea area.

were nevertheless civil servants in positions subordinate to those of various administrators frequently gave rise to friction. The staffing of the service demanded a degree of creativity. Not all of the engineers by a long way were in the possession of a diploma from Delft. Many were locally recruited from the military ranks or were drawn from supervisory circles. In order to be considered for an engineering rank this group of candidates had to sit exams before an engineering committee chaired by the chief engineer.

The ambitions of the engineer De Bruyn

Not long after the Bureau of Public Works had been created the director, Uhlenbeck, received (in 1858) a memorandum on the expansion of East Indian Public Works from engineer H. de Bruyn. De Bruyn was a man with vision. Having been educated at the Military Academy in Breda he had gone on to make a career for himself in the Soerabaja division where major river projects had proceeded to be executed under his supervision, such as the diversion of the Solo estuary and the distribution and channelling of the water of the Brantas. Uhlenbeck respected the plan but asserted that he could not pass it on to the governor-general. When, however, De Bruyn visited the Netherlands and brought the memo to the attention of the Minister for the Colonies, J.J. Rochussen, the latter was very impressed. Unfortunately, though, he estimated that the time was not yet ripe for such plans. Nevertheless Rochussen came to view De Bruyn as a valuable advisor. One of the central themes in the recommendations provided by De Bruyn was that more attention should be paid to practical know-how. The young engineers graduating from Delft in ever-larger numbers lacked, to his mind, the necessary practical skills and were not familiar with the very specific geographical circumstances. He therefore lobbied for study trips to be organised to destinations such as the Po Plain in Italy and the floodplains of the Loire.

To that end, De Bruyn actually organised a study trip to Italy in 1860 and he took with him two young engineers from Delft (S. Westerbaan Muurling and A. van Lakerveld). In 1861 he sailed to the East Indies with three engineers (two of whom were T.J. Stieltjes and J. Dixon). The idea was that the engineers could do research into ways of improving transport facilities. De Bruyn would accompany them and instruct them. Hardly had De Bruyn set foot on what was to him familiar East Indian soil when he was engaged to help with flood problems. On 13th April 1861 De Bruyn succeeded Uhlenbeck as director of Public Works. His vision in the field of water management was quite staggering. He soon stipulated the necessary size of the service: 181 engineers, 30 architects and 515 supervisors.

On 26th December 1864 Public Works was expanded as planned but the growth was especially in the higher ranks and not, as De Bruyn had foreseen, in the lower ranks. The director of Public Works was given:

- two 1st class chief engineers and two 2nd class chief engineers
- eight 1st class engineers, twelve 2nd class engineers, twenty 3rd class engineers and twenty aspirant engineers
- a number of architects and overseers.

Not all of those engineers were in the possession of a diploma from Delft but within ten years that was to become one of the basic requirements. The expansion of Public Works was, in part, a consequence of new technological developments. As of 1863 the mining

Table 1. **List of names of Dutch East Indian Public Works directors**

G.H. Uhlenbeck	Nov	1854	
H. de Bruyn	April	1861	
P.J.G. Beyerinck	Jan	1868	
W.H.F.H. van Raders(baron)	Febr	1870	
H.de Bruyn	Aug	1874	

H.J. Bool	June	1877	(former secretary)
T.C.J. Kroesen	April	1879	(former auditor's office administrator
H.L. Janssen van Raay	Dec	1884	(former postal services director)

M.J. van Bosse	June	1889	(former head of harbour works)
G. van Houten	Sept	1892	
J.E. de Meyier	June	1898	
H.P. Mensinga	Oct	1901	
A.P. Melchior	June	1905	
W.B. van Goor	Aug	1908	
J. Homan van der Heide	Febr	1911	
P.J. Ott de Vries	Sept	1914	
A.G. Allart	July	1920	
J.W. de Bruijn Kops	May	1922	
J. Blackstone	May	1925	
E.H.M. Uljee	Sept	1927	
J.A.M. van Buuren	May	1930-Dec 1933	

Source: Ravesteijn, *De zegerijke heeren der wateren. Irrigatie en staat op Java, 1832-1942*, appendix K.

industry fell under the bureau's sphere of influence, as of 1865 steam engineering and, as of 1862, telegraph and postal services but much of this was to be of short duration. As a result of expansion and reorganisation instigated by the colonial administration in 1866 when general management departments were created and Public Works was made a department in its own right (i.e. the Department of Civil Public Works, abbreviated to BOW: Burgerlijke Openbare Werken) the mining sector disappeared. It was transferred to the new Department of Education, Service and Industry. Later (in 1907) the Department of Government Companies became responsible for the post, telegraph and telephony services. In 1875 BOW received a new offshoot: the State Railways. At the same time, plans for the new harbour in Batavia were in full swing and so BOW's interest in harbour works also intensified.

Engineers versus clerks

The list of names of Public Works directors (see Table 1) demonstrates that they succeeded each other at quite a speed. The average term of office was less than four years. Not infrequently they were forced to leave due to illness or because of the psychological pressures, as was the case with De Bruyn. On 10th January 1868 De Bruyn went off on sick leave. This might well have been linked to the fact that there were permanent tensions within the administration between engineers and the clerks, between the technicians and bureaucracy. It was notably the Inland Administration representatives, the residents who held all the power in the regions, who felt that their power was being undermined. It was a conflict that continued against the backdrop of modern colonial state development in the East Indies where the government started, around 1870, to withdraw from the economic arena so that colonial exploitation automatically fell into the hands of private enterprise while at the same time the state continued to present itself as a champion for business and for the people. In the long-term it was a trend that would help the engineers but it was not a smooth path.

In 1874 De Bruyn once again became director. In the interim period he had been chairman of the committee for the design of irrigation rulings and a member of the Batavia seaport committee. After De Bruyn's second period in office the Public Works directorate fell into 'administrative' hands with the appointment of H.J. Bool. By Royal decree of 25th November 1877 it was determined that BOW engineers must follow the directives laid down by the Regional Administrations. The residents gained more power and bookkeeping surveillance was considerably stepped up. Three chief engineers were sent packing, which led to much upheaval within the corps. T.C.J. Kroesen introduced economy measures and under H.L. Janssen radical cutbacks were seen in 1885.

Meanwhile the way affairs were being run was coming under sharp attack, not least from the engineer C.L.F. Post. A former East Indian Public Works engineer, he wrote a book[13] in which he denounced the intervention of the 'non-experts' with various technical affairs. The book in question was even debated in parliament. The government authorities were in a difficult position: on the one hand the engineers were right, the East Indian public works situation desperately needed to be improved, both for the local population and for the agricultural enterprises but on the other hand the colonial administration was making a loss and cuts needed to be made. Sound council was scarce but it did exist in the form of the engineer praised at the beginning of this chapter, De Gelder. After his successful harbour works in Batavia he became the instigator of the new East Indian Public Works department ruling that was introduced in 1885 and which, despite intrinsically linked cutbacks within the department, placed engineers in the position they had so long wanted to be in.

The public works ruling of 1885

The Royal decree of 10th February 1885 precipitated sharp Public Works personnel reductions, especially in the higher ranks. The body of engineers was virtually halved when numbers were reduced from 75 to 39. The numbers of architects and supervisors fell from 164 to 159, which was a relatively minor reduction. In October of that same year further developments followed in the shape of the new public works ruling. The department was divided into the General Division including the Technical Office and seven regional Public Works divisions (five of which were on Java) and the Regional Division where predominantly architects and supervisors were active and where the residents had the final word.

What was new at the time of this ruling was the establishment of what was known as the Irrigation Brigade consisting of some four engineers headed by a 1st class engineer (W.F. Heskes). This service unit was given an important mission. BOW would first of all have to improve existing irrigation facilities. In addition to that the densely populated areas that lent themselves to it would have to be irrigated and finally 'waste ground' would also have to be taken in hand. The approach adopted was amazingly academic. The brigade was required to measure, observe, test and experiment. Suggestions could also be put forward but after that the actual work had to be handed over to the General Division which, for those purposes, was given access to its own specialised Technical Office. Apart from being weighty, the Irrigation Brigade's mission was also politically sensitive because of the direct link between irrigation and public affluence. The consecutive chief engineers of the Irrigation Brigade were: W.F. Heskes, J. Nuhout van Veen, A.P. Melchior*, C.W. Weijs, P.J.G. Beyerinck, J. Homan van de Heide*, J. Haringhuizen and J.H. Thal Larsen (* later also BOW directors). At times the brigade was a little overzealous in its approach 'by exaggerating the theoretical side of things while paying too little attention to smaller works'.[14] It was therefore not long before the BOW director intervened by making the head of the brigade director of a new branch of the Technical Office (known as Irrigation and Drainage) so that the Public Works hierarchy could once again be perpetuated. Under BOW director Van Bosse, irrigation affairs started to expand in another respect. As of 1888 Irrigation Divisions were created according to river catchment areas: 1888 Serajoe, 1892 Brantas (engineer C.W. Weijs), 1892 Demak (engineer J.C. Heyning), 1907 Pekalen, 1908 Pemali-Tjomal, 1909 Madioen, 1916 Tjimanoek. In 1890 a General Irrigation Plan was drawn up for the whole of Java which comprised 19 projects. The underlying idea was, in both cases, to achieve more cohesion and mutual cooperation.

Personnel

Despite the relatively attractive salary, the incentive amongst Dutch engineers to travel to the East Indies in the mid nineteenth century was not that great. The lack of cultural activities, the unhealthy climate and the minimal medical facilities in those parts were undoubtedly contributory factors. Moreover at that time there was also plenty of work for engineers in the Netherlands. In order to make the prospect of going to the East Indies more attractive to engineers a ruling was introduced in 1874 whereby upcoming Delft engineers could be enticed to take the plunge because of the favourable conditions. They were, for instance, offered a study allowance and the opportunity to be seconded to a major European works project for one year after completing their studies. It was a ruling that only existed for four years and which only produced 19 East Indian engineers, though it included such outstanding talent as the likes of A.G. Lamminga.[15] After the turn of the century it even became necessary to engage foreign engineers. In 1920 the foundation of the Technical College in Bandoeng offered some solace during the latter days of the colonial era (see Table 2).

Railways

Transport by rail offered certain definite advantages for the transportation of products such as sugar at the time of the Cultivation System and coal during a later phase. But who was

Table 2. **Number of civil engineers employed by the Dutch East Indian government and, after 1854, by East Indian Public Works**

Year	Delft engineers	Foreign engineers	Bandoeng engineers	Total
1829	3 à 4	-	-	-
1829-44	5	-	-	-
1844-54	10	-	-	-
1854	35	-	-	-
1878	60	-	-	60
1888	53	-	-	71
1898	92	-	-	92
1908	109	-	-	109
1913	155	15	-	170
1918	181	20	-	201
1923	195	56	-	251
1928	207	17	30	254
1930	204	14	45	263
1932	191	14	47	252
1934	154	10	38	202

Source: *Limburg, De toekomst der academisch gegradueerden.*

to pay for the expensive laying of the tracks? Whilst private companies experimented with small local lines the Minister for the Colonies sent two respectable engineers (L.J.A. van der Kun and D.J. Storm Buysing) out to the East Indies in 1859 to investigate the matter. They were unanimous in their view that only the state could create railway lines in Java. It was not, however, until 1877 that the proposal was passed in the Dutch Lower House (the Second Chamber). Between 1877 and 1879 an engineer by the name of D. Maarschalk was responsible for the first railway line connecting Soerabaja with Malang. Formally he worked under the auspices of Public Works but in practice he operated very autonomously. As far as BOW was concerned this was a completely new field and the department was not exactly equipped with the relevant expertise. This was also apparent from the fact that after having completed the project Maarschalk returned to the Netherlands where, under the Ministry for the Colonies, he set up a Technical Bureau.

It was obvious that all rail construction activities were directed from the Netherlands. The overviews of the government funding pumped into BOW at that time clearly reveal that the money expended on railway and tramline construction was rapidly predominating (see Tables 3 and 4). In 1906 the State Railways became even more distanced from BOW. In 1909 the railways was made an independent service within the Department for Government

Companies. Within that department there was a new technological terrain that was gaining ground: waterpower. With a view to the need to provide electricity for the railways the State Railway Company's waterpower bureau and the Electrical Engineering branch of the Government Companies department created a new Waterpower and Electricity Service in 1917 and the headquarters was in Bandoeng.[16]

Decentralisation and specialization

The intensification of government activity around 1900 precipitated considerable bureaucratic corps expansion in the colonies. As a result of the Decentralisation Act of 1903, 32 local authorities (communes), 15 regions and 6 sections of regions were given their own councils, administrative powers and necessary financial independence. Many matters which had previously been the business of Public Works became – where possible – locally determined and executed. Decentralising meant bringing decision-making and the powers to decide closer to the people and that, in turn, meant that more projects could be realised more rapidly. The multi-layered bureaucratic structure did, however, also precipitate more procedural steps and delays.

Under the director, W.B. van Goor, BOW was broken down into a number of specialised services. Departments that had been reorganised or newly created took over tasks which had previously been the domain of BOW. As has been mentioned, the railways was transferred in 1909 to the Department of Government Companies. As of 1924 sanitation fell under the Department of Education, Service and Industry (the Public Health Service). BOW also saw various expansions:

1908 the Bureau for Public Buildings with its own Technical Office.
1911 the Technical Office for Harbour Works.
1912 the Technical Office for Bridges and Roads (as a sequel to the Road Traffic Inspectorate, established in 1908).
1915 the Technical Office for Sanitation Works.

In this way the former Technical Office was ultimately divided into five (technical) sections: Irrigation and Drainage, Buildings, Bridges and Roads, Harbour Works, and Sanitation Works.

During these decades the labour market was expanding rapidly for engineers. With the arrival of the motorcar kilometres of asphalt had started being laid. Ethical Policy and the growth in trade stimulated the construction of schools, official residences and offices. Urbanisation prompted city expansion, drinking water provision, sewerage works and measures taken to combat bandjirs. After Java had been tackled irrigation technology would have to spread systematically throughout the various Outer Regions (the islands outside Java and Madoera). After the success seen in Tandjok Priok, various other cities started to create harbours of their own so that they could link up to the existing network of harbours and scheduled services. The railways no longer confined itself to freight transport but instead became increasingly important in the area of passenger transport, which demanded certain levels of speed and comfort.

One continual point of attention and debate pertained to the scale best adopted for the effective tackling of projects. In the case of harbour activities the authorities experimented with contractors in the hope that this would help to speed things up but that was unsuccessful. In a drive to raise efficiency much use was made of standard design in both the irrigation and building sectors. In the case of functional irrigation works that was not a problem at all but in the case of housing, government buildings and other representative structures this policy was criticised by 'thoroughbred architects'. In view of the fact that the realisation of large-scale public works demanded big investments and big risks, engineers were frequently brought in to research issues and provide recommendations.

Generally speaking the 1900-1920 period was a golden age for engineers stationed in the East Indies. They were in no way hampered by the First World War. In fact very much to the contrary sectors like the railways and the coal mining industry flourished. The establishment of the Technical College in Bandoeng and the General Engineering Congress of 1920 were the crowning glory of the work of engineers in the East Indies.

The dismantling of BOW

The year 1925 saw the further decentralisation of administrative matters when provinces were created and, in the process, waterways and waterworks were delegated to provincial public works services. It was just a few of the large-scale irrigation networks that remained with BOW. The transferring of personnel to the provinces was a laborious process and before the operation had even been fully completed the BOW department was ordered by the government to set up an investigation into the efficiency of the provincial public works services. By the early thirties, the time when all of this was taking place, all kinds of economic measures were being proposed. Decentralisation became a covert way of introducing cutbacks. Alongside of irrigation, bridges and roads switched to provincial authorities and in 1935 that was followed by buildings. Already in 1924 Sanitation Works had been transferred to the Public Health Service (under the Education, Service and Industry department) where further cutbacks proceeded to be made. Indeed it was with a view to economising that on 1st January 1934 the new Department of Transport and Public Works was created which incorporated BOW and Government Companies. Shortly afterwards the Buildings section and the Bridges and Roads section were merged and personnel reductions followed. After the Second World War, the Department of Transport and Public Works was renamed the Department of Public Works and Reconstruction.

Calculation, planning and rationalisation

Many of the complaints voiced in the area of public works could be directly or indirectly traced to a lack of control often in connection with finances. In the case of the less successful projects it gradually became more customary to afterwards (sometimes as a result of questions raised in the House) set up parliamentary committees to investigate matters. As state intervention intensified, so the demand for preliminary control increased. By working on precisely calculated and reliable figures people endeavoured to produce plans that were above all else realistic. Obviously this was something that suited the engineers down to the ground as long as the calculations were confined to their own fields of expertise. However, it became more and more the task of Public Works to completely calculate the plans in an integral and all-embracing fashion before finally drawing the relevant conclusions. Eventually it was no longer the case that the engineers were able to independently determine everything. Agricultural, sanitary and socio-economic facets also came into the picture and thus also the voice of administrators, agricultural experts, economists and other developers.

Regarding the irrigation works, it was only after many exploratory studies had been completed that the above-mentioned General Irrigation Plan for the whole of Java was launched. The Rentability Committee (i.e. Profitability Committee) that had been created in 1893 subsequently adopted the habit of assessing the separate plans according to economic feasibility. The agriculturalists, later engineers, who had been educated at the Agricultural College in Wageningen formed an influential group within the irrigation sector. In 1905 the Department of Agriculture became their headquarters. Irrigation plans were linked to crop plans and to that end agricultural research was carried out in Buitenzorg where the country's plant centre was situated. There attempts were made to discover what constituted the optimum crops and the best cultivation methods. Increasing use was made of various research methods which meant that all the following laboratories came into the picture: the Hydrodynamic Laboratory (1927, Bandoeng Technical College), the Provincial Hydraulic Laboratory in Semarang (1930) as well as the Soil Mechanics Laboratory (1930, Bandoeng Technical College). The General Water Ruling, drawn up in 1936, with its precise regulatory stipulations represented for some the height of scientific planning. For yet others it represented the height of technocratic interference.

When it came to the matter of road construction the data issuing from the Metal Research Laboratory (1908, converted in 1912 to the Laboratory for Materials Research) was particularly influential. In particular it was information concerning the qualities of the new material known as asphalt that was required. From 1913 onwards road network plans were drawn up that were backed up by traffic projections and statistical research. In the case of the sanitation works, it was the Laboratory for Technological Hygiene and Sanitation (1935) affiliated to the Public Health Service that was able to provide the necessary scientific bases for the various plans and projects. Where drinking water supplies were concerned which came under two different departments, a Central Committee for Drinking Water was established in 1936 in order to encourage cooperation between the two factions.

With the installation of the People's Council in 1918 and the rise of further political input on the part of the population, control from those quarters also started to become apparent. In short, the promotion of prosperity was no longer a convenient slogan that was readily believed. It was something that had to be proved and made convincing. Calculations and calculators had become indispensable. The economic crisis of 1921-1924 and later the structural and prolonged period of economic crisis and depression of the nineteen-thirties not infrequently gave such calculation work negative connotations.

The civilisation offensive

Established in 1854, the development of Dutch East Indian Public Works ran more or less parallel to the rise of modern imperialism where the process of colonial state formation came to form an extension of state support for private enterprise and the protecting of the interests of the population. We say more or less parallel because there were always many

Table 3. **Relative government public works expenditure and spendings in other policy areas in the Dutch East Indies (in percentages)***

Policy areas	1871	1881	1895	1905	1913	1921
Justice	4.2	4.4	4.9	4.5	3.2	2.3
Finance	7.8	8.9	11.7	17.5	19.6	17.0
Tax authorities	*3.1*	*3.1*	*2.8*		-	-
Pensions etc.	*3.6*	*4.4*	*7.6*	*8.3*	*5.6*	*4.7*
Inland Administration	45.3	32.0	22.9	20.1	12.2	9.8
Administration	*7.3*	*6.4*	*9.2*	*7.9*	*3.8*	*1.8*
Cultures	*30.1*	*19.7*	*6.0*	*4.0*	-	-
Education	1.7	2.4	3.4	3.7	4.1	3.5
European	*0.7*	*1.0*	*1.7*	*2.0*	*1.7*	n.a.
Javan	*0.3*	*0.8*	*1.0*	*1.1*	*1.9*	n.a.
Healthcare	0.8	1.3	1.6	1.7	1.9	1.7
Mining/industry	4.6	4.2	6.1	7.5	3.4	2.5
Public works**	8.5	17.1	20.4	18.5	37.7	39.2
Railways	*0.7*	*6.8*	*8.1*	*8.9*	*11.3*	*16.4*
War department	19.2	22.0	21.5	21.7	12.3	11.5
Marine department	7.9	7.7	6.0	5.6	5.6	4.4
Interest etc.	-	-	1.4	-	-	8.1
Total	100.0	100.0	100.0	100.0	100.0	100.0

* For the years 1871-1895 the figures are estimates and for 1921 they are provisional.

** Apart from expending money on irrigation works, it was also spent on roads, bridges, public buildings and government companies such as: the railways, and the postal and telegraph services.

Source: Booth, 'The evolution of fiscal policy and the role of government in the colonial economy', 215, 224.

potentially disruptive factors looming on the horizon; factors varying from crop failures to famines[17] and economic depression and from unrealistic expectations to fraud and failure, arrogance, incompatible temperaments and personal protectionism. War and the politics of opportunism were also capable of turning the tables.

The rise of East Indian Public Works was coupled with an emancipation struggle on the part of civil engineers. They were up against the clerks, the non-technical civil servants involved in administration. De Bruyn gave the organisation a push in what was, for the engineers, the right direction. It was especially Bool who knew how to curb the power of the engineers by bringing in a Public Works ruling that was favourable for the administration and unfavourable for the engineers. De Gelder saw to it that things finally turned in favour of the engineers. Even though later they would have to relinquish something of their newfound power, a technocratic line was mapped out with the introduction of the 1885 ruling that would never more be strayed from but which, with the arrival of Ethical Policy, would only be reinforced. This obviously had consequences for the creation of public works. In the area of irrigation a real civilisation offensive had been mobilised. The harbour in Batavia was improved but new harbour works, both in Batavia and elsewhere, would soon follow suit. Rail construction took off at quite an accelerated pace. (See Tables 3 and 4). In general it may be asserted that the process of modernisation in the East Indies, as well as the evolution process of present-day Indonesia, which not only included the infrastructure but also the economy and the region's politics, had been given quite an impetus. In the coming chapters we shall see where all of that would lead.

By the end of the colonial era much had been achieved but there was still also much to be done. An overview created on the occasion of the tenth *dies natalis* (or foundation day) of the Bandoeng Technical College in 1934 illustrates this very point (see Table 5). It must be noted that in that year some 30 percent of all Dutch civil engineers were offered employment in the East Indies but the ratio of seven civil engineers to every million inhabitants in the colony as opposed to 115 to the same number in the mother country is certainly staggering and gives an idea of the immensity of the scale of the public works still to be realised in that sprawling archipelago. All the development work therefore continued, even after independence. Some chapters (especially the ninth one) will show just how colonial planning came to play a special part in that particular process.

Table 5. **Overview of the civil engineers working in the Netherlands and in the Dutch East Indies in 1934**

Civil engineers	**Numbers (rounded off)**	**%**	**Per million inhabitants**
In the Netherlands	920	70	115
In the East Indies	420	30	7
Total	1340	100	20

Source: 'Viering van de 10e dies natalis der TH Bandoeng'.

Table 4. **Inflation-corrected government spending devoted to public works in the East Indies (in millions of guilders)**

Year	Irrigation works	Harbour works	Post and transport	Rail and tramways	Year	Irrigation works	Harbour works	Post and transport	Rail and tramways
1871	0.1	-	0.1	-	1906	0.5	0.1	0.8	7.5
1872	0.1	-	0.0	-	1907	0.3	0.1	0.3	6.0
1873	0.2	0.1	0.1	-	1908	0.5	0.5	-	8.0
1874	0.2	-	0.0	-	1909	0.2	0.6	0.5	6.9
1875	0.8	-	0.0	0.5	1910	0.2	0.7	0.4	9.2
1876	0.9	0.1	-	4.0	1911	0.3	1.7	1.5	14.3
1877	1.2	5.7	0.0	3.8	1912	0.3	4.0	1.8	16.6
1878	0.9	0.9	-	3.3	1913	0.3	7.7	3.9	38.8
1879	0.3	4.0	0.0	6.8	1914	0.2	13.0	2.0	39.4
1880	0.4	3.4	-	8.1	1915	0.2	13.0	1.5	34.7
1881	0.4	2.7	-	7.9	1916	0.2	10.2	2.5	32.9
1882	2.0	2.1	-	8.1	1917	0.2	8.1	4.8	30.0
1883	1.7	1.7	0.0	5.8	1918	0.4	10.0	6.1	26.0
1884	1.3	1.5	0.1	7.1	1919	0.3	13.1	8.1	37.0
1885	1.1	1.3	-	8.0	1920	4.7	20.3	11.0	54.4
1886	1.1	0.4	-	6.5	1921	5.8	37.1	15.9	69.6
1887	1.1	0.4	0.1	4.1	1922	3.8	25.1	7.9	64.7
1888	1.4	-	1.1	7.1	1923	3.2	15.6	2.6	54.8
1889	1.6	0.3	0.0	6.2	1924	3.5	4.4	1.3	22.8
1890	1.6	0.3	-	7.1	1925	3.9	2.0	1.2	27.1
1891	1.8	0.1	-	9.1	1926	3.8	3.5	6.6	16.7
1892	1.5	4.0	0.8	8.9	1927	3.1	2.3	3.0	15.1
1893	1.9	0.4	-	6.8	1928	3.3	1.9	4.0	21.6
1894	2.9	0.6	-	6.6	1929	3.7	2.1	6.0	19.8
1895	3.2	0.3	-	8.8	1930	3.2	2.8	6.2	14.8
1896	3.4	0.1	-	8.9	1931	2.0	0.7	2.9	12.5
1897	3.0	-	0.3	9.4	1932	0.9	0.1	1.4	2.2
1898	3.4	-	-	14.0	1933	0.7	0.1	1.5	0.9
1899	1.2	-	-	10.7	1934	0.4	0.1	1.2	0.3
1900	0.8	-	-	10.5	1935	0.6	0.2	0.7	0.4
1901	0.8	0.2	0.8	9.3	1936	0.5	0.2	0.8	0.4
1902	1.3	0.3	-	16.4	1937	0.5	0.2	3.1	1.0
1903	1.1	0.2	2.2	19.6	1938	0.4	0.6	1.2	1.7
1904	1.2	0.3	-	9.5	1939	0.6	2.0	1.3	2.3
1905	0.8	0.1	1.2	9.3	1940	0.6	1.9	1.3	-

Source: Mansvelt and Creutzberg, *Changing economy in Indonesia. A Selection of statistical source material from the early 19th century up until 1940, volume 3*, Table 2.

When, in about 1920, roads were constructed in Bali that was still done in the 'old' way.
Many labourers were hired to lay stones,
probably intended for the subsurface. The
whole road would then be finished with a
layer of chippings or aggregate. Small dykes
were also created alongside of the roads.

CHAPTER 2 **THE ROAD TO A NEW EMPIRE**

Road construction organisation and techniques

MARIE-LOUISE TEN HORN-VAN NISPEN

Herman Willem Daendels (1762-1818), governor-general of the Dutch East Indies (1808-1811).

The history of road building in the Dutch East Indies goes back further than 1800. Alongside the old locally created routes, the United East India Company constructed a number of roads between important posts, and in areas where such connections were perceived as being strategically opportune. Connections were created between, for instance, Batavia, Bantam and Buitenzorg. These routes were not always completely passable for people and horses. They were grassed over roads, unsurfaced cart tracks or footpaths. From travel journals and reports dating from the end of the eighteenth century one can deduce that generally speaking the roads were either badly maintained or not maintained at all. Footpaths in woods were overgrown, various stretches of road were only accessible in the dry season and any kinds of cross-river links were completely lacking. It thus took eight days to travel from Batavia to Tjipanas, partly by buffalo cart and partly on buffalo-back. Tjipanas, high in the mountains of the Preanger Regencies, was a convalescent home for sick 'Company Servants'.

In the nineteenth century, roads were of strategic and economic importance. The first half of the century saw much activity in the area of road construction. As the century progressed, so the demand for transport and means of transport increased. The rise of the car and especially of goods vehicles at the start of the twentieth century necessitated a totally different, more systematic approach to road construction. The years between 1915 and 1920 saw a clear turning point, both from organisational and technological points of view. Road network plans began to take shape and to be implemented. Within the Dutch East Indies the political and economic centre of gravity was in Java. It was that power which ultimately dictated the extent and quality of the road network. By the end of the nineteenth century, the road network in Java was already quite extensive. Of all the other islands, it was notably in Sumatra that the quality of the roads was, for the time, quite good.

THE 1800 TO 1870 PERIOD

Organisation

The coming into office, in 1808, of the governor-general, H.W. Daendels, heralded a change in Javanese road building policy. The reasons for this were military and economic. Java had to be prepared to defend itself against the British, all of which meant that it had to be possible to speedily move and supply the relevant troops. Others who also benefited from the good road networks were the civil servants with their reports and the cultures entrepreneurs who were becoming increasingly important. Whilst journeying through Java in May 1808, Daendels drew up a plan to construct a road between Buitenzorg and Karangsemboeng (in Cheribon). Several days later he decided to also create a west-east connection between Anjer and Panaroekan. As existing *desa* roads were improved and linked up to the network, so large sections of the various routes were rapidly completed. By contrast, road building in 'desolate mountainous' parts, coastal marshlands and across inland seas was laborious, made many ill and claimed a lot of lives. The roads in the Preanger Regencies, for example, had to cross the slopes of the Goenoeng (mountain) Megamendoeng and the Goenoeng Masigit (between Tjiandjoer and Bandoeng); in Cheribon there was the swampy area to be negotiated. According to the governor-general's resolution, work on the road between Buitenzorg and Soemedang should commence in the 'coming dry season as soon as the inhabitants' had 'finished the coffee picking, transporting and rice harvesting'.[1]

It was the head of the engineering corps who was put in charge of the entire project. He was able to call upon a number of military engineers. With the exception of the structural works and a few complex parts of the route, most of the construction workers deployed were unpaid labourers, thus 'not an expense to the state'. For the sections of the route around Soemedang and Cheribon, the road workers were paid. The 1000 workers who worked in Cheribon received '4 two-and-a-half guilder coins per month as well as 4 gantangs of rice'. The Soemedang road workers received 2 two-and-a-half guilder coins and 3 *gantangs* of rice.[2] A certain percentage was *boeijangers*, that is to say, labourers without a family. They were also deployed along other route sections where it would be determined, per section, how many 'heads' would be deployed and just how much money was available. The workers were paid in rice and salt. The supervisory work was done by a number of European 'invalids' who received 24 five-cent pieces a day.

By mid-1809 some 1000 km of road or '300 hours of travelling time' had been completed. The travelling time from Batavia to Soerabaja had been considerably reduced, both for postal services and for passengers. Before all the road building was undertaken, it used to take three weeks in the wet season and two weeks in the dry season for letters to reach their destination whilst in the dry season passengers had to allow a month's travelling time. (In the wet season 'travelling was deemed impossible for white people'.) Once the new road connection had been created it was possible to establish regular postal services with stagecoaches leaving from Batavia and Soerabaja twice a week. Letters arrived within six to seven days and passengers were able to travel the distance in nine to ten days, depending on the time of year. Express post could be delivered within four to five days.

The Great Post Road leading from Buitenzorg to the Preanger Regencies (West Java). In the middle of the road one can see the 'post' where the horses would have been changed. The dykes skirting the road are also clearly visible.

In various respects, Daendels proved to be quite a supporter of French methods. Apart from getting the idea to create road connections from the French, he also copied the French idea of changing horses. Along the new postal route horses were therefore stabled at 'stations' created at 8 to 9 km intervals so that not only the stagecoaches but also people using horse-drawn carriages (or travelling on horseback) could hire horses or change them.

Apart from the Great Post Road, a coastal route was created between Batavia and Cheribon, a connecting north-south military road was built (to Soerakarta and Djokjakarta) and various major inland routes were created. Horses were once again stationed along the sections where roads had been improved so that carriages drawn by teams of four or six horses could comfortably reach their main town destinations. It was not, however, easy to negotiate every incline in the mountainous regions. Not infrequently the roads had longitudinal slopes of 1:7 or 1:8. At such junctures two, four or even eight buffalos would be hitched up just to pull the coach up the hill and invariably the coachman and passengers would have to walk alongside. Going downhill things went 'at a terrifying speed even with blocks on the brakes' and the coachman would have to muster all his skill to avoid having an accident. Otherwise the roads were, according to Daendels, 'comparable in their sheer perfection to French roads' whilst according to yet others they were comparable to 'kolf courts' (i.e. very flat).

Daendels reports that apart from being used for military ends these roads were also used for non-military traffic. Stagecoaches, private carriage owners and civil servants all made regular use of the roads. For the ordinary Javanese population, though, they were not accessible, neither for transporting their cattle, nor for their transport carts (i.e. *tjikars*). For those particular transport purposes separate cart tracks were created alongside the postal routes, which were not surfaced and did not provide bridges.

It was Daendels who had given the start sign for the construction and improvement of the road network. Afterwards, the job commenced by him continued to be pursued with such verve and on such an extensive scale (with the help of unpaid labour) that the governor-general decided, in 1835, to put a stop to all road-building activities and to declare that no new roads should be created without his prior written consent.[3] Some 20 years later, in 1853, it was decided that no more separate roads for the transport of goods should be constructed and in 1857 the government decided that the postal routes should be opened up to all traffic, provided that the vehicles first met a number of requirements. Nevertheless, since there were already a number of cart tracks running parallel to the post roads, people continued to make use of those alternative routes, including the fords in rivers.

With the mid-century blossoming of all the cultures activities came a growing need for transport facilities. The various agricultural enterprises used a combination of properly surfaced and unsurfaced roads to transport and distribute their products. When it came to the matter of public road transport, endeavours were made to expand the range of modes by introducing camels, donkeys and elephants but that was not a great success. In 1862 one Dutch parliamentarian even asked why lama transport had not been tried![4] (Ultimately the big solution would come in the form of the railways.) At that same period people were also forced to concede that the Great Post Road of which Daendels had been so proud did not fulfil the new military requirements. It was therefore decided that 'Java should be provided with a road that was strategically better positioned and more suited to military purposes'.[5]

That road would follow a southerly route and would pass through such places as Djokjakarta and Surakarta. In 1854 a start was made on that particular road and 20 years later various parts of it were completed.

On the other islands all the various road networks were created later than those of Java. Generally speaking it was military operations and the 'increasing domination of the inland areas' that gave rise to such road-building projects. For instance, at the beginning of the nineteenth century two main roads were created on the island of Sumatra leading from Padang to inland destinations. One of the roads passed through the Anei crevice while the other went through Tikoe to Loeboek Basoeng and Matoea. The road that passed over the gorge was very suitable for passenger and goods transport by horse but only partially suited to carriage transport. The other road was more all-round serviceable and linked up with a big connecting road between Fort de Kock and the Bondjol valley.[6] The main roads ran parallel to the mountains. A smaller network of roads was created by connecting up the main roads with various side and transverse roads. Part of the network led from the main road and through the mountains to the coast. In the Outer Regions, the 1840 to 1870 period was marked by political restraint but on the island of Java road building simply continued apace.

Technology

Governor-general, Daendels determined that where possible the Great Post Road should be two Rhineland measuring rods wide (circa 7.5 m). Every 400 Rhineland rods (circa 1.5 km) there would be a roadside post, not only to indicate distances but also to identify road maintenance sections.[7] One important requirement was that the road should also be passable in the wet season, both for carriages and other vehicles. That meant that the road had to be surfaced and that a number of large structural works had to be created. The available military engineers visited the relevant spots to determine just where and how the road should be constructed. As was customary, the road route was divided into sections and each engineer was made responsible for one particular section.

Though Daendels had determined that the bigger roads should be around 7.5 m wide they often ended up 10 to 12 metres wide. What was used for the surfacing was rubble, gravel or a combination of both. It was only in the built-up areas of larger cities that actual paving was used. The roads were given no sub-surface because up until 1857 no heavy traffic was allowed to go on the main post roads. Alongside the roads, small dikes were formed to mark them off from the remaining landscape.[8]

A comparison with the Netherlands

Up until the time of French rule, road construction had not really featured as a major issue. In 1811, however, Napoleon came up with a road plan which involved extending all the main roads leading from Paris to all corners of the empire and that, of course, included the Netherlands. To a large extent, King Willem I took over the plan and adapted it to the situation of the day. Up until around 1840 many large roads continued to be constructed and surfaced which meant that the resultant effect was a good network of through roads for that particular period in history. After 1840 the government turned its attention to waterways and railways. In several provinces, though, road construction work did continue so that extensive networks emerged.

THE 1870 TO 1920 PERIOD

Organisation

After 1870 roads started to appear in the Outer Regions, again with multi-purpose goals. What took precedence was the political and military interests or, as the engineers F.G. Dumas and J.H. Blok of the Bridges and Roads section of the Department of Civil Public Works put it: 'in order to retain power it [was] imperative that the inland areas be easily accessible'. In the second place what also had to be considered was economic interest: people wanted the cultures to flourish and transport facilities to be available. Finally there was the consideration that had more to do with freeing regions from their isolation.[9] In many places the new roads provided connections between the industries and the railways while in other places they made it possible for companies to be set up in parts that would otherwise be difficult to access. The rise of tin mining on the islands of Banka and Billiton made it essential for roads to be constructed there and new roads were also built in Bali, Lombok and South Celebes. Invariably such construction work would be stimulated by local Residents, just as happened around 1870, for instance, at the instigation of the Resident of Palembang in Sumatra. At the other end of the spectrum in the old East Coast Residence, in Sumatra, road building did not start until a relatively late stage for the simple reason that the rivers there were perfectly navigable so the alternative transport facilities sufficed.

Towards the end of the nineteenth century the roads in Java were divided into various categories: the large post roads, the large inland roads and the less important inland and *desa* roads maintained by unpaid labourers. It was not, however, a differentiation that was generally made. The circular letter sent out by the director of Public Works in September 1887 also differentiated between three types of roads: the main post roads with a lot of traffic, the less important routes, and the mountain roads and 'less important inland roads'. Sometimes the categorising was done according to whether the roads were or were not maintained by unpaid labourers and sometimes the criterion was whether or not the roads were surfaced.

On the island of Java, goods transport took up a central position. Often transport came in the form of two-wheeled freight carts pulled by water buffalo. Especially in the rainy seasons the roads would be badly damaged by such carts and would thus sometimes become unusable. At such times everything possible would be done locally to restore things to normality. It clearly emerges from the ordinances of 1901 and 1905 that the damage caused by freight cars was considerable. The 1901 ordinance established that *desa* roads should be viewed as 'traffic means opened to the public by the municipalities' and that such thoroughfares should be open both to pedestrian traffic and light vehicles but not to heavy traffic. If such roads were damaged by transport systems privately organised by various enterprises, then the companies themselves would be answerable for the damage. The ordinance did not apply to damage caused to roads built on the basis of unpaid manpower or to structural works. All

Not only Java but also Sumatra was active in the creation of roads. As East Sumatra possessed good navigation routes, road construction came to that area later. The waterways did link up with the country roads. Round about 1878 a section of the road in the residence of Djambi ran parallel to the Batang Hari. Apart from there being a connection between the road and the river (a path with steps) one can see a post for the changing of horses.

In North Sumatra, in about1910, people started working on a road that was planned to connect Gajo land with the northern coast of Atjeh. The negotiation of the mountainous area necessitated much excavation work. It became one of the few roads to provide access to that particular region.

of that was further stipulated and detailed in the 1905 ordinance. Damage caused to roads constructed by unpaid labourers by 'private agricultural or industrial enterprises' had to be compensated 'provided that such damage was greater than that which, in view of the local conditions, would have been caused anyway by normal vehicular traffic'.[10] It is not hard to imagine that this instantly gave rise to discussion as to what constituted 'general' or normal traffic and what fell into the 'other' traffic category. The ordinance was reversed in 1906. Still, the matter of who should pay for the damage done to roads built by unpaid labourers remained and the notion of what could be seen as 'normal' and 'exceptional' traffic remained unresolved.

The introduction of bicycles and cars to the East Indies towards the end of the nineteenth century gave a whole new dimension to the 'good roads' concept. It also made it necessary for road traffic to be regulated. Locally, certain stipulations were introduced in relation to traffic and it became customary to drive on the left. The first rulings designed to regulate vehicular traffic were introduced in 1900 and everything was laid down, from maximum speeds to driving licences; in 1910 came rulings pertaining to bicycles. Both sets of rulings were to constitute the beginnings of road traffic legislation.[11]

It was the introduction of decentralisation, established under the Decentralisation Act of 1903 and the 1905 Decentralisation Resolution, which was to lead to the creation of town councils and regional councils. At that point the responsibility for the managing and maintaining of the post roads and inland roads was transferred from the central powers to the local authorities. It was simultaneously determined that new roads 'and all the accompanying works' such as milestones and bridges would be funded by the regional or local authorities. For the first few years after the introduction of decentralisation it was still possible to make use of unpaid labour except for in some of the leading cities and in the Residence of Batavia. Gradually, though, where the upkeep of roads in Java was concerned, the phenomenon of unpaid labour was phased out until, in 1916, it was completely abolished. On all the other islands road maintenance continued to be predominantly based upon unpaid manpower.

As of the beginning of the twentieth century, road-building activities once again intensified and that was especially due to all the other demands being placed upon roads. The Outer Regions were also placing more and more demands on the Public Works department. Especially on Sumatra there were extensive plans to create and improve road infrastructures. In order to establish some sort of unity the Road Traffic Inspectorate was established in 1908, a move which also involved appointing a traffic routes' inspector for the Outer Regions. In 1912 the Inspectorate became the Bridges and Roads section of the Public Works department.

The East Indian government gave the Road Traffic Inspectorate the task of drawing up road plans for Java and Sumatra. What was missing in the Javanese road network were several important links and a number of bridges. Public Works designed a general road plan for Java and afterwards the heads of all the Regional Administrations were asked to provide detailed road plans for their particular regions. The total cost involved in executing the general plan was estimated at around 8 million guilders and another 35 million for the various regional road plans. In Sumatra there were a number of separate local and regional road networks. The challenge facing the people there was thus to link up those networks by creating a number of large inter-regional connections. The result would be one interconnecting and virtually entirely surfaced road between Kota Radja and Padang that would pass through Medan, Sibolga and Fort de Kock. From Padang there would then be a connection with the South Sumatra road network. Bridges included, the cost of all of this was an estimated 60 million.[12] The actual implementation of the Javanese road plan commenced in 1914. In that same year the Sumatra plans were produced and very soon afterwards work also commenced there. A road plan was also to ensue for Borneo and in the post-1914 period roads suited to modern vehicular traffic were also constructed in the Minahassa area, in Timor and in South Celebes. The budgeting for the respective expenses to be incurred by all those plans was as follows: 5 million guilders for the Minahassa area, 38 million for Borneo and 5.5 million for South Celebes.[13]

Technology

At the end of the nineteenth century (in 1893) the road networks of Java and Madoera comprised 3,300 km of large post roads, 6,600 km of main inland roads and 10,500 km of minor inland roads maintained by unpaid labourers and an unrecorded number of kilometres of minor inland roads known as 'desa roads not maintained by unpaid labourers'.[14] Ten years later the annual report published by Public Works made mention of some 23,793.5 km of post and inland roads in Java. What was recorded for the Outer Regions in the Public Works records of 1903 was almost 13,000 km of roads.[15] The precise status of these roads was not, however, mentioned.

The determining and plotting out of the various road routes was no easy task, particularly in the mountainous areas. In practice, the final choices made did not always turn out to be the best ones. Sometimes it was the case that at the time of construction alternative routes were simply not possible. What therefore often happened was that after a certain period of time the roads would be re-routed or one or more ancillary roads would be built alongside existing roads. A circular letter issued by the Inland Administration director on

21st December, 1880 (no. 12770) expressed the wish that an end be put to this situation because of the 'great pressure placed on the population that was expected to carry out such unpaid labour'. In future it would be the responsibility of the engineer employed by Public Works to determine the new road routes on the basis of 'relevant topographical maps'.[16]

The most common road width in Java in the nineteenth century was 10 to 12 m, some 5 to 6 m of which would be surfaced. The gradual reduction in unpaid labour in Java and Madoera, leading later to the abolition of such labour, was something that had slowly made road construction and maintenance considerably more expensive. For that reason the director of Public Works proposed in 1887 – by means of a circular letter (15th September no. 8030A) – that there should be three standard road widths: 5.50 m for the main post roads, 4m for less important roads and 3m for mountain roads and other inland routes. 'In the interests of safety' it was decided that the roadsides should still be flanked by small dykes. The depth of the hardened surface varied from 6 to 20 cm including or excluding the under layer. The exact depth chosen depended very much on the volume of traffic and the maximum weight of the vehicles using any one given route.

Until 1920 use was made, where necessary, of natural stone for the under layer as that was a commodity that was readily available. Alternatively coral or river stones were sometimes used. On the whole, though, it was concluded that the underlying surface was solid enough. On top of the under layer along the staked out route the road surface would then be cambered. Outside the cities such surfacing consisted of hard core or gravel which may or may not have been cemented together. Within the various cities asphalt was often used. The material used to surface the roads would then be compacted with road pounding machines or steamrollers; it was believed that simply allowing carts and cars to level out the surface was rather old-fashioned. [17]

As far as road maintenance was concerned, the director of Public Works had stipulated in his 1872 circular precisely how that was to take place. Each day a number of 'military conscripts' would be responsible for checking their respective stretches of road. Any holes that appeared in the surface or gullies that were created by carriage wheels had to be immediately filled up. It was also important to make sure that rainwater could easily be drained away. Major maintenance work would be carried out 'in the usual fashion'[18] or, in other words, by means of the annual resurfacing projects.

One important facet of road construction in the East Indies was the matter of how steep an incline in a road could be. What was taken as the maximum was a slope of 1 in 20 with, in exceptional circumstances and for the shorter distances, 1:15 as the limit. In mountainous regions it was not uncommon to come across even steeper inclines. At that time the steepness of Dutch roads was usually between 1:35 and 1:40. As was mentioned in congress reports of the day, Public Works engineers were not, however, bothered by the steepness of some East Indian roads. The stone-based surfacing remained in place even after heavy storms 'provided that the cambering was adequate'. The fact that heavy freight carts were able to negotiate slopes that even proved challenging to cars remained a wonder: 'one needs a good driver and a good vehicle to climb these slopes' it was claimed.[19]

As of 1912 Public Works had its own Materials Research Laboratory where, for instance, the material used for the surfacing of roads could be investigated. Tests were carried out to establish just how great the load created by freight traffic was. Similarly, various types of asphalt were examined.[20] In the East Indies naturally occurring asphalt could be found in Palembang, Cheribon and on Boeton. It should also be mentioned that as of 1895 asphalt was used that actually constituted a by-product of the local oil industry.

A comparison with the Netherlands

On a national scale, it was the rise of the bicycle and the car that made roads the focus of interest around the turn of the century. Dutch motoring organisations (such as the ANWB and the KNAC) started to protect the interests of cyclists and motorists. In 1915 endeavours to do something to improve roads at national level floundered in parliament. As far as the Directorate General of Public Works and Water Management (Rijkswaterstaat) was concerned 'the roads question' was just one of many topics on the agenda. A number of provinces did devote attention to road construction. It was not, however, until the end of the twenties that a systematic approach was adopted to road matters and to the possible ways of financing infrastructure.

THE PERIOD AFTER 1920

Organisation

After the First World War vast numbers of American lorries started being imported into the East Indies for the purposes of transporting the products generated by the various burgeoning enterprises (cultures, industries). Heavy loads then proceeded to be transported at relatively high speeds, all of which proved disastrous for the roads which had originally been constructed to take light loads transported at a slow pace. It therefore became urgent to adapt the roads to the new transportation requirements and so the road issue was placed on the East Indian agenda. But, just like in the Netherlands, the permanent shortage of funds for road maintenance proved to be a problem. Up until about 1916 it had been possible, just about everywhere, to make use of forced labour which meant that all the road building and road maintenance costs could easily be curbed. After 1916, though, the cost of labour became an expensive item on all the relevant budgets. It gave rise to questions such as: should the road issue be centralised or decentralised and should it be just the government that involves itself in such matters or should private parties (and enterprises) also be involved.

In May 1920 there was an international General Engineering Congress in Batavia and road construction was one of the topics of discussion on the agenda. According to the engineer, R. van Sandick who reported on that part of the congress in *De Ingenieur* (The Engineer), the whole issue was badly covered. Only a handful of engineers had submitted 'preliminary recommendations' despite the 'expensive and extensive experiences of East Indian Public Works'.[21] After all the discussions those involved concluded that having a road conference would be extremely useful as it would give everyone the opportunity to talk about various experiences and to establish, in view of the increasing volume of traffic and weight of vehicles, just what stipulations should be laid down for roads.

With its Bridges and Roads section and its Materials Research Laboratory, Public Works seemed, to all intents and purposes, to be well equipped to execute the road plans. In

In the 1930s the Dutch East Indian Roads Association possessed a test circuit. It was situated in the grounds of the Bandoeng Technical College. Different types of surfaces were tested on the 175 m long and 5.5 m wide track. Two 'vehicles', a lorry and a two-wheeled goods cart (i.e. grobak), were used to test the road.

addition to that the directors of regional or municipal works in larger areas and bigger cities who were expected to be involved in road construction and maintenance were engineers. The cooperation and division of tasks between Public Works and local services was not always perfect. In 1920 the newly opened Technical College in Bandoeng provided courses in surveying and in road and water engineering in order to better meet the needs of engineers. Four years later the first wave of civil engineers, twelve in total, graduated from Bandoeng. In 1921 the Materials Research Laboratory was moved to the grounds of the College and the laboratory director was made extraordinary professor.

At the instigation of the Technical College, a Dutch East Indies Road Congress was held there in 1924. Simultaneously a roads exhibition and various demonstrations were organised in the Bandoeng Exhibition Hall. It was decided that during the congress on roads only technological and 'technological-economic' matters should be addressed 'so that laymen (civil servants, members of local councils etc.) would also be able to benefit from the event'.[22] The kinds of issues on the agenda were therefore: construction and maintenance, administrative and financial control, funding, the draining and the surfacing of roads, urban roads, rails in roads and pipelines under roads. The issue of roads in Outer Regions was also considered. One of the results of the congress was that a battery of road-related nomenclature was established. It was, however, more difficult to get the participants thinking along the same lines in other areas of road construction. Their work terrain was so divergent that they were forced to employ different methods and to use different materials. One of the congress objectives was to create an overview of all those aspects.[23]

One concrete consequence of the congress was the establishment of the Dutch East Indian Road Association. It was an association that employed engineers who carried out research, did experiments and published various findings affiliated to the field of road construction. For those purposes they first had at their disposal a 'test road' and later an experimental station, also known as the Road Construction laboratory. The association's committee, known as the Road Council, was broad-based and also had a representative in the Netherlands. The association had its own independent sources of funding so that every five years it was able to organize a road congress. [24]

When, as of the mid-twenties, the regions were converted into provinces, all the roads that had been under the control of the regions fell under provincial administration. Each province was given its own provincial state water service which, in turn, comprised a number of smaller district services. The responsibility for bridges and roads was passed on to the district services. At the same time, the less important roads remained under the control of municipalities and *desas*. By the year 1934 it was the governor-general who had supreme control of provincial public works affairs. In that same year the Public Works department was changed to the Department of Transport and Public Works and not long afterwards the Bridges and Roads section and the Buildings section were merged.

The sharp rise in traffic in the post-1920 period made it necessary to adapt all the legislation relating to traffic. Up until then land, water, rail and tram routes had been dealt with quite separately, as though they were all totally unrelated. In the early twenties, though, people began to realise that the sharp rise in vehicular traffic was having repercussions for the railways and for tram routes and that the 'traffic issue' thus needed to be viewed in a broader context. It was for that reason that in 1925 a Motor Traffic Committee was established: to inventorise the numbers of lorries and busses on the roads, to advise the government on the matter of how lorries, busses, trains and trams could best meet the transport demands, to see how the safety of passengers could be protected and to decide how users of public roads might contribute to the financial costs. The committee brought out its report in January 1928.[25]

Meanwhile the original Motor Ruling of 1917 had already been amended in 1927; the need for a more uniform approach to road traffic made it undesirable to postpone things any longer. In 1933 a Road Traffic Ordinance was drawn up which, in the interests of implementation, was accompanied by the Road Traffic Statutes and the Road Traffic Resolutions. All of this was essentially based on the committee's report. Everything to do with road transport became organised, from traffic rules, vehicle requirements and driving licences to public transport and regulations concerning the actual roads themselves. With a certain degree of regularity, road traffic legislation was also subsequently adapted to suit new traffic situations. Licensing systems were introduced for private transport and for freight transport that were executed at both local and regional levels as the licences were granted for relatively short sections. The ordinance introduced a transport obligation, tariff controls and the obligation to transport post. Passenger transport by road was organised on a small-scale and was chiefly in the hands of locals. (Unlike in the case of the railways which were operated by large enterprises and were viewed as 'foreign'.[26])

One of the outcomes of the Road Ordinance of 1933 was a Road Traffic Committee, set up in that same year, and a number of Transport Committees. The Road Traffic Committee was partly comprised of civil servants and partly of association representatives who protected the interests of road users. It became their task to advise the governor-general and the different departments on various aspects of traffic affairs. The committee had to produce annual reports on its activities. The transport committees were established in conjunction with freight transport. It was possible for the governor-general to set up transport committees for each of the various regions. Apart from having to remain up-to-date on developments in the goods transport sector it was also their task to produce recommendations on the granting of permits and all related official objections and, furthermore, to mediate between the different transport companies.[27]

In 1938 the Road Traffic Ordinance, the Road Traffic Statutes and the Road Traffic Resolution, drawn up by the Transport and Public Works department, were drastically amended. In the new legislation, which became effective on 1st January 1939, all kinds of definitions were laid down such as: for public transport vehicles, for parking-linked terms and for road maintenance authorities. Different driving licence categories were introduced and duty and off-duty stipulations were laid down for lorry drivers. Rulings were also drawn up concerning the withdrawal or confiscation of registration numbers and driving licences, and uniform traffic regulations were introduced.

Similarly, as of 1st January 1939, records started being kept in connection with all kinds of possible statistics relating to such matters as: traffic accidents, bicycle and vehicle numbers and the numbers of individuals (broken down according to country of origin) in possession of bicycles, vehicles, lorries and hand-carts. In the case of motorised vehicles it was possible to trace the statistics back as far as 1917 as ever since that date registration numbers had been issued. From the relevant statistics it could be seen that on 1st January 1939 some 37,500 private cars were registered in Java and Madoera and 14,115 on all the other islands. In addition to that there were 1,769 omnibuses in Java, 7,473 in the Outer Regions, 6,779 lorries in Java and 6,081 lorries elsewhere in the region.[28]

When it came to the matter of financing road construction and road improvements, taxation proved to be a good source of income. In 1908 taxes started being levied for petrol which meant that the existing taxes already being paid for petroleum were extended to include petrol. In 1921 a differentiation was made between petroleum and petrol, all of which culminated in a sharp raising of taxes for the latter.[29] The first motor vehicle taxes started being charged in 1932 in the Outer Regions and in 1933 in Java and Madoera. At the end of 1934 a new motor vehicle tax ruling was implemented that applied to the whole of the East Indies.[30] Other good sources of income for the central state emanated from the import duties paid on cars, the legal dues for driving and other kinds of permits and all the local taxes paid on bikes, cars and so on.

Technology

Whereas in 1915 the traffic routes inspector for the Outer Regions had asserted that car traffic should adapt to the existing road network, the Public Works director presumed in 1920 that it was time to make all the roads suitable for car and freight traffic. In his presentation of that same year to the Dutch Road Congress he remarked that in the East Indies it was really necessary to create new road profiles. The points that needed to be addressed most urgently were: road widening, the extending of bends and the reducing of the number of bends. The biggest problem, though, was the quality of the material used for surfacing.

As far as road-building technology went, the East Indies also saw a major turning point around 1920. Around that time the rapid growth of car and freight traffic necessitated rapid road improvement: roads needed to be made more even and, for freight traffic, the metal surfacing needed to be reinforced. In the case of Java, this largely involved improving existing roads but in the Outer Regions it often involved the building of new roads. Thanks to the availability of new knowledge and new materials, such as asphalt, it was possible to meet this growing demand for infrastructure. The conclusions drawn during the Dutch East Indies Road Congress on all the various topics did lend a certain degree of unity to the construction process.

The first phase of the process, involving the measuring up and staking out of the route, did give rise to the predicted problems in mountainous and very overgrown areas. Apart from needing to be equipped with their measuring instruments the surveyors also needed quite a team of assistants to clear the terrain and hold the beacons. Surveyors were not, however, always available. Especially in the Outer Regions it was often 'a civil servant or a cultures employee ... who had to map out a route'. For that particular category of people and for the 'aspiring road constructor' the Bridges and Roads section of the Public Works department had compiled a special instruction manual. The manual not only listed the necessary equipment but also detailed the desired work method and provided important calculations. None of the inclines along the route could be too steep or the bends too sharp. It was also stated that it was preferable to embed a road in a hillside rather than build it on banked up earth in view of the danger of subsequent subsidence in the wet seasons.[31] The road constructors did their best to avoid having to cross rivers. At the same time as plotting out a route it was also necessary to have an indication of the total road length and the required road depth. A careful estimating of the road dimensions was not only important for knowing precisely how much land had to be acquired but also for the making of concrete plans and for budgeting.

Around the year 1920 the substantial demands being made upon road surfaces by freight traffic were soon translated into new road profile demands. The width of the hardened surface area on the main roads in Java was increased to 5 to 6 m whilst in the Outer Regions, where traffic was less heavy, a width of 4 m was regarded as acceptable. In line with the newly laid down stipulations for road cross-sections, only roads in raised areas or alongside of gorges would be given small dykes on the downside but elsewhere such dyke structures had to be removed to improve road drainage. Since nearly all the roads were flanked by such characteristic East Indian dykes their removal constituted a fairly major operation. On either side of the road metalling, roadside ditches were dug to take the excess water that flowed off the hardened surface. That also explained why the roads were cambered. With the introduction of asphalt, though, cambering became dangerous because of the slip factor in wet weather and so consequently the verges and ditches became a hazard. Such roads thus started being constructed with less cross-sectional sloping.

In the case of the main roads in the cities it was maintained that 1:30 was sufficient and for main roads outside the cities the standard incline was 1:20. The most extreme slope in the city was 1:25 and outside the city areas 1:15. Such a maximally steep slope could only continue for 300 m and needed to be interspersed with 'rest sections' or strips of road at least 50 m long with 1:200 slope factors.[32]

In general, the minimum radius for bends was lower in the East Indies than in the Netherlands in the period around 1920. Endeavours were made to keep the radius to 20 m, the minimum being no more than 10 m. The Dutch East Indies Road Congress determined that the norm to be adhered to should be a minimum radius of 40 m on level terrain, of 25 m in hilly parts, of 15 m in mountainous areas and of 10 m in 'extremely mountainous' parts. It should not be forgotten that in mountainous areas there was always a very high ratio of bends. There it would not be unusual to have as many as 30 bends per kilometre. The adapting of the road profile and notably the reducing of the number of bends in any one route was a major operation and thus, inevitably, an expensive business.

As of roughly 1920, the road constructors also started to devote more attention to road subsoil. Investigations were carried out to see which roads required a fortified substrata consisting of large stones and where an under layer or foundation could be created in an alternative way. Along the roads where less traffic was expected the typical subsoil would consist of sand, clay and possibly other locally available materials such as coral chalk stone (or *karang*). In the 1920 *De Waterstaats-Ingenieur* (The Public Works Engineer) journal there was a debate about the using of *karang*. It was believed that the coral chalk that came straight from the bay of Batavia was not particularly suitable but that the petrified coral type was suitable.

From the thirties onwards more care was given to the steam rolling of the under layer and tests were carried out in which an asphalt emulsion was used as a 'soil stabiliser'.[33] On top of that would come the hard surface and/or the final covering layer. The thickness of the hard surface and the metalling layer would all depend on the anticipated volume and weight of traffic.

After 1920 the kind of surfacing used on the outlying roads on a massive scale was asphalt. It was a by-product of the oil industry and since there was such an industry in the East Indies 'petroleum-based asphalt' was experimented with at quite an early stage in the history of road building. From the mid-twenties onwards there were many types and brands of asphalt available on the market. The natural asphalt present in various regions in the East Indies was only used to a limited degree and then it was often mixed with petroleum asphalt. When opting to finish a road with asphalt there was the choice between a simple surface hard core layer (quick, inexpensive) or a deeper road-metal layer (reliable but more expensive). At the Dutch East Indies Road Congress it was possible to find proponents for both methods but no final conclusions were drawn. It was concluded that more research needed to be done into the two techniques. In various areas test strips were subsequently laid where different foundation layers and surfacing materials were experimented with. For instance, the *tijkar* lane of the Soerabaja-Porrong road was divided into various sections for a total distance of 14 km and different types of asphalt were tested out. A complete report of the results was published in *De Ingenieur in Nederlandsch-Indië* (The Engineer in the Dutch East Indies) by the engineer, H.W. Pareau-Dumont, in 1939. What emerged was that in practice there was a preference for the quicker and cheaper kind of road surfacing. Between 1925 and 1935 some 10,000 km of hard core road was thus asphalted in that particular fashion.[34]

The kinds of roads frequently constructed in the Netherlands with clinkers or cobbles and the new-style cement-concreted roads were never or hardly ever seen in the East Indies. The types of stone naturally occurring in the East Indies were not sufficiently 'durable' and clinkers could not be fired at high enough temperatures. The roads would definitely be worn down too quickly by the freight carts with their iron wheel-rims. The Dutch East Indies Road Congress proposed that it would be wise to continue with the testing of different surfaces provided that the 'sub-surfacing was adequate'. Because of the more porous type of stone, cement-concrete surfacing could not be made durable enough. Portland cement would have produced a better type of cement-concrete but that kind of cement was very expensive in the East Indies. In Deli (Sumatra) cement-concrete had been used for the cycle paths alongside of the main roads but there, of course, wear and tear was not such an issue. Generally speaking it was found that cement-concrete was too heavy and too expensive to use on the main roads.

During the International Road Congress held in Scheveningen (the Netherlands) in June 1938 the desirable 'driving-surface features' of a modern road were described as even, anti-slip and 'sufficiently light-diffusing to make good visibility possible in artificial light'.[35] The nobleman and engineer, C. Ortt, was present as representative for the Dutch East Indian Road Association. Later on in that same year, when talking on the same subject, he mentioned that in the East Indies it was especially the anti-slip feature of road surfacing that was concentrated on. To that end, a special sort of cold pliable petroleum asphalt material was available (known as cutback asphalt). There was still discussion about the desired degree of evenness of any road surface. The kinds of factors that needed to be taken into consideration were the speed and shock absorbing qualities of the vehicles. The Roads Association had a special vehicle that was able to gauge the evenness of surfaces. The final requirement, regarding 'the light-diffusing factor', was viewed as less important as there was very little evening and night traffic in the East Indies.[36]

In view of the great changes seen in road building practices in the East Indies in the years after 1920 engineers were convinced by the end of the thirties that they were well on the way to resolving all the road issues. Even though estimates pointed towards a predicted 50 percent traffic volume increase, it seemed that continuing in the same fashion whilst differentiating between slow and fast traffic, was the best policy. By 1938, the 1913 road plan for Java had almost reached completion. By then some 23,000 km of surfaced roads had been created, the Sumatran road network was 10,000 km long and Bali had 800 km of road suitable for cars. The total amount of surfaced road in the whole of the East Indies was estimated at some 70,000 km.[37] By East Indian standards the roads in Deli were particularly good. The main roads were flanked on either side by cycle paths. There, 84 percent of all the roads had asphalt surfacing as opposed to 50 to 70 percent of the provincial roads in Java.[38]

Conclusion

The economic depression of the thirties that hit the East Indies very badly, also affected road construction. It was only on the islands of Sumatra and Borneo that roads were still constructed and improved; in Java and Madoera that only still happened 'at detailed' level. The Second World War and especially the Japanese occupation of the East Indies did little to help the traffic and infrastructure situation. In 1945 most of the Europeans were able to pick up where they had left off and so endeavoured not only to regain authority but also

Modern road construction in Sumatra round about the year1930, near to a Javanese settlement. A steamroller was used to compact the surface layer. This would possibly have been topped with a layer of asphalt.

to reinstate law and the plantations and to repair the neglected roads. The re-establishing of transport means was seen as the key to economic recovery but the political climate was rapidly changing. During the political upheavals that took place between 1947 and 1949 many of the roads in Sumatra and Java (especially in the West) were made impassable or were undermined. For Indonesia which gained its independence in 1949 that meant a bad start at least as far as road infrastructure was concerned. By 1965 as much as 45 percent of all the roads were considered unusable and only 5 percent qualified as good. In the mid-eighties it was asserted that road traffic was still badly hampered by the poor state of repair of the roads. By then the total road distance in Indonesia was estimated to be 220,000 km.[39]

A comparison with the Netherlands

After the First World War road users started placing increasing pressure on the government (through the motoring organisations known as the ANWB, the KNAC and through the KIvI), as did road constructors and municipal authorities, to create new roads and to make all the existing roads suitable for vehicular traffic. The first Dutch Road Congress was in 1920 and at that point the entire road traffic issue was examined. What did become immediately clear was that a structural approach to the matter would not only require large financial investments but also a different kind of organisation and different materials. One of the consequences of the congress was the establishment of the Dutch Road Congress Association (the NWC).

The Road Taxation Act, adopted by the Dutch parliament in 1926 was what provided the financial framework for road construction and improvement. By 1923 the Public Works department had set up a road improvement district (which originally consisted of one individual) to study the problems and bring out recommendations. The first National Road Plan was presented in 1927 and provincial road plans were also created during the same period. In order to design and build the many bridges that featured in the National Road Plan, a special Bridges Office was set up at the Public Works department. In 1927 the State Road Structures Laboratory was created so that the materials used in road construction could be studied and, not least, the different types of asphalt used.

In the Netherlands the kind of road profile adhered to was selected by the state road planning organisation and the provincial road planners. In the first National Road Plan it was standard to have cycle paths lining all the motorways. In North Brabant they had already started creating cycle paths alongside major roads by the beginning of the twentieth century.

Up until the time of the Second World War, clinkers and cobbles were the main kinds of surfacing material used in the Netherlands. After the mid-twenties it became common to use asphalt but not on a very large scale. After 1927, roads in the east and south of the country were also sometimes made of cement-concrete. Especially in the west of the country, the combination of the peaty subsoil and the heaviness of the traffic accounted for much subsidence and for fracturing in the smooth and sealed off road surfaces. It was clear that extra attention needed to be given to the substrata of roads.

In the Netherlands, too, the width of the hardened surface was also the subject of debate. By the end of the nineteenth century the width had been reduced, almost everywhere, to 2.5 to 3 m. In the National Road Plan of 1927 the new standard width quoted for surfaced roads was 5 to 6 m. As of the mid-thirties Public Works started launching a network of motorway plans which included dual carriageways and flyovers. Most of these plans were not executed until after the Second World War.

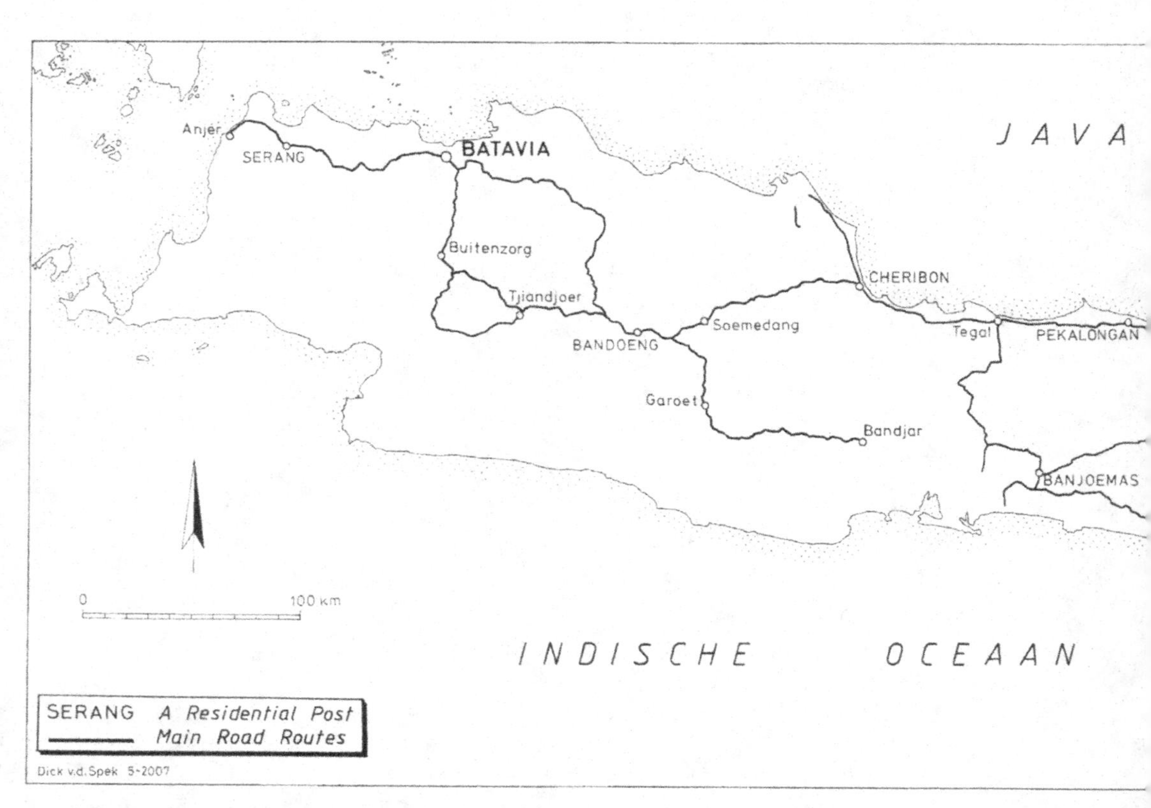

Map of the main roads in Java and Madoera c.1940.

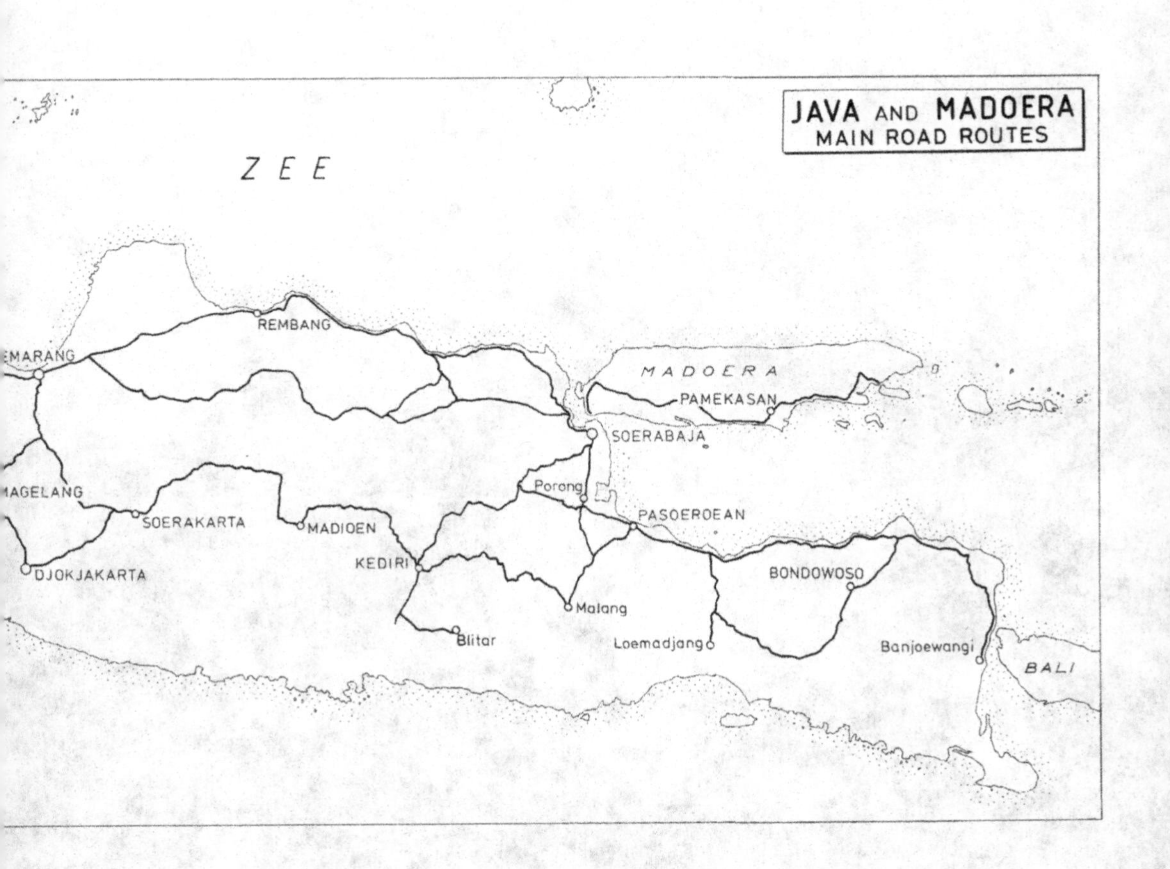
JAVA AND MADOERA
MAIN ROAD ROUTES
ZEE
REMBANG
MADOERA
PAMEKASAN
SOERABAJA
MAGELANG
Porong
PASOEROEAN
SOERAKARTA
MADIOEN
KEDIRI
BONDOWOSO
DJOKJAKARTA
Malang
Blitar
Loemadjang
Banjoewangi
BALI

A Netherlands East Indian Railway Company train destined for Solo at Semarang station (Central Java), photographed in the eighteen-eighties.

CHAPTER 3 **THE LOCOMOTIVE OF MODERNITY**

Building the network of railways and tramlines

AUGUSTUS J. VEENENDAAL, JR.

The Dutch East Indies before the first railways

Though, officially, it was virtually the whole of the Indonesian archipelago that constituted the Dutch East Indies, all actual influence – in terms of Dutch rule – was predominantly confined to the island of Java once the British had handed back the area in 1816. Elsewhere, it was just the odd trading posts and administrative areas that were retained by the Dutch. Self-government exercised by indigenous monarchs was the chief form of administration and as long as things more or less complied with the demands of the Dutch high command in Batavia everybody was satisfied. It was only in the Moluccas, notably on the island of Banda, a vital pepper producing area, and Banka, where even then tin was being mined, that there was clear evidence of Dutch rule. After a rising led by local rajahs, Palembang (in Sumatra) was finally conquered in 1825 whilst another local rising organised by the Padris culminated, in 1837, in West Sumatra's submission to Dutch rule.[1] The politics of abstention which had previously taken hold in the era of the United East India Company was once again, of necessity, revived. Above all else, what needed to be carefully guarded was the finances. The Javan war of 1825 to 1830 had cost a tremendous amount of money and had been entirely financed by the Netherlands, itself a country burdened by massive national debt. Between 1817 and 1829 the mother country had run up debts to the tune of 40 million guilders in conjunction with its administration and control obligations in what was usually termed a 'profitable colony'.[2]

It was thus essential for the East Indies to at least become financially self-supporting and to generate a surplus in order to contribute to the Dutch national budget. Johannes van den Bosch, who was made governor-general in 1830, conceived a plan for government-imposed cultivation of the specific crops that were then in great demand on the world market. Those crops were sugar, coffee and indigo. The Cultivation System need not be explained in any great detail here.[3] It is sufficient to indicate that in financial terms it was a resounding success. Java soon became the cork upon which the Netherlands floated so that before long all the Dutch state debts were cleared by the positive East Indian returns. Eventually the surplus was big enough to fund major projects such as the post-1860 creation of the Dutch State Railway network and projects that focused on connecting Amsterdam and Rotterdam with the sea.[4]

Transport in the East Indies

Obviously all the increased economic activity on the island of Java brought with it greater need for transport routes. Above all else, it was the sugar refineries that rapidly multiplied and expanded their capacity. Between 1830 and 1840 the total exports rose – in terms of weight – from over 36 million kilo to more than 166 million and it was a trend that would only continue.[5] Everything that was produced had to be transported by oxcart along primitive roads. The only properly surfaced road of anything other than local significance was the Great Post Road traversing the entire length of Java. It had started being built under governor-general Daendels and was finished in 1816 during the interim period of British rule. Local and regional roads linked up Djokjakarta and Soerakarta, the main cities of the Principalities, with the rest of the island.[6] Even though the more or less completely surfaced

Great Post Road existed, the journey from Batavia to Soerabaja still took nine days or more. A system of *pasanggrahans* or simple government-supported guest houses, made the journey bearable for travellers.[7] Though the situation on Java left much to be desired, conditions elsewhere in the archipelago on other islands were much worse. There were hardly any roads and where there were roads they suffered badly because of the frequent torrential rains.

Interinsular traffic

In such a sprawling profusion of islands as the Dutch East Indies, transport between the various islands was obviously extremely important. The government created a navy so that messages and government officials could be taken from one island to another and shipping could be protected from piracy. Later, after reorganisations in 1861, this navy was officially named the Government Navy and given other additional duties.[8] This semi-military navy consisted largely of lightly armed ships able to navigate the farthest corners of the archipelago. After 1848 steamships gradually took over the role of the schooners and sea-going proas.

The Government Navy ships were not designed for civil commercial shipping ends. Apart from the local proas, it was ordinary sailing ships that were deployed for those purposes. They provided more or less regular services between the main ports. The government recognised the need for such regular steamship services and so in 1850 a contract was signed with the former naval officer, W.F.K. Cores de Vries, to the effect that two shipping lines would be equipped with steamships and that in return each line would receive a fixed subsidy for the services provided.[9] In 1863 a new contract was signed with the Netherlands East Indian Steamshipping Company, a British company, despite protests from Dutch entrepreneurs who had requested an extra one cent subsidy per sailed sea mile.[10] This Netherlands East Indian Steamshipping Company met its contractual obligations reasonably well but gradually it was realised in Dutch and East Indian circles that in times of international tensions it might not be safe to leave such vital services in the hands of foreign partners. The politics of establishing Dutch authority in every corner of the archipelago was obviously something that demanded a reliable Dutch shipping enterprise for the transportation of troops and equipment. The final upshot was the establishment, in 1888, of the Royal Netherlands Packet Shipping Company (Koninklijke Paketvaart Maatschappij, KPM) and the closing of a new contract with the East Indies government for a large number of lines. On 1st January 1891 the KPM put into service its 13 new ships plus a further 16 purchased from the Netherlands East Indian Steamshipping Company.[11] Over the course of time, the KPM proved to be a reliable and efficient company serving the whole of the archipelago. Because of its very nature, the KPM was in no way a rival for the East Indian railway and tramway companies. Indeed, it was rather a complementary form of transport, like for example in Sumatra and Belawan where the Deli Railway Company (Deli Spoorweg Maatschappij, DSM) started of its own volition to improve harbour facilities in the eighteen-nineties. The warehouses built by the DSM were gratefully rented by the KPM. There were several exceptions though, like in the case of the Soerabaja-Semarang-Tandjong Priok line that met with heavy competition from the railways along that particular route.[12]

The first railway projects

The success of the Cultivation System led to ever-increasing sugar and coffee production levels but the problem was that it was incredibly difficult to transport everything from the Principalities (i.e. the main production area in Central Java) to the harbour of Semarang. Between 1833 and 1840 transport costs tripled and so it was not surprising to see that the government became interested in introducing various measures.[13] For a while, camel transport was experimented with but as the animals could not adapt to the Javanese climate that idea was soon dropped. Similarly, it was found that Brabant donkeys were also unsuited to the climate.

In the Netherlands, steam trains had been in use since 1839 and that was something that had not gone unnoticed in the East Indies. People started toying with the idea of creating a railway network in Java. For the time being, however, it seemed that steam locomotion was not an option and invariably it was animal traction that was put forward as the alternative. Already in 1841 the company of Dixon & Co. in Amsterdam requested permission to lay a railway line between Semarang and the inland Principalities.[14] It was to be horses or buffalos that would provide the traction. As Dixon asked for a five percent interest guarantee from the government his request was rejected. Other applications, including an amended one from Dixon, all came to nothing.

Discussions concerning the permissible level of state intervention invariably became rather heated, just as in the mother country. Unlike in the Netherlands, though, where it was not until 1860 that railway construction would be taken over by the state, there were many in the East Indies who believed that private investors would not be sufficiently interested in railway construction.[15] Already in 1842, in other words shortly after Dixon's initial application, it was decided by Royal Decree on 28th May that 'an iron road should be laid from Semarang to Kedoe and the areas known as the Javanese Principalities in order to ease goods transport in wagons drawn by buffalo...'.[16] Undoubtedly the intentions were good but as no candidates came forward who were interested in exploiting such an idea the whole plan just fizzled out. The debate on the pros and cons of government intervention did continue but nothing concrete came out of it all. Meanwhile the governor-general had given the army officer, G.H. Uhlenbeck, orders to consider plans for a route between Semarang and the hinterland.[17] Uhlenbeck's report, which was published in 1844, was extremely negative: he even referred to it as an 'ineffective enterprise'. Another army officer, D. Maarschalk, launched a project in the eighteen fifties at the request of the East India Company for Administration and Annuities that wanted a railway line between Batavia and Buitenzorg, but the official application was rejected.[18]

For a long time nothing whatsoever materialised but the tides were turning. In 1859 the Minister for the Colonies ordered that all existing projects and reports be handed over to a couple of experts: L.J.A. van der Kun, the chief inspector for Dutch Public Works who was one of the few real Dutch experts in the field of railway construction, and D.J. Storm Buysing, the chief engineer at Dutch Public Works. Their conclusions were clear: only the state would be able to construct railways in Java. Providing subsidies for private entrepreneurs in the form of an interest guarantee was out of the question because then the state would only suffer the disadvantages and not any of the possible advantages of having a railway. It all amounted to an echo of the debate going on in the Netherlands where the minister in charge, F.A. van Hall, had opted for state-backed construction work and had rejected interest guarantees for private enterprise on the grounds that such a set up would create too much uncertainty for the state.[19] Van der Kun and Storm Buysing also recommended that from the government angle the most urgently required routes should be investigated and determined as soon as possible. Amazingly, the Minister for the Colonies instantly followed up the recommendation by appointing the artillery officer, T.J. Stieltjes,[20] and the Amsterdam industrialist, John Dixon, advisors for technological affairs in the East Indies. Both men instantly left for the East Indies.

Upon arrival, together with several others including the engineer N.H. Henket,[21] they formed a Transportation Committee, which speedily and professionally got down to the job in hand. At the end of 1862 Stieltjes and his colleagues presented their report in which a number of rail links were recommended.[22] Slightly earlier, in January 1862, a Dutch group consisting of W. Poolman, S. Fraser and E.H. Kol had submitted a request for permission to create the Semarang-Soerakarta-Djokja rail link. This request had also been submitted to Stieltjes, who had reacted pretty negatively. The new governor-general, L.A.J.W. baron Sloet van de Beele, did however decide in favour of the group's proposal but obviously on the understanding that it first be approved by the minister and the Dutch parliament. In the past there had been a clash between Sloet and Stieltjes when, in 1858, the former had been one of the men to apply for permission to construct a section of the railway network in the Netherlands. At that time it had been Stieltjes who had forcefully opposed the plan. It was partly thanks to Stieltjes that the request had then gone on to be rejected in the Lower House and clearly this was something that Sloet had not forgotten.[23] Now that the ball was in Sloet's court Stieltjes was dismissed but he got his revenge by publishing a series of pamphlets in which all the mistakes, misleading information and incorrect figures produced by the government were outlined.[24] The campaign was successful because the minister's proposals in the Dutch Lower House to grant Poolman c.s. their requested permission met with considerable opposition. Only after the route proposal had been amended and the requested interest guarantee had been curtailed did the Chamber grudgingly agree to the proposal. On 6th June 1863 permission was officially granted and the franchise agreement published in the Dutch 'Bulletin of acts, orders and decrees' (*Staatsblad*).[25] Poolman c.s. immediately handed over the rights to the Netherlands East Indian Railway Company (Nederlandsch-Indische Spoorweg Maatschappij, NISM) that had been incorporated on 27th August 1863 with the seat of the company in The Hague.

Despite this apparently positive gesture there was still a great amount of dissent. Certain people were clearly not convinced that it was necessary to have any railway lines at all in Java. One anonymous army officer attached to the East Indian army wrote a paper in which he queried many points in the report written by Stieltjes.[26] What was much more controversial was the pamphlet written by the former government official, J.D. van Herwerden, in 1863.[27] He maintained that building railway lines in Java would not be in the interests of the local population or the European crop cultivators, who would be able to look for alternative ways to transport their products. From the point of view of the defence of the country, the railways would also be useless because a potential enemy would soon be able to disrupt the infrastructure by, for instance, blowing up bridges. The first war in which the railways had played a significant part, namely the American Civil War, was clearly not something that Van Herwerden had taken into consideration. He furthermore maintained that it would be technologically virtually impossible to create railway connections in Java, especially in view of the heavy rains of the west monsoon season that would soon wash away the tracks. In the face of all these negative predictions the NISM nevertheless went ahead with its plans.

The Netherlands East Indian Railway Company

The NISM managed to capture the interest of the former military engineering officer, J.P. de Bordes, in the line construction plans. De Bordes was no stranger to the railway world as, since 1860, he had been secretary of the Railway Construction Committee in the Netherlands. In that position he had supervised all the preparations for the construction of the state-backed lines.[28] Before leaving for the East Indies in the summer of 1861 Sloet van de Beele had been on the same committee, which just goes to show how really small the Dutch and East Indian railway worlds were and how very much intertwined.

Even before De Bordes had left for the East Indies the first problems had already surfaced. The shares of the new enterprise adding up to a total of ten million guilders were supposed to be sold by the General Company for Trade and Industry established in Amsterdam, a daughter of the French Crédit Mobilier company.[29] This did not, however, go as smoothly as had been presumed. After a year, there were still some 7000 shares that had not been disposed of and the General Company had moreover got itself into trouble over its handling of the shares of the Netherlands Company for the Exploitation of State Railways and other large enterprises like the Netherlands East Indian Trading Bank. At the end of 1864 the whole business folded up. The director, Alexander Mendel, ran off to London after it emerged that he had endeavoured to prevent the catastrophe by involving himself in various extremely dubious transactions. The business was saved by the intervention of the French mother company. Small wonder, therefore, that in popular parlance it came to be nicknamed the Company for Fraud and Industry.[30] The NISM shares, amounting to around 60 percent of the total and still around six million too short, were transferred to a French group. A bond loan to the tune of four million was placed with an English syndicate under terms that were highly unfavourable and by the end of 1868 things had become so desperate that the NISM directors were forced to turn to the government for support.

The precise details of the financial support offered by the state do not need to be given here. New interest guarantees, interest-free advances and loans followed in rapid succession and the Second Chamber, where Stieltjes – by then a member of parliament – played a big part, was permanently in the opposition. With great difficulty the Minister for the Colonies finally managed to push through a package of support measures in 1869 with a slim majority of 37 to 31 votes.[31] Still that was not enough as the Djokja connection would not be completed until July 1872 whilst the Willem I branch was scheduled for completion in May of the next year. The revenues from traffic thus remained disappointing for the time being and the government and the agricultural enterprises in the Principalities once again had to dig deep into their pockets to keep the whole project afloat. It was only after the lines had been completed that the NISM gradually started to recover from its financial difficulties. Meanwhile it had become very evident that there was a great need for a reliable transportation system to convey all the sugar from the Principalities to Semarang.

Part of the government support for the NISM consisted of an obligation to also construct a Batavia-Buitenzorg line along a route that had been previously plotted by Maarschalk and which had subsequently been detailed by Henket. Permission to construct this line was given in 1865 and by the end of January 1873 it was operational.

NISM construction work

De Bordes did not take his new task lightly. Indeed, on the way to Java he took the opportunity to inspect a newly opened real mountain railway in Austria, the Semmering line, and then in the British East Indies he went to look at a second such line. The terrain in Java was partially level or at the most slightly undulating but part of the route, especially the sections between Soerakarta and Djokja and the line that branched off to the Willem I fort, were proper mountain routes.[32] More than 100 bridges were required, some of which would need to be fairly big, such as one with two 20 m spans and another that would have to be 28 m long. In some places the required earthworks was heavy, such as where soil had to be dug out for a length of 600 m and to a depth of 19 m. Elsewhere embankments had to be built and heights of 17 m or 18 m were no exception. In the branch line leading off to fort Willem I it was even necessary to create a 162 m long viaduct. In order to erect the iron bridges De Bordes deployed two bridge wagons so that long sections could be transported to their destination and subsequently eased in place on temporary supports. In that way the work of erecting and riveting the bridge sections on the spot could be kept to a minimum which, because of the shortage of skilled personnel, would otherwise have taken months. The bridge wagons themselves were twin-axled with one long articulated carrying base upon which the bridge sections could be rested. They were of Belgian design, built by Evrard in Brussels. At first, imported creosoted pine sleepers were used but it was soon discovered the indigenous ironwood or djati (teak) sleepers were more suitable.[33]

De Bordes had to operate mainly with local workmen, sometimes as many as 9000, and just a handful of European employees. Everything was new and previously untested, the men were unfamiliar with the work, the climate was exhausting and yet, despite all of that, he managed to make reasonable headway. At first the engine drivers and conductors all came from the Netherlands, but by 1868 the first native engine driver had been trained and was already working. The stone for bridges and buildings and the wood for sleepers could be found locally but all the iron had to be shipped in from Europe. Rails came from England and Belgium, most of the iron bridge sections came from the Netherlands and the steam locomotives were from England and Germany. De Bordes, who was familiar with the Dutch railway system, opted for the 1435 mm gauge which was the standard European width. He did not, however, see the completion of his line. After a dispute with the directorate he felt obliged, in 1870, to hand in his notice.[34]

The first locomotives were ordered from the Borsig factory in Berlin-Tegel and were delivered in June 1865 where they went into service as NISM nrs. 1-2.[35] After that Borsig did not supply any rolling stock to the East Indies but it did deliver to the Holland Railway Company in the Netherlands. The next consignments of locomotives arrived in 1866 and subsequent years and came from the Beyer, Peacock & Co. factory in Manchester, the same factory which, in 1863, had supplied the first locomotives for the Netherlands State Railways. The NISM engines were produced in two different models, one with three coupled axles and one with two coupled axles and one carrying axle under the firebox. Both types had separate six-wheel tenders. Both were also the factory's standard models and so similar specimens could be found all over England and in all countries supplied by the English market. By 1875 the NISM had a total number of 21 engines in service.[36] The first carriages were four-wheeled, like those then common in the Netherlands, with individual compartments and doors to the side. Even the freight trains conformed more or less to European norms. In that way, seeing the NISM trains and tracks was just like experiencing the Dutch railway system in a tropical environment. At first, there were three passenger trains between Semarang and Djokjakarta and just one train per day on the Willem I branch line.[37]

The battle of the gauges

As has been mentioned, the NISM adhered to what was termed in Europe the 'standard' rail gauge of 1435 mm or 4 feet and 8.5 inches. However, just because this was seen as the norm in Europe it did not mean to say that it was automatically the most suitable gauge for colonial railway lines. In the Netherlands for that matter, the tracks had originally been broad gauge but things were later changed just because neighbouring countries such as Belgium and Prussia used the standard gauge. The final sections of the Holland Railway Company's broad gauge tracks did not disappear until 1866.[38] The advantage of the narrow gauge track was that it was cheaper to construct because tighter curves were possible and less ground area was required for tracks and embankments, but it limited the capacity.

One instance of a successful narrow gauge railway with steam traction was to be seen in Wales where the Festiniog Railway had been specially constructed for the purpose of transporting slate from inland quarries to the harbour of Portmadoc. The line opened as early as in 1832 and it had an extremely narrow gauge of two feet (610 mm). Originally the empty wagons had been towed uphill by horses but in 1863 the first steam locomotives were introduced. All over the world this drew the attention of engineers and the Festiniog Railway may be said to have had quite an impact on the development of narrow gauge railways.

In the British colonies, railway construction had begun much earlier than in the Netherlands East Indies. In Australia various of the territories had had railways since 1855, some were standard gauge (e.g. in New South Wales) whilst others had broad gauge tracks (like in Victoria and South Australia). In Queensland which was thinly populated and covered a huge area, it was decided in 1865 that the narrow gauge system would be cheaper. What was selected was the English width of three foot and six inches, which had also been successfully used for about three years by a number of Norwegian lines.[39] Tasmania, Western Australia and New Zealand rapidly followed suit so that the narrow gauge railway soon became a very appropriate way of accessing less highly developed areas with difficult terrain and providing cheap rail links that nevertheless met the traffic requirements.

In the British East Indies things developed differently. There the first line had already opened in 1851 and curiously enough the rail gauge chosen there was five feet and six inches, a width that was not even common in England.[40] By 1869 this broad gauge network covered 4241 miles. However, as it was soon discovered that such broad gauge lines were often too expensive to build in relation to the expected volume of transport, the cheaper solution of creating narrower gauge lines was opted for. In this particular case metric dimensions were selected, the assumption being that the British East Indies would very soon be switching to the metric system.[41] That did not happen, though, which meant that the only thing metrical about the railway network was the gauge of the tracks; everything else remained in feet and inches.[42] By 1879 the narrow gauge network had no less than1700 miles of railway.

The three foot and six inch rail gauge quickly came to be known as the Cape gauge because in the Cape Colony and later in the whole of South Africa a huge interconnecting network of these dimensions was to be developed. In fact, this gives a rather inaccurate view of things because the first railway lines in the Cape Colony and Natal were constructed in standard gauge and it was not until 1872 that the decision was finally taken to construct any future lines based on the narrow gauge system of three foot, six inches.[43] Elsewhere in the world as well, people were busy experimenting with narrow gauge railways so that in Newfoundland and South America relatively large networks sprang up that were all narrow gauge. In the North American state of Colorado, a number of lines were constructed in a gauge of three feet (910 mm) like, for instance, the Denver & Rio Grande Railroad, the first few sections of which were entirely financed from the Netherlands and so were very familiar in Dutch engineering circles.[44]

These developments had not gone unheeded in Dutch engineering circles and they were actively discussed. The Minister for the Colonies, who was apparently not sure about the wisdom of agreeing to the standard gauge for the NISM, decided, in March 1869, to appoint two experts by the name of J.A. Kool and N.H. Henket to report on just what was the most suitable rail gauge for Java. The latter was familiar from the Transport Means Committee but Kool was new to the East Indian railway scene. Originally he had been a military engineer but in 1845 he had exchanged the military life for civilian life and had become an engineer on the Aachen-Maastricht railway. In 1860 he was made the supervising engineer for the Maastricht-Venlo state line and, as of 1863, he had been put in charge of the entire southern network to be constructed by the state.[45] Kool and Henket did not delay with the publishing of their report which came out in September 1869. Narrow gauge tracks of 1000 to 1067 mm were emphatically recommended because of being cheaper to lay and more than adequate for the kind of transportation demands anticipated. The Semarang-Principalities line that was under construction could remain a standard gauge line but it would be best for the Batavia-Buitenzorg line, for which the NISM had gained a concession, to be changed into a narrow gauge line.[46] Naturally Stieltjes once again got involved in things by recommending an even narrower gauge of 760 mm. He had visited the Festiniog Railway in October 1873 and had been impressed by just what was possible with that railway.[47] A rail width of 760 mm – two foot and six inches – was, he maintained, really wide enough for present and future Javanese transportation requirements. Fortunately the government opted for the Cape gauge because in time, as traffic on the railways rapidly increased, 760 mm would really have been too narrow.[48]

Stagnation and discussion

Apart from the rail gauge issues, Kool and Henket had also given their opinions on further railway requirements in Java. They put forward a number of line suggestions amounting, in total, to the creation of some 950 km of track. What they proposed would effectively link up all the main towns and harbours. What they did not convey was how the project should be tackled: if the state was not interested then private enterprise could be given interest guarantees. For further recommendations their report was passed on to a new committee consisting of De Bordes and Kool and Henket themselves together with a number of East Indian civil servants.[49] The committee arrived at the conclusion that two east-west mainlines, as proposed by Kool and Henket, would be rather excessive and that it was the southern line that was needed most urgently. Only De Bordes formed the minority by opting for a northern line and by recommending 'his' own standard gauge. The committee unanimously decided that the preference should lie in private construction and exploitation and that possibly an interest guarantee should be given for the capital outlay required for construction. In the East Indies itself it was very much believed that private parties would never be able to set up a cohesive railway network. It was for that reason that both the army commander and the director of the Civil Public Works department were more interested in the idea of state-backed construction work.

This therefore formed yet another impasse and one which, this time, would be resolved by the Minister for the Colonies, P.P. van den Bosse, who served in the third liberal Thorbecke cabinet. On 6th November 1871 he put forward a parliamentary bill in which it was proposed that the state should fund some four railway lines. The bill led to a flood of protests, especially as Van den Bosse was personally known to be a champion of private initiative. Even when the minister amended his plans the opposition did not abate. Ultimately the decision was postponed because the cabinet fell in July 1872. Van den Bosse's successor, J.D. Fransen van de Putte, himself a former East Indian sugar plantation man, instantly withdrew his predecessor's bill.

The situation very much resembled the state of affairs of two years before. There were not a huge number of applications from the private sector and because of the uncertainty surrounding the way in which the government should support the private sector, even the minister was loath to issue concessions. He first wanted certainty about the financial side of things which was why he turned to a new committee, composed of people from the financial world, for advice. The banks represented were: Lippmann, Rosenthal from Amsterdam, the Rotterdamsche Bank, the Rotterdamsche Handelsvereeniging (i.e. Trade Association) and the Bank für Handel und Industrie in Darmstadt.[50] The well-known businessman, L. Pincoffs, was the committee chairman and Stieltjes was the committee's secretary. In September 1873 this new Financial Committee produced a prototype franchise of the sort that could be granted to private companies, in which the conditions for the state were so objectionable that the minister did not even consider entertaining the idea. One major problem was the unfamiliarity with the terrain through which the envisaged railway lines would have to pass. Because of this the committee wanted to divert all the possible risks to the state. Increasing numbers of politicians, even previous supporters of privately funded railway projects, became convinced of the notion: 'better to have state-funded railways than no railways'. Fransen van de Putte thus wanted to propose that the state take responsibility and that it should start with a line going from Soerabaja to Pasoeroean and on to Malang (115 km) since people were knowledgeable enough about that area and that route. In the end, as J.H. Geertsema's cabinet was voted out in August 1874, it was his successor, W. baron van Goltstein, who finally came up with a bill proposal along those lines in November 1874.

In the Second Chamber things were not easy for the minister. The many politicians who were basically opposed to a state-run railway company raised all the old arguments and the earlier-mentioned former resident, Van Herwerden, was also galvanised into action again. In his opinion the state should not expend funds on something that would, in the end, only benefit private businessmen and planters who were, apart from anything else, often also foreigners into the bargain:

> *Let us once again be warned! The opening up of Java to people from everywhere by providing a railway network will soon mean that we will be flooded with foreigners. Very soon, and faster than one might imagine, it will lead to a loss of sovereignty in the East Indies.*[51]

One or two individuals even argued that the government would not be capable of coping with major projects. Since the creation of the Great Post Road nothing of importance had, after all, been accomplished in the East Indies. In the end, the bill was passed by a majority of 44 to 21 votes in the Chamber but only after Van Goltstein had been forced to promise that the construction work would be entirely state-based and that the exploitation of the line would definitely be left open-ended. It was, therefore, a repetition of the 1860 debate relating to the bill proposed by Van Hall on state-backed railway construction in the Netherlands. At last a decision had been taken: the 'State pleasure route' could be embarked upon.[52] The relevant law was published in the Dutch Bulletin of acts, orders and decrees on 6th April 1875. The former military engineer, D. Maarschalk, who had previous railway experience, was the man put in charge of the job. Officially he fell under the supervision of the director of Public Works but in practice he had a very high degree of freedom.[53]

The creation of State Railway lines

Maarschalk's project actually began at a very inconvenient time. The sugar crisis, a catastrophic fall in the price of cane sugar on the world market which badly hit a large number of East Indian companies and even badly damaged the East Indian banking system, started in the eighteen-seventies and continued for a long while. What especially accounted for the sudden drop in price was the increasing competition from beet sugar that was being cultivated in Europe. Ultimately it was to lead to a permanent fall in the price of sugar.[54]

Despite all these setbacks Maarschalk speedily got on with the job in hand. The Soerabaja-Pasoeroean line was opened in May 1878 and by July 1879 it had been extended as far as Malang. Maarschalk, always much in favour of narrow gauge railways which, by then, had come to be known as the standard track gauge for the East Indies, was good at economising and remaining within the budget. The railway line was simply constructed but had been laid out with a view to the possibility of later needing to be extended and reinforced. It was maintained that rails weighing 27 kg per meter would suffice for the seven light 2-4-0 tank locomotives that only weighed 14.5 tons and were supplied by the English company Fox Walker & Co which had its factory in Bristol. For the Bangil-Malang section characterised by steep inclines, the same factory supplied five six-coupled engines that were not much heavier than the other type but which, with their total weight available for adhesion, were able to provide the required power.[55]

Regarding the system of working the lines, no official pronouncement had ever been made by the legislator, but as the States General had silently granted the money needed to operate the lines, the minister presumed that the railway lines laid by the state would automatically also be operated by the state. Certain opponents maintained that in that way state working had, as it were, been smuggled in but, like it or not, that was the situation.[56] Once the line had been completed, Maarschalk left for the Netherlands on vacation where he immediately set up the Technical Bureau that was affiliated to the Department for the Colonies. That Bureau, comparable to the Crown Agents in England, drew up and prepared specifications for all the materials that needed to be supplied and took care of the requesting of quotes for the delivery of all the rail-related paraphernalia, including the actual rolling stock. Officially the Bureau was declared to be a temporary arrangement but it lasted for a long time and did much useful work.[57]

Meanwhile work had started on the building of the Buitenzorg-Bandoeng line plus the Sidoardjo-Madioen-Solo line which connected up with the NISM's broad gauge line leading to Djokjakarta. Because of the gauge difference it was always necessary for passengers to transfer and for freight to be transferred, all of which remained time-consuming and inconvenient. This was especially a problem when, much later, the state railway line between Djokjakarta and Tjilatjap on the Javanese south coast was completed. The existence of the NISM line between Batavia and Buitenzorg also constituted a problematic element when it came to extending the network, even though it conformed to the East Indian standard rail gauge of 1067 mm. Negotiations with the NISM finally led in 1877 to an agreement where it was decided that Batavia-Buitenzorg would be purchased by the government whilst the NISM would be given permission to extend its lines in the Principalities. This led to a fierce debate which was so violent that it precipitated the Second Chamber's rejection of the relevant bill in 1878. A second attempt floundered in 1881 in the Upper House.[58]

All the while, the expansion of the state railway network under Maarchalk's successor, H.G. Derx, continued unabated until financial constraints, caused by the disappointing prices of tropical products and the costly Atjeh war led to the temporary halting of all new construction work.[59] In those years a fundamental discussion once again started up surrounding the desirability of state intervention in economic life in general. The Cultivation System was gradually phased out to be replaced by private agricultural enterprise on a big scale.[60] Naturally this gave rise to further questioning concerning the state's operating of the railways. The fairly favourable outcome of the exploitation of the new state lines only served to fuel the discussion pertaining to the desirability or undesirability of having a network that was owned and operated by the state. The NISM first offered to only take over the best lines but when that offer was badly received the company immediately offered to take over all the railway lines including those that still had to be constructed by the state. This also met with heavy protest, both in the East Indies and in the Netherlands, and even though the governor-general, O. van Rees, the East Indian Council and the Minister for the Colonies, J.P. Sprenger van Eyk, declared that they were in favour of selling out, the Lower House was not enthusiastic. As slowly the financial skies started to clear, the discussion gradually calmed down.

It was during those times of uncertainty about the future of the State Railways (Staatsspoorwegen, SS) that the chief inspector of the railway's independent status was also abolished. As of 1888 he came directly under the director of Civil Public Works, and all of the railway's departmental managers were no longer answerable to the chief inspector but rather to the Public Works director. It was a strange kind of organisation in which the chief inspector no longer had any control over how services were operated. The final upshot was a slack, not very alert and detached kind of management. In practice that translated into indifference on the part of personnel, a lack of contact between the various service departments and unclear responsibilities which negatively influenced the service offered to the travelling public. Despite the many complaints aired in the East Indian press it was not until 1906 that an end was finally put to this difficult situation and the chief inspector, though still subordinate to the director of Public Works, was reinstated as the proper head of the railway company. In 1917 that dependence on the director of Public Works disappeared altogether when the State Railways was once again made an independent branch of government service.[61]

By 1894 the first link between the Western and Eastern lines of the state rail network had been created but the lines still did not form a unified whole. Any passenger arriving at the harbour of Tandjong Priok and wishing to journey on to Soerabaja had to do that in the following way: first he had to take the state line to Batavia, there he had to change to the NISM line to Buitenzorg, after which the state took over as far as Djokja. At Djokja he had to switch to a broad gauge NISM train as far as Solo after which the State Railways would transport him to Soerabaja on a standard gauge line. The journey took 32.5 hours in total and included an overnight stay in Tasikmalaja and a night in Djokja. After 1896 the travelling time was shortened so that only one overnight stay was necessary, in Maos, where the State Railways had a hotel specially built for that purpose.[62] Obviously though vastly improved, the whole journey still remained roundabout and time-consuming but for freight the problems were much greater as everything had to be transferred a number of times. Unfortunately the hotel in Maos soon got a bad reputation as the comfort and service was not reputed to be all that brilliant. Admittedly a third rail had been laid along the Djokja-Sola section so that both gauges were catered for but it was not until 1905 that through passenger trains were able to make use of that track and even then they had to travel slowly because of all the many stations and stops on the way.[63]

By this time the state railway network on the island of Java had grown, in 1900, to a total length of almost 1600 km and various lines were still under construction. The isolated NISM line between Batavia and Buitenzorg still presented problems. Even though the adherence to a uniform rail gauge had facilitated smoother connections, the state of the railway lines in and around Batavia still left much to be desired and, understandably, with the prospect of being bought up by the state, the NISM was not particularly keen to put energy into improving its network. Negotiations with the NISM resulted finally in 1909 in a parliamentary bill to purchase the line for a good ten million guilders but the proposal was rejected by the Second Chamber. Only after the price had been reduced to 8.5 million in exchange for better conditions for the NISM did the House agree to the transaction. By then it was 1913.[64]

Since there was, to his mind, a need for a clear plan concerning the necessity for train and tramlines in Java, the Minister for the Colonies, W.K. baron van Dedem, published such a railway plan in 1893. It had been largely compiled by the chief inspector of the State Railways, J.K. Kempees, and it included proposals for a number of lines to be constructed by the state. There was, however, a difference of opinion about the status of the northern line (Cheribon-Semarang) which, according to Kempees, should be made a main rail connection for military and economic reasons. In that way a second west-east line would be created, but the minister found that unnecessary, maintaining that there a tramline would suffice. All the lines would either be state or privately financed and, where necessary, there might also be a certain amount of state subsidising.[65]

Private rail and tramlines

Once the Semarang-Principalities and the Batavia-Buitenzorg lines had been completed the NISM had gradually recovered from its financial difficulties. The government's advances could be paid back and it was even possible to pay out dividends. Though the NISM itself had little interest in expanding its network, there were others who believed that the railway company's positive results were reason enough to seek new franchises.

Since 1867 the first partially successful horse-drawn tram had been in service in Batavia. The rail gauge had the unusual dimensions of 1188 mm. Steam traction had been considered but had been rejected as too dangerous for the remaining traffic which was

A Javanese State Railway locomotive at Batavia South station which belonged to the former Batavian Eastern Railway line company.

Poerwodadie Station (Central Java) which belonged to the Semarang-Joana Steam Tram Company, photographed around 1905.

why the Netherlands East Indian Tramway Company (Nederlandsch-Indische Tramweg Maatschappij, NITM) that was established in 1881, decided to use fireless locomotives which would be provided with fresh steam at two loading stations positioned at each end of the line.[66] They remained in use until the company merged with the Batavia Electric Tramways[67] in 1930 and until the time of the electrification of the former NITM company, in 1933.

What was of more importance was the incorporation of the Samarang-Joana Steam Tram Company (Samarang-Joana Stoomtram Maatschappij, SJS) in 1881 which wanted to lay a line between both places and a series of local lines in Semarang. The first idea was to make the tracks 914 mm wide but fortunately the plans were amended to 1067 mm before building work began. By 1884 the planned network, covering 87 km, was fully operational with the familiar square, European-style steamtram locomotives and lightweight rolling stock. Along these kinds of lines, often laid on the sides of existing roads, speeds were slow

The Semarang-Cheribon train. On its route this train passed no less than 27 sugar refineries.

yet fast enough to meet the transportation requirements. For rural access, notably for the local population, companies like the SJS were invaluable.[68]

The success of the SJS also encouraged others to go ahead and seek franchises. The East-Java Steam Tram Company (Oost-Java Stoomtram-Maatschappij, OJS) followed suit in 1886 with city lines in Soerabaja and a short line in Modjokerto. There was also the Serajoe Valley Steam Tram Company (Serajoedal Stoomtram-Maatschappij, SDS) that had a couple of lines around Poerwokerto in Central Java, was 90 km long and was started up between 1896 and 1900. All these lines, with the exception of the SDS, which was more akin to a local line, retained the character of the steamtram era with their light infrastructure and small rolling stock.[69]

As the existing tramway regulations of 1885 were seen as too restrictive for the new steamtram transportation requirements a special committee was created in 1891, in true Dutch style, which was expected to draw up a new set of regulations. In 1893 the new regulations were enforced despite complaints from various quarters that they might be too liberal. Indeed that proved to be the case because the whole thing led to a veritable explosion of new requests for tramway franchises. Numerous companies, both existing ones and new ones, requested and obtained franchises for new lines, most of which were intended as subsidiaries for existing lines.[70]

One of the first to benefit from the new regulations was the Semarang-Cheribon Steam Tram Company (Semarang-Cheribon Stoomtram-Maatschappij, SCS) which opened its new lines between 1897 and 1901. These lines totalled a length of 323 km and included the Tegal-Slawi-Balapelang line that had been taken over from the unfortunate Java Railway Company (Java Spoorweg Maatschappij, JSM).[71] The SCS was a tramline that ran more or less on its own tracks and had more the character of a main railway, though at first the rolling stock used was lightweight. In the 1893 regulations it was stipulated that it should be possible for mainline stock to also run on the new tramlines so that through freight transport services could be provided. One of the bigger new companies was the Madoera Steam Tram Company which, between 1898 and 1913, constructed a network that was more than 220 km long on that small neighbouring island of Java. Even the old NISM got a new lease of life and created a number of railways but especially tramways in its traffic area of Central Java. Some of the lines were broad gauge but most, like the line between Semarang and Soerabaja, were built to the standard East Indian gauge.[72]

When it came to transport on the new tramlines, the old square four-wheeled locomotives manufactured by Backer & Rueb in Breda and Beyer, Peacock & Co. in Manchester no longer met the requirements and the bigger companies, like SJS and SDS, had already invested in three and even four-coupled engines for goods transport before 1914. After 1912, with the opening of the Tjikampek-Cheribon state line, the SCS became a link in the Batavia-Semarang connection. As increasing numbers of sugar refineries sprang up along the existing line it became necessary to bring in more and more heavy engines. Between 1908 and 1913 Hartmann-Chemnitz delivered no less than 27 modern four-coupled locomotives with separate tenders that had a weight in service of 16.8 tons. After the First World War the volume of transport increased to such a degree that those engines were no longer able to cope with the express train traffic catered for in conjunction with the State Railways. It was for that reason that in 1922 some 19 real express train locomotives appeared on the rails, large 4-6-0 machines with separate tenders that could run at 75 km/hr. By that time the tracks had been reinforced and, where necessary, straightened and made suitable for speeds of up to 85 km/hr. Inevitably the steamtram character of such trains had disappeared altogether.

As new lines opened and traffic on the existing lines increased, so too did the need for new locomotives. In order to provide express trains for the wide gauge track, the NISM switched in 1902 – earlier than in the Netherlands – to six-coupled engines with 4-6-0 axle configurations and an in-service weight (just the locomotive) of about 50 tons. In the end, the NISM owned 14 such locomotives and after 1923 the fleet was boosted with another four stronger engines. For the fully-loaded sugar trains Hartmann supplied, in total, eight heavy (66 tons) 2-8-0 machines in the post-1912 period which did the job perfectly. Towards the end of the thirties the NISM, just like the State Railways, was busy planning to raise the speed of passenger trains to 100 km/hr. Obviously this meant that once again new engines would have to be purchased and after the sale of Netherlands Railways' locomotives fell through, an order was placed with Werkspoor for four heavy 4-6-0 engines. Because of the outbreak of war those locomotives were never produced.[73]

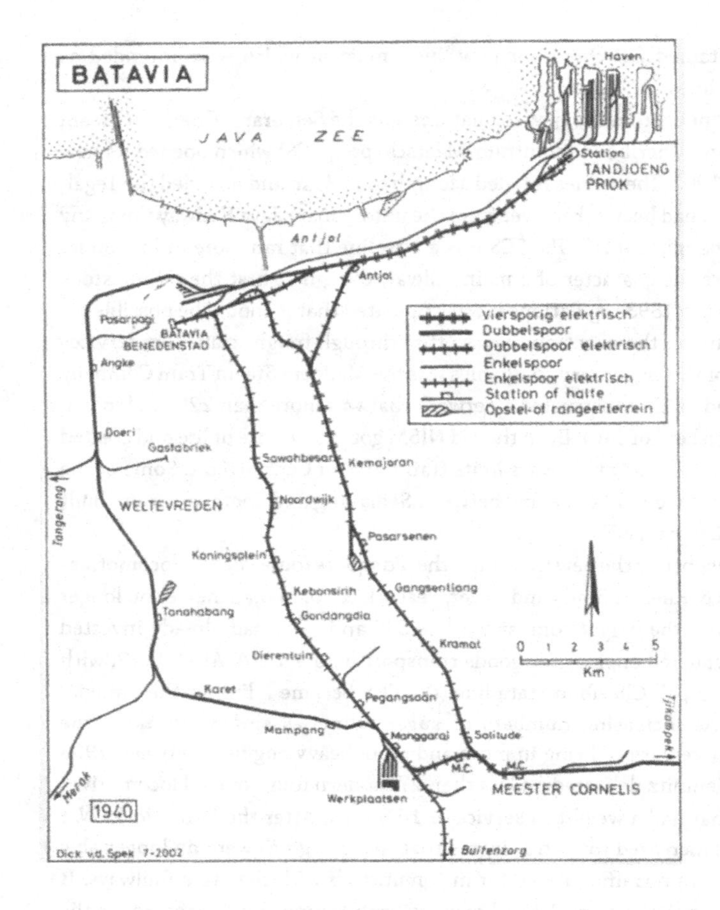

A map of the Batavian railway network round 1940.

For the NISM's new standard gauge railway lines (1067 mm), especially the connection between Semarang and Soerabaja that was initially operated as a tramway, it was mostly tank locomotives that were deployed with two and three coupled axles and they were built as 0-4-2, 0-6-0 and 0-6-2 engines by Hartmann as of 1898. Later, between 1908 and 1911, it was Werkspoor that was to deliver another 15 even heavier 2-6-2 tank locomotives. In view of the fact that in 1903, when the line first went into operation, the travelling time between Soerabaja and Semarang, with a transfer at Goendih, took as long as twelve hours and ten minutes, the NISM decided to reinforce the tracks and improve the traction. To this end ten 4-6-0 engines with separate tenders were ordered from Beyer, Peacock & Co. in 1912. The reason why so many were ordered was because they could also be used on the Batavia-Buitenzorg line but as that line was sold off to the state in 1913, that part of the plan never materialised. As those engines were well suited to the task in hand, another 20 were built after the First World War by Henschel-Kassel, Werkspoor and Beyer, Peacock & Co. with just slight modifications to various details. In time, the Soerabaja-Semarang travelling time became considerably reduced, especially after the opening of the Semarang-Gambringan standard gauge line, thus making it no longer necessary to transfer. After the line had been upgraded to a 2nd class railway line in 1928, the journey only took just over five hours. Once various improvements had been made (including to the locomotives), new rolling stock had been introduced and the line had been re-classified (in 1937) and upgraded to a 1st class railway line (suited to greater axle load and higher speeds than a 2nd class railway line) the journey from Soerabaja to Semarang took only four and a half hours.

If one was to take into consideration all the privately owned companies then one could also include the numerous factory lines belonging to sugar enterprises and other kinds of plantations dotted around Java and Sumatra. Some of these had networks, usually with 600 to 700 mm gauges that were dozens of kilometres long because they linked the cane fields with the factories and then connected up with the public rail and tramlines. In that way the Djatiroto sugar company (Trading Company Amsterdam – Handels Vereniging Amsterdam, HVA) in East Java had a 161 km-long network. Usually these lines did not cater for the transportation of passengers but the company's own personnel and labourers would be carried in primitive carriages. The Forestry department, a government company, also had numerous – sometimes very extensive – narrow gauge networks all over Java and Sumatra. At first traction was usually provided by animals but quite soon different kinds of light steam locomotives were deployed instead. In the pre-1940 era the Djatiroto factory had no less than 80 locomotives in service.[74] Factories like Orenstein & Koppel-Berlin, J.A. Maffei-Munich and DuCroo & Brauns in Amsterdam/Weesp supplied hundreds of engines to the East Indies for such purposes.[75] Also on Sumatra, especially in Deli, various of the HVA's palm oil plantations had interconnecting networks with 700 mm gauge with more than 100 steam locomotives in service.

The State Railways in the twentieth century

The story of privately owned companies brings us up to the twentieth century, thus returning us to the biggest railway company in Java, the State Railways. Apart from the newly constructed lines (to be dealt with in more detail later on) there was a small private company begun in 1883, the Batavia Eastern Railway Company (Bataviasche Ooster Spoorweg Maatschappij, BOS), which operated a line that led eastwards from Batavia to Krawang and was later taken over by the state.[76] This railway line, totalling 63 km, was opened in stages from 1887 onwards and formed the start of the Preanger state line going to Padalarang, a real mountain railway that opened in 1906 with its numerous spectacular bridges and viaducts and a tunnel at Sasaksaät that was almost 1 km long. Because of the BOS takeover the lines leading to Java's western extremity could also be linked up to the Tandjong Priok harbour line and the rest of the state network. Another important connection was the Tjikampek-Cheribon line that was 137 km long and reached completion in 1912. Since the Semarang-Cheribon Tramway Company had meanwhile opened a tramline between both cities, a kind of through line skirting the northern coastline between Batavia and Soerabaja was created. It alternated between being a main railway line and a tramway which thus made it temporarily of less importance to through traffic. The proper second connection between both cities did not come until 1916 when the Cheribon-Poerwokerto-Kroja state line was completed, with the 156 km new line linking up with the existing line from Djokja. As the track between Djokja and Soerakarta with a third rail was not suitable for high speeds a free connection between Djokja and Solo on the 1067 mm gauge, running parallel to the NISM broad gauge line, was finished in 1929. As this line was completely separate from the other line it was more suitable for high speeds than the existing track with its third rail.[77] The laying of that line roughly marked the finish of the state railway construction activities. By 1925 the state had thus created or taken over a total length of 2740 km of track in Java for a total expenditure of some 414 million guilders.[78] Of this entire network no more than 173 km was double-track and that was chiefly in the areas around Batavia and Soerabaja. The whole network was kept running by almost 40,000 employees, including the state-run lines in Sumatra and the Atjeh tram network. Asian personnel were very much in the majority with, for instance, the locomotive and train personnel consisting almost exclusively of Javanese people. It was only in the higher echelons, both on the technical and the administrative side that the European personnel component predominated.

Competition from road transport

Just as with other railway companies everywhere else in the world, in the twenties and thirties of the twentieth century the State Railway company in Java began suffering from increasing competition with road transport.[79] The road network was gradually expanding and improving under the careful supervision of the government. By the end of 1938 Java had a road network that consisted of 9000 km of asphalted roads, 14,000 km of otherwise hard surface roads and 4000 km of unimproved roads, in other words, sufficient to put internal combustion vehicles firmly on the transportation map. Elsewhere in the archipelago the situation was less favourable, though it should be mentioned that in that same year Sumatra had 8000 km of surfaced roads.[80] Energetic entrepreneurs threw themselves wholeheartedly into the business of transporting goods and passengers, especially where the short and medium distances were concerned. Cheap lorries and buses began to pinch more and more trade from the railway market to such an extent that on Java the SS started to devise ways to put an end to all of that. It seemed that one good solution might reside in the electrification of a portion of the network.[81]

Already before the First World War the first optimistic reports on the feasibility and desirability of electric railway lines in Java had been published, but due to a lack of funds that was as far as it had got. In 1917, however, a new report came out that had been compiled by P.A. Roelofsen, chief of the State Railways' hydropower bureau and later director of the Service for Hydropower and Electricity.[82] His report recommended that the network around Batavia should be electrified and that simultaneously a number of hydropower stations should be built in order to generate cheap electricity.[83] It was a plan that was well accepted so that already in 1919 work began on the construction of the first hydroelectric power station, the Oebroeg station in the river Tjitjatih, to the south of Buitenzorg. Under the supervision of the engineer G. de Gelder, who had already gained experience with electric railway networks in South America,[84] the lines in the Batavia area were rapidly electrified using the same system of 1.5 kV direct current that had been adopted in the Netherlands.[85] It was due to a lack of funds that work on the line leading to Buitenzorg was temporarily postponed but by 1925 electric trains were running from Tandjong Priok to Batavia and on to Meester Cornelis. Just as in the Netherlands, the trains used consisted of light motor carriages and coaches supplied by Beijnes in Haarlem and Werkspoor in Utrecht, but unlike in the Netherlands heavy electric locomotives were also ordered for the through train services. In the Netherlands all long-distance and international trains still ran with steam, even where the tracks had been electrified. The seven required locomotives came from Germany and Switzerland but two were jointly manufactured by Werkspoor in Amsterdam and Heemaf in Hengelo and they were the first electric locomotives to be constructed in the Netherlands. In Hengelo Heemaf even built a small test track so that the locomotives could first be tried out.[86] The motor carriages were very much like heavy American interurbans, with Westinghouse or General Electric electrical fittings, automatic couplers and air brakes.[87] The delayed Buitenzorg line was finally supplied with electric overhead wires in 1929 and required more motor carriages and a further six locomotives. By then some 125 km of railway lines were operating on electricity.

The express train service between Batavia and Soerabaja had always been given priority in the minds of the SS directors and the notion of creating a one-day connection between the two cities was one that had been in circulation for a long time. It was not until 1917, though, that the SS had its very own line via Cheribon which was shorter than the line that went via Bandoeng. The separate Solo-Djokja standard gauge railway line finally reached completion in 1929. On 1st November 1929 the first One-Day Express train started running between Batavia and Soerabaja and gradually the travelling time of 13.5 hours for the 825 km was reduced to 11.5 hours at an average speed of 80 km/hr, thus making that train one of the fastest if not the fastest in the world on 1067 mm gauge tracks.[88] At least it was quite a difference when compared to the journey time of 32.5 hours of 1894! The rolling stock for this train was already available: comfortable air-cooled, bogie carriages, 18.5 m long, including the restaurant and mail carriages. They were built by the Dutch factories Beijnes, Werkspoor and Allan. Traction came in the form of 4-6-2 express locomotives of the 700 series, delivered in 1910-1911 by SLM-Winterthur and Hartmann. They were two-cylinder engines with a total in-service weight of 88 tons, though the axle loading did not yet exceed the 10-ton mark. Even though at first it was not expected that the trains would travel faster than 90 km/hr it emerged from later test runs that these locomotives could easily reach 127 km/hr. Another twenty heavier engines (of the 1000 series) were supplied by Werkspoor in 1920, once again in the form of a 4-6-2 (i.e. Pacific) but this time as four-cylinder compound engines with an in-service weight of 109 tons and an axle loading of 12 tons. This impressive locomotive was designed by the head of the State Railways in Java, W.F. Staargaard, I. Franco, professor of mechanical engineering at the Technical College in Delft, and by Werkspoor. As a young engineer F.Q. den Hollander was also involved in the construction side. These Pacifics tended to be deployed for the One-Day express trains which were regularly loaded to more than 400 tons, and were able to reach average speeds of 100 km/hr on the level sections. During test runs much higher speeds could be reached but complaints about the unstable motion at such speeds remained. The One-Day trains were a resounding success and provided a good alternative to the daily flight connections which had already been established between the two cities.[89] In 1929 a single Batavia-Soerabaja journey in a One-Day train cost 37.50 guilders for a first class ticket and 24.70 for a second-class ticket. The KNILM Fokker aeroplane did the journey in a mere five hours with an in-between stop in Semarang but it cost 115 guilders just for a single journey.[90] In addition to the One-Day train a night express service was also laid on as of 1936 which, even though it may have been slower, was equipped with comfortable sleeping cars and soon became very popular. Up until 1918 there were no night services because of the inadequate security and the danger that large animals such as buffaloes and elephants may stray onto the tracks: few of the lines were fenced off.

Between Soerabaja and Malang (96 km) the competition with the motorbus was considerable and to put up some kind of resistance the SS introduced another five express trains on that line in 1934 which were known as the Fast Five. Among the citizens of Soerabaja Malang, located high in the hills, was a popular destination, hence the fierce competition between transport modes. For those trains the State Railways used the existing heavy 4-6-4 (79 ton) tank locomotives of the 1300 series that had been built in 1921-22 by Henschel-Kassel. They took an hour and a half to get to Malang. For the last 49 km leg of the journey, from Bangil onwards, two engines would be used to haul the train uphill in double traction. The new trains were a great success and so in no time passenger numbers went up by 90 percent, which meant that the service had to be expanded. Improvements to the 1300 locomotive series resulted in a travelling time of one hour and twenty minutes, provided that the train was not heavier than 75 tons. From 1938 onwards twelve trains per day traversed that route. It was no longer the Fast Five but rather the Express Malang train service.[91]

Between Batavia and Bandoeng the Fast Four trains, also introduced in 1934, covered the 175 km distance in two hours and 45 minutes, making three stops on the way and thus doing the journey an hour faster than in the past. That journey also included a mountainous stretch (for 56 km) through the Preanger, between Poerwakarta and Padalarang where there were sharp curves and steep inclines. There, too, it was thanks to the 1300 series locomotives that much time could be saved. An added attraction was the fact that some of those trains went on to Tandjong Priok on days when a mail boat was due in from the Netherlands (usually on a Monday or Friday) or due to leave for the Netherlands (on the Wednesday). It was by taking such innovative measures, making such technological improvements and being alert to public demand that the State Railways in Java was able to retain its share of the market and even, to a significant extent, expand it.

Railway technology in Java

From the technical point of view the Javanese railways, both the state-run component and the private lines, were very much based upon what was customary in the Netherlands. Ultimately most of the engineers in charge of things had been educated at the Delft Polytechnic where the majority of their Dutch counterparts had also been educated. There was furthermore a degree of interchange between the East Indies and the mother country, like in the case of J.L. Cluijsenaer who will come to play a part later in this story. At first the East Indian railway network was much lighter and less elaborate than the Dutch network for the simple reason that there was much less traffic in the colony. It was quite normal to find there single-track lines, light rails and few to no signalling systems. In the Netherlands, by contrast, there were already a lot of double tracks and heavy rails. In addition to that block

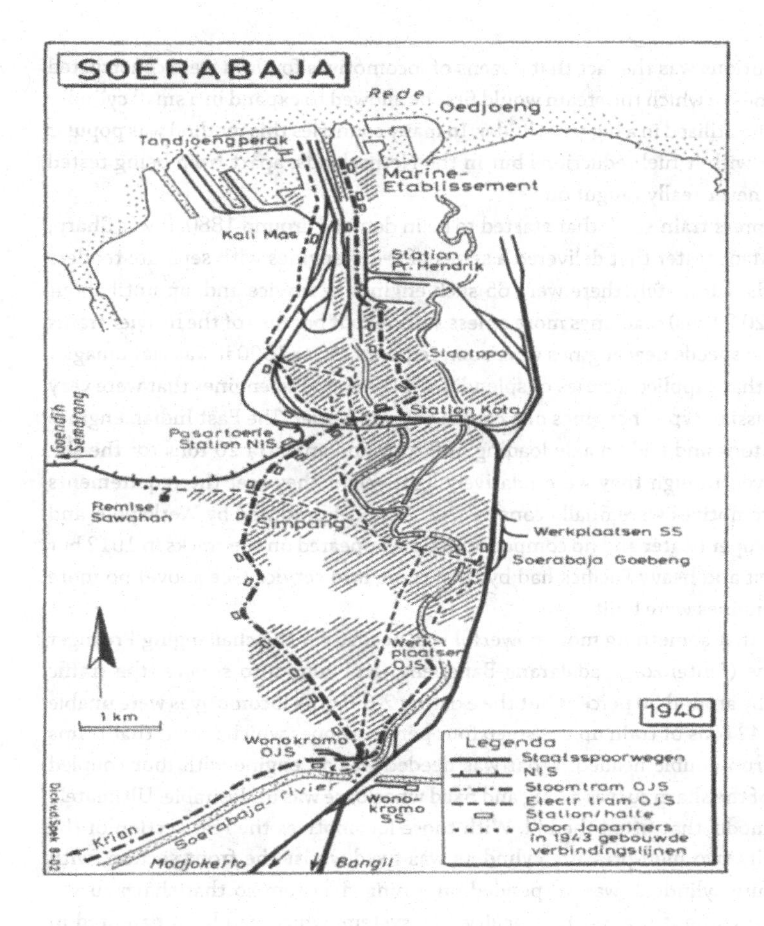

A map of the railway lines and tramlines in Soerabaja in the year 1940

signalling had also been introduced. For the relatively low travelling speeds and the light rolling stock of the time the very basic Javanese set-up sufficed. In contrast to in the mother country, a central buffer system was introduced with automatic coupling so that carriages could easily be linked together. It was a Norwegian system that had been devised by the engineer C.A. Pihl and was improved upon in Java in a slightly heavier form. For a long time the system worked reasonably well but it was not adequate for smooth journeys at higher travelling speeds so, for instance, with the One-Day service carriages the European system with side buffers and screw couplings was introduced.

As the volume of traffic increased, so it became necessary to bring in heavier locomotives and to make the tracks heavier as well. The light four and six-coupled tank-engines that were no heavier than 25 tons and could go no faster than 30 km/hr – with the exception of a few series that could get up to 60 km/hr on the level – no longer fitted the bill and so were banished to branch line services.[92] Most of the engines had been supplied by German and English manufacturers as there was, at that time, no locomotive-building industry in the Netherlands. It was not until 1898 that Werkspoor started to build mainline locomotives.

What was rather curious was the fact that dozens of locomotives for Java were constructed as compound engines in which the steam would first be allowed to expand in a small cylinder before going on to be utilised in a bigger cylinder. In many countries this method was popular because it brought with it fuel reductions but in the Netherlands, apart from being tested out a few times, it never really caught on.

For the real express train stock that started to be in demand around 1880, it was Sharp, Stewart & Co. of Manchester that delivered a series of 2-4-0 engines with separate tenders from 1880 onwards. Ultimately there were 65 such engines in service and, up until about 1900, these light (20.2 tons) machines more or less sufficed but because of the heavier trains and faster travelling speeds new engines were soon required. After 1900 it was Hanomag of Hannover-Linden that supplied a series of splendid 4-4-0 compound engines that were very inspired by the Prussian type of engines of a couple of years earlier. The East Indian engines just weighed 31.2 tons and had an axle loading of 9 tons and an extra 20 tons for the six-wheeled tender. Even though they were relatively lightweight they met the requirements and so 44 such locomotives were finally constructed. A model produced by Werkspoor and Hartmann, with a super heater and no compound action, appeared on the tracks in 1912 but as Winterthur's fast and heavy Pacifics had by then come into service (see above) no more than eleven such engines were built.

It was obvious that something more powerful was needed for the challenging Preanger line. Since that line (Buitenzorg-Padalarang-Bandoeng) had gone into service, the traffic demand had risen by around 50 percent but the existing 2-6-0 tank locomotives were unable to haul more than 42 tons of train up the steep four percent slopes which meant that trains usually had to be run double-headed. What was needed was an engine with four coupled axles but because of the sharp curves a long and fixed wheelbase was undesirable. Ultimately it was the Mallet model that fitted the bill. With those locomotives the rear section of the driving gear with its two high-pressure cylinders was fixed whilst the front section, with the two low-pressure cylinders, was suspended on a hinged system so that sharp curves could be negotiated without too much difficulty. The system, which had been patented in the eighteen-eighties by the Franco-Swiss engineer Anatole Mallet, had found widespread use in Europe since the nineties, both for narrow and standard gauge tracks.[93] In later years the Mallet locomotive was to take off 'big time' in the United States but the first SS engines in Java were of more modest proportions. Hartmann, virtually the State Railways' exclusive 'by Appointment' supplier, built the first 0-4+4-2 Mallet tank locomotives for Java in 1900. They had two fixed and two articulated driving axles and a carrying axle under the engine driver's cab.[94] The axle loading was still little more than 9 tons and the weight in service amounted to 42 tons. The locomotives met the exact requirements of the day and so, up until 1908, Hartmann and Schwartzkopff of Berlin produced some 16 such engines. Once again, though, the demand for heavier rolling stock in conjunction with rising passenger numbers soon put pressure on the service.[95] In 1904 Hartmann therefore developed a much heavier engine, a 2-6+6-0 tank engine weighing 60 tons. Between 1904 and 1909 Hartmann and Schwartzkopff supplied 23 such engines and after that Werkspoor built a further ten in 1911 of a slightly heavier type. They were good engines that hauled everything that was hitched up behind them at the maximum permitted speed of 50 km/hr on the Preanger line, but there was one problem: the flexible steam pipes between the high-pressure and swivelling low-pressure driving wheels were difficult to keep taut.[96] This was not something that was peculiar to Java because all over the world engineers were wrestling with the problem of how to seal off these steam pipes but nobody had yet found an answer.

Just to take some kind of action, in 1910 the SS of Java decided to try out a locomotive with six fixed coupled axles. The Austrian engineer, Karl Gölsdorf, had proven in practice that an engine with six coupled axles in one frame where a number had lateral play could provide the flexibility required for sharp curves. In 1912 Hanomag was thus given the order to design a 2-12-2 tank locomotive for the Javanese network. What emerged was a most impressive engine weighing almost 80 tons that turned out to be much more powerful than the 2-6+6-0 Mallets. Admittedly the sharp curves of the Preanger line were negotiated without too much difficulty but the wear and tear on the tyres of the front and rear driving wheels was so great that they had to be replaced every 6000 kilometres which meant that it became preferable to deploy those engines on lines other than the Preanger line. Nevertheless, between 1912 and 1920, 28 such 'Javanics'[97] were constructed by Hanomag and Werkspoor for use on the Preanger line, for use elsewhere in Java and for the coal transport lines on the west coast of Sumatra.

Transport demands continued to increase and for the mountain routes something more powerful than the 2-6+6-0 Mallets or the Javanics was soon required. After the experiences with the 2-12-2 type it was evident that more fixed axles could not be used and so the decision was taken to revert to the Mallet locomotive of the past. As the First World War was by then fully underway the usual German factories were no longer in a position to supply rolling stock and so an order was placed with the American Locomotive Company. In 1916 that factory supplied eight 2-8+8-0 engines with separate tenders within the space of six months. They were the largest locomotives that the State Railways had ever owned. Their in-service weight was 133 tons, their adhesion weight 84 tons and they had a maximum axle loading of more than 11 tons, all made possible by the fact that in most of the places where those locomotives were used the rails were heavier[98]. In 1918 twelve slightly heavier locs were supplied by ALCO and afterwards, between 1923 and 1925, Hanomag, Hartmann and Werkspoor built 20 slightly modified and improved engines. It was found that the American locomotives were somewhat too slow for express services on the mountain tracks whilst the European models had no trouble getting up to the higher speeds and so they were used on passenger trains.

As the axle loading could not be any higher than 11 tons on certain lines, the State Railways ordered 30 lighter 2-6+6-0 Mallet locomotives in 1927, 16 of which were to come from Winterthur and 14 from Werkspoor. They weighed a 'mere' 110 tons in service and were generally satisfactory.[99] Apart from 20 2-6-2 tank locomotives from Hohenzollern in Düsseldorf and Borsig, intended for the lighter passenger trains that were delivered in 1929 and 1930, the 2-6+6-0 Mallets were the last locomotives to be ordered by the State Railways. The world economic crisis precipitated such a fall in traffic that part of the older locomotive fleet was even scrapped, whilst a number of the bigger and more modern machines were temporarily taken out of service but stored for possible later use. By June 1941 the State Railways in Java had 525 engines; some of the very oldest of which were only represented in a mothballed form.[100]

There were four well-equipped workshops for all the maintenance that had to be done on this fleet of engines, Manggarai (Meester Cornelis), Bandoeng, Madioen and Soerabaja-Goebeng and two smaller but outdated workshops in Poerwordjo and Djember. The workshop in Manggarai that was opened in 1920 was designed according to the newest trends and equipped with the most modern facilities and tools.[101] The older series of engines tended to be wood-fired but later models ran on Sumatran coal or oil.

Generally speaking, it can be said that the image of the State Railways in Java was very Dutch, which was not surprising when one bears in mind how Dutch the whole character of the managing of the business was. It was only the impressive Mallet locomotives that were not required in the flat Dutch countryside and were therefore unknown in the Netherlands. Through all the international contacts and the level of familiarity with specialist literature, foreign inventions and innovations were usually implemented relatively quickly, generally simultaneously in the Netherlands and the East Indies. There was just one striking example of where the East Indies was clearly ahead of the mother country. In France the engineer André Chapelon had been busy experimenting with steam locomotives since the twenties in an endeavour to make them more efficient by streamlining the internal steam channels, improving the fireboxes and the exhausts and optimising the balancing of the drive. The results were spectacular: not only was the pulling power of the engines that had been renovated in that way considerably increased but the fuel consumption was, at the same time, considerably reduced and the engines became much quieter which inevitably meant less wear and tear on the moving parts. In the Netherlands his work was, of course, known but for some inexplicable reason little or nothing had been done with that knowledge. At the time the man in charge of traction for the Netherlands Railways, W. Hupkes, was more preoccupied with new diesel traction developments and his assistant, P. Labrijn, the man responsible for the construction of steam locomotives, was clearly not very interested either, though certain of the existing NS locomotives could well have benefited from Chapelon's kinds of modernisations. In the East Indies, by contrast, there was a great deal of interest in what Chapelon was doing. Ways and means were found to modernise the newer locomotives so that they were quickly suited to hauling trains such as those used for the Fast Four and the Fast Five services. The steam channels were thoroughly renovated, the combustion and exhaust systems improved and the driving gear was re-balanced, all in accordance with the upgrading method dictated by Chapelon.

The 4-6-4 tank locomotives of the 1300 series were the first to be taken in hand in the way described above. After that the 900 series engines were completely overhauled. Some 42 of those 2-8-0 locomotive types had been supplied by various factories between 1914 and 1921. They were powerful but slow engines suited to freight service in level countryside. As powerful machines were required for the Buitenzorg-Soekaboemi mountain line a number of those locomotives were taken out of the mothballs in 1938 and given the Chapelon treatment. It was certainly a successful action. Not only was the in-service speed raised to 75 km/hr from the previous 50 km/hr, but the adhesion weight was increased with 2 ton of iron blocks so that the locomotives became suited to the mountain lines. At the new higher speed they also ran extremely quietly. In test runs they even effortlessly got up to speeds of 130 km/hr! The 2-8-2 tank locomotives of the 1400 series that had been built by Hanomag and Werkspoor in 1921/22 for the slower freight services (hence the lack of attention to the balancing of the driving wheels) underwent a similar metamorphosis to the above-mentioned engines. In 1938 they were rebuilt and, once again, the results were spectacular. Speeds could be increased to 70 km/hr from 45 km/hr which meant that the travelling time between Bandoeng and Soekaboemi could be considerably reduced. As the firebox and the exhaust had been improved, coal consumption fell in spite of the fact that the engines were being worked much harder. With their small 1100 mm driving wheels they still ran quietly at speeds of more than 80 km/hr thanks to being better balanced. Other locomotives were subjected to similar treatment in which attention was especially paid to small but important adjustments to the balancing, the exhausts and the fireboxes. In that way the 2-12-2 Javanics were able to reach speeds of 75 km/hr instead of the 55 km/hr of the past.[102] Thanks to such attention to detail and the introduction of the most modern European techniques, the State Railways possessed a fleet of locomotives at the time of the outbreak of war which, though largely antiquated, was relatively well adjusted to the traffic demands of the day. The number of passenger carriages was also adequate but there, too, the number of really modern steel carriages equipped with the latest innovations such as air conditioning was much too small.

All in all, it can be concluded that by 1942 the well integrated network of railways and tramways that had been developed in Java met the requirements. Private and state-run companies complemented each other well and the rolling stock could be fairly easily used on all the lines. It was just the broad gauge tracks of the NISM that created an obstacle in what would otherwise have been a truly unified network throughout the whole island. It was the Japanese occupier, who was not hampered by shareholders in the Netherlands, who later narrowed the lines down to the East Indian standard gauge.

Railway lines in Sumatra

For much of the nineteenth century the islands outside Java remained the government's poor stepchildren. Indeed, it was for good reason that they were termed the 'Outer Regions'. It was only in odd spots that there was some kind of Dutch authority and any kind of familiarity with the inland regions of bigger islands like Sumatra and Borneo was virtually non-existent. The commercial exploitation of the riches beneath the earth's surface was more or less confined to the tin mining activities on Banka and Billiton while the production of spices in the Moluccas remained important. Gradually it became clear in Batavia and The Hague that there were many possible competitors around and that the Dutch colonial state could only be safeguarded if authority was properly exerted throughout the archipelago. This therefore constituted externally stimulated colonial expansion but as soon as the natural resources of some of these outlying possessions became known the pressure exerted could also be termed internal.[103] Railway construction in these Outer Regions was therefore directly influenced by the developments just sketched above as will emerge from the following sections.

The Atjeh Tram

The first railway line to be laid outside of Java could be termed a war child, originally without much economic use. Atjeh, situated in the northernmost part of Sumatra, had always been

The heaviest steam locomotives in Java were the 1 D+D Mallet locs in the 1209-1220 series which had a duty weight of 137 tons. They were built in America and the Javanese State Railway Company found that they were best suited to the mountainous terrain of the Preanger (West Java). The locomotive photographed here is the 1209 seen in Bandoeng in 1939.

an extremely independent part of that island which did not bend easily to Batavian rule. Vague agreements with Britain also hampered any decisive attempts to curb piracy and other detrimental practises until, after the Sumatra Treaty of 1871, the Netherlands was finally given a free hand throughout Sumatra.[104] A first expedition designed to subject Atjeh to central rule was undertaken in1873 but met with little success. A second expedition, undertaken in that same year, was a success as the *kraton* of Kota Radja was wrested from the hands of the sultan, but elsewhere in Atjeh the opposition remained strong. It culminated in endless guerrilla warfare with the Dutch troops withdrawing to the main town of Kota Radja. It was very difficult to transport all the provisions from the Olehleh road to Kota Radja, which was why in 1874 the decision was taken to build a harbour jetty in Olehleh with a railway line leading to Kota Radja.[105] The whole decision-making process was made flexible by military necessity so that in 1874 the proposal was approved. By the end of 1876 the line, which was a mere 5 km long, had been completed. It adhered to the by then standard East Indian track gauge of 1067 mm. The whole project was lightweight with small four or six-wheeled locomotives and light carriages but it served the purpose.[106]

For present purposes it is not necessary to trace the events of the war and how it progressed but once the railway fell under civil administration a line extension, to as far as Glé Kambing, was recommended using a narrower 75 cm gauge so that the existing military roads could be followed. The relevant project plans were drawn up by the captain of the corps of engineers, Wouter Cool.[107] Due to the opposition put up by the people of Atjeh only 8 km of the planned route was actually laid but in 1884 the original Olehleh-Kota Radja line was converted to the narrower gauge. All further rail construction work was predominantly dictated by the changing military strategies so that a belt line around Kota Radja was laid with a few radials that was altogether 39 km long. Military factors made it once again necessary, in 1890, for the network to be placed under military control and extended. Seulimeum, situated some 45 km away from Olehleh was reached in 1898 and under the governorship of J.B. van Heutsz it was decided that the line along the coast should be extended to link up with the Deli Railway Company (DSM). The construction work took a long time. It was not until 1917 that the 511 km line to Pangkalan Soesoe on the Strait of Malacca was ready for use. The DSM (see below) linked up two years later from a southerly direction but using the East Indian standard gauge of 1067 mm. This meant that the last section, from Besitang to Pangkalan Brandan, was three-railed. In 1916, as proof of the peace-making that had been done, the Atjeh Tram was transferred from the War department to the State Rail and Tramway services. From then on it came to be known as the Atjeh State Railways (Atjeh Staats Spoorweg, ASS).

The first locomotives for the 75 cm track were supplied by Hanomag in Hannover. They were small 2-4-0 tank locomotives of no more than 10 tons and ultimately twelve were put into service. Two of the locomotives were built in 1904 in the workshops of the Atjeh Tram in Kota Radja and Sigli and that was no mean task[108] The heavier six-coupled engines were delivered in 1898 by Hanomag and Werkspoor but the problem of the small space for wood fuel became more and more of an obstacle as the network extended. This problem could be partly rectified by building separate tenders for water and wood in the own workshops. For the 20 km mountain line in the Seulimeum area, special Mallet locomotives were ordered from Esslingen in 1904. They were six hyper modern 0-4+4-0 engines, still constructed as tank engines and as the driver's cab proved to be too cramped and too hot, the engines were lengthened by one metre, which meant that the coal supply could also be increased. Since the axle loading was somewhat on the high side an extra axle was placed beneath the new, enlarged cab.[109] In that way they were turned into 0-4+4-2 engines weighing just over 31 tons. For the long and fairly level line along the coast something bigger and faster was needed than the 0-6-0 engines. That was to lead, in 1922, to the introduction of two fine 4-6-0 Werkspoor locomotives with separate tenders. They were intended for the ASS 'express trains' even though the maximum speed was not much more than 35 km/hr. For freight train purposes, DuCroo & Brauns and Hanomag each supplied six 2-8-0 locomotives with large bogie tenders in 1930. These were to be the last locomotives ordered by the ASS. After 1912 most locomotives switched to wood burning, though sometimes when the supply of firewood was insufficient, coal was used instead.

In the thirties, rail traffic declined sharply as competition with road traffic grew. With the narrow tracks and their light construction it was hard for rail speeds to be increased and so road traffic (cars and lorries alike) stole a sizeable portion of the market. A journey from Medan to Kota Radja of some 650 km with DSM and the Atjeh Tram took two days with an overnight stay in Lhokseumawe. It was suggested that the tracks could be reinforced or converted to the standard East Indian gauge but ultimately those options were rejected on the grounds of expense. The idea of discontinuing the service was also seriously considered but it was social and military factors that tipped the balance in favour of continuation. A night train service was even laid on, in the first place to transport cattle to Medan but a passenger carriage was also hitched on behind. The goods wagons were usually four-wheeled and had a 4 to 5 tons capacity. Later wagons were equipped with bogies and could carry twice as much. The first passenger carriages were four-wheeled and later six-wheeled with movable axles. Even later still, after the completion of the entire line, comfortable bogie carriages appeared on the scene which closely resembled Dutch tram carriages such as those used by the Geldersche Steam Tram, which also ran on 75 cm tracks. All in all, the Atjeh Tram did what was expected of it, first in terms of military requirements and later, after pacification, in terms of developing the region. Towards the end of the period discussed here the tram had, however, virtually outlived itself. With the arrival of asphalt, lorries and buses it became clear that roads were better suited to meeting the transportation demands.

The Deli Railway Company

Whilst everywhere throughout the Outer Regions the presence of the Dutch – not to mention Dutch administration – remained a rare phenomenon for a long time, there was one region in Sumatra that had attracted the attention of Dutch planters early on. That region was the Sultanate of Deli, situated on Sumatra's north-eastern coastline where the soil turned out to be extremely fertile and particularly suited to the cultivation of tobacco. One planter entrepreneur, J. Nienhuys, had already started cultivating tobacco there in 1863.[110] His agricultural experiment was a success and soon (in 1869) the Deli Company was established. Before long it turned into a large and influential enterprise for the production of tobacco, rubber and tea.[111] The success of the tobacco growing combined with the virtual total lack of effective Dutch rule in those parts soon led to all kinds of complications. The planters resolved the chronic shortage of labour by contracting Chinese and Javanese labourers, who invariably lived and worked in deplorable conditions. It was only towards the end of the nineteenth century that government intervention on a large scale put an end to the worst social abuse.[112] At the same time, massive profits were being made but it was the planters and traders who were benefiting.

The prosperity in Deli and the surrounding area rose even more when, in 1883, the tobacco planter A.J. Zijlker struck oil in the Langkat Sultanate between Deli and Atjeh. As a result, the Royal Company for the Exploitation of Petroleum Wells in the Dutch East Indies (better known as Royal Dutch) was established in 1890 and it took charge of oil production. After a number of difficult years the company finally extended its business to the whole of Sumatra and elsewhere in the archipelago and the world. Ultimately it succeeded in destroying the monopoly of American Standard Oil.[113]

The Atjeh tram station (North Sumatra), in about 1920.

With such spectacular regional developments going on the inadequacy of transport facilities obviously became a big drawback. There were hardly any roads and the rivers were generally not easily navigable. Little wonder, therefore, that people's thoughts turned to tram or rail transport possibilities. The first franchise was requested in 1881 by a certain De Guigné but it was rejected by the government on the grounds that he was not Dutch.[114] Immediately afterwards a request was submitted by J.T. Cremer, Deli Company administrator and successor to Nienhuys, to create a connection between Belawan, the 'harbour', and Medan, the chief town in the sultanate.[115] Initially it was thought that the tramline would run on 70 or 75 cm gauge tracks but eventually it was the standard East Indian gauge of 1067 mm that was selected. The franchise was for a Belawan-Medan line (23 km), a line going from Medan to the east (11.6 km) and a westward line from Medan to Bindjei (20.9 m). In contrast to the state-run lines of Atjeh and the western coast of Sumatra this was a purely private affair. The initial outlay of 2.6 million guilders was fairly easily raised, especially by the East Indian cultures companies and half a million poured in from the Netherlands.[116] Nobody thought for one minute that the state should be involved in the whole project, nor was a subsidy from the government considered. The Deli planters were keen to keep things in their own hands. The harbour works in Belawan were also created and operated by the railway company, which was something exceptional for the East Indies. The actual construction work was not easy. The coastal strip was marshy and unhealthy and it was difficult to construct a bridge over the Deli River to the island where Belawan was situated, because there were no solid layers of hard sand for the foundations. In the end, the bridge was built on iron screw piles that were driven 15 m into the ground but even then it did not rest on a solid basis. The bridge with its 20 spans, each 18.5 m long, was the last section of the first line to reach completion in 1888.[117] The bridge did not need to be replaced until 1932 so, in other words, the engineers had done a good job. From the economic point of view, the new company was a success. In 1888 permission was obtained to extend the coastal line further from Medan in a south-easterly direction to Perbaoengan through a region where a large number of new agricultural companies had been established. By 1890 that section had also been completed. In the west, the line was extended from Bindjei to Selesai, again for the benefit of several crop-growing concerns in the area.

After 1890 there was a slight recession in the tobacco-growing industry which was chiefly caused by tariff measures that had been taken in the United States. Because of that railway expansion stagnated for a while but after 1897 improvement once again became evident, mostly because by then the coffee and tea growing businesses were also flourishing. After 1898 the existing lines were extended, in the north towards Pangkalan Brandan (1904) and in the south as far as Tebing Tinggi with a branch line, a real mountain line, leading to Pematangsiantar (1916). Rail construction work then slowly continued from Tebing Tinggi in a south-easterly direction until, in 1937, interrupted by the world economic crisis, Rantau Prapat was finally reached. By then the network was 553 km long.

As Belawan left much to be desired as a seaport, because of the shallowness of its harbour mouth, many were in favour of seeking a different suitable spot elsewhere. Aroe Bay, to the northwest of Belawan, seemed to be a more suitable place for a deep-water harbour. A specially created East Indian state committee declared in 1909 that it was in favour of creating a harbour in Aroe Bay but the various interested parties – especially the agriculturalists and the DSM – did not go along with the plan. The distance from Medan, the administrative and economic centre of Deli, was considered to be too great. It just so happened that at that time two Dutch experts, J. Kraus, a professor from the Technical College in Delft, and G.J. de Jongh, director of Civil Public Works in Rotterdam, chanced to be in the East Indies in connection with advisory matters relating to harbour construction. Going on the available information, they maintained that Belawan could be retained and that with the help of modern floating hopper dredgers it should be possible to dredge the river mouth until the desired depth was attained. It was also their contention that the government should take over the exploitation of the harbour from the Deli Railway Company.[118] The East Indian government agreed to those proposals and in 1912 the harbour works were sold to the government for half a million guilders. The experiments with a modern silt dredger were most successful and Belawan was soon converted into a real ocean harbour. In 1921 the first Dutch mail boat, the 'Rembrandt', owned by the Nederlands Company, sailed into the harbour and moored there.[119] Despite the fact that the choice had fallen with Belawan, the

DSM still responded to the request for a line to Aroe Bay via Besitang where a link up was made with the Atjeh Tram. Together with this Atjeh Tram connection it was finally possible, by 1921, to reach the harbour of Pangkalan Soesoeh.

When DSM tracks started being laid it was presumed that train speeds would be low and train weights modest; the highest permitted axle loading was 10 tons. The maximum speed was fixed at 50 km/hr and the tracks were laid with light rails and without a proper ballast bed. Gradually the transportation demands increased which meant that tracks had to be reinforced, bridges rebuilt and safety increased by introducing fixed signalling systems. Express trains were laid on between Belawan and Medan which travelled at 70 km/hr and had comfortable bogie carriages.

The first locomotives were real contractors' machines but the Deli Railway Company immediately had four and six-coupled tank locomotives constructed by Hohenzollern that performed well on most level tracks.[120] However, the fast increasing transport volumes, resulting especially from the palm oil and rubber plantation production that gained momentum after 1900, made it necessary for new and heavier locomotives to be introduced. Six-coupled tender engines of 0-6-4 and 2-6-4 axle formations were produced by Hartmann and Werkspoor after 1900. Those locomotives weighed around 50 tons and had a maximum speed of 70 km/hr which was enough for the new express trains. For the mountain line leading to Pematangsiantar heavier (60 tons) engines were needed, four of which were produced by Werkspoor in 1917 and had 2-8-4 axle configuration with four coupled axles. The last locomotives to be newly built were ten 2-4-2 tank locomotives specially ordered for the express trains. They were produced by Hanomag in 1928 and exactly fulfilled the requirements. Even though there was an abundance of petroleum in the area, all the Deli Railway Company locomotives used wood as fuel. The DSM even had its own forests especially for that purpose. Alongside of these 'ordinary' steam locomotives the DSM also experimented with a Sentinel steam carriage that did not meet the requirements at all, and with a number of petrol-engined carriages constructed by Renault for the less intensively used branch lines. At the beginning of the Second World War, in 1941, the Deli Railway Company had more than 61 steam locomotives, 10 motorcars, 223 passenger carriages (most of which were eight-wheelers) and 1867 (chiefly four-wheeled) goods wagons, some of which, especially the tank wagons, were owned by companies like Royal Dutch Shell and Good Year.[121]

DSM's financial results were generally positive, except for in the years 1930-1935 when losses were incurred. Dividends were frequently paid out to the shareholders, except of course in the Depression years.[122] In the company's best years, 1928 and 1929, just over a million tons of freight and almost seven million passengers were transported. The crisis years saw a halving of that transport volume but by 1941 the economy had picked up so that a respectable 875,000 tons of goods and 4.5 million passengers were conveyed by rail in that year alone.[123] The manpower employed by the DSM in 1928 to facilitate all of these transport services consisted of 154 European and 3718 Asian employees. By 1939 the personnel numbers had dropped considerably but there were still 72 European and 2000 Asian employees on the payroll.

All things considered, the Deli Railway Company was an essential part of the development of Deli and its various agricultural companies. One consequence of that dependence was, of course, that the DSM results became closely allied to the prices of the tropical products on the world market. Nonetheless, the DSM management succeeded in transforming the company into a modern railway concern that measured up to the best comparable companies in other tropical regions.

The State Railway line on Sumatra's west coast

Apart from the capital of Padang, the fairly inaccessible regions of Sumatra's west coast had long remained unexplored. In the inland region the centre of Dutch rule was located in Fort de Kock. In the Padang Highlands of the Ombilin River the mining engineer, W.H. de Greve, discovered good quality coal in accessible fields in the year 1869. As there was a scarcity of fuel in the East Indies it was obviously not long before people came up with the idea of exploiting those coalfields. Up until then coal had been imported from Australia, South Africa and even from Britain which had made it a very expensive commodity. Sumatran coal promised to be much cheaper but there was still the question of how to transport the mined coal. Steep inclines would have to be negotiated to get the coal to the quayside in Padang. Another possibility was to transport it to a point on Sumatra's eastern coast which would be closer to Singapore. While searching for a suitable route De Greve actually drowned and so all the plans were temporarily halted.[124] In the inland areas there was hardly any Dutch presence and the topography was pretty much unknown. Private parties submitted franchise applications but even they were uncertain what to expect and the interest guarantee requested of the government was blocked by certain fundamental and practical objections.

It was not until 1873 that the matter was resumed. It was the engineer J.L. Cluijsenaer who was ordered by the government to investigate the possibilities for laying a coal line to the coast.[125] His first idea was to create a fairly direct connection between the coalmines to Brandewijn Bay and Padang via the Soebang Pass. Though short (96 km) it would have been a winding and steep route.[126] In a second report he discussed the possibility of creating a 115 km branch line that would lead to Fort de Kock.[127] On returning to the Netherlands and contemplating again the steep inclines that would be needed if the line were to lead to Brandewijn Bay, Cluijsenaer started to think of other possibilities. In a third report published in 1878 he described the possibilities that a new technique seemed to offer.[128] The Italian engineer, Thomas Agudio, had devised a system whereby trains negotiating steep inclines could be actually pulled uphill with cables. His basic idea dated from 1863 but it had not been tried out until 1875 when a first trial was carried out with the Mont Cenis railway. Though the system seemed to work Cluijsenaer hesitated to recommend that kind of traction for the Ombilin railway line.

What appeared to be more promising was the invention of the Swiss engineer, Nikolaus Riggenbach, who had opened Europe's first rack or cog railway in 1873 – what Cluijsenaer called a rod rail line – the 7 km-long Vitznau-Rigi railway line. Shortly afterwards a number of other similar lines were built in Germany and Switzerland which meant that, unlike with Agudio's idea, Riggenbach's was actually successful in practice. In his report Cluijsenaer therefore ventured to cautiously recommend the rack rail system for the Sumatran coal line, but on a completely different and in fact much longer route which would keep to much of the originally plotted route and would pass Padang Pandjang and go on to Fort de Kock. From

Belawan station (Sumatra) which belonged to the Deli Railway Company photographed soon after its opening in 1888.

Padang Pandjang the steep Anei gorge could then be followed to descend to sea level using the rack rail system. The whole railway line would be some 150 km long.

Despite the optimistic prospects of the proposal no proper decision was made due to the fact that the previously described discrepancies between those who were for and against state exploitation served to paralyse the various relevant decision-making processes in the East Indies and the Netherlands. After pressure had been brought to bear by different members of parliament including J.T. Cremer, the Minister for the Colonies, J.P. Sprenger van Eyk, finally submitted a proposal for the creation of a state-funded coal line along the lines described by Cluijsenaer in his final plans.[129] J.W. IJzerman, an engineer for the State Railways in Java who had assisted Cluijsenaer with his previous plans, was given the job of exactly plotting out the route.[130]

IJzerman succeeded in manoeuvring himself into an almost inviolable position which was quite something in an age when the head of the State Railway Company in Java had actually been made subordinate to the director of Public Works. Evidently he enjoyed the support of the Minister in The Hague who simply ignored the authorities in Batavia.[131] IJzerman basically stuck to the route that had finally been mapped out by Cluijsenaer, not via the Soebang Pass but taking the roundabout route via Padang Pandjang and making use of two rack railway sections for the steep and rugged Anei gorge and for an area to the east of Padang Pandjang. A large modern coal loading facility was created in deep water in Brandewijn Bay which, for the occasion, had its name changed to Koninginnebaai (Queen's Bay) for the coaling station at Emma Harbour. The total length of the Emma Harbour-Padang Pandjang-Sawah Loento (i.e. the coal mines) line was 154 km. Some years later the Padang Pandjang-Fort de Kock-Pajakombo branch line was constructed which was 52 km long.[132]

The whole project took quite a long time, especially because of the difficult section through the steep and narrow Anei gorge. Three tunnels had to be drilled, one of which was more than 800 m long and just in the Anei gorge alone, eight bridges had to be constructed which involved endless digging and banking up. IJzermann went for the standard East Indian gauge (1067 mm), despite all kinds of recommendations that 750 mm and sharper curves would have been better. In retrospect it emerged that he had made the right decision as heavy coal trains were able to travel along the 1067 mm tracks but would not have managed on the 75 cm tracks. The rack rail system devised by Riggenbach and proposed by Cluijsenaer was adopted by IJzerman.[133] In 1891 the line leading from the coast to Padang Pandjang went into service but a massive *bandjir* in the Anei river destroyed a kilometre and a half of track in 1893, together with a couple of bridges. The repair work took months and cost more than half a million guilders. The whole line, including the coaling station at Emma Harbour, was finally reopened in 1893 and the last four or so kilometre section leading to the coal washing installation in Sawah Loento, including another tunnel of almost a kilometre long, followed in 1894. The whole construction job, including the Pajakombo extension, had cost nearly 20 million guilders thus amounting to 94,000 guilders per kilometre, which in view of the difficulty of the terrain was not really so exorbitant.[134]

The construction of this heavy mountain line had been a great technological feat and one that had certainly drawn attention in the entire railway world. The coal trains, consisting of hyper modern four-wheel iron self-discharging hoppers were generally hauled with double traction and were often also pushed along the short incline outside Padang Pandjang.[135] The maximum train weight on the steepest sections was originally no more than 105 tons but when two trains were linked together with one locomotive at the front and one in the middle, the total weight could be doubled. On the adhesion sections the maximum permitted weight was 600 tons and at the beginning and end of the rack rail sections a lot of shunting had to be done.[136] For the adhesion sections 30 2-6-0 37-ton tank locomotives were purchased that were built by Esslingen whilst, up until 1905, Esslingen produced 33 engines for the rack railway sections, all of which were tank locomotives with 0-4-0, 0-4-2 or 2-4-0 axle configurations that weighed just under the 30 tons.

Despite the railway line's relative success, continuous attempts were made to find other ways of transporting the coal to the coast. In conjunction with the steepness of the hills the transport capacity was really too small to fully exploit the potential of the mines. There was always the nagging thought that the original route leading over the Soebang Pass might have been better, though there the inclines would definitely have been steeper. In order to have done with the endless misgivings the Minister for the Colonies ordered, in 1906, that

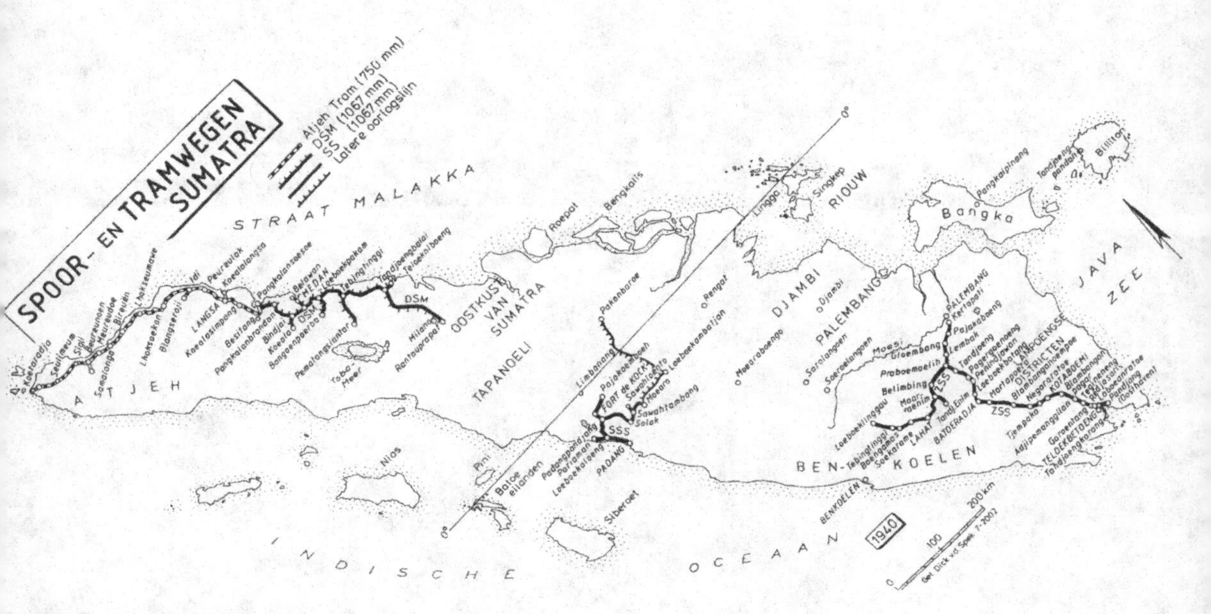

A map of the railway lines and tramlines in Sumatra in 1940.

a full investigation be made into the desirability and feasibility of creating a line over the Soebang Pass. Cluijsenaer and IJzermann were among the committee members and so it was not surprising to see that the committee unanimously rejected the Soebang route as unnecessary and much too expensive. Small improvements made to the existing line would be sufficient to increase the transport capacity.[137]

In its recommendations the Soebang committee had also presumed that heavier rack locomotives would be introduced on the existing railway. Two 0-8-2 type engines with four coupled axles and a rear carrying axle with a service weight of 50 tons were supplied by Esslingen in 1913. The Swiss factory, Winterthur, supplied a similar engine in the same year. In practise those modern engines proved able to operate just as easily on the adhesion sections as on the rack rail sections. The First World War heralded a decrease in the supply of imported coal to the East Indies which meant that the Ombilin mines were stretched to the limit. More rolling stock was urgently required but because of the war the even more powerful engines that had been ordered were not delivered until 1921. In total 22 heavy 0-10-0 engines, each weighing 53 tons and equipped with five coupled axles, were supplied by the same factory. Those engines were able to pull 190 tons on the steepest rack rail sections. In practice one locomotive would be placed in front of the train and another at the rear. By then the trains consisted mainly of bogie coal wagons with 24-ton loading capacities. The year 1930 turned out to be a top year for the Sumatran State Railway (Sumatra Staatsspoorweg, SSS): the total amount of coal transported was over 640,000 tons but later it dropped to just over half that amount, largely because of the economic crisis but also because of the

A State Railway coal train crossing the spectacular bridge over the Anei crevice situated on Sumatra's western coast.

competition with the Boekit Asem government coal mines in South Sumatra after 1925. The demand for bunker coal also fell in connection with the gradual switch to diesel engines in the shipping sector. By 1942, however, the SSS still had 31 adhesion and 25 rack locomotives in service, 136 carriages and 1033 goods and coal wagons.[138] Little else than coal was ever transported along that route. The line passed through a relatively thinly populated region with few or no large plantations and the coffee and tea cultivation which had once existed there in the nineteenth century had all been wound up in the crisis period. What still needed to be transported could easily be transported by lorry. What did remain was the demand for passenger trains so that people could be taken to the markets. With the rising losses of the thirties the idea of closing down the mines and terminating rail transport was seriously contemplated but mainly because of social considerations the plan was never implemented and everything continued as usual.[139] The gradual easing off of the consequences of the economic crisis saw to it that after 1937 the losses of the thirties were slowly turned into modest profits so that discussions revolving around closures gradually ebbed away.

The South Sumatran State Railways

At first South Sumatra was slow to develop which meant that for a long time no railway services were required in that area. Already in the nineteenth century there had been plans for a railway network, most of which had been launched by private parties. Every time, though, those kinds of plans had met with funding problems and refusal on the part of the government to offer support. Here, again, it was apparent that it was really only the state that could afford to fund such construction work.[140] Apart from anything else this coincided with the government's plans to encourage people from vastly overpopulated Java to migrate en masse to South Sumatra which was so thinly populated. The first colonies were formed in 1905 and by 1910 some 4500 Javanese had moved to the Lampong districts.[141] It was not until 1908 that a first survey was carried out in connection with plans to lay a railway line from Oosthaven (Eastern Harbour), just across from Java, that would go to Palembang via Prabamoelih and would have a line branching off from Prabamoelih to Moeara Enim for the transportation of the coal which had been discovered there in substantial quantities.[142] It had moreover been discovered that in Djambi and the Palembang region there were considerable oil reserves under the ground, reason enough to seriously start thinking about constructing railway connections. The construction work began in 1911 both at the Eastern Harbour end and in Palembang but it was not until 1927 that the two places were linked up. By then 527 kms of line were in use. The 135 km line between Lahat and Loeboek Linggau went into service in 1933. The branch line leading to Moeara Enim plus the short line from there to the Boekit Asem mines had already become operational in 1919.

Traffic on the South Sumatran State Railways (Zuid-Sumatra Staatsspoorwegen, ZSS) developed speedily so that before very long a net profit of 3.7 percent of the initial outlay of approximately 58 million guilders had been recovered which was not bad going, all things considered. The line was also operated very simply and cheaply, as a tramline with maximum speeds of 30 km/hr, even though the heaviest stock from the Javanese State Railways was used. Coal traffic was important, ultimately 400,000 ton per year was transported and for that purpose special bogie wagons with removable containers were purchased. After the

The arrival of the railways and the tramways also gave the native population the chance to venture beyond their familiar villages and towns.

outbreak of war in Europe in 1939 the production figures of the mines soared spectacularly which meant that on a daily basis the ZSS was transporting 4000 ton of coal, that is to say, roughly eight train loads carrying 480 tons all of which amounted to 1.4 million tons on an annual basis. That was not a bad record for a single-track 'tramline' with a maximum axle load of 10 tons, steep inclines to cope with and various sharp curves.[143] In the meantime, the maximum speed for passenger trains along certain sections had been raised to 59 km/hr so that the 400 km or so between Kertapati (near Palembang) and Panjang (the Eastern Harbour) could be covered in a mere eight hours.

In spite of their extremely simple infrastructure the South Sumatran Railways were able to contribute significantly to the development of the area and they formed an important component in the politics of the government that was directed to the opening up and populating of the island which was, at that time, virtually uninhabited.

Celebes

Except for the islands of Java, Madoera and Sumatra there was just one other island in the East Indian archipelago where railway lines were built. For the islands of Borneo, Celebes, Bali, Lombok and numerous others, many plans were drawn up but because of a lack of funds and uncertainty about the economic usefulness of the lines, little came of all the wonderful plans, with one exception.[144] The 47 km long line in South Celebes from Makassar (Pasarboetoeng) to Takalar was opened in 1922 but the planned continuations in a northerly direction towards Maros and Tanete never reached fruition even though the necessary funds had been set aside. The short but serious economic crisis of 1921 meant that the plans were postponed before finally being abandoned altogether. The existing line, which was especially intended for goods transport and had been constructed in accordance with the standard gauge norms for the East Indies had, it is true, been simply operated as a tramline but it did not completely meet expectations. Eight years later it was closed down.[145]

Epilogue

In the Dutch East Indies the railways and tramways played an important part in the economic development of Java and Sumatra. Not only was the production of export crops such as sugar, tobacco and coffee thus furthered on a far greater scale than ever before, but the development of new export products such as rubber and petroleum was also made possible. It was not only European big business that benefited from the new transport means but also the indigenous population that came to appreciate the comforts of the steam tram and the steam train. Many of the transportation companies laid on extra services on market days to take people to the market places where they could then sell their chickens, goats and anything else they wanted to trade. Where the luxury trains such as the One-Day service were chiefly intended for European and Chinese passengers, the local inhabitants made grateful use of 3rd class travel where the tariffs were relatively low and the comfort reasonably good.[146]

After the Second World War and the ensuing period of unrest leading up to the transfer of sovereignty, the existing railway network was intensively used. However, there was extremely sharp competition from road traffic which meant that very many of the more rural lines had to be closed down. Certain parts of the formerly Dutch railway network are still being used today, with new rolling stock and equipment. The train still meets a very real need but hardly any new lines have been constructed since the time of the transfer of power.

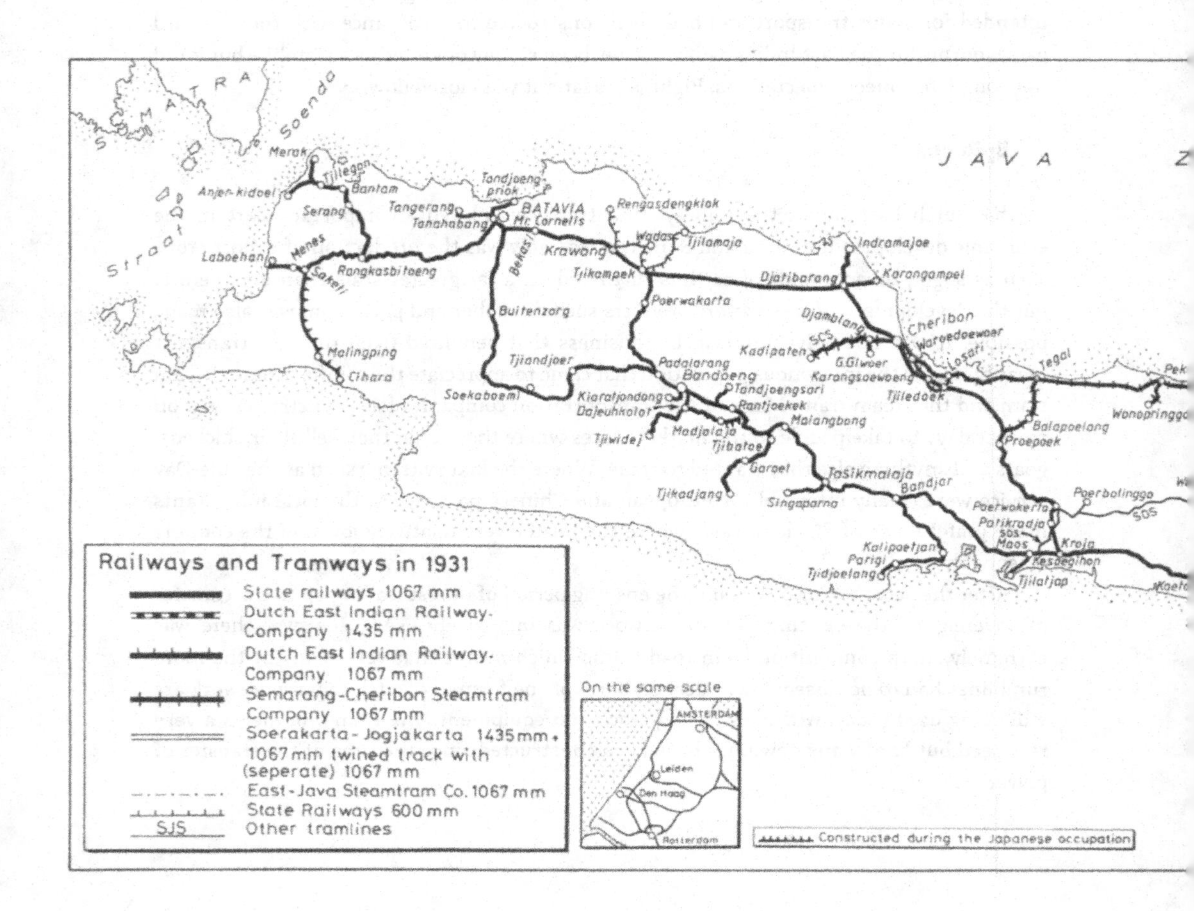
SUMATRA
Soenda
Straat
JAVA
Merak
Tjilegon
Bantam
Anjer-kidoel
Serang
Menes
Laboehan
Saketi
Rangkasbitoeng
Malingping
Cihara
Tangerang
Tandjoeng-priok
BATAVIA
Tanahabang
Mr. Cornelis
Bekasi
Buitenzorg
Tjiandjoer
Soekaboemi
Rengasdengklok
Wadas
Tjilamaja
Krawang
Tjikampek
Poerwakarta
Padalarang
Bandoeng
Kiaratjondong
Dajeuhkolot
Tjiwidej
Madjalaja
Tandjoengsari
Rantjoekek
Malangbong
Tjibatoe
Garoet
Tjikadjang
Singaparna
Tasikmalaja
Bandjar
Kalipaetjan
Parigi
Tjidjoelang
Djatibarang
Indramajoe
Karangampel
Djamblang
SCS
Kadipaten
G.Giwoer
Cheribon
Waroedoewoer
Karangsoewoeng
Losari
Tjiledoek
Tegal
Balapoelang
Proepoek
Poerwokerta
Patikradja
SDS
Maos
Kroja
Kesoegihan
Tjilatjap
Poerbolingga
Wonopringgo
Railways and Tramways in 1931
State railways 1067 mm
Dutch East Indian Railway-Company 1435 mm
Dutch East Indian Railway-Company 1067 mm
Semarang-Cheribon Steamtram Company 1067 mm
Soerakarta-Jogjakarta 1435 mm + 1067 mm twined track with (seperate) 1067 mm
East-Java Steamtram Co. 1067 mm
State Railways 600 mm
SJS Other tramlines
On the same scale
AMSTERDAM
Leiden
Den Haag
Rotterdam
Constructed during the Japanese occupation

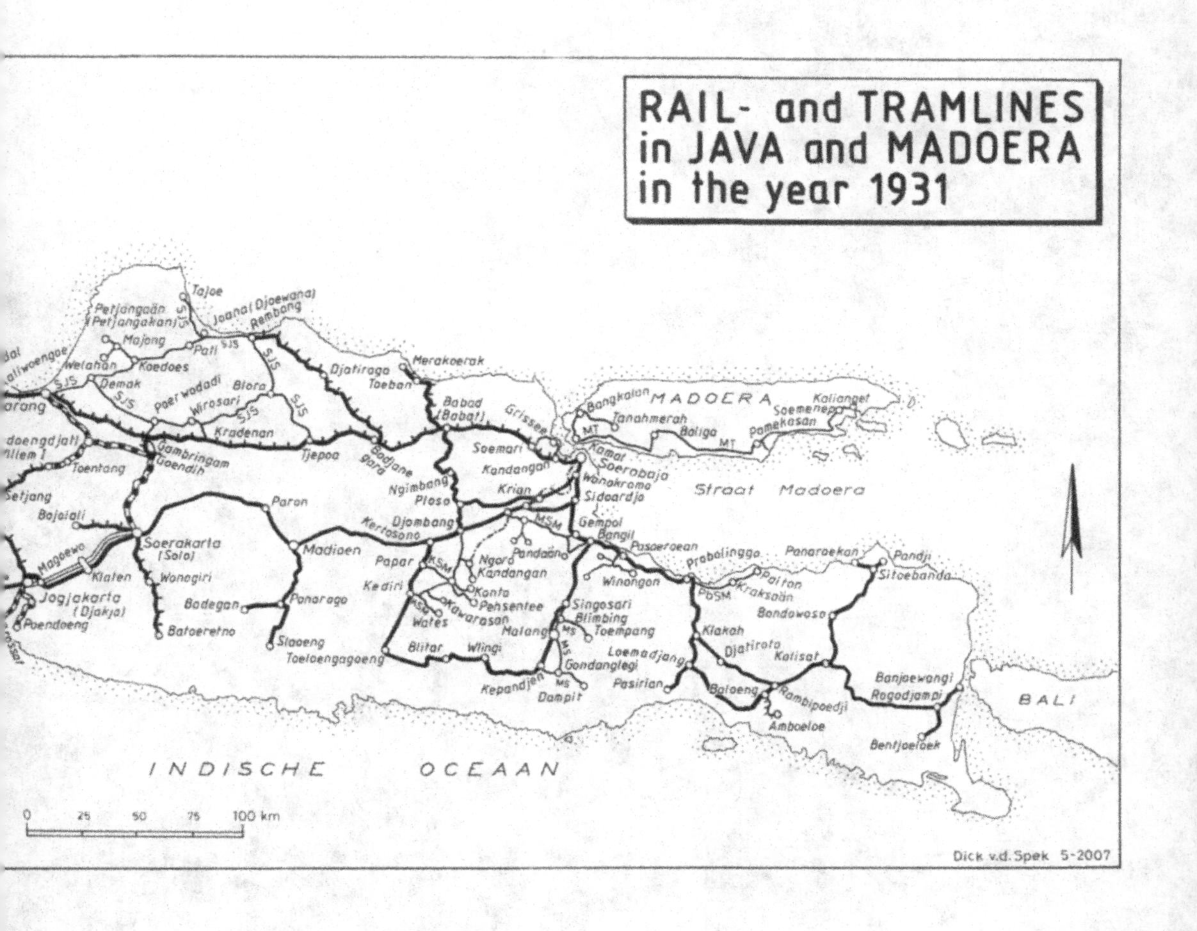
RAIL- and TRAMLINES
in JAVA and MADOERA
in the year 1931
Tajoe
Joana (Djoewana)
Rembang
Petjangaän
(Petjangakan)
Majong
Pati
SJS
Welahan
Koedoes
Demak
Poerwodadi
Blora
Wirosari
Kradenan
Djatiroga
Merakoerak
Toeban
Gambringam
Goendih
Toentang
Setjang
Bojolali
Tjepoe
Bodjane
Gara
Babad
(Babat)
Grissee
Soemari
Kandangan
Ngimbang
Ploso
Krian
Paron
Djombang
Kertosono
Soerakarta
(Solo)
Madioen
Magoewo
Klaten
Wonogiri
Jogjakarta
(Djakja)
Poendoeng
Badegan
Ponorogo
Batoeretno
Slaoeng
Toeloengagoeng
Papar
KSM
Kediri
Ngoro
Kandangan
Konto
Pehsentee
Kawarasan
Wates
Blitar
Wlingi
Kepandjen
Malang
Dampit
Gondanglegi
Singosari
Blimbing
Toempang
Pandaän
MSM
Gempol
Bangil
Winongan
Pasoeroean
Sidoardjo
Wonokromo
Soerobaja
Kamal
MADOERA
Bangkalan
Tanahmerah
MT
Balige
Pamekasan
Soemenep
Kaliangel
Straat Madoera
Probolinggo
PbSM
Kraksaän
Paiton
Panaroekan
Pandji
Sitoebanda
Bondowoso
Klakah
Djatiroto
Kalisat
Loemadjang
Pasirian
Baloeng
Rambipoedji
Amboeloe
Banjoewangi
Rogodjampi
Bentjoeloek
BALI
INDISCHE OCEAAN
0 25 50 75 100 km
Dick v.d. Spek 5-2007

The Blang Me Bridge in Atjeh (North Sumatra), which was seriously damaged by a bandjir in 1902.

CHAPTER 4 **FREE EXTENDING OVER THE BANDJIR**

Building bridges for roads and railways

MICHEL BAKKER

Assisted by: Jan Arends, Hans Brinkhorst, Ben Coelman, Auke Kingma, Hein Klooster, Jan Kuipers, Jan van Loenen, Henk van Maarschalkerwaart, Kees van Meijgaard and El Ypey [1]

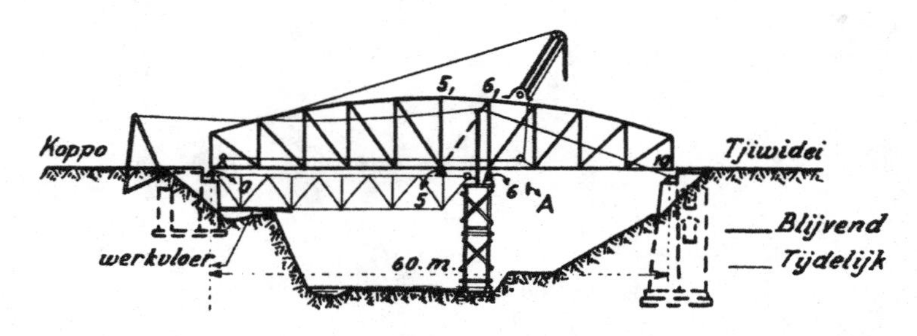

The recycling of a bridge that once spanned the Tjitaroem river at Krawang and was later used for the Koppo–Tjiwidej tramline, not far from Bandoeng (West Java), 1933. Assembled according to the free extension method.

At the end of the nineteenth century the construction of bridges in the Dutch East Indies gave rise to more problems than the building of bridges in the mother country. In the East Indies it was not uncommon to have to construct bridges over deep ravines which made it necessary to erect towering supports. There was another problem though. When the Dutch railway network was being constructed it was some years before people dared to build bridges across the big rivers. It was feared that drifting ice would destroy the supporting pillars. In the East Indies Dutch engineers were confronted with comparable problems, albeit of a slightly different kind. Heavy *bandjirs* transporting mud, sand, shingle, pebbles and boulders through the *kalis* could very rapidly destroy everything in their path.[2] Such sudden high water levels always came as the result of heavy rainfall generated by tropical rains in the river basins. In no time flotsam and jetsam would build up around pillars and trestlework or the bridges themselves, thus rapidly leading to blockages and damage. It was not uncommon for bridge sections or sometimes even entire bridges to be washed away. In efforts to curtail the damage, bridge watches would be organised whenever that was deemed necessary so that those on duty could keep the bridges clear.

Bridges had to be constructed for the railways and for road transport purposes. In the East Indies the construction of roads and railway lines in many mountainous areas was hampered by the vast numbers of streams and rivers that flowed criss-cross through the landscape. The building of the required bridges was therefore an expensive occupation which forced people to impose restrictions. On the whole, the large bridges built for the railways tended to be steel truss bridges. Smaller bridges invariably consisted of steel girders or wooden joists. In order to restrict the total length of the span, piers and bents would often be introduced. A good example of this style of bridge was the type that was made of steel horizontal sections mounted on screw pile bents. At the end of the nineteenth century, the volume of motorised traffic was not yet great. For rural populations it was therefore still often possible to make do with bridges that were made of bamboo or which had been constructed by the local people and were made of rattan. In the dry seasons it was sometimes possible to cross the rivers at given fordable spots. Around 1900 there was some 20,000 km of surfaced postal and country roads on the islands of Java and Madoera. Those roads had around 250 stone arched bridges with spans of more than 10 m, 1500 smaller arched bridges and 10,000 bridges constructed with iron girders and wooden decks.

The way in which bridges were mounted was something that had to be adapted to East

Indian circumstances. Whenever it was impossible to create temporary supports because the bridge stood too high above the riverbed and the support pillars were too expensive or because it was impossible to erect structures in the river due to the risk of *bandjirs*, then the free extension or cantilever method would be adopted. The system of main girders and the shape of the different parts would then have to be such that this method would be practicable. Another possible way of assembling a bridge without extra vertical support systems was by devising a suitable sliding across method. In this way the bridge would be mounted on the access point of one of the abutments. Upon completion, the entire bridge would be pushed along its longitudinal axis before finally being secured in position. The method was suitable for girders on more than one support point. What was known as a 'launching nose' was required to ensure that the stresses arising in the girders when they were extended to their maximum would not be too great. With this method constructional stresses would occur in the assembly phase that were just as great or perhaps even greater than those ensuing from the design proper which meant that it was necessary, when designing any bridge, to account for that and to adapt the design and construction accordingly.

In the following sections more details will be given on bridge building in the Dutch East Indies. The consecutive facets to be considered will be railway bridges and traffic bridges. In the case of railway bridges, attention will first be given to the design, fabrication, transport and assembly facets. After that a brief sketch will be given of developments in the field of all the different kinds of iron and steel bridges constructed on the island of Java between 1862 and 1940. Brief attention will be devoted to several bridges built on the island of Sumatra. A similar sketch will be given concerning road bridges with regard to the use of reinforced concrete as a building material and the building of concrete bridges. Subsequently the all-important 'own weight' of bridges will be discussed. In the case of traffic bridges attention will also be devoted to those built with wood or bamboo, and suspension bridges will be considered. A large part of the chapter will be devoted to the examining of various particular bridges so that different authors can each give their views whilst making sure, each time, to enumerate what is characteristic about any particular bridge. These bridge examples will be given in chronological order. The famous bridge builders will come into sharper focus: Professor J.H.A. Haarman, Professor P.P. Bijlaard and the engineer W.J. van der Eb. The first mentioned builder will be dealt with in connection with the standard arched bridge discussed in the section on iron and steel bridges. The next two builders will come into the picture when the bridges over the Kali Progo and the Tjisomang viaduct are examined. Finally, the aspect of the special role of military engineering in general and the Corps of Engineers in particular with regard to bridge building and repair will be highlighted.

RAILWAY BRIDGES

The designing, manufacturing, transporting and assembling of bridges

Building materials and guidelines

Railway bridges tended to be made of steel. Though admittedly stone was a good and durable material it was not sufficiently suited to the large and high styles of bridges that needed to be built for the tram and railway lines. Steel was durable and strong but it had to be imported. In the East Indies there was no iron and steel industry and so all the material had to be pre-fabricated and transported ready for assembly. The chance of defects, component loss, material shortage and damage was thus great and could easily lead to difficult situations at critical moments. The process of compensating for mistakes was time-consuming. In other words, it was important to find suppliers that were good and reliable. Despite all these pitfalls, steel emerged as the perfect material in the whole story of the rise of tramways and railways in the East Indies and later also in the period of heavier road vehicles.

Definite directives had to be given on stress loads and permissible stresses. In 1868 a circular sent out by the director of Civil Public Works outlined certain guidelines. The effective load per m^2 of bridge surface should lie between 350 and 400 kg. Moreover, calculations had to include a moving load comparable to an 'eight pound canon' towed by a load hauler. That amounted to a weight of 6 ton on two wheels where the track width was 1.478 m. With traffic bridges in factory areas it was safer to calculate on the basis of double-axled wagons where the axles were placed at 3 m intervals and the wheel axle load was 3 ton. In the case of heavier loads – should that indeed occur – it would have to be easy to quickly adapt the bridge to meet the requirements.

The permitted stress for *djati* (teak) wood was set at 110 kg / cm^2, both for the pull and push tensions in the direction of the grain. In the case of drawn steel, pressure and tension were set at 1200 kg / cm^2, while 800 kg / cm^2 was considered to be acceptable for shear stresses. In the case of drawn iron 600 kg / cm^2 was adhered to for tension, pressure and shear stress. No precise stipulations were laid down for the actual material itself.

In 1878 Mr. J.K.E. Triebart[3] lobbied to have a number of standard bridges imported in order to build up a stock. He accepted the fact that such a system would require about twelve percent more material. On the basis of the above-mentioned principles he then went on to calculate just how big the steel constructions would have to be for a number of different bridge structures between 12 and 24 m.[4] In those days the dimensions of the products fabricated in the rolling mills were not that great. Long girders would therefore automatically involve many weldings. Sheet material had a maximum width of 146 cm and was generally 2 to 3 m long. Angle iron and strips could be anything up to 8 m or more long. The steel cross-girders and main girders were made up of sheet and angle iron and rolled, double, T-profiles were used for the longitudinal girders. Many interconnecting links were required to create a girder from the available sheet material. The girders were formed by joining together the various lengths with rivets.

Whether or not the ideas put forward by Mr. Triebart were actually implemented remains unknown. What we do know, is that in the later construction of bridges for busier lines they were purpose-designed by the relevant department to fit the given individual circumstances.

Design and manufacture

The creating of designs for steel bridges has frequently been a topic of debate. In the early days it was a job that was frequently left to the factories themselves. In that way it was possible to easily integrate the rapidly changing innovations of production technology.

Nevertheless, it was not long before the department's own designs were opted for so that a degree of standardisation could be achieved.

It is not clear how the required bridges were contracted out or put out to tender or precisely how contracts for orders were drawn up. In the case of the bridges constructed for one of the main lines, a kind of categorisation was created by the directorate. This had to do with the construction of the Goendih-Soerabaja line, a franchise that went to the Dutch-East Indian Railway Company in 1900. That particular project is described in detail by B.M. Gratama in *De Ingenieur* (The Engineer).[5] The contracts went to German, Belgian and Dutch factories. Factory capacity was something that greatly influenced choice. Giving all the jobs to one manufacturer would have led to everything being in the hands of a single German company and that was not desirable. It was felt that it was important to also involve Dutch companies.

When it came to the project in question it was not possible to order everything all at once which was why the bridges were ordered in 50 to 350 ton consignments. Two Dutch companies (709 ton), two Belgian firms (641 ton) and three German companies (974 ton) were involved in the project. The Dutch companies were Pletterij Enthoven and Penn & Bauduin. For this particular project the price of processed steel amounted to about 0.20 / kg including around 0.03 / kg for the sea freight costs (in Dutch guilders).

The bridges were supplied in the 1898/1900 period. It was the golden age for the European steelworks, both for the rolling mills and for bridge manufacturers. Companies had trouble meeting their orders and everywhere there were capacity problems in the factories, all of which led to erratic delivering and to roughshod workmanship. At the rolling mills the material had to be inspected – charge branded and strike marked – but nothing much came of the strike trade marking intentions.

Thomas iron was accepted for the rolling of steel and, for a small portion of the work, Siemens Martin iron was used. The requirements for the rolled materials met with the Technical Conditions laid down by the Ministry for the Colonies with just a few minor amendments.

The bridges were made from profiles that met with the *Deutsches Normalprofilbuch für Walzeisen* (Aachen 1897). All the double T and U iron profiles and most of the L profiles were manufactured in the large German rolling mills. The sheet and universal steel, and a portion of the L profiles came, for the most part, from Belgian factories. A certain amount of the Belgian steel originated from blocks derived from French rolling mills.

The individual girders were riveted together in the factories but the interconnecting of the girders used to form entire bridges was something that occurred on-site during the assembly phase. In the factories the bridges would always be pre-assembled though before being shipped overseas.

Much of the rust-proofing of the components occurred with the help of acid baths. After then being neutralised in calcium baths the components would be daubed with linseed oil. The neutralising of the acid in the calcium bath was something that had to be strictly monitored. The fluid in the calcium baths frequently needed to be changed and it had to be stirred. Cockerill experimented with mechanised anti-rust methods but the desired results were not achieved. After the parts had been assembled the completed structures would then be given two coats of paint in order to prepare them for transport and long-term storage.

After the final assembly stage, the paint would then be scraped off and the finished bridge properly painted.

The riveting together of the component parts to form the final bridge was heavy work requiring a high degree of precision. For that reason, much careful attention was devoted to the accurate preparing of the various elements. The holes had to be well formed, well trimmed and slightly submerged on the outer edge. The holes in the parts that needed to be connected would be bored on top of each other. The results of hand riveting could be superb but true craftsmen were few and far between which was why the preference often lay with mechanical (i.e. pneumatic) riveting. Just to give an indication of how riveting gangs worked: an average team could do 200 rivets a day or perhaps as many as 300 rivets on a good day under optimum conditions.

Around the year 1900 the normative load would have been a four-axled locomotive that would have had an axle-load of 8.4 ton and the axle intervals would have been: 2.37 m, 1.35 m and 1.575 m. The wind pressure was estimated to be 150 kg / m^2 for a bridge loaded with a train and 300 kg / m^2 if the bridge was empty. The train speed at that time was 25 km / hr and – according to H. Gerber, see below – the mobile load had to be multiplied by a factor of 1.5. The permitted stresses varied from 800 to 1200 kg / cm^2. No cambering would be applied in order to ensure that the assembly fit was as close as possible. The bridge's appearance was thus of secondary importance.

Transport and assembly

Another bridge, the assembly of which has been extensively described, was situated along the Cheribon-Kroja line and crossed the Serajoe river.[6] The bridge was delivered in 1914/1915 by Pletterij, formerly L.J. Enthoven & Co. in Delft and met the accompanying design requirements. Even though the quotation was slightly higher than the bid put in by a foreign firm the job went to Pletterij in the interests of protecting the domestic market.

The total weight of the bridge was some 670 tons. The bridge sections arrived in Tjilatap by steamship on 25th September 1915. The ship which was carrying 6000 consignments was unloaded in no less than 24 hours. The load was transported to the construction site by a steamtram from the Serajoe Valley Steam Tram Company. That company's line was also going to use the new bridge which meant that the materials could be delivered straight to the construction site. Half of the material had to be transported to the other side of the valley. To that end a special cable-lift was constructed with the aid of wooden towers. The parts that needed to be hoisted were pushed onto a beam plateau until they were under the hook. After that they were raised and transported with runner lorries that carted an average of 60 tons a day. Assembly according to the free extension method made use of a gantry crane that would be situated on the already assembled section. The constructors worked towards the middle of the bridge gradually assembling from both banks. The final middle section of this particular bridge was put in place on 18th December, within three months of the components having being shipped in. At the end of December the first train rolled across the bridge. The assembly of the whole bridge, which weighed 670 tons, had cost 48,580.00 guilders or 70.50 guilders per ton of bridge.

The bridge sections were pneumatically riveted together. There were three installations

The in 1919 prefabricated assembly of a railway bridge in the Hollandsche Constructie workshops in Leiden (the Netherlands).

on either bank, each with four hammers. The riveting equipment was supplied by Pokorny & Wittekind in Frankfurt am Main. It consisted of a 16 hp petroleum engine that powered a compressor which then compressed the air to six atmospheres in an air-chamber. The engine, the compressor and the air-chamber were all mounted on a solid base. The base had wheels under it so that it could easily be moved along the rails. Personnel from the State Railways' (SS) Construction and Bridge Building division were responsible for the assembly work. The personnel who were drawn from the local population had practiced working with pneumatic hammers during a previous assembly job and so possessed the necessary experience.

Ultimately all the work was done within the company. The Indo-European personnel and the native workers were internally trained by a State Railways' construction bureau. The native workmen, many of whom had already been employed by the company for ten years, were first taken on for a daily wage of 0.30 cents as ordinary labourers (carriers) and would then be trained in riveting, forging, machine operating, etc. As their skills increased so their wage would gradually be raised. Those who were able to read the types and numbers of the parts would be given special increments. In that way some of the natives eventually learned how to write their names or even became literate. At first they were educated by the Indo-European operators but later the cleverer Javanese workers took over that task. It was obvious that this particular organisation cultivated mutual understanding between European and native personnel which, in turn, led to feelings of solidarity. Such solidarity was conducive to the work atmosphere which meant that the assembly work could ultimately be completed twice as quickly as scheduled.

Developments emanating from World War I

As time passed the Dutch steel industry, which had seen rapid growth between 1900 and 1915, gradually started to dominate the market where the supplying of bridges to the Dutch East Indies was concerned. Pletterij Enthoven, Penn & Bauduin, Begemann and other firms as well supplied between them tens of thousands of tons of steel. War broke out in 1914 and as of 1916 it was no longer possible for goods to be transported overseas from Europe. There was still, however, a great demand for road and railway bridges. The government was therefore forced to place orders with local companies that were not particularly suited to the task or which simply did not have the capacity to meet the orders.

These were companies that were more oriented to maintenance than to new construction work. Endeavours were made to forge links with countries that could be contacted in the outside world such as Japan and America. The war changed things which meant that post-1918 it was no longer so much a matter-of-course for material to be supplied by the Netherlands. Despite the shortage of skilled and schooled personnel, local production resources started to be stimulated more. Other countries also started to see the market potential in the Dutch East Indies.

The government had a special purchasing office for the Dutch East Indies, the Technical Bureau at the Ministry for the Colonies in The Hague. The bureau was started up in 1880 by David Maarschalk who had been made responsible for the creation of the first Javanese railway link for the State Railways and it continued until around 1920. When the various companies tried to resume business after the war they all went through that Technical Bureau. What soon emerged in the East Indies, though, was that other trends were developing, trends that were focused on independence. Already before the war several metal companies had been established in the East Indies. These were companies that were predominantly preoccupied with maintenance work (for the sugar industry, irrigation works, the electronics branch). A number of steel companies also started to investigate the possibilities of setting up business in the East Indies.

The endeavour to set up business

The Dutch Construction Workshop in Leiden, the Dutch Factory for Mechanical and Railway Materials 'Werkspoor' in Amsterdam and the Royal Factory for Carriages and Rail

Wagons, J.J. Beijnes in Haarlem all made serious attempts to settle in the Dutch East Indies. In February 1921 several gentlemen were sent out to the East Indies to investigate the possibilities for starting up a consortium.

It was discovered that orders had been placed with maintenance workshops that simply did not have the capacity to meet them in time. The State Railways' own production workshop had already been extended. Meanwhile a German consortium had also been following the developments in the East Indies and had already paid a visit to the authorities and put forward a proposal to establish a factory in Java. The governor-general had given the German businessmen his full sympathy and support.

The Dutch businessmen were also welcomed with open arms, received full co-operation and were advised to start up a company as soon as possible. The government made it clear that they would prefer to have Dutch industrialists operating in the East Indies. It was also highly likely that the Dutch government would fully co-operate by making concessions for the first few years. Upon returning to the Netherlands, plans were thought through, reports were written, negotiations were continued and state backing was sought. In December 1921 all of that culminated in a request being put to the Minister for the Colonies for orders and work to be guaranteed in the event that the relevant companies should decide to set themselves up in the East Indies. Such a guarantee had already been discussed with the directors of the State Railways and the Central Bureau.[7] The guarantee would be for the delivery of 80 passenger carriages, 150 goods wagons and 1500 tons' worth of bridge construction work.

In the meantime, thanks to an increase in the provision of raw materials, the price of steel constructions dropped sharply. In 1920 it was possible to earn 250 guilders per ton but in the course of 1921 that had fallen to 120 guilders per ton. Another problem was the many strikes in the Dutch metal industry. One development took the form of the building of a factory by the above-mentioned German consortium, the Rhein-Elbe Union. The *Algemeen Handelsblad* (General Trade Journal) newspaper of 24th August 1921 reported that work had started on the construction of a huge machine factory in Cheribon.

In September M.C.E. Bongaerts, a member of the Lower House and commissioner for the Dutch Construction Workshops, put a number of questions to the Minister for the Colonies relating to the activities of the German consortium. From the reply given by the Minister, S. de Graaff, it became apparent that already in November 1920, during an audience with the governor-general, promises had been made regarding the German consortium. For further elaboration people were referred to the delegated member of the Committee for the Development of Factory Industry in the East Indies. At that time a job had been commissioned and the Dutch manufacturing branch had not even been given the chance to compete for the tender. In November of that year the Technical Bureau had been asked to quote its prices but just before the tenders were received the government gave the pending assignment to the German company. The governor-general, the lawyer D. Fock, then declared that he felt bound to accept the decision that had been made. In subsequent years there would never again be any occasion when other firms in the East Indies or in the Netherlands would be placed at such a disadvantage.

The following year, in 1922, word was received that the German consortium was struggling to meet its obligations. The affair also had repercussions for personnel. The director of the Technical Bureau, which had changed its name to the East Indian Central Acquisitions Service, resigned. For the Dutch consortium, which was in the process of being set up, he had been one of the most important contacts.

Ultimately the Dutch consortium initiative was not pursued. In the Dutch Construction Workshop archives everything suddenly dried up. The consortium never came into being.

Permanent iron and steel railway bridges, 1862-1945

In the Dutch East Indies railway and tramlines started being constructed in the eighteen-sixties, first on the island of Java. At first there were only private railway companies, like for instance, the Dutch-East Indian Railway Company that was established in 1863. By 1875 the State Railways Company had been formed. After decades of expansion in Java, Sumatra and several other islands, all further railway construction work finally ground to a halt in the nineteen-thirties at the time of the great economic depression.[8]

It was especially the increased axle loads of locomotives that led to the overall gain in train weight. The regulations surrounding the calculations for bridge construction therefore had to be perpetually adjusted.[9] Just to illustrate this point, a record is given below of how that trend progressed:

Year	Axle load	Load per metre of rolling stock	
	locomotive	locomotive	wagons
1878	9 ton	5.0 t/m	2.6 t/m
1907	12 ton	4.7 t/m	3.2 t/m
1911	13 ton	5.6 t/m	3.3 t/m
1917	16 ton	8.75 t/m	5.0 t/m
1921	20 ton	8.75 t/m	5.0 t/m

The way in which iron and steel bridge construction developed did, in broad outline, mirror the way in which it progressed in the Netherlands. Until around 1900 it was foundry iron that was used to create bridges but after that period molten iron was used. That material was later called 'molten steel' before eventually just being referred to as 'steel'. Even from the point of view of bridge type (or construction form) there were parallels: in the case of smaller bridges it was the full-bodied type that was common whilst for bigger bridges it was trellis bridges (until circa 1875) and later truss bridges that were seen. The latter type was created with arched or straight edges. When it came to the matter of the way in which the construction work was completed there were, however, great differences when compared to the Netherlands. As was indicated at the beginning of this chapter, construction methods had to be adapted to the East Indian circumstances.

Foundry iron trellis bridges

Examples of foundry iron trellis bridges constructed in the second half of the nineteenth century are these:

The reinforced bridges over the Tjitandoei river (West Java). Reinforced by means of series of arches and upright supports placed beneath the bridge.

- The spanning of the river Tjitaroem along the Padalarang-Buitenzorg railway line with three lattice bridges, each of which was 54 m long.
- The spanning of the Tjitaroem along the Meester Cornelis-Tjikampek line. The bridge was purchased by the State Railways from the private Batavian Eastern Railway Company. The bridge had to be reinforced to take the heavier trainloads.
- The spanning of the river Tjitandoei near to Tjiamis on the Tjibatoe-Bandjar (Preanger) line. The bridge comprised three 62 m long spans. In 1936 the bridge was reinforced by placing arches on uprights. This made the bridges identical to the Langerse type (named after the Austrian engineer who had developed that particular style not long before). In Europe bridges were also being reinforced in a similar way. After the 1930s it was a style that was often adopted in new bridge building.

Railway bridges of the post-1906 era

After the period around 1900, iron production methods had progressed to such a degree that it was possible to create molten iron on a big scale. Molten iron was superior to foundry iron and so it was not surprising to see that it was exclusively such molten iron that came to be used for rail bridge construction.

There were also other developments going on at that time in the world of bridge constructing. The trellis girder bridges were disappearing and were being replaced by truss bridges which were more open. Molten iron arched bridges were also appearing on the scene. Many of these advancements were directly linked to new discoveries in the field of applied

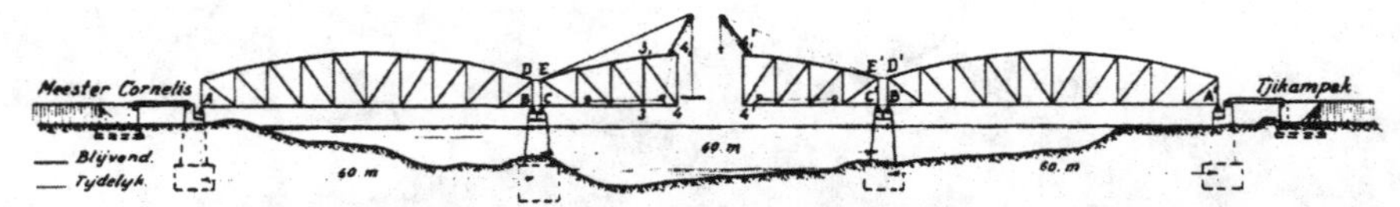

FIG. 1. OUDE BRUG OVER DE TJI TAROEM IN DE LIJN MEESTER CORNELIS—TJIKAMPEK.

The new 1906 bridges over the Tjitaroem river, not far from Krawang (West Java) on the Meester Cornelis–Tjikampek line which consisted of three iron truss bridges with arched upper edges and spans of 60 m.

mechanics, such as the theory relating to the bruckling of beams. An example of such a developmental step is illustrated by the above-mentioned spanning of the Tjitaroem river along the Meester Cornelis-Tjikampek line.[10] In 1906 the original trellis girder bridges were replaced by three molten iron truss bridges, each with a 60 m length span. The main spans were made of single trusses with a curved upper edge, pressure loaded vertical sections and diagonal lengths that predominantly took the tensile forces.

In 1918 it was decided that it would be desirable to double the railway line at that point. In order to anticipate future reinforcement requirements it was determined that the bridges built in 1906 should be replaced and that the whole project should involve six new bridges designed to cater for double-track requirements. The project was rounded off in 1922. The style chosen for the new bridges was arched bridges with tension bars fabricated in molten iron. Experience had been gained with this type in 1915 and, meanwhile, it had become the standard accepted design. Over the course of time it was introduced in many different places.

Bridge examples

The iron bridge over the Kediri river

In view of the fact that for quite some time there had been a need for a sturdy bridge over the Kediri river at the town of Kediri for traffic passing over the Great Post Road between Soerabaja and Madioen in East Java, it was determined by means of a government resolution of 16th May 1854 that in accordance with a certain army captain's design a stone arched bridge should be constructed. The construction work was scheduled to commence in 1855 and the total cost was estimated at 128,891.00 guilders. In the course of that year the chief engineer at the Public Works division in Soerabaja protested extensively to the director of Civil Public Works about the bridge construction plans. His main objections were these:

- The large river discharge, especially at high water, made it necessary to construct a bridge that would provide as little as possible resistance to the periodical huge volumes of water. Not only would such a bridge cause the water level to rise but, because of the sandy riverbed, the acceleration of the current between the piers would also lead to an undermining of those same piers.
- Moreover, the process of blocking off and keeping dry the foundation pits for the piers could give rise to major problems because of the shifting riverbed.

The development of bridge constructions: three generations of bridges over the Tjitaroem near to Krawang on the Meester Cornelis–Tjikampek line. Left, the original horizontal trellis bridge. In the middle, the 1906 truss bridges with their curved upper edges. To the right, the new 1922 arched bridges with their horizontal girders. The old truss bridges that became available in1906 could be recycled elsewhere. They were used for the Koppo-Tjiwidej tramline situated near to Bandoeng that was used for light transport (Bijlaard, Vrije Uitbouw, uitgevoerde werken, 12).

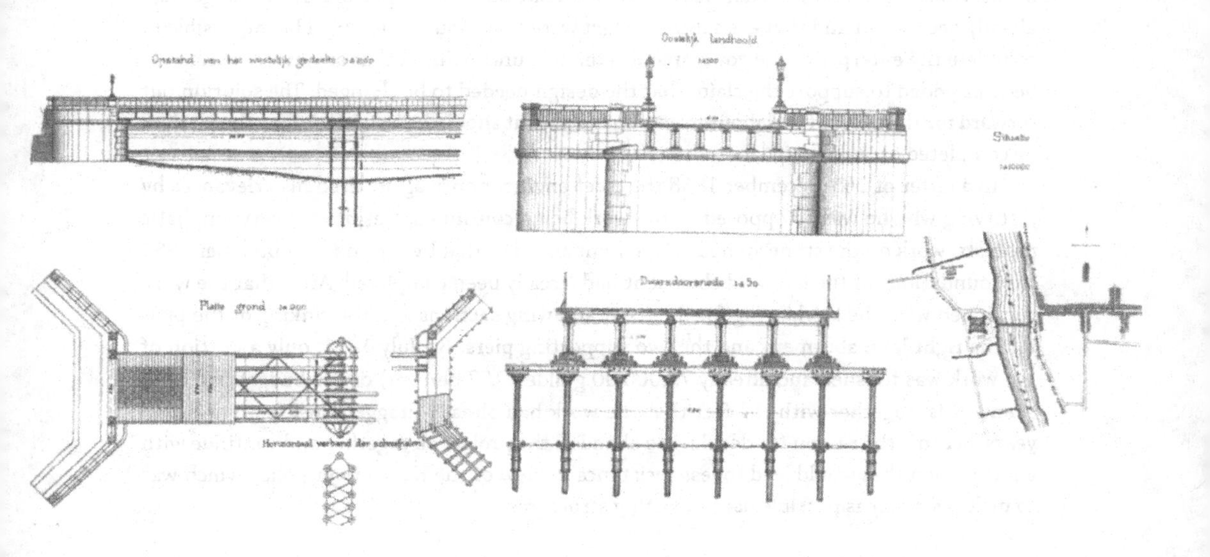

The design for the iron bridge over the Kediri river at Kediri (East Java).

The iron bridge at Kediri, 1869.

- A design and budget were submitted for a bridge with iron box girders laid across piers with iron screw piles that were intended to replace the heavy stone construction. In that way all the mistakes made in the first design could be carefully avoided.

Ultimately the objections were rejected and instructions were given to start with the construction of the stone arched bridge. The problems encountered already by the digging of the foundations for the left abutment confirmed the previous opinion of the chief Public Works engineer who thus felt obliged to again write a letter (dd. 28-06-1858) reiterating his previous reservations. What was also pointed out was that 40 percent of the budget had already been spent and that even if the budget were to be doubled it would be impossible to complete the enterprise. The government remained unconvinced that any evidence had yet been provided to support the claim that the design needed to be changed. The solution put forward for the lack of free labourers was that a permit should be issued to allow the work to be completed on an unpaid labour basis.

In a letter of 16th December 1858 the chief engineer once again aired his grievances by justifying why he was so opposed to the work being continued. Despite his very emphatic protests, work on the stone arched bridge went ahead so that by the end of September 1859 the foundations of the left land abutment had already been completed. After that the work continued with the bricklaying for the front and wing sections and the sinking of the piles for the right land abutment and the two supporting piers. By July 1861, only a portion of the work was finished and already 73,000.00 guilders (57 percent) of the budget had been spent. This, together with the fact that the work had already dragged on for a number of years, became the reason for deciding to abandon the project all together and continue with another plan that would lead to less resistance in view of the Kediri river policy which was to make as much as possible use of existing structures.

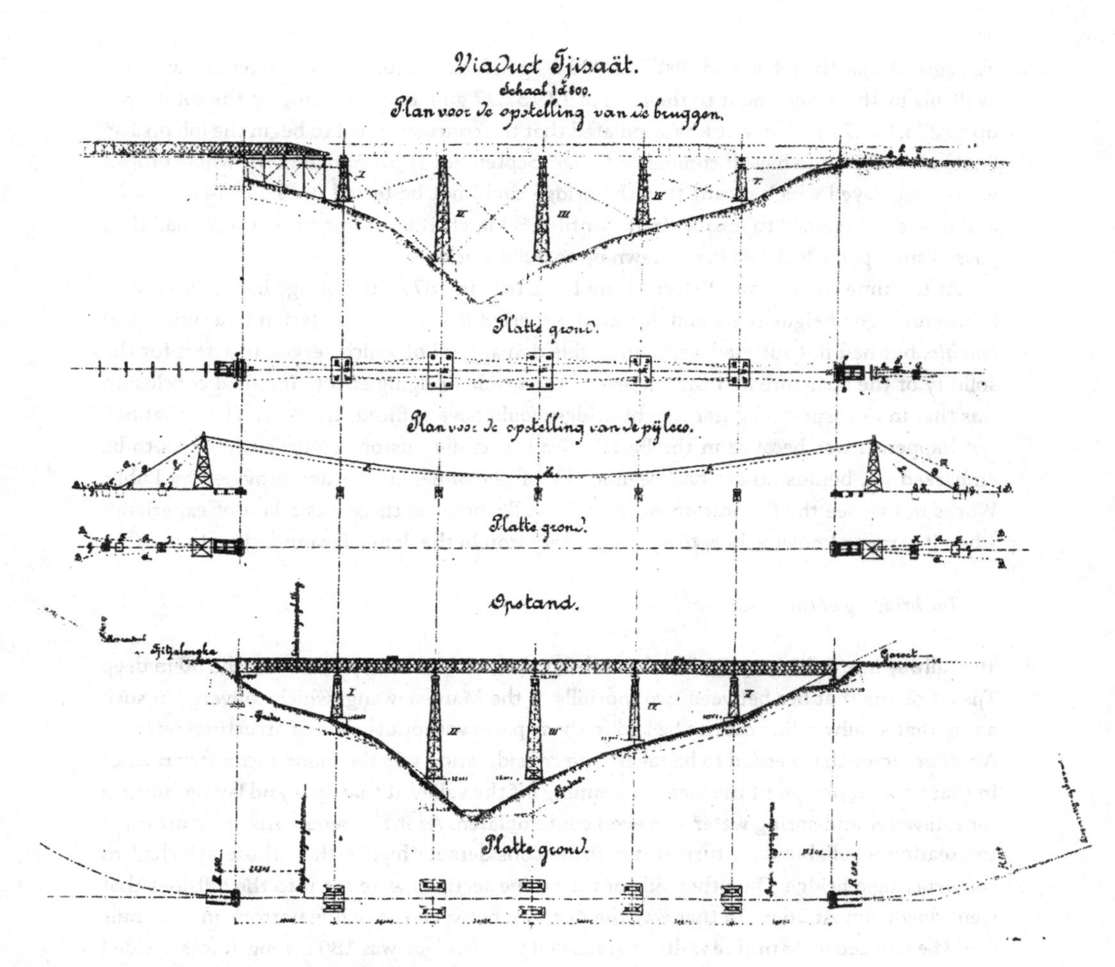

The Tjisaät viaduct (West Java): the bridge configuration plans (scale 1 : 800), approximately 1888.

A new design was developed by the engineer in training, S. Westerbaan Muurling, and submitted on 1st May, 1862. In essence it consisted of iron girders mounted on trestlework with iron screw piles and the total cost was estimated at 230,825.00 guilders.[11] In the accompanying memo describing the plans one of the points mentioned was that though few iron screw piles had been used in the East Indies, sufficient experience had been gained elsewhere to have confidence in the system. It was furthermore the case that these types of piles would not narrow the river's profile in any way and so would not hinder the water flow. The already partly constructed left land abutment was also integrated into the design, together with the already half sunken piles for the land abutment on the right-hand side.

Tendering on 27th April 1863 in Batavia failed, as did attempts to draw up a private agreement on 30th December 1863. On 31st July 1865 the plans were finally approved and it was agreed that the job would cost 212,000.00 guilders. On top of this, material was made available by the government to the tune of 13,152.27 guilders thus bringing the total costs up to 225,152.27 guilders. It was stipulated that the contractor had to begin the job on 18th September 1865 and have it completed by 18th September 1867. In reality the whole project was very delayed which meant that the bridge could not be tested until 11th March 1869 and it was not opened to normal traffic until 18th March 1869. Some fifteen years had thus passed since plans had first been drawn up to build a bridge.

At the time when J. van Velzen wrote his article in 1877[12] the bridge had already been in use for a good eight years and during that period it had been subjected to a number of *bandjirs* but had not suffered any appreciable damage, all of which served to attest for the solidity of the structure and the success of the whole bridging effort. The final conclusion was that in retrospect a lighter weight bridge would have sufficed. The reason why that had not happened was because in the East Indies it was not customary for such works to be embarked on, because there had been a lack of personnel and means provided by Public Works to oversee the fabrication work and, finally, because there was a lack of experience when it came to knowing how to preserve sheet iron in the damp Javanese climate.

The bridging of the Tjisaät ravine

The railway line linking up Tjitjalenka and Garoet (in West Java) passes over the 80 m deep Tjisaät ravine that lies between two foothills of the Mandalawangi which converge in such a way that a railway line cannot be laid in those parts without the aid of structural works.[13] Another factor that needed to be taken into consideration was the many secondary ravines in that area. At one point the idea of damming off the valley at that spot and laying culverts for rainwater and spring water was even contemplated. As it happened, the costs attached to creating a reliable dam turned out to be considerably higher than those attached to constructing a bridge. On either side of the ravine sections were cut into the hillsides that went down almost 26 m. In that way the distance between the ravine bottom and the rails could be reduced to 48 m. The width that needed to be bridged was 180 m long. It was decided that the job should be done with two 90 m bridge sections. What turned out to be the most economical solution was to half extend each bridge over four supporting uprights. Apart from the central pillar it was therefore necessary to construct four in-between columns. The uprights were made of iron and were placed at 30 m intervals. A special casing form was selected for the two abutments. Just like the lower sections of the supports, they were masoned. The masonry consisted of quarried stone cemented together with ordinary local mortar consisting of equal portions of lime, sand and brick dust. Four bluestone supports were submerged into the lower pillar sections for the iron uprights. Two anchor bolts were then passed through these supporting stones or pillow-beres which were anchored to the masonry by means of an anchor plate. The anchor bolts served to fix the upright sections or columns of the iron pillars in place.

Each pillar consisted of a lower frame and an upper frame with four stanchions between the two parts that were interconnected by cross bars. Between the link-ups both horizontal and more or less vertical stabilizing wind braces were introduced. Most of the fields contained two crossing diagonals. Only the fields of the highest two pillars that were

After the major bandjirs of May 1899 the old bridge over the Moedjoer (East Java) was almost completely destroyed.

completely perpendicular to the bridge axis had two diagonals that crossed over each other. Every stanchion was made up of four quadrant profiles with plates between the flanges. The respective heights of the iron pillars were: 12.78 m, 31.13 m, 31.13 m, 19.09 m and 11.98 m.

For the mounting of the pillars, use was made of a cable-lift that was spanned above the ravine. Large sections such as the uprights and the upper and lower frames were suspended from the cable from the sides and also eased into place by the cable. Smaller components were taken down the slopes and were then hoisted into place with the help of the cable-lift. This method proved to be much more efficient than fully manoeuvring with the cable-lift. Even the smallest components were taken down the slopes into the ravine below. Those parts would then be taken up by local workmen by means of bamboo ladders lent against the pillars.

As the quadrant profiles were of a limited length, the uprights had to be composed of a number of sections. In that way each pillar, depending on its final height, easily consisted of two to four sections. In order to weld the parts together, a close-fitting iron cylinder would be passed between the quadrant profiles of the uprights and affixed with specially fabricated bolts. The end of the screw thread part of the bolt would be flattened and have a hole punched through it. In that way the bolt could then be hung on a line. A section comprising four uprights, the cross connections and the stabilising parts would be assembled before starting on the next section. The cylinders would then be affixed to the upper edge of the four uprights so that the components could be slid over the cylinders to start on the next section of the pillar. Finally one of the labourers would climb to the top of the highest columns and would lower down the bolts on lines. When welding, an iron hook would be used to pull the string and thus also the bolt through the connecting holes. With the help of a spanner the nut would then be screwed onto the bolt after which the projecting lip would be removed. In order to ensure that the bolts could not slip off, the screw thread would be levelled off.

After all the components had been properly assembled, the upper pillar frame had to be fitted onto the column. It turned out that with one of the taller piles the upper extremity was tilted so that the upper frame did not fit. All the horizontal cross connections then had to be undone. With the help of four tension stays the uprights were then pulled into the right position so that the upper frame could be fixed in place. The connections that had been undone could then be done up again.

Because the ravine was so deep the option of building assembly scaffolding was too expensive, hence the reason that it was decided that the bridge should be built on the land to either side of the ravine. The two bridge halves were built as double trellis girders with parallel upper and lower edges and the rail deck on top. Each bridge was made up of 30 sections, each 3 m long. The height of the girders was 3.2 m. The sections were connected by means of overlapping diagonals. The second set of overlapping diagonals extended over half the next section length so that in that way the middle of the rectangular sections were connected from the sides. The whole assembly of the bridge went without a hitch. It was assembled on the Tjitjalenka bank side where the space was limited. In order to be able to construct a significant section in one go, a 25 m long jetty was built out from the land abutment and a section of the ravine slope was carved out. In that way it became possible to build a bridge section of almost 74 m all at once. For assembly purposes a kind of nose was fixed to the end of the bridge that was 13 m long. It was there to ensure that the bridge could be properly pulled over the supports.

After the first bridge section with the nose construction had been completed that section would then be dragged over the third column with the help of a winch at a rate of 13 m per hour. The next section of bridge would then be assembled and once that was ready it could be pulled as far as the fifth column. Finally the last section was assembled and the whole bridge was pulled into position. The nose construction could then be removed and with the help of screw jacks the bridge was lowered onto its supporting pillars. The land abutments were built up to the required height and the space between the sides was filled with sand.

The first train was able to cross the bridge on 10th January 1889. Some 355 tons of iron had gone into making the bridges and 196 tons into the columns. The whole project had cost 147,703.00 guilders. The assembly of the iron columns and the bridges and the laying of the rails had taken just over two months without having to work any night shifts!

The renewal of the railway bridge over the Moedjoer river

Between the places Klakah and Pasirian in East Java there was a railway connection which, though not of great economic importance, was nevertheless viewed as essential. In the Pasirian area the railway line crossed a number of rivers located to the south of the Smeroe volcano. Around the year 1900 the volcano was still active. Before 1898 much of the ash and lava erupted from the volcano was transported away by the Boesoek Sat, a river that flowed westwards from the crater and through a ravine. The mountain river's estuary was situated in a plain and that was where all the transported volcanic material was deposited. The railway track which, at the time of its construction, had led over a rise soon found itself in a dip.

There were a number of other ravines and mountain rivers situated to the southwest of Besoëk Sat. One of those rivers was the Moedjoer which, until 1898, was a fairly calm river. The railway line traversed that river by means of a railway bridge with a span of some 30 m. During its design phase the dimensions of that bridge had been liberally estimated.

In the year 1898 the transportation of volcanic ash and lava suddenly switched from the Besoek Sat to the Moedjoer. The explanation for that sudden change of route was probably to be sought in movements in the crater area leading, in turn, to the partial blocking of the Besoek Sat. The Moedjoer which, up until then, had had a *bandjir* width of less than 20 m suddenly turned into a wild flowing river with a *bandjir* width of more than 120 m. The water transported mud, sand, gravel and pebbles. In places where the river shelved less steeply all those materials sank so that the riverbed simply became higher and higher. At the point where the railway bridge crossed the river, huge volumes of water were forced between the pillar openings and the bridge that had, by then, become far too small. During a *bandjir* in October 1898 it was not long before one of the railway line's horizontal support systems collapsed over a 100 m length. After the *bandjir*, the dam was repaired and reinforced. As it was generally believed that this would only constitute a temporary solution a new bridge was immediately designed with a total gap area of more than 100 m.[14]

The earthen dam repair work had, indeed, been very short-lived. As soon as another *bandjir* came it was very quickly destroyed again. Obviously the traffic over the bridge could not be obstructed for too long which was why it was decided that a temporary bridge should be erected in line with the old one. The temporary bridge was supported with wooden bents, the lower side of which was only dug in to a depth of 60 cm because of the shortage of time. It was anticipated that the mud and sand deposits would automatically further bank up the 60 cm. The temporary bridge was 80 m long and the wooden supports were placed at 5 m intervals. In May 1899 there was a series of heavy *bandjirs* which, contrary to all expectations, brought very little sand and mud. This meant that cracks developed in the riverbed and the foundations of the temporary bridge were laid bare. Seven of the bridge supports were washed away and were later found in the sea. The temporary bridge was immediately repaired and protected with a light dam of *brondjongs* created upstream of the bridge. *Brondjongs* (gabions) are circular baskets made of woven bamboo. They were fixed in place with dyke protection poles made of bamboo and filled with boulders. Ultimately the dam created new problems because downstream of it a deep-water cushion (natural stilling basin) developed so that the dam soon threatened to collapse. The solution was therefore sought in reinforcement and level raising, a solution which only gave rise to the need for more reinforcement and dam raising after the next *bandjir*.

Meanwhile, from the Kontoh river, three truss bridges with parabolic arched upper edges had become available. It was estimated that those bridges would be suitable for recycling for the Moedjoer, an operation that would not only require two abutments but also two in-between piles. In view of the potential problems and the great danger precipitated by *bandjirs* it was determined that the job should be completed as quickly as possible. Petroleum lamps were therefore hung up so that the construction work could also continue in night shifts. The lamps, which provided sufficient light, were connected to a network of lines laid in the bridge building area.

It emerged that it was virtually impossible to dig foundation pits for those particular supports and the job would, moreover, have taken too long. It was therefore decided that a sunken pit foundation should be created. The sunken pit consisted of a wooden ring beneath which a T-profile was mounted that was to serve as the cutting iron. Ring-shaped stones were also cemented to the wooden ring. The wooden ring and the lowest levels of the stonework

The assembling of the bridge span.

The drawing across of the bridges.

sloped away conically on the inner side. The connections between the wooden rings and the masonry were realised by means of iron anchors that were 1 m long. Once the sunken pit had been built up 2 m high the outside of it was plastered over on the outside so that the sliding edge was as smooth as possible.

The sinking was achieved by extracting the earth on the underside. As the ground in that area contained great quantities of gravel and pebbles it was largely removed with the aid of baskets and as much as possible of the work was done in the dry season. Centrifugal pumps powered by traction engines were used to keep the inside of the pit dry. Only where there were proper sand layers could a dredger be used. The welling up of the ground water meant that the volume of sand and gravel that had to be shovelled away turned out to be much greater than just the content of the pit. When the friction between the sides and the surrounding earth became too great welding sheets, rails and other heavy material were placed on top of the pit. Where there were proper sand layers the friction could be reduced by quickly pumping the pit empty. If boulders around the pit created too much friction then allowing the pit to fill up with water could help to resolve the problem.

The pits were sunk to depths of around 6 m below the riverbed surface. With the help of wooden boxes suspended from ropes, concrete would then be deposited in the bottom of each pit to a depth of around 2 m. Once the concrete had hardened the pits could then be pumped empty. Hardly had the final box of concrete been poured in when a huge *bandjir* approached at an amazing speed. In no time everything was covered in a layer of sand which rose, in some places, to a height of 3 m so that all that could be seen of the traction engines was the steam pipes protruding above the sand. As the concrete had been laid in all the pits it was fortunately relatively easy to empty them again. The masonry in the pits was then completed and the piers and abutments built up.

Meanwhile the bridges which had been transported to the spot in a dismantled state were assembled on the new stretch of line. Between the abutments and the piers assembly bridges made of the trunks of coconut trees were built and two slide rails were placed on top. The next step involved drawing and easing the spans into place.

Whilst the new bridge was being constructed the temporary bridge had to remain intact. However, the above-mentioned temporary *brondjong* dam had to be removed before the spans of the new bridge could be installed. On the one hand it meant that sand could again

The new bridge over the Moedjoer, 1903.

The completed bridge over the Serajoe river (Central Java), 1915.

fill the holes created in the riverbed beneath the temporary bridge which consisted of the former original bridge and its elongation whilst on the other hand the temporary bridge was exposed again to the danger of destructive *bandjirs*. Fortunately only some of the bents of the temporary bridge subsided during the instalment of the new spans and it was easy to quickly set them straight again so that the rail traffic was not disrupted for too long.

The standard arched bridge with a tension bar and engineer Haarman

In 1913 Professor J.H.A. Haarman gained a PhD for his research into 'angle iron brackets'. In his dissertation he focused on the angle iron of longitudinal and diagonal girders. In practice, such joints were prone to fracturing. In dealing with the problem he emphasised the fact that one of the consequences of allowing iron to become molten is that the stresses lessen and he alluded in that connection to the extra reserves of static, indeterminate structures. In the speech that he gave on the occasion of the third anniversary of the Technical College in Bandoeng on 30th June 1923 that was an important topic of debate.[15] By then his theoretical knowledge in the field had led to a professorship.

At that time Haarman was also the head of the Construction and Bridge Building division of the State Railways company. In the early days of the SS it had been the company's own staff who, up until around 1910, had been responsible for the designing of railway bridges. After that German manufacturers were hired to do the job for a short time. Round about the year 1913 the SS decided that it should once again be responsible for the design work and

The new Blang Me Bridge whilst still under construction in 1930.

for the assembling of bridges.[16] The fabrication of bridge components was something that occurred in the factories of construction workshops and invariably they were situated in the Netherlands. At the time Haarman was busy working on the design of various bridges for the Cheribon-Kroja railway line.[17] The arched bridge he designed to span the river Losarie which dissected that railway line was one with tension bars. The bridge's shape deviated slightly from the norm: the bar of the lower edge of the truss bridge first commenced at the third intersection of the horizontal tension bar so that the arch forces were more evenly distributed over the upper and lower girder of the arch. That also turned out to be a more favourable position for the placing of the transverse links so that greater cross stability could be achieved. For the stresses created by the structure's own weight the system was statically determined by the introduction of a hinge at the top of the arch.

All these innovations proved to be extremely advantageous when it came to the matter of assembly and free extension mounting because through the static determining of the structure it became possible, during the construction process, to very accurately calculate the iron bar forces and thus avoid any undesired stress distribution.

The bridge proved very suitable for absorbing any pressure forces created during free constructing and so from then on it became a standard model that was frequently copied. Obviously, though, in this particular instance it had to be fully adapted to the river Losarie's bridging requirements.

An example of a place where a similar bridge was built was across the Serajoe river along the Cheribon-Kroja railway line. That bridge comprised a central 90 m arch that was flanked on either side with spans that were each 60 m long. The 60 m spans were subsequently linked up to the land abutments by sections that were 25 m long. The respective 90 m and 60 m spans were of the arched type with a tension bar (i.e. the Haarman-style). The 25 m spans were truss bridges with parallel edges. Both the side spans extended three metres over the central opening. At the end of the overhanging sections there were the fitted central spans that had been introduced in a hinging fashion. In that way, what was in effect created was the Gerber system, named after the German engineer, H. Gerber, who had patented it in 1866. In addition hinges were also placed at the centre of each arch in the three linked together bridges. The whole project was completed in 1915.

The bridge over the Blang Me river

At the time when the tramway was laid in Atjeh there were hardly any roads which was why it was often the case that bridges were constructed for mixed transport means so that trams, ox carts and pedestrians could all make use of the same bridge. In that region all the rivers look like small, innocent, idyllic little streams in the dry season but when the rainy season comes they often turn into wide, wild, streaming torrents easily capable of washing away any structure that is not constructed sturdily enough to withstand the forces. Clearly, therefore, it was necessary to devise foundation structures that were much more solid but not too expensive.

One of the very obvious structural materials to use was wood, a material that was in abundant supply. However, at the beginning of the twentieth century it was not possible to construct wooden spans that were such that the river piers could be omitted. Pole trestles were built from the long trunks of the betel palm and were bored into the riverbed with a propeller blade until they hit a stable, weight-bearing level. The trestles were erected at 8 m intervals and then a bridge was built on top of them using single beams. What emerged was that those structures were much too flimsy to withstand the pressure of water in the monsoon season and the torn down trees and other driftwood transported by the river. So it was that in 1902 the three Sigli-Lhoseumawe tram line bridges over the Kroëng (river), including the bridge in the Blang Me area, were all destroyed during flooding. It was then decided that the main river should be bridged with truss girders so that a greater spanning distance would be made possible. The main span of the Blang Me Bridge thus became 49 m. For the approach bridges the cheaper construction consisting of trestles mounted on screw piles was maintained. The total length of the bridge was 77 m.

The piers on which the truss girders were supported required heavier foundations and so stone piers were built. In places that were susceptible to *bandjirs* it became the policy to always in future go for trusses on piers option. On the landside of the stone piers trusses were mounted on screw piles wherever that was deemed necessary. It was thus possible for the water to flow past the stone piers on the landside.

Despite all these sensible measures the bridge was once again destroyed in 1916 and so it was decided that the next time it should be lengthened by adding a further 50 m span and another three 8 m overland bridges on screw pile bents. Even that was not sturdy enough because once again, in 1926, that bridge was also washed away. In 1930 the bridge builders constructed a main span that was 100 m long, thus making it the longest bridge in the East Indies at that time. That bridge also almost got destroyed. During the building process the water rose so high that it washed over the decking of the bridge while it was under construction and 40 m of the temporary tram and road traffic bridge was wrecked. The double-tracked main span went into service on 10th July, 1930. The iron truss with the curved upper edges was built by Werkspoor Amsterdam at the request of the State Railway service. For the assembly work, 73 labourers who were familiar with the type of work were taken to Atjeh. Since the bridge had to be used for trams and other traffic the entire car deck was provided with an asphalt layer over the full width, also between the rails. On the new bridge two busses could easily pass each other.

The Tjisomang viaduct and engineer Van der Eb

In conjunction with the fact that a certain amount of soil subsidence was apparent in a particular area, it became necessary to divert a section of the railway line between Bandoeng and Batavia. This also involved constructing a new viaduct over a deep ravine. Round about the year 1930 it was the engineer W.J. van der Eb, who had been employed by the State Railways' Construction and Bridge Building division since 1926, who came up with plans for a bridge over the Tjisomang ravine.

What was selected for the viaduct support was three tall trussed uprights in combination with two stone piers and two abutments. Four steel truss bridges were then mounted on the stone and steel supports, each of which had a length (or span) of 37.50 m. Small 13 m long bridges were introduced on top of the piers so that connections could easily be formed with the referred to truss bridges. The spans between the abutments and the stone piers were

The bridging of the Tjisomang ravine (West Java), 1932. Assembly taking place with the aid of an 'assembly nose'.

respectively 3 m and 21 m. The deepest point of the ravine was approximately 95 m below the upper edge of the rails. The entire bridge was fabricated from trusses with raised track surfaces.

What was perhaps most striking was the height of the trusses, it was 9.38 m in which the height to span ratio (h/l) was 1/4. What had emerged from earlier studies, carried out by the State Railway's construction bureau, was that with certain bridging structures the weight of a truss diminished as the height increased until an approximate h/l relationship of 1/6 was achieved. When the height was further increased so that the h/l was made equivalent to 1/4, the weight would remain virtually constant. By making the horizontal sections higher, the height of the uprights could be kept down. This was all made possible by the fact that there was a raised track surface and no construction height restrictions. As the truss girders were so high the edges became light in relation to the rest which meant that there was less sideways rigidity. In order to give the bridges sufficient sideways stability the bridge width was extended to 4.17 m which amounted to a 1/9 part of the total span.[18]

The bridge components were assembled with the aid of a frontal extending crane that belonged to the State Railway Company. For the free extension mounting of the first bridge, use was made of an assembly nose. The actual manufacturing was contracted out to three different companies: the Dutch-East Indian Industry in Soerabaja, the Union in Bandoeng and the Braat firm in Soerabaja. Constructors were specially brought in from Europe and the whole job was completed in 1932.

The completed Tjisomang viaduct, 1932.

Between 1932 and 1940 Van der Eb was also busy replacing a number of old bridges that were no longer suitable for the heavier rail traffic. It was also a time when people were beginning to question to what extent the foundry iron previously used in bridge building was still suitable in connection with its weakening over the course of time. Ultimately this was to lead to the decision to systematically replace all the old bridges by introducing bridges with modern designs that were made from molten iron.

Around the year 1937 Van der Eb carried out detailed research into the fatigue phenomena of wrought iron and molten iron. He studied the breaking point of such material when subjected to a number of changing loads under stresses that were smaller than the fracture strength of the material. It was a phenomenon that had been examined by the German engineer, A. Wöhler, in about 1860. Van der Eb's approach to the problem was different. He presumed that with metal fatigue the limit of the available resilience capacity is actually exceeded. It was a brief mention in a publication by the Russian professor, N. Streletzki, that gave him the idea to correlate the amount of energy consumed with the gradual reduction in the flexibility of the material over the course of time. It occurred to him that it could be a way of determining the degree of fatigue. In that connection he researched the elasticity of a large number of test samples obtained from the material derived from dismantled old bridges. It was partly the development of the tensile testing curves that indicated to him just how the elasticity had declined.[19]

After the Second World War, Van der Eb left for the Netherlands where he first found employment with TNO and carried out research into steel and concrete constructions. He then moved to the State's Public Works' Bridge Building bureau where he designed the large bridges over the river Meuse in Venlo, the Merwede Bridge in Gorkum, the bridge over the Rhine in Rheenen, the Wilhelmina Bridge over the Meuse in Maastricht, the Van Brienenoord Bridge in Rotterdam, the bridges across the IJssel in Rheden, the bridge over the Haringvliet at Numansdorp and finally the bridge over the St. Annabaai in Willemstad in Curaçao. Van der Eb was also the man who designed the steel structure for the flood barrier in Krimpen aan den IJssel and for the drainage sluices in the Haringvliet.

The bridge over the Kali Progo on the Bandoeng-Djokja railway line and engineer Bijlaard

Professor P.P. Bijlaard joined the Construction and Bridge Building division at the State Railways company in about 1923. Before going on to discuss the design and construction of the bridge over the Kali Progo that was constructed in 1931, an overview will first be given of Bijlaard's earlier activities. Around 1923 he was involved in the dismantling of the bridge over the Tjitaroem on the Meester Cornelis-Tjikampek railway line that had originally been constructed in 1906. While working on that particular project he became acquainted with the reverse process to free extension mounting. He was afterwards also involved in the recycling of the old disassembled bridges that were once again deployed in free extension methods for the Koppo-Tjiwideij tram line.

In 1926 he was put in charge of the assembly of the bridge over the Way Kommering on the Palembang-Telok Betong line in Sumatra.[20] The job involved the building of three truss bridges, each with a 61.5 m span. Two of them could be directly mounted on the scaffolds under the bridge whilst the third was built according to the free extension method. Another bridge construction project along that line was the bridging of the Way Oempoe in 1927 consisting of an arched bridge with a tension bar (following the Haarman method) that had a 60 m span. That was followed by another truss bridge with a 20 m span. In 1927 he also moved to the Bandoeng Technical College where he was made a professor.

Whilst Bijlaard was working in the Construction and Bridge Building division he was perpetually preoccupied with the theoretical background to bridge building. One of his areas of focus was, for instance, the problem of secondary stress.[21] Secondary stresses are the extra stresses that come to the fore during calculations when simplified representations of constructions are replaced with those that correspond more closely to reality. A good example is the bending stresses that emerge alongside normal stresses in the diagonals and edges of trusses if, instead of presuming that the joints are hinged, one presumes they are fixed and if one also takes into account the rigidity preventing the bending of the bars and beams. During the assembling of free extension constructions considerable forces can be manifested in the trusses that are under construction and the secondary tensions can play quite a big role. The aim is always to minimise the secondary stresses by restricting shape changes through the application of all kinds of devices and to restrict the formation changes by regulating the jacking up sequences.

The assembly of the Way Oempoe Bridge along the Palembang-Telok Betong line (Sumatra), 1927. The construction material was supplied from the right, from the Telok Betong side.

The completed bridging of the Kali Progo (Central Java) in 1933. Seen from the pendulum construction side.

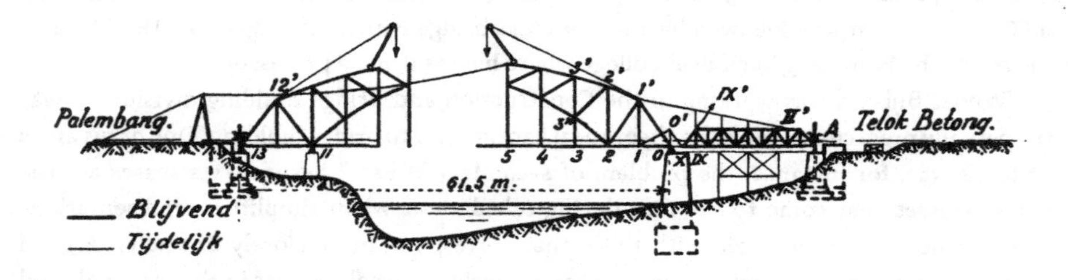

Bijlaard also studied the problem of bridge fatigue and conducted in that connection a number of laboratory tests. In so doing he arrived at the conclusion that secondary stresses could indeed affect the lifespan of bridges. What he emphasised was that construction methods should minimise secondary stresses. Bijlaard was keen to point out that in the event of fatigue the magnitude of secondary stress should not become greater than 1/4 of the primary stress.

Another subject that captured Bijlaard's interest was the economic side of trusses. As soon as such structures become very tall the weight factor decreases but there is a certain height which is, in that respect, ideal. Previously, proportions where the height was equivalent to 1/8 of the span were the norm. After further research and discussions with the professor mentioned later on in this chapter, N.C. Kist of the Technical College in Delft, Bijlaard arrived at the conclusion that from the economic viewpoint a ratio that was between 1/5 and 1/8 was the most favourable.[22] One important ground rule was that the slant of the diagonals should not diverge too greatly from the favourable slant of 45 degrees. In order to ensure that the fields did not become too great and that the buckling length of the diagonals and the pressure-loaded edges was not too unfavourable it was believed to be advantageous to subdivide the trelliswork segments.[23] The first SS bridge to be built in that way according to the new revelations was the bridge over the Kali Asem for the Klaka-Pasirian line that was designed by Bijlaard in 1927 and consisted of a truss bridge with a 61 m span. The truss was 10 m high and was divided up in such a way that the diagonals adhered to the 45-degree constraint.

In the continued endeavours to achieve an as great as possible saving of weight and thus also of bridge expenses, Bijlaard designed a new type of bridge for the Kali Progo which adhered to a new system based on the principle of a truss girder with a pertinent height/bridging of 1/6. The relevant span was 96 m and the height 16 m. The truss was furthermore provided with a subdiving of the diagonals and verticals. What was unique about that system was the application of two slanting hinged supports at the abutments. The entire structure was statically determined and stabilised by means of a stabilisation rod introduced on one side and supported by a land abutment. In that way a special shape for a three-hinged frame was obtained. Apart from saving weight through the selection of the earlier-mentioned truss system, what the introduction of the hinged supports mainly achieved was the effect that the lower and upper edge bars of the truss were relieved of the load. Bijlaard calculated that ultimately that would result in a weight saving of around 30 percent of the entire bridge weight.

One of the disadvantages attached to subdividing the truss in that way was that the secondary stresses that were introduced turned out to be greater than in the case of a single truss. In order to minimise the secondary stresses, Bijlaard decided to give the subdividing bars a preliminary length change which was such that after assembly certain bars would already be pre-bent but would lose such curvature when maximally loaded.[24]

Finally the hinged support would be attached to the abutment by the stabilisation bar. In that way a kind of collar arm was created upon which one of the extremities of the truss could rest. In view of the fact that apart from being subjected to vertical stress the abutment also undergoes a tilting moment the whole fixing had to be securely done. The collar arm could also serve as a point from which to mount the bridge by the free extension method. At the other end of the truss the free hinged support served as a point of support. The bridge sections were fabricated in the Netherlands by the Royal Dutch Machine Factory, previously known as Begemann, in Helmond. After having been manufactured the main girders were completely laid out in the factory so that all the connection holes could be bored. Once that had been done the whole bridge was test assembled on the factory grounds before being taken apart and shipped to the East Indies. The assembly work started in 1932 in the workshop and was conducted by the SS's own personnel. The whole job was completed in 1933 and in the same year all the test runs were done as well.

TRAFFIC BRIDGES

Materials

There was one golden rule that applied to all building projects of times past and that was that builders were always inclined to use materials that were readily available in the immediate vicinity. In the East Indies that was, of course, wood. Before the dawn of the industrial age wood was the material that was usually used to build bridges in the East Indies. Indeed, the bridges built for the oldest tram and railway networks were also fabricated from wood. Wood came in all shapes and sizes but the best and strongest was *djati* (teak). Because of the ever-increasing demand for teak it was not long before it became scarce and therefore expensive. Soon alternatives were sought in the form of stone, steel and later concrete.

Originally the East Indian road traffic bridges were built for much lower traffic class bridges than those listed in the Regulations for the Design of Steel Bridges (Voorschriften voor het Ontwerpen van Stalen Bruggen, VOSB) adhered to in the Netherlands which were namely: class 30, class 45 and class 60 (in tons). Depending on the local circumstances and the anticipated traffic developments, a certain taxation system was laid down for the road class divisions according to the relevant load patterns. The wooden bridges in Southwest Celebes were, for instance, geared to a maximum weight of 7.5 tons or sometimes a mere 4.5 tons for just one lorry on the bridge at a time. That weight was much less than the maximum load laid down in the VOSB regulations but the traffic intensity was also much lower. The ordinary traffic often consisted of people with carts pulled by oxen, buffalo or horses. It was also common for notices to be placed on bridges, as was usual in the case of military bridges, to the effect that no more than one vehicle at a time could cross the bridge or that the distance between any two vehicles should be greater than the length of the bridge span so that there could never be more than one car on a bridge at any given time.

In an extensive article the engineer R. Roosseno Soerjohadikoesoemo[25] wrote about the wonderful advances being made with reinforced concrete for the purposes of bridge building for road traffic in the 1933 to 1940 period in the East Indies. He was undoubtedly a very enthusiastic engineer who had been able to develop thanks to the possibilities opened up by the rise in the volume of road traffic. Up to a point his article was 'instructive' and had been written to share with his colleagues the knowledge and experience that had come with new construction methods. At the time of writing Roosseno Soerjohadikoesoemo was a 2nd class engineer affiliated to Public Works and in 1940 his services were made available to the provincial public works department for East Java where he was attached to the office of the second district in Kediri.[26]

From an article published by Professor Kist about the own weight of East Indian and Dutch bridges it was concluded that up until 1928 the railway bridges in the East Indies were predominantly made of steel. In the Netherlands it was the building of road traffic bridges that took precedence after 1928 but notably in the case of the bridges that spanned the big rivers, it was steel that was often still used, especially the bridges that featured in the tables referred to by Kist. In the East Indies funds were released in the thirties from what was termed 'rubber money' and the 25 million funds. Various large bridges were constructed in Java, Sumatra and Borneo, many of which were made of reinforced concrete. Roosseno Soerjohadikoesoemo's first question was therefore: why did the bridge builders choose reinforced concrete? That was the prelude to a treatise that could well have been used as course material in a technological education course and which, in broad outline, still holds today – especially for countries that find themselves in comparable situations. The main arguments he levelled were these:

A structural work made from reinforced concrete has more of a permanent character than one that is made of wood or iron.

Whilst the annual maintenance expenses for wood and iron structures are considerable, provided they are well constructed, reinforced concrete bridges require virtually no maintenance.

The consideration that a relatively large portion of the construction expenses, such as the materials and the wages, remain in the country and is even directly linked to the population can be another good reason for deciding to select concrete.

The costs of any structural works depend in the first place on the availability of good natural building materials such as sand, gravel, fresh water and framing wood. Throughout the entire East Indian archipelago the basic cement and reinforced iron prices were the same, which meant that it was only the transport costs that varied.

Before Roosseno went into more details concerning the design possibilities opened up by reinforced concrete bridges, he summed up the strong and weak points of wood and steel usage for bridge building in the East Indies.

In the East Indies there were a number of good wood types that were perfect for bridge building. Wood was the obvious material for small and temporary structures. As construction iron was hard to come by in the 1914-1918 period the Public Works department focused at that time on wood as the obvious solution.

When building a wooden truss the hardest part of the job is the creation of the joints. In conjunction with the differing traffic loads it is important to make sure that the hinge points are true hinges. Though it is possible to create hinged joints in wooden trusses it was really only feasible to do that with the help of iron. This meant that for a bridge with a 20 m span, some 140 kg iron was required per m^2 of wood. It turned out that the most suitable type of truss for those kinds of bridges was the Howe-joist style, a parallel beam structure with double diagonals and verticals that were provided with tension bars so that the diagonals could be pre-tensioned. In that way the entire structure became much sturdier. In order to make wooden bridges more durable under East Indian conditions the construction has to be properly protected against sun and rain. The best way to achieve that was by building a roof or awning above the bridge.[27]

Structures fabricated from iron (or steel) are extremely light. The material is homogeneous and reliable and it is possible to apply low safety coefficients. The lifetime of such bridges is, however, very dependent on their maintenance. Sea air will, for instance, rapidly corrode them. For a number of reasons: material efficiency, ease of transport and easy on-site assembly, these bridges tend to be predominantly built as trusses.

The disadvantage of reinforced concrete in the case of bridge construction is the tremendous weights involved. Later what is known as the 'own weight' phenomenon will be discussed in greater detail. What will be demonstrated here is the fact that the relative own weight factor is much more disadvantageous in the case of concrete than where steel and wood are concerned. In the case of a 20 m bridge span that is 4.30 m wide and suitable for class two traffic[28] the rest load / mobile load ratios are as follows:

Wood	43 ton / 36 ton = 1.2 : 1
Iron (or steel)	36 ton / 36 ton = 1.0 : 1
Reinforced concrete	108 ton / 36 ton = 3.0 : 1

This, then, is just one of the reasons why any reinforced concrete bridge always automatically falls into a higher load category. Retrospective reinforcing of concrete bridges is always much more complicated than in the case of any other kinds of materials.

Foundations and constructions

Roosseno Soerjohadikoesoemo wrote that there were generally three kinds of foundations that were suitable for bridges: foundations directly built on soil or subsoil, foundations in pits or caissons and foundations built on reinforced concrete piles. The first type was the cheapest but in connection with local situations not feasible everywhere. If the solid stratum was somewhat deeper but not too deep then the second method would be recommended. One disadvantage was that the pit wells were very heavy and therefore difficult to transport whilst lowering them into place was something that required skill derived from practice. If neither of these methods was possible then people were forced to go for the final and most expensive option. Only if the posts could be kept 'free' by projecting above the water level would it possible to keep the costs down but then one would have to take into account the risks of collisions or *bandjirs*.

The concrete piles would be preferably driven home but in the soft soil of Borneo adhesive piles had to be used. The constructors endeavoured to enhance the adhesion factor by introducing Takechi collars but there was disagreement about the effectiveness of that. The pile load was generally between 10 and 30 tons. In areas where the ground was soft, the road slope leading up to the abutments could exert horizontal and riverwards loads on the pillars and thus cause damage. The answer to that problem was to lengthen the bridge thus reducing the approach incline. During *bandjirs* the river would create erosion holes around freestanding piles, which would normally be repaired through the dumping of quarry rocks.

In his article Roosseno also provided a description of different types of reinforced concrete bridges, and he also gave a number of practical instructions and tips. Again, that section would not be out place in a present-day course book on hydraulic engineering.

Plate bridges

The advantages that were listed were: the minimal construction height (1/17 to 1/23) of the span, little soil displacement in the feed-on roads, simpler form work and reinforcing. Moreover, it was not usually necessary to have transverse force reinforcing. A disadvantage was the tremendous own weight, particularly where continuous plating was concerned where the pressure on the central support points was great. The economic limit was spans that were around 10 m long where the required reinforcement amounted to around 150 kg / m^3 concrete.

Horizontal beam bridges

This type of bridge consisted of a plate supported by beams which together formed a structural entity. The size of the horizontals was between 1/9 and 1/11 of the bridge span itself. In the case of bridges with a number of supports that would drop to 1/11 to 1/14. Most of the Public Works horizontal bridges with a car deck width of 4.50 m would have two to three main beams. The advantage of this system was that the surface plate fulfilled three functions: it served as a pressure flange for the TT beams, as a supporting structure for the mobile stresses and as a load dispersal construction over the main horizontals. Sometimes there would be no cross supports and just a few which partly had a load dispersal function while at other times there would be many cross supports so that the bridge surface would be divided into square sections that spanned in two directions and the covering plate could be made thinner, the minimum requirement being 12 cm! Such horizontal bridges were used in the East Indies up to spans of 20 m. Elsewhere in the world, though, a surface plate would be introduced beneath and between the horizontal components which meant that a box girder would be formed that could be used for spans up to 60 m.

Girder bridges with sectional steel reinforcing

In certain special circumstances these types of bridges could constitute an attractive option especially because construction-technically it would then become possible to simplify the support system for the formwork.

Girder bridges with shored up uprights

This was a very economical variation on the types given above but only practicable if the subsoil could effectively absorb the horizontal reactions. Up until 1940 it was a variant that was little used in the East Indies.

Clamped arches and vaulting

These kinds of constructions could only be used in cases where the foundation soil was good and where there was sufficient freeboard (i.e. height between the water level and the soffit of the bridge). Roosseno Soerjohadikoesoemo recommended that the relationship between the arrow (i.e. the height of the bridge) and the span should be as great as possible with such bridges because in that way the arched forces would decrease but the lead-up road should not thus be steeper than 1 : 20.[29] If the arch was well modelled, 70 to 80 kg of reinforcement material per m^3 would usually be sufficient.

This aqueduct over the Kali Baroe, Banjoewangi (East Java) constitutes a good example of a bridge type with a raised driving surface, 1940.

A type with a low-lying driving surface as exemplified by this arched bridge over the Ajer Klingi, Palembang (Sumatra), 1940.

Rod stiffened arched bridges

> Where the foundation soil was good and the freeboard sufficient, it would be logical to select a stiffened rod arch bridge with a raised car deck, in other words, predominantly in mountainous areas, whilst in the case of wide rivers where there was the risk of *bandjirs* and the freeboard was small the arch would be constructed above the car deck (arched bridge with stiffened rod arching and a low-lying car deck).

At the end of the article Roosseno reminded his readers once again of the most important conclusions and listed a number of companies that had been involved in the building of various bridges and which are still in fact known today in the Netherlands. He also noted that the GBV NI 1935 (Gewapend Beton Voorschriften voor Nederlandsch-Indië or Reinforced Concrete Regulations for the Dutch East Indies of 1935) offered too little footing for the calculation of point loads on the floor plates and that the regulations therefore needed to be supplemented.[30]

The own weight of East Indian and Dutch bridges

In two related articles Professor Kist[31] gave, with the help of other experts from the Netherlands and the Dutch-East Indies, a comparative overview of the bridges constructed for single and double railway lines and for ordinary traffic. The bridges that were especially mentioned were those with a span of 25 m or more which meant that bridges made of wood or concrete were simply not considered. What is particularly interesting in the articles is the comparisons that are drawn between the different types of bridges and the analyses that are made explaining why one particular constructional style was more successful in

The bridge at Batoeradja (Palembang) over the Ogan (1940) constitutes an extremely elegant design. Strictly speaking it is not a stiffened arch bridge but rather a truss girder bridge with a parabolic upperside. The way this links up with the bridge's peripheral sections makes the whole structure eminently practical and elegant.

the Netherlands whilst another caught on better in the East Indies. The Dutch national character was not in any way renounced because when drawing comparisons use was made of the 'economy factor f'.

It would appear that up until 1912 there were a number of permanent appendices that accompanied the yearbook of the Royal Institute of Engineers and that appendix P contained a 'frequently consulted listing of the own weights of Dutch bridges'. It is easy to understand that when people had to communicate without the telephone or the e-mail in an era when technological development progressed in fits and starts, there must have been a need to just sometimes compare notes with the results and ideas of others, especially for engineers stationed in remote outposts and working in a fairly solitary fashion. In both his articles Kist opted to present the salient characteristics of Dutch and East Indian bridges by giving extensive tables in which 10 to 15 prominent design principles were highlighted.

In *De Ingenieur* of 1928, 28 single-track bridges in the East Indies, 5 then recently constructed single-track bridges in the Netherlands and 3 road traffic bridges in the Netherlands were compared. In the 1937 article the bridges that had been built in the 1928-1937 period were listed and that was 13 single-track bridges in the East Indies, 5 single and 3 double-tracked bridges in the Netherlands plus 15 bridges for ordinary traffic in the Netherlands and none in the East Indies, and all in steel. One might therefore conclude that up until 1928 the East Indian railways was busy expanding in all directions but that after that year a more limited number of bridges sufficed and then only single-track bridges. In the Netherlands, by contrast, the railway network's main lines had been firmly established by 1928.

Another thing that furthermore seems to be apparent was that in the East Indies there was no demand for long-distance road traffic. The referred to road traffic bridges in the

Netherlands in 1928 were bridges that were necessary in conjunction with hydraulic works, such as the bridge over the Wessem-Nederweert canal and the one linked to the canalisation of the Meuse at the weir in Grave. In 1937 things were very different! By then the State Road Plan was being implemented and spectacular bridges were being constructed over the big rivers. At that time railway bridges were only required for new navigation routes. Most of them were still single-tracked but in the Randstad (i.e. the urbanised west of the Netherlands) there were also double-track bridges.

In 1928 and 1937 Kist explained how he had obtained all the data on which his comparisons were based. In fact it is amazing that it was possible to find so much information on the 1911 to 1928 period. Perhaps people were not, at that time, so concerned about the expenses attached to keeping so many metres of archive material or perhaps it was the case that those involved regarded such details as essential and therefore worth keeping.

The 'economy factor f' referred to in the introduction is a coefficient which according to theoretical deduction only depends on the truss plan for the main girders.

$$f = \frac{Q}{Q + Pv + 9\ M/l} \times T/l$$

There:

- Q is the weight of the main girders in tons
- Pv is the remaining bridge weight
- M is the moment of greatest possible mobile load in the middle of the bridge
- l is the span length in metres
- T is the average permitted tension stress
- f is the economy coefficient which depends solely on the truss plan[32]

The f factor was used both to make estimates during the design phase of the own weight of various horizontal sections and to test the efficiency of given detailed designs.

When comparing rail and traffic bridges the loads also play an important part which is why it is useful to maintain these basic principles. In the case of very big spans the load class was not of such great importance but in the case of spans up to 50 m it was particularly the weight of the locomotive that was crucial. What was also very important was the type of transport that used the line: heavy freight traffic in difficult to negotiate mountainous areas was something very different from light tram traffic that travelled along coastal plains.

Another noticeable point is that in 1928 issues of weight and economy were permanent considerations of importance. Steel was expensive and labour relatively cheap in the initial building phase: little was, however, said about the long-term maintenance costs. In the East Indies weight was more important in view of the fact that transportation to the assembly point could not usually go via water, like in the Netherlands, and the reaching of construction points in mountainous or difficult to access locations constituted an important cost factor. Every kilo counted!

What emerges from the 1937 table is that there were a number of bridges in the Netherlands that were less economical. The arched bridges with full bodied girders and tension bars, like the one in Vianen (which still exists but may have to be dismantled) and the one in Hedel with economy f of respectively 3.50 and 3.63 compared to the norm of 2.6 to 2.9 for truss bridges were therefore 'expensive' but very attractive in the landscape. Similarly, the old Moerdijk Bridge with its diamond shaped truss structure was not very economical either with its f = 3.04 but it was, on the other hand, easy to maintain. The Dutch bridge builders looked enviously to the extremely economical road bridge in Nijmegen, f = 1.99, and the Progo Bridge in the East Indies, f = 1.95, but because of the horizontal bearing forces those types of bridges could only be constructed in places where the soil on which they were built was capable of absorbing such horizontal forces. Clearly that was more likely to be the case in mountainous areas than in a country like the Netherlands with its young sedimentary deposits. Indeed, with these types of bridges factor f calculations are not really fair, as the complicated foundation scenarios will often partially cancel out the savings made in the superstructure.

The bridges that were really uneconomical were the 'Langerse boog' (Langer arch) type that were f = 4.42 and 3.41 respectively. There the horizontal forces were internally absorbed by the main girders / tension bars so that the foundations were not horizontally loaded. These types of bridges were aesthetically pleasing in the landscape and must certainly have been easier to maintain than the truss bridges.

The qualitative improvements in the type of steel that was becoming available also, of course, influenced the efficiency of the various types of bridges. The f formula that was compensated by introducing the strength via T led, in turn, to a comparing of similar kilo factors. The effects of buckling or folding in higher quality steel were not, for instance, discounted; neither were the possible effects of higher processing costs or such like taken into consideration. Despite everything, however, the systematic approach did at the time provide good chances for all concerned to make some interesting comparisons. There are also a number of interesting conclusions to be drawn by historians, not only in terms of technology but also concerning the circumstances under which bridge builders in the East Indies had to practice their profession and the extent to which they did, in effect, stand-alone.

Bridge examples

The bamboo bridge over the Tjitandoei in Indihiang

Bamboo is a strong and elastic material. *Bamboo in Building Structures* by Jules J.A. Jansen provides a number of facts and figures that help to give one a rough idea of various essential details:

1. the volume weight (density) of bamboo is 600 to 700 kg/m^3;
2. its strength qualities depend very much on moisture content;
3. when the bamboo shoot (stalk) reaches its third year the cell walls thicken; that is, when it is at its maximum strength;
4. the pressure strength is around 70 N/mm^2; for buckling calculations the phenomenon of crooked shoots will have to be taken into consideration.

Selamatan on the occasion of the opening of a bridge in the Preanger (West Java), probably at Tjikampek, c.1905.

When bamboo is bent the hollow inside part changes shape, it flattens out which means for one thing that tensile stresses develop perpendicular to the fibres. In combination with the shear stress in the neutral fibres that leads to splitting. Hence, a low shear value is assumed which is around 2 N/mm² instead of at least 7N/mm² in 'pure' shear tests. The calculated value of the bending strength lies at more than 80N/mm². The elasticity module is over 20,000 N/mm². The tensile strength is so high that it was hard to determine during tests: the low pressure strength perpendicular to fibre combined with the shear strength, which is also low, is what makes it impossible to introduce the tensile strengths.

Even the engineering journals of the day devoted attention to bridges made of bamboo though admittedly it was a subject that was given less space than the issue of iron and concrete bridges.

> *The replacing of temporary bridges with permanent ones has led to the slow disappearance of very interesting structures and all entirely at the instigation of departmental heads who carry out the jobs without the technical assistance of the local population. Though they were structures that might have consumed a large amount of material and labour resources as well as requiring a great deal of upkeep they were, nevertheless, evidence of the ingenuity of the natives and for many years they had met the transport needs.*[33]

One particularly striking bridge made of bamboo was the one over the Tjitandoei at Indihiang on the border of the Preanger and Cheribon regencies which was built in 1889 at the instigation of the *loerah* (*desa* chief) by natives.[34] Even then there were few bridges of that type in use though it was reported that the traffic on that particular bridge was busy.

The bridge was 33 m long and 3.75 m wide. The four shoring constructions, hereafter called 'props', that were required consisted of six to eight sticks of bamboo that were 8 to 10 cm in diameter. Every 75 cm the bamboo tubes were tied together with string made from *idjoek*.[35] The bound together props were interlinked by horizontal and diagonal bamboos. The horizontal placed at the top of the props served to reinforce the vertical framework that supported both the car deck and the *atap* roof (i.e. made of Nipa palm leaves). Those vertical sections would cause bending moments in the props which was why the props were also linked to poles that were attached to the verticals in order to minimise the bending moments. It was established that the roof would definitely catch the wind which meant that the bridge could noticeably bend in a sideways direction. A double thick *sassak* (i.e. made-to-measure woven mat) that was laid over the framework served as a (car) deck.

Bamboo bridges required a great deal of maintenance and only actually lasted for five or six years but they were cheap. It only cost about 300 guilders to make the bamboo bridge over the Tjitandoei, in other words it cost less than 10 guilders per metric length of span. Even at the time it was built that would not have been expensive. The materials would have cost around 190 guilders and the labour about 80 guilders. The bridge was constructed in 14 days.

The bridge over the river Ajer-Silau on the road from Tandjong-Balei to Deli

That was a bridge that was of great importance to the tobacco plantations in that area. For many years before the bridge was built there had just been a small ferryboat connection. The ferry had been the cause of much frustration in the busy tobacco-harvesting season. In fact the ferry formed part of the main connection between Tandjong-Balei and Deli (on the Sumatran east coast) and the government had decreed that the link there had to be improved. Early in 1894 the authorities therefore gave permission for a permanent bridge to be built over the Ajer-Silau just slightly up river from the above-mentioned ferry crossing point.[36]

The bridge consisted of two iron truss spans, each 30 m long, which rested on one land abutment and two masonry piers. On the left bank there were two further spans, each 10 m long with iron beam girders that rested on a bank pile and two pillar trestles. The whole bridge was 4 m wide from side to side and thus had a total length of 80 m. Before the building proper commenced a temporary wooden bridge was erected that later also came in useful for the building and the assembly of the iron superstructure. The abutments on the right bank and river piers were built on masonry pits, mounted on *djati* wood pit rings.

The pits, the under edge of which lay 3 to 4 m below the riverbed, were filled with concrete and linked on the upper side by a built up vault. The abutment pits were kept dry by means of pumps and subsequently filled while dry. Where river piers were concerned, the lower section was laid in concrete under water and the rest was likewise built 'in dry conditions'. When it came to sinking the pits for the river piers the builders had difficulty

A model of a bamboo bridge dating from the 1930s.

excavating the riverbed. At first the sand was dug out with spades then later labourers dived under water and brought up the sand in buckets. As that latter method was seen as very inconvenient the rest of the job was completed using four revolving buckets that were used day and night to dig further and finish the job. In total 16 m³ of sand was excavated every twenty-four hours.

When sinking one of the pits for the mid-river piers the builders hit a tree trunk when they had gone down about 1.5 m. It turned out that the only way to dislodge the wooden obstacle was with dynamite, so four sticks of dynamite were used. After the explosion the pit was found to be cracked in a couple of places which meant that four braces then had to be attached to the sides that were fixed in place with the help of swivel bolts. In that way the cracks were reduced from around 60 mm to around 5 mm. The remaining cracks were then filled up from the outside and inside with Portland cement.

The screw piles of both the piers that were needed for the ten metre span were sunk to about the same depth as the soffit of the foundation of the flow piers. A lot of bricks were required to build up the piers, about 650,000 in all, which were then made on site. There were two stone ovens available, each of which could fire 40,000 bricks. A horse-powered brick machine was brought in from Singapore which was able to produce 5000 bricks a day with the help of just one horse. It turned out that the machine was excellent for the job of mixing the materials but not for the shaping of the bricks. That final stage was done by hand with the help of wooden moulds. For the mixing of the mortar the machine was powered by two small Siamese oxen. The building bricks, which measured 200 x 100 x 50 mm, were of a good quality and were manufactured for a minimum price, especially when compared to, for instance, the 'Waal river bricks' made in the Netherlands.

Through the intervention of the Ministry for the Colonies the iron superstructure of the two 30 m truss spans was delivered by the German company Krupp. The remaining ironwork was drawn from 'the country's supplies in Batavia'. The total costs for the whole

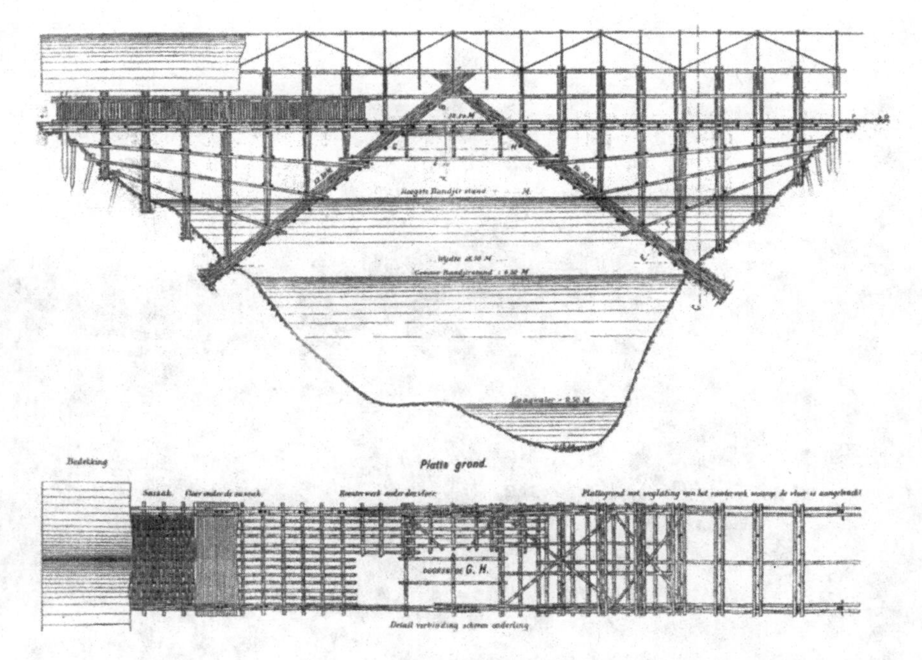

A bamboo bridge that was constructed over the Tjitandoei on the road that led from Indihiang to Tjiamis and Cheribon, approximately 1893.

bridge amounted to approximately 54,000 guilders. The bridge was opened to traffic on 8th October 1896 after first having been named the Wilhelmina Bridge by the wife of the chief administrator of one of the local tobacco companies.

Wooden bridges in Southwest Celebes

The traffic in Southwest Celebes was not as heavy as in Java. In the thirties of the 20th century the population was around the two million mark and there were some 2000 km of roads, 1500 km of which fell into the classes III, IIIA and IV. The remaining 500 km in thinly populated areas where there was not very much traffic were class V roads. There both narrow and wide waterways were bridged with temporary wooden truss bridges.[37] In the thirties of the 20th century a portion of the temporary bridges that fell into the higher class road bracket started being replaced by permanent bridges. The limited means in combination with the then plentiful supply of good construction wood led to a situation in which the routes that were not so intensively used were given wooden bridges with large spans.

Up until 1928 bridges with spans of 14, 18, 24 and 30 m were constructed as Howe-style bridges (i.e. made of truss with parallel lower and upper edges) for one lorry of 7.5 ton and a car deck width of 3 to 3.3 m. These bridges were made of good hardwood and, provided that they were well maintained, could last for 25 to 35 years.

Alongside these wooden truss bridges built in Howe style, truss bridges with curved upper edges were also constructed in the years running up to 1941, the main reason being

Bridge over the Ajer-Silau on the road between Tandjong-Balei and Deli (Sumatra), 1896.

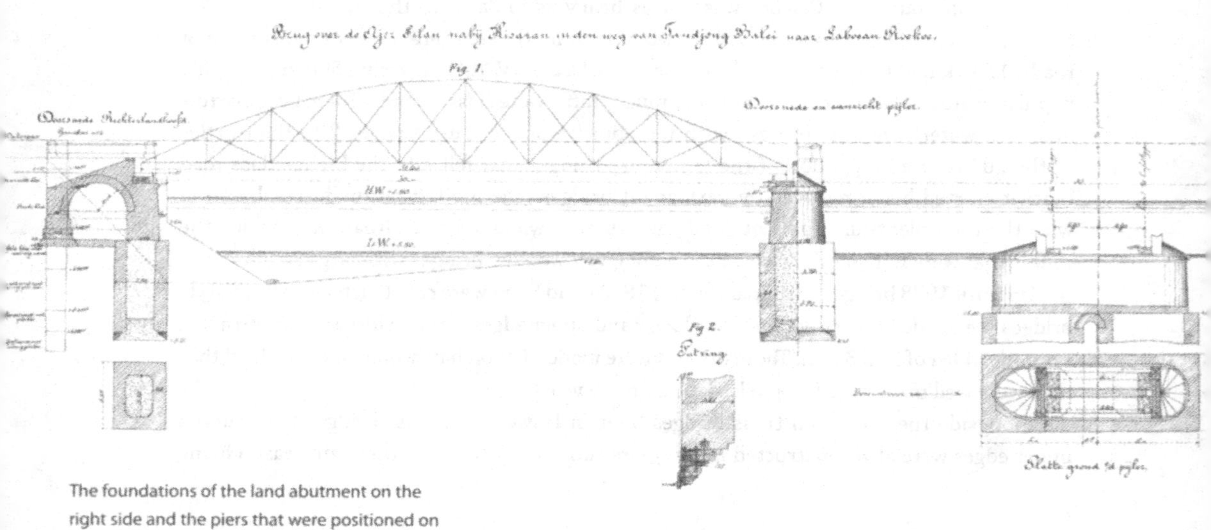

The foundations of the land abutment on the right side and the piers that were positioned on masoned pits and wooden well rims.

The 'horse power brick machine' from Singapore.

that it meant important savings could thus be made in the ever more scarce supplies of hardwood.

In the case of small spans of 11 to 15 m, bridges were built along standard designs and were prefabricated in the Landschap sawmill in Raha on the island of Moena. After the wood had been sawn and prepared the bridge would then be assembled before being taken apart again for transport to the place where it was to be finally constructed. In the case of the eleven-metre bridges, single shoring was used and for fifteen-metre bridges the shorings would be extended.

Individual designs adapted to the particular situation and possibly even to the locally available wood dimensions were created for the bridges that required larger spans. A good example of such a bridge was the Tandoeng bridge. In the vicinity where it was constructed there just happened to be 20 x 24 cm wood available in lengths of 15 to 16 m. The available types of soils and subsoils made it possible to absorb horizontal loading. It was therefore feasible to create a double shored up suspension bridge with a 30 m span. The circumstances were not always that favourable. Sometimes, depending on the timber available, bridges would be made with parabolic-shaped upper edges, stiffened tension bars and iron hanging rods.

The arched bridges had an upper edge that was made of double wooden pressure beams which were cold fitted at the joints. The verticals consisted of thin steel drawing rods and the double tension bars were stiffened by placing on top of them a lattice girder that simultaneously served as the handrail. The transfer of forces from the hanging rods to the arch passed through a hardwood block via an iron plate that had been forged in shape and there was another small plate for the nut. The whole construction would then be finished with a zinc plate to make sure that no water could get into that important joint.

In that way bridges could be built with spans of up to 48 m. With any length that was greater than that the narrowness proportions between the width of the car deck and the

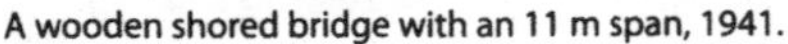
A wooden shored bridge with an 11 m span, 1941.

A wooden shored bridge with a 15 m span, 1941.

height of the main girders would rapidly become such that the stability would be hard to guarantee only by stiffening. For the larger spans, the solution was usually sought in suspension bridges with wooden reinforcing beams. The kinds of wood sorts used were especially *bajam* from South Celebes, teak from the island of Moena and *bangkirai* from Kendari (Southeast Celebes). The wood was often obtained by mediation on the part of the State Forestry Service. Great care would then be given to selecting wood that had been ringed beforehand, displayed no defects and was of a sufficient age. It was therefore absolutely acceptable to adhere to the same tensions as those which, via the European standards, applied for oak and beech. Though all of this sounds reasonably safe, if the spans had been any greater then the matter of the joints would have become critical and the craftsmen who were able to do that work properly were not always available.

In contrast to the views put forward by Roosseno Soerjohadikoesoemo in *De Ingenieur in Nederlandsch-Indië* (The Engineer in the Dutch East Indies) in 1940, the engineer W. Paardekooper questioned whether the covering of the bridge with some kind of roof or awning did not simply provide an illusion of certainty regarding the bridge's lifetime (see above). Apart from anything else, the kinds of bridges that he had been building in Celebes with vaulting or parabolic girders would have been virtually impossible to roof over in any way. Hence the reason that he preferred to cover the joints with zinc plating, pay attention to precision in the details, use a wood preservative and carry out perpetual maintenance.

If the solid ground layer was deep below the surface or if there was a danger of cracking and subsidence, the foundations for wooden bridges would often take the form of pit foundations with concrete pipes of about 1 m in diameter filled with minimum cement-concrete. Whenever the soil was soft, pile foundations with wooden piles or reinforced concrete pillars would be introduced. In many places in Celebes the soil is often gravelly and sandy. There circular reinforced concrete piles of about 40 cm in diameter would be spouted into the ground, levered into place and driven. The piles would then be provided with a collapsed old water pipe that would lead to the steel pile-shoe. Water, air, or a combination of both, was used for spouting.

A bridge with a parabolic arch and a stiffened tension bar that has a 24 m span (Noni Bridge), 1941.

A stiffened rod arch bridge with a 32 m span (Pembaoeang Bridge), 1941.

The strengthening of the reinforced concrete girder bridge over the Tjimanoek at Leuwidaoen

What this particular bridge building example will show is that even with a sophisticated bridge building project unexpected things can happen. It will also show how, in the East Indian situation, matters were tackled and resolved in a responsible way. As it was proving to be insufficiently strong to support the increasingly heavy traffic, the wooden truss bridge over the Tjimanoek in Leuwidaoen (Garoet) needed to be replaced.[38] The new bridge that was designed was a reinforced concrete girder bridge with a theoretical span of 28.80 m consisting of two main girders with, on top of them, the bridge car deck. Together the horizontal girders and the car deck functioned as T-horizontals.

In the nineteen-thirties it was not yet customary to construct bridges of these dimensions in reinforced concrete. In the Netherlands the approach to the bridge over the river Noord in Hendrik-Ido-Ambacht was under construction and there the set up was similar: six horizontals were placed side by side and a concrete surface was drawn over them but the span of that continuously constructed bridge was approximately 13.50 m and it was then the longest beam bridge span.[39] Also in the approach viaducts constructed as plate viaducts leading up to the bridge over the Oude Maas (i.e. the old branch of the Meuse) in Dordrecht most of the 55 spans of the, in total, 1000 m long viaducts were between 16 and 22 m long with two exceptions in the continuous constructions where the length was 40 m.[40]

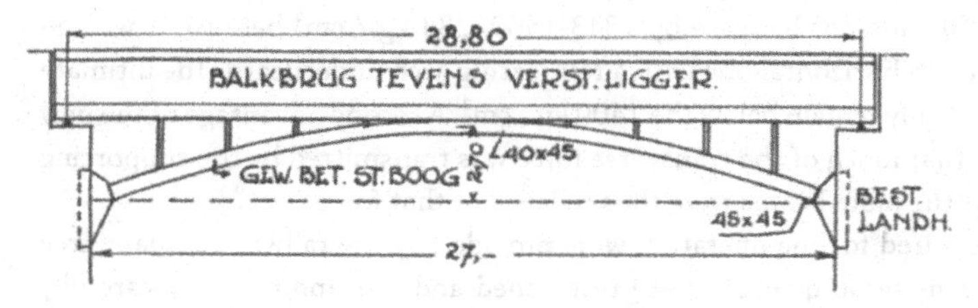

The reinforced bridge in its final state, 1939.

What happened to this respectable bridge over the Tjimanoek? During construction the plan had been to first cast the lower section (2.2 m) of the two 2.5 m high main horizontals on a scaffold that was to support the formwork for the two beams, the idea being that afterwards, using the same scaffolding, the remaining beams would be put in place and the floor would be poured. Two days after the main horizontals had been cast a *bandjir* struck, sweeping away the scaffolding under the bridge but not the bridge flooring. After the first shock it was discovered that the formwork was still there but that it was hanging from the incomplete beams that were two days old. What was the explanation for that? In those days the calculating of reinforced concrete proportions was based on the n-method in line with the elasticity theory but a fictitious ratio of 5 was taken between the elasticity modules of steel and the concrete whilst in the fracture stage that was supposed to be around 15.

At that time a new collapsing method[41] had been adhered to and it was discovered that under stress circumstances a safety margin of 1.04 was present. It was sheer luck therefore that everything was still more or less in place. The concrete pressure stress would have been crucial in the first couple of days but the unexpectedly good behaviour was attributed to the compressing of the concrete by vibration, not a generally applied principle in 1939. After further maturing of the concrete the carrying strength could even increase but then it would be the reinforcement steel that would be of crucial importance. The bridge finishing included the pouring of the last section of the girders and the flooring plate. Upon completion the tension in the steel could rise to 1560 kg / cm^2 and that was far above the normally tolerated stress level.

If bridges finished in this way were also subjected to traffic loads then the steel tension of a completed bridge could go up to 1843 kg / cm^2 which is very close to the liquefaction limit, the risk being that the concrete would fracture so that, in time, the reinforcement steel rods would rust. Unburdening the hardened beams by replacing and restressing the supporting scaffolding was not simple and there was the risk of another *bandjir*. As there was ample freeboard between the river and the bridge and the abutments were in an excellent condition (as was the carrying capacity of the soil) it was possible to introduce stiffened support arches under each of the horizontals.

The supporting arch was made of reinforced concrete that was provided with stiffly constructed reinforcement. The stiff reinforcement material could easily be cheaply made from excess material, in this case 5 m long cross-girders from an old railway bridge. At the very top of the concrete arch, space had been made for two 60 ton jack screws that would give the arches the right stress, thus simultaneously reducing the stress in the reinforcement of the main beams and bringing it down to 1216 kg / cm^2 from the already mentioned 1560 kg / cm^2. On top of this one still had to account for the load stress created by passing traffic. The steel tension would thus not increase by 1843-1560=283 kg / cm^2 because it was no longer a beam bridge but a horizontal one with a reinforcing arch underneath. The ultimate tension would thus probably remain below the 1400 kg / cm^2. A second advantage of the new construction was also that much of the transverse force was transmitted by the supporting arch which meant that the main beams were thus relieved of that force.

The screw jacks required for the operation were provided by the railway company free of charge. The various material qualities were researched and the findings very carefully noted by the Materials Research Laboratory of the Bandoeng Technical College. It may be

The suspension bridge over the Tjimandiri near Palaboehan Ratoe on the Pasawahan-Palaboehan Ratoe road, Preanger, c.1920.

concluded that here creative and brave use was made of the available technical knowledge and all the means available and that the accident caused by the *bandjir* was resolved in a completely logical way.

Suspension bridges

Unlike in the Netherlands it was customary in the Dutch-East Indies to build suspension bridges. For pedestrians that was quite soon an attractive solution, but also for road traffic. As the volume of traffic was then low this type of bridge often emerged as the most economic solution. The first kind of suspension bridge to appear was the 'sagging' or cable suspension bridge where the weight was entirely borne by the cables. There were a number of quite important disadvantages attached to this kind of bridge like, for instance, the limited traffic capacity allowed and the pronounced vertical displacement created by the traffic. This last factor directly affected the steel fatigue aspect, gave rise to massive maintenance costs and created an unpleasant sensation for those crossing the bridge.

When the volume and weight of road traffic started to increase it became necessary to switch to 'stiffened' suspension bridges with truss girders in the plain of the cables. After the building of the first few such bridges the designs underwent developments until a more or less standard design was established. The first stiffened suspension bridge was completed in 1917 (see below). A number of these bridges are still in use and remain as silent witnesses to the expertise of the designers of their day. They led to extremely elegant structures demonstrating an efficient use of materials. The achievement is that much greater if one considers that the designers in question were not able to fall back on Dutch know-how. A good example of a suspension bridge that still exists today is the bridge near to Pelaboehan Ratoe in West Java.

Fairly extensive reports on a number of the bridges can be found in the various technical details of the designs and assembly processes. Just three of those bridges are described in brief here below.

Bridges on the Loemadjang-Pasirian road (East Java)

The two bridges in question were the first to be constructed as stiffened suspension bridges. The original wooden bridges were destroyed by the consequences of the Smeroe volcano's eruption of 1909 and therefore needed to be replaced. At first the bridge design was based on the 'sagging' suspension bridges that had respective spans of 120 m and 50 m. The building work started in 1913 but during the construction work one of the abutments of the smaller bridges was washed away. It was then decided that the spans of both bridges should be 120 m and that the design should be based on the 'stiffened' type. Despite this rather drastic amendment to the plans the bridge's substructure was still completed in 1915. The two bridges were fully operational by 1917. Tabulated below are just a few of the characteristic features and statistics:

- Width between the handrails: 3.4 m with a wooden car deck and pedestrian areas.
- Traffic load: 350 kg / m^2 and a concentrated load of 7.5 tons with an axle weight of 5 tons.
- Cable supports on the abutments: stiff hoists with rolls for the cables.

Note that in the standardised versions these hoists were replaced with trestlework that hinged underneath without rollers.

- Cables: 2 x 4 cables 52 mm in diameter of the closed type and not prestretched (though that was considered) with adjustment equipment (both mechanical and hydraulic) in the anchor blocks.
- Cable heads: locally cast with an alloy consisting of tin, antimony and lead.
- Wind bracing: a K-beam in the lower edge with four wind bracing cables to the banks of the river.
- Components of the stiffening girder: maximum length 6.24 m and a maximum weight of 1.78 ton.

- Joints of the stiffening girders and hoists: riveted in the factory, linked with fitting bolts on site.
- Constructor: Dutch East Indian Enterprises.
- Cable supplier: Felten and Guilleaume in Cologne.
- Costs: the total expenses of the bridge over the Kali Moedjoer amounted to 83,364 guilders.

Note that these bridges were identical to the illustration on pages 46 and 47 of the bridge over the Air Tetap on the Manna-Bintoehan in Benkoelen (Sumatra) of around 1918.

The bridge over the Kroeng Geumpang (Atjeh)

This bridge was constructed in approximately 1930. Up until that time the transportation of goods required by the military garrisons in the region was provided by elephants. That was an expensive and time-consuming form of transport and so the creation of a road and bridge connection was viewed as essential. The main span was 44 m and the width of the bridge 2.50 m. It is perhaps worth mentioning that the stiffening girder was entirely made of wood.

The pedestrian bridge over the Malili river (Celebes)

This pedestrian bridge, completed in about the year 1933, was what was known as a 'sagging' type. What was most apparent about the design was the fact it made use of leftover material kept in storage. Here are just a few of the features and statistics:

- Width: 0.90 m between the handrails.
- Load: 80 persons.
- Bearing cables: fabricated from remains, linked together by means of safety clamps.
- Other materials: pylons and deck made of ironwood, the hanging components and rails were made of galvanised telephone wire that was 5 mm in diameter.
- Costs: the total costs amounted to 625.00 guilders excluding the excess materials.

Note: the ironwood parts were supplied and constructed by unpaid labourers.

The technical activities of the military troops 1946-1949

After the invasion of Pearl Harbour the East Indian government declared war on Japan on 8th December 1941. On 8th March 1942, after the Java Sea battle and nine days of fighting in Java, most of the Dutch East Indian forces capitulated. Northern Sumatra followed suit some three weeks later by which time Borneo was already occupied. Up until the time of the American invasion, part of (Dutch) New Guinea remained free of Japanese occupation. As the East Indian forces had withdrawn it had been partly a strategy of 'scorched earth politics' that had reigned – especially in the case of factories, refineries and such like – but despite that very little damage of any significance was done to the infrastructure in Java, partly because the occupation had been so rapid. The maintenance of the infrastructure was, however, neglected during the time of Japanese occupation but when the allied forces and the Dutch returned in 1945 it was discovered that the infrastructure had not been significantly disrupted. The return simultaneously marked the beginning of an armed conflict between the Netherlands and Indonesia that was to last until the time of the official transfer of power on 27th December 1949. The highlights of the struggle were the first Police Intervention that lasted from 21st July 1947 until 4th August 1947 and the second Police Intervention from 18th December 1948 until 31st December 1948. In the interim period a guerrilla war raged on. During that period of conflict it was the Indonesian republican forces that caused a great deal of damage to the infrastructure. Particular at the time of the Police Intervention many bridges were destroyed and roads were made impassable in all kinds of ways.

Especially in the more mountainous areas there were many culverts under the roads. These were attractive objects for the Indonesian forces to blow up thus creating gaping holes in the roads. Often temporary bridges would then be laid across the caverns in the road or they would be made passable with the help of mechanical equipment. At a later stage new culverts were created which were actually more suitable for all the heavy traffic than the original ones would have been.

Bridges that were made of steel horizontal girders mounted on screw pile trestles were destroyed by planting bombs behind the abutments and then detonating them. The abutments would then be rendered unusable and the steel girders would be bent or fractured. Where possible, the steel girders would afterwards be welded together again or a central pile would be introduced to make the span smaller so that the shortened horizontals could be reused. Once the abutments had been repaired and the girders had been put in place a teak wood car deck would then be laid.

The large steel bridges that the Indonesians tended to destroy were the road bridges for the simple reason that most of the Dutch military troops relied on road transport for all their manoeuvres. Lacking motorised transport the Indonesians themselves had to rely more on rail transport and so naturally they tended to leave the railway bridges intact. In certain cases and for some forms of transport the military sometimes managed to make the rail bridges suitable for combined road and rail traffic.

When the large steel bridges were destroyed it was often the case that one of the piles or abutments had been blown up, thus forcing the bridge to automatically collapse into the river. A few contorted steel bars then had to be straightened or strengthened and the support system temporarily restored so that the bridge could be returned to its original position. In a number of cases it was also necessary to first create a Bailey bridge which could then be used as a kind of scaffold and working platform for the assembly of a new permanent bridge. Whilst all the construction work was continuing traffic still had to be able to pass over the Bailey bridge. Once the new bridge had been completed it would similarly be used to then dismantle the Bailey bridge.

In the areas under Dutch control the repairing of all the damage done to bridges was the responsibility of the military troops which needed to be available in considerable numbers. In Java the Dutch military forces consisted of a number of divisions: the A division on East Java that will not be further discussed, the B division in Central Java and the C division or '7th December division' in West Java. The divisional commanders had at their disposal a military commando that led and co-ordinated the activities of the military troops stationed there which formed part of the Royal Dutch East Indian Army (Koninklijk Nederlands-Indische Leger, KNIL) and the Dutch Army and maintained the lines of contact with the civil authorities. The restoration of the various infrastructures that had been damaged involved intensive co-operation with the Public Works department, the State Railway Company, the General Dutch East Indian Electricity Company and companies such as Braat and De Vries Robbé.

Some of the repair work done in such constellations actually constituted prime examples of ways in which in exceptional circumstances, such as in war situations or in the face of natural disasters, it soon becomes possible to meet the needs of the population in general and of people needing to make use of any kinds of public facilities.

Perhaps one of the most spectacular repair jobs was the restoration of the supply pipe for the hydraulic power station in Djelok. When a bomb that had been placed under the pipe was detonated, all the water flowed out of the riveted together steel pipe that was around 2 m in diameter and the upsurging pressure completely flattened the pipe. The consequences of that deed were considerable for Central Java and so repair was extremely urgent. A number of metres of the flattened pipe were taken off to De Vries Robbé where the sections were repaired by means of welding. A new concrete pressure pipe was cast and the uprooted abutment was repaired. The State Railway Company took responsibility for the transportation and the Public Works department helped with the supplying of materials.

In much the same way all the repair jobs that needed to be done to roads, railways and water supply installations formed part of the second-line duties of the army, backed up of course by civilian experts from the relevant companies. The way in which duties were divided between civilians and military personnel was often dictated by the degree of safety in any given area. In 1946/47, for example, the repairing of the main Buitenzorg-Bandoeng road that crossed through the Poentjak pass turned out to be a combined effort undertaken by the army and Public Works. During the operation the less safe sections were taken care of by the military troops.

The army's main task was to facilitate the mobility of its combat forces. During combat actions, in other words during the first-line duties, their main tasks were to clear mines, dispel road blocks made of trees – what was termed chopping – sometimes involving trees that were more than a metre in diameter, remove all kinds of obstacles, fill or bridge holes in roads, complete emergency repair work to roads and bridges and create military bridges. In order to complete all those projects what was especially required was the kind of manpower and equipment that could be transported in convoys during any military operation. Specially prepared bridge material such as Bailey components could only be transported to a limited extent. Other essential materials could often be found on site by for instance excavating, breaking down existing structures or demolishing material from ruins. The vast majority of the material required to build bridges was kept in storage behind the battle lines in military depots consisting of warehouses and workshops.

Apart from storing Bailey material all kinds of material that could be used for semi-permanent bridge construction work was stored in those depots. This included, for instance, material required for steel bridge repair work that had been obtained from work done in co-operation with Public Works and construction companies. In the workshops everything that the troops could possibly need at the various job sites and which had to be purpose-built because of not being otherwise obtainable was manufactured. To take one example: when it emerged that wood could not be obtained through the usual channels it had to be sawn to the desired lengths in the army workshops and stored in a warehouse. In West Java

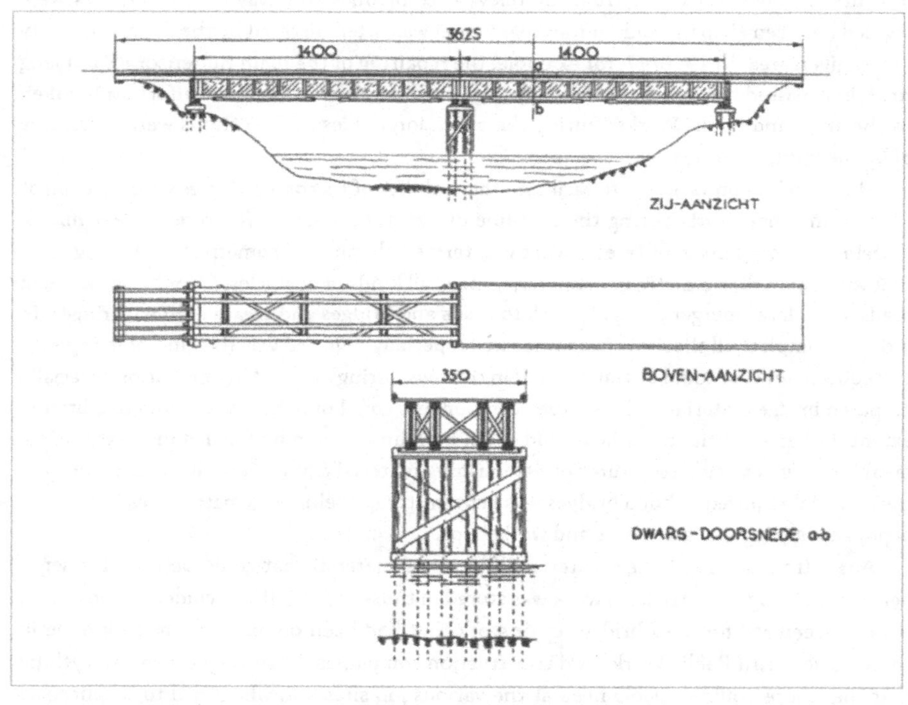

Bridge over the Tjisankoej not far from Kamasan (West Java), around 1947.

a military depot was created at Tjimahi in Bandoeng. In Central Java army depots were created in Semarang, Tegal and Poerwokerto after the first Police Intervention. In turn, those depots were supported by the general Army Depot situated in Meester Cornelis.

The available amount of Bailey material was relatively small in relation to the demand. For that reason it was necessary to improvise in all kinds of ways and save the required material when the advance routes opened up. All the Bailey bridges that had been created were as quickly as possibly replaced by permanent structures. As Public Works still hardly possessed the means to carry out such projects it was a task that was entirely left to the military troops. In that way more and more Bailey material was thus made available for ensuing campaigns.

The new permanent structures that were erected were often steel girder bridges or wooden bridges. The bridge built over the Tjitaroem at Tjilampeni was 64 m long and was suitable for traffic class 9 (i.e. ton) traffic. The wood used to construct the bridge was the standard sawn wood obtained from the military depot. The locally available teak was extremely well suited to the task in hand. A slanting bridge over the ravine near Soekoredjo was made of the same type of wood. One of the disadvantages attached to such wooden bridges was that the free span is so small that the bridges become very vulnerable during bandjirs.

Just to give an indication of the quantities of Bailey material involved here are a few facts. Up until 24th February 1948 Bailey bridges with a collective length of 1200 m were constructed by the army in Central Java including the 87 m bridge at Boemi Ajoe, which was the largest Bailey bridge in the whole of Java. In mid-May 1948 the military depot in Semarang had 500 m of class 12-18 Bailey bridges. As gradually more material was accumulated and new supplies were brought in the depot had 1000 m of Bailey bridge material by December 1948 in readiness for the second series of Police Interventions. In the first ten days of that campaign the army stationed in Central Java constructed no less that 600 m of Bailey bridges.

In West Java as well the military had its hands full with the task of repairing all the damage caused to infrastructure by the Indonesians. Because of the task of the Dutch troops and the style of combat of the opposing forces the emphasis for the army came very clearly to rest on bridge building. The area for which a battalion of C division '7th December' was technically responsible was actually larger than the land area of the Netherlands and it contained literally hundreds of bridges. After the first campaigns mounted by the police each military field company controlled an area that was roughly the size of an average Dutch province. It was a time when the military field companies and especially the various military depots were pushed to the limits to rectify all the damage that had been done. Together with Public Works and the Railways some 600 bridges were constructed with a total length of 3500 m and damaged road sections were repaired. When it came to the matter of repairing the roads, as much as possible use was made of local labour. It was the kind of support that was especially useful for the repairing of roads along mountain routes that had sometimes been dug up for considerable distances.

Finally, we shall give just a few examples of the bridges that were constructed by the field companies in West Java with the support of the depot companies. On the road between Bandjaran and Soreang there was a bridge that had been constructed on steel girders at Kamasan over the Tjisankoei and consisted of two 12.5 m spans. Shortly before the first

The influence of East Indian bridge building techniques in the Indonesian era: the concrete Radjamandala Bridge (West Java) designed by Professor R. Roosseno Soerjohadikoesoemo and completed in 1979.

Police Intervention the bridge had been completely destroyed by dynamite. The steel girders were completely contorted, both abutments had been utterly destroyed and the pier had been damaged. The pier and abutments were afterwards replaced by teak wood trestles on driven piles. For the actual bridge section, eight 14 m long and 1.25 m high wooden closed side horizontals were fashioned by the company's staff in the depot workshop.

Not far from Batoedjadjar on the Tjimahi-Goenoenghaloe road a Bailey bridge was built over the Tjitaroem that had a raised car deck. The bridge was 41 m long and it served to replace the destroyed central span of the existing bridge. The special parts required for that irregular Bailey bridge design such as half panels and wind bracings were, once again, produced in the depot workshops.

Over the Tjitandoei near to Indihiang on the road to Bodjongdjenkol there was a truss bridge of the type that was very common in the East Indies. The bridge was 3 m wide and had a span of 34 m. During the first Police Interventions it was very badly damaged by explosives. The final sections were severely skewed and the entire bridge had been twisted on its lengthwise axis by 45 degrees and was 100 cm out of line with the central axis. Both the abutments had been completely destroyed. In order to restore the bridge, both the end sections had to be regarded as lost which meant that the available remaining span was reduced to 27 m. That difference in bridging length therefore had to be compensated by the new abutments. The experts at Public Works estimated that the resultant narrowing of the river flow profile would be acceptable. The Public Works department also provided support during the reconstruction work and took responsibility for the building up of the new abutments.

Between 1910 and 1920 the Dutch East Indies seemed to be caught up in a veritable port fever. Throughout the archipelago new breakwaters, quays and landing stages were being constructed at a tremendous pace. The East Indies had some 500 harbours, eight of which were large and twelve of which qualified as medium-sized; together they constituted part of the international shipping route network. In those days the actual developing of harbour works was quite a recent phenomenon. Until well into the nineteenth century, ports had really been little more than simple natural anchorages nestled in the shadow of a bay, reef or island and invariably situated near to a river estuary, where boats could simply drop anchor. The only known harbour works were long piers that jutted out into the sea to ensure that the navigation channels remained deep enough. In the last quarter of the nineteenth century, though, there was evidence of a turnaround and the construction of Tandjong Priok, the new Batavian harbour, could be seen as the breakthrough. Tandjong Priok, the first ocean port of the Dutch East Indies, was entirely designed on the drawing boards of colonial administrators and engineers. The construction of Tandjong Priok was to mark the beginning of a change in attitude towards the purpose of harbours and it was to herald a period of active intervention in harbour construction works throughout the East Indies.

It is curious to note that in the historiography of the Dutch East Indies and Indonesia there is no research into harbour affairs. In the maritime tradition it is the matter of ships and shipping companies that tends to be exclusively considered and what tends to be central to economic research is shipping movements and trade. Even in historical studies into transport systems, harbours remain strangely unmentioned or are, at best, simply viewed as a fact of life. If one bears in mind that at the end of the nineteenth century there was strong public interest in harbour development then this is, indeed, a curious state of affairs. Politicians and administrators, trade associations, ship owners and, not least, engineers all became involved in the debate which was, to a significant extent, conducted publicly via the press.

Within the context of a project known as The Java Sea Region in an Age of Transition 1870-1970, three main studies into the harbours of Tandjong Priok, Semarang and Soerabaja are currently being carried out. The studies focus on the infrastructural, technological and economic developments in the Java Sea region which is seen as the central maritime zone within the Indonesian archipelago.[2]

The issue that remains central to this chapter is the development of harbour works in the Dutch East Indies in the nineteenth and twentieth centuries and the part played by Dutch East Indian engineers. Particular attention will be given to the history of Tandjong Priok as it was the harbour that was central to the rise of East Indian engineers and the Department of Civil Public Works (Departement van Burgerlijke Openbare Werken, BOW). In conjunction with the development of an East Indian harbour system the creation of other major harbours such as Soerabaja, Semarang, Belawan, Padang and Sabang will obviously be described as well but merely in broad outline.

Detailed technological discourse will not be entered into here, tempting though that may be in a study of engineering works, as the main aim of this book is for it to be especially a descriptive analysis of the development and implementation of technology within the historical-social framework.

This introduction will first be followed by a brief explanation of the concepts used in this exploration of harbour history, together with a number of details relating to the context within which that same harbour history was formed. The remainder of the chapter comprises two parts. In the first part it is the rise of the engineer in harbour construction and management that is central. There I shall start by explaining what kind of a state the Dutch East Indian harbours were in during the first half of the nineteenth century. I shall then go on to discuss the conflict that emerged around 1870 concerning harbour improvement, especially where the harbour of Batavia was concerned. Afterwards I shall sketch how the works in Tandjong Priok were executed and how the East Indian engineer was very much in the forefront. The final paragraph of that section will present a brief overview of the physical and economic situation in relation to all the remaining East Indian harbours round about the turn of the century. The second part of this chapter will deal with the emergence of the professionalisation of the engineer and the harbour system. What will be central in that part of the chapter will be the physical and organisational side of the reorganisation of East Indian harbours as prescribed in 1910 by the Kraus-De Jongh committee. I shall consider those recommendations and discuss their precise relevance to the East Indian harbour system. We shall see how in the space of just one decade it was possible to witness the centralisation of harbour policy under the Department of Public Works. Several giant leaps will then be taken so that harbour history after 1920 can be described. The chapter will conclude by placing East Indian port history within the wider framework of technological development and state formation.

Background

In the seventies (of the 20th century) the British transport geographer, J. Bird, introduced the 'Anyport model' that traced the historical development of harbours in phases. Bird described port evolution from the angle of the changing facilities offered by any harbour, ranging from rudimentary provisions for storage, loading and unloading to large-scale reception installations for tankers or bulk carriers.[3] The British-oriented model was subsequently frequently used in similar port studies and elaborated and modified in different variants. Another phase-model for port development is the one that comes to us from the French geographer, A. Vigarie, who based his notions on trading activities and economic cycles. Vigarie concluded that over the course of time the port tends to move from its original spot, usually in the centre of the city, to a location where there is more space for deeper basins and to thus develop the capacity to receive larger vessels. Only a section of the urban apparatus follows the harbour to its new destination and in the search for space the harbour gets cut off from its original location.[4]

In his study of the harbour and the city, H. Meijer takes the function of the harbour as a starting point for a phased model of harbour evolution including: the entrepôt port, the transit port, the industrial port and the distribution port. It is not only the character of the harbour that changes but also its location in the city and even the structure of the city itself.

Traditionally harbours have always been places where the city meets the international transport network. The entrepôt harbour was usually situated in the heart of the city and the quays where goods were loaded, unloaded, stored and traded constituted a part of the urban public space. For political and military reasons the city would monitor that junction

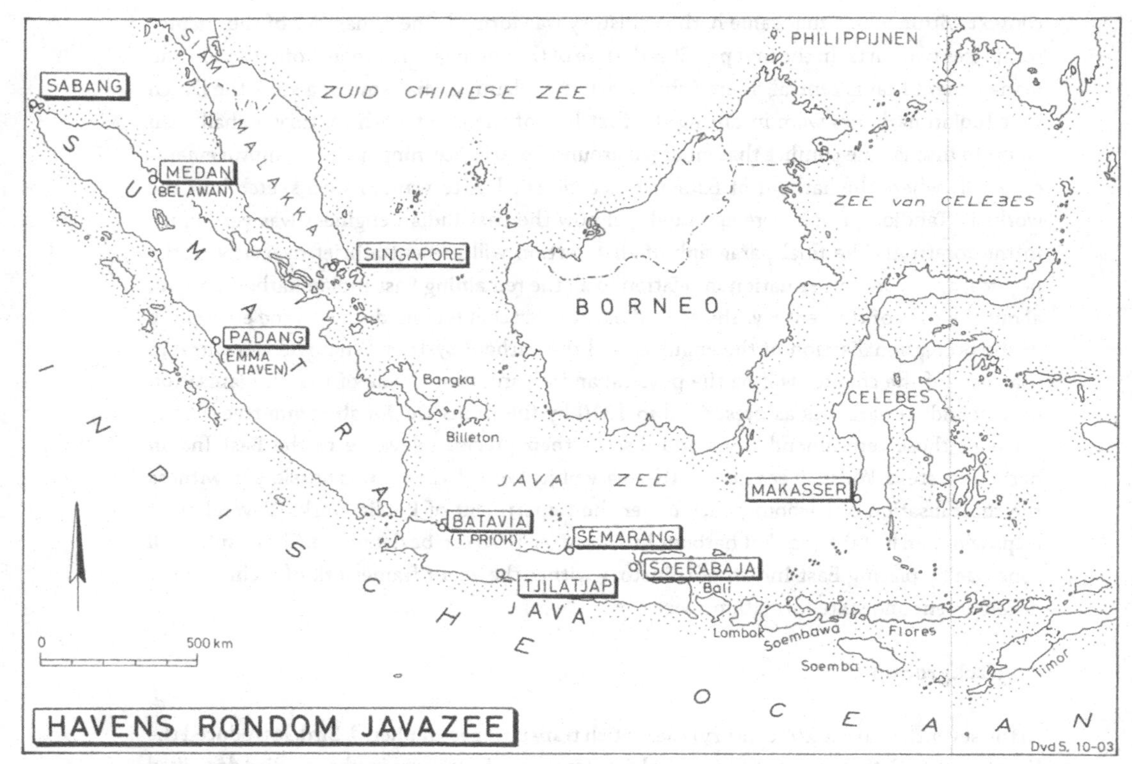

A map of the main harbours in and around the Java Sea region.

in many respects and a certain degree of protection would also have to be guaranteed. The harbour quay operated as an intermediary between the international transport and trade network and the city's local political and commercial network and was ruled by an urban elite. Many important administrative, trading and religious institutions would be situated on the quayside. In the nineteenth century, influenced by the industrial and transport revolution, there was a fundamental transformation in the position and function of the city and the port. In the first place there was the rise of new transport networks devoted to the transporting of raw materials and products between the new industrial centres. In that way the need arose for a new type of harbour – the transit harbour – that did not so much form a centre for international trade as rather a transhipment point. In that way the traditional interweaving of international and urban infrastructure was, to a large extent, lost. Apart from anything else, the established urban trading elite was not intimately linked to the emerging transport economy so that its grip on the harbour weakened. The increasing dimensions of steam ships, the links with the railways and the altered nature of port activities (transhipping instead of entrepôt requirements) were the factors that determined the needs for the new port situated on the city's periphery. Freight would then find its way to various destinations, including the city, along the new infrastructures of roads, canals and railways. This was the phase, then, when the functions of the city and the harbour became detached from each other. In the twentieth century this trend continued and the harbour turned into an industrial and more or less autonomous locus outside the urban boundaries. Industries would be established in or in close proximity to the harbour area where goods would no longer be just transported but also produced. In the most recent period, the second half of the twentieth century, the modern harbour has turned into a distribution centre; into a junction for logistic organisation and telecommunication. In that way the harbour is therefore winning back its place in the urban landscape.[5] This most recent harbour phenomenon does, however, extend beyond the scope of this chapter.

The way in which colonial harbours are to be typified diverges from their Western counterparts because in many ways their development adhered to a different pattern. Colonial harbours often formed part of a maritime and land infrastructure that had somehow been superimposed upon existing trade and transport patterns. Colonialism was indeed the whole raison d'être for colonial harbours and so understandably they were typically harbours that were geared to export activities.[6] Much the same applied to the harbours of the East Indies.

In the first half of the nineteenth century trade and shipping in the East Indies was dominated by the Dutch Trading Company (Nederlandse Handelmaatschappij, NHM), an enterprise that had been established and controlled by the government. The Cultivation System, introduced in 1830, which compelled Javanese farmers to grow for export was responsible for a sharp rise in the export of products from Java. For some four decades that Cultivation System was to result in an annual colonial balance surplus.[7] The trading of the Javanese products emanating from the Cultivation System was entirely in the hands of the NHM, which only used ships that sailed under Dutch or Dutch East Indian colours. According to estimates, between one third and two-thirds of the ships that sailed to Java were actually chartered by the NHM.[8]

That closed system which accounted both for the cultivation of export crops and for the transport thereof under government supervision, produced good results but began, after a while, to show signs of decay. At first, the system certainly stimulated navigation and shipbuilding activities but its weakness was that it did little to encourage modernisation. Because of the certainty of the freight contracts there was less incentive to increasing the speed of goods transhipment in the harbour or the sailing time to the Netherlands.

The advent of steam power was to herald radical changes in shipping. Steam ships were suddenly no longer dependent upon wind power which meant that fixed routes could be adhered to with fixed navigation schedules. The transition from sail to steam was not very radical in the Dutch East Indies. It was in 1825 that the first steam ship actually entered into service in the East Indies but it was also to remain the only steam ship until 1840. Until the sixties of the nineteenth century sailing took precedence. On the oceans it was only after 1870 that steam power came to the fore. The Dutch fleet that sailed to the East Indies gives us a good overview of the development. In 1860 a mere 12 of the 392 East Indian sea-going vessels were steam ships. By 1880, though, that number had grown to 43 as opposed to 368 sailing ships. In 1910 some two-thirds of the entire fleet (161 ships) was steam-powered, whilst there were still 88 rigged ships that sailed to the East Indies. The shift becomes more sharply accentuated if one takes as the basis for one's calculations the volumes of freight transported: in 1860 steam ships were only responsible for 4.7 percent of the total freight volume but by 1880 that had risen to 28.7 percent. In 1910 steam transport accounted for 85 percent of the total East Indian shipping volume.[9] The reason why the volumes of freight shipped had risen so sharply was because of the introduction of steel to the ship construction sector.

The opening of the Suez Canal in 1869 represented a turning point in steam navigation. By linking together the Mediterranean Sea with the Red Sea the canal literally halved the travelling distance between Europe and Asia but it was only accessible to steam ships. Because of steam power, ships no longer had to depend on the wind which meant, in turn, that strict schedules could be introduced. This meant that different working methods had to be adopted on the actual ships and in the harbours. Shipping companies wanted to ensure their ships of moorings in the harbours. With the increasing size of ships came also the need for deeper harbour basins and longer quays. Other things that were introduced were new mechanical devices such as steam cranes, winches and transporters, all of which made it easier for freight to be loaded and unloaded faster and more safely. It was at that time that stevedores first put in an appearance on the quayside.[10]

It was not only technological advances in navigation that were to dictate the pace of change in East Indian harbours but also infrastructures on dry land where steam power had initiated a revolution. The railways provided access to an expanding hinterland which, in turn, sought shipment opportunities for its products via the ports. The backdrop to the East Indian harbour history was the political and economic integration of the East Indies. Throughout the nineteenth and twentieth centuries the colonial government did all in its power to integrate the East Indian archipelago into one cohesive political and economic system. In the vast archipelago steam shipping was – notably as of 1870 – an important factor in state formation as J. à Campo shows in his monumental study of the Royal Netherlands Packet Shipping Company (Koninklijke Paketvaart Maatschappij, KPM).

Operating according to scheduled steam services between the East Indian islands, the KPM transported mail and freight and provided a means of transport for colonial civil servants and military troops. Even though the KPM was a privately-run company the packet boat services were initiated, stimulated and supervised by the colonial government. The expansion of steam services was closely allied to the military and administrative expansion set in motion by the colonial government after the second half of the nineteenth century. The growing KPM network also linked up with the integration of the Outer Regions (or Outer Islands) into the colonial economy and the world trade network.[11]

The colony's economic development reflected two simultaneous processes that served to reinforce each other. In the first place, the colonial government adhered to an active economic policy which, after 1870, provided much more space for private initiative. In the second place, as a result of rapidly expanding export production (including coal, tin, rubber and tobacco), the economic importance of the Outer Islands also grew fast but all the freight and finance flows were channelled to Java, thus reinforcing its central position in the colonial system. According to H. Dick, the colonial administration succeeded in turning the archipelago 'inside out' if one bears in mind that up until that point in time the Outer Islands had really been more oriented towards other South-East Asian countries than towards Java. In that respect, the economic integration of the East Indies was to lay the foundations for the later national economy of Indonesia.[12]

In this whole process the colonial government ascribed a key role to the maritime sector. Especially in the post-1890 era, KPM's shipping lines, and transit and tariffs policies were all geared to channelling freight transport through Dutch East Indian ports, reinforcing a national 'element' in the colonial economy. The underlying motives were not only economic but also political: it all served to confirm the external sovereignty of the colonial state. In geographical terms, the scheduled packet shipping services turned the East Indies into a close-knit geographical entity[13] in which the ports were the main nodes of the network. After 1870 the colonial government endowed the ports with renewed significance.

J.A.A. van Doorn once described the late-colonial history of the East Indies as a 'colonial project'. He recognised in the Dutch colonial regime a kind of instrumental rationality that was based on cost-benefit analysis. The Netherlands operated as an active and interventionist 'project organisation' keen to give shape to an artificial, colonial society. Particularly when contemplating railway and irrigation development, Van Doorn saw the engineer as one of the project leaders.[14] When Tandjong Priok was constructed the ports became an instrument within the colonial project. Whilst for the greater part of the nineteenth century colonial governmental measures with regard to harbour affairs were more or less ad hoc, port development was to gain a more programme-steered character after 1880. The ports constituted a prime terrain for the engineers to conquer and were thus to leave their technocratic mark on port development and management.

Natural harbours and rudimentary harbour works

Up until the mid-nineteenth century the East Indian archipelago consisted mainly of natural harbours formed in a river mouth or creek. If the river mouth was navigable then ships requiring little draught could sail to the trading place. Where it was not possible to sail inland, ships would moor on a roadstead that often provided shelter and was often situated behind a reef, peninsula or island.

In some places more elaborate harbour works were constructed with two piers jutting out into the sea from the coast. Between the piers a navigable channel was maintained, the so-called harbour channel. In view of the shallowness of the waters of the Java Sea coastal zone, such piers were particularly suited to Java's northern coast.

As early as in 1634 two 800-metre long harbour piers were constructed in Batavia. Due to the perpetual silting up and shifting of the coastline[15], the dams had to be continually lengthened so that by 1817 they were at least 2 km long. By 1874 the harbour dams projected some 4 km into the Java Sea which meant that they had been lengthened by no less than 1875 m in less than 60 years or, an average of 32 m per annum.[16]

There were other harbours in Java, such as Soerabaja, Semarang and Pasaroean, which had a harbour channel situated between two piers that connected with the river. The formation of sandbanks in the inlets in front of the entrance to a harbour canal or in the river mouth was an acute problem for shipping and one that was frequently alluded to in the various reports on Javanese harbours. Large and medium-sized vessels, such as the European East India ships or the Chinese junks would drop anchor on the roadstead whilst smaller vessels, the so-called lighters or loading *perahus* (proas) would then ferry freight and passengers to and from the quayside where, in favourable circumstances, they were able to moor on simple mooring posts or landing stages. In the Straits of Madoera, not far from Soerabaja, the canoe-like *kromans* plied the waters. In Semarang the proas were known as *banting* and in Batavia they were called *tjoenia* or *sampan*.[17]

Similarly around all the other islands away from Java the various harbours were nothing more than natural mooring spots. Makassar, for instance, which constituted the geographical port between the west and the east of the archipelago had a splendid anchorage that was sheltered by several coral islands. There, too, proas were used to provide transport between the deeper mooring spot and the quay. Even Ambon, which since the early days of the United East India Company had been the trading centre for the Moluccas, had its own natural bay where ships could safely anchor.

In the early days, the Dutch colonial government had no particular port policy and government intervention was mainly confined to maintaining the depths of the harbours and providing customs facilities in the harbours that were open to international shipping routes.[18] In the larger ports there were always two customs offices known as the 'Kleine Boom' and the 'Groote Boom' (i.e. the smaller and the bigger customs posts). The Kleine Boom was the first harbour office that passengers had to report to and it was the point where the proa skippers had to hand over the ships' papers. At the Groote Boom the goods would be stored in customs sheds so that the import and export levies could be determined. The customs officers fell under the Department of Financial Affairs whilst it was the harbour master who was responsible for the overall supervision of the harbour and he was a Naval representative. The harbour master had to regulate shipping traffic and, in day-to-day terms, he was responsible for the physical upkeep of the harbour.

Round about the year 1850 the physical state of the harbours was exposed to much criticism. At that time shipping movements and international trade were increasing fast and in the harbours that was leading to congestion, both in the actual harbours and at the customs offices. In the East Indies a heated debate flared up because, according to critics, harbour affairs had been neglected. Already since 1839 the government had been levying taxes of five percent on import and export rights specifically in order to be able to maintain and improve the harbours. According to the *Tijdschrift van Nederlandsch-Indië* (Dutch East Indian Journal), up until the year 1861 that tax alone had brought in a sum of 6,210,670 guilders for the inland revenue but despite that the maintenance of the harbours had been so badly neglected that trade and shipping activities were leading to great dissatisfaction and shipping was even becoming perilous.

> *The neglecting of the harbours of Java not only contravenes the direct interests of the government which is itself a major cultivator and trader in the East Indies but it also contravenes the interests of trade and industry in general which are suffering from the problem and are repressed by it and, ultimately, it also constitutes an obvious injustice.*[19]

The discontent surrounding East Indian harbours was also fed by the success of Singapore on the other side of the Java Sea. That port city, established in 1819 by Thomas Raffles, had managed to overshadow Batavia as the main transit harbour and trading centre in South-East Asia. Singapore had three main advantages: firstly there was its strategic position on the southern entrance to the Straits of Malacca, the gateway to the Indian Ocean, the

Java Sea and the Chinese Sea. In the second place, Singapore was able to benefit from the Malaysian peninsula as its productive hinterland. In Singapore, regional transport routes linked up with international shipping routes. In the third place it was an excellent natural harbour which, in combination with its several deep sea-docks, was able to provide efficient and cheap loading and unloading facilities. Apart from these three major assets Singapore was able to profit from its status as a free port which meant that ships were not required to pay any dues or levies.[20]

In order to find an answer to the competition emanating from Singapore the East Indian administration decided to also give a number if its harbours similar free port status and the one that was by far and away the most important in this respect was Makassar.[21] Apart from extending harbour dams and dredging the harbour channels, the East Indian government did nothing to improve the physical state of its harbours. The continual silting up and formation of sandbanks served to greatly hamper shipping movements, notably in the Batavia and Semarang areas. The harbours were so difficult to access that freight traffic between the mooring spots and the harbours regularly came to a standstill. During the wet west monsoon season, which lasted from December to February, there were always a lot of breakers around the sandbanks, which made it virtually impossible for the proas to shuttle back and forth. On such days the harbour master would raise the blue flag to indicate that shipping movements to the boat anchorage point had been completely halted. It was quite common for the newspapers to carry reports of proas that had nevertheless braved the sea only to have capsized and in extreme cases crew members sometimes lost their lives beneath the waves as well. In the dry monsoon period the channels were so shallow that even sloops, which required a minimal draught, were unable to ply the waters.

In Batavia the sorry state of affairs surrounding the harbour there came to light in a rather painful way when in 1861 the British viceroy, James Elgin, got stranded in the mud of the harbour channel. Before the eyes of the entire reception delegation Elgin was finally brought ashore with great difficulty. As this incident was given some prominence in the East Indian press, even the Dutch Lower House in The Hague started to intervene in the harbour question. Indeed, in the very same year, the governor-general also got stranded in the harbour channel while endeavouring to go ashore in Batavia.[22] After that incident the harbour channel was thoroughly excavated with mechanical dredgers and by hand. By 1864 the channel had been made so deep that there was even space for a sizable Danish steam barge to drop anchor in the channel.[23]

It was not only the physical state of the harbours that gave rise to concern but also the organisational structures that created a barrier to trade and shipping. The turnaround time for ships – in other words the space of time within which a ship could be loaded and unloaded – depended very much upon the nature of the available proa fleet. Up until 1854 the proa fleet was strictly controlled by the government, in much the same way that it had been in the time of the United East India Company. After 1854 the proa services were gradually privatised. Several Batavian trading firms banded together to establish the Batavian Proa Ferry Service and so took over most of the registered proas in the city. The Proa Ferry Service was the first private enterprise to provide links between the city and the coast. The ferry company also deployed the Tjiliwong, a small and shallow steamship, for passenger services. In 1861 a second transport service began to compete in the form of the

Batavian Tjoenia ferry. For many years both companies continued to operate an oligarchic-like regime: they successfully warded off new rivals, perpetuated an artificial scarcity and thus made it possible to charge high fares. Ship owners and other traders were compelled to remain with them because the alternative option of organising their own proa fleets with suitable and big enough crews would have consumed so much time and money that it would not have made sense to take the loading and unloading into their own hands.[24]

In the newspapers there were many complaints about the long waiting times at the proa ferries and about the transporting of goods that was so careless that frequently consignments ended up in the sea or disappeared in other ways. By the same token, the passenger transport provided was not exactly comfortable. The Tjiliwong steam ferry lacked the necessary safety requirements for passengers. The Batavian press, which was showing increasing interest in harbour matters, surmised that the problem of the poor service offered by the proa companies was a direct result of a lack of competition.[25]

All the customs formalities also added to the delays as was borne out by the long rows of proas moored at the Groote Boom each day. In Batavia – like in Semarang and Soerabaja – the customs offices had a shortage of inspectors and storage room. The proas were kept waiting a long time before they could finally sail further with their freight. In the roadstead, the ocean-going vessels were, in turn, kept waiting by the proas. A *djoeragan*, a proa skipper, described in the newspaper *Bataviaasch Handelsblad* (Batavian Trade Journal) how busy he had recently been clearing customs: the paper work, the checks and the estimating of the value of the freight had, he reported, all taken days.[26] The director of the Department of Financial Affairs readily admitted that the customs office had a shortage of space but he also suggested that the traders really exaggerated the situation and that because of their inefficient work approach they were also partly responsible for all the delays.[27]

Between 1850 and 1870 the harbour question regularly came to the fore. One time the harbour in Batavia would be at the receiving end of all the criticism and the next time the harbour of Semarang would come under attack. The Semarang Chamber of Commerce repeatedly appealed to the authorities to urgently improve the harbour which had become very silted up. However, as the *Java Bode* (Java Post) newspaper rightly asserted, for decades the colonial government had only wanted to patch things up.[28] It was clearly evident that the harbours of Java would have to be improved and so it was that already in the fifties the first plans were drawn up. In that era of Cultivation System agriculture and credit balances the colonial government kept state investments down to a minimum which meant that no fundamental harbour improvements could be made. Maintenance was restricted to the deepening of harbour channels and to the extending of piers. The former marine officer, P. Bruining, whose plans for the harbour at Batavia had been rejected characterised the colonial government as one that constituted nothing more than 'a calculated status quo'.[29]

Princely intervention

It is true to say that by the 1860s the Java harbour problems had found their way onto the political agenda, both in The Hague and in Batavia but all the plans put forward were abandoned because the colonial government was not prepared to make the required substantial investments. In 1871 the harbour question was suddenly given a heavy political slant because of princely intervention. That was the year when the Dutch Steam Shipping Company (Stoomvaart Maatschappij Nederland, SMN), established a year earlier, started up its first scheduled steam service between the Netherlands and Java. The main problem confronting that shipping company at that time was lack of adequate harbour facilities, both in the Netherlands and in the East Indies. Within a matter of a few years the accessibility of the harbours of Amsterdam and Rotterdam was improved thanks to major large-scale engineering projects such as the construction of the North Sea Canal (for Amsterdam) and the New Waterway (for Rotterdam). The SMN noticed that any time gained during steamship voyages was rapidly lost through the slow unloading of the ships in the East Indian harbours. It was not just the case that much time was lost transferring the goods to the small proas but all the bureaucracy surrounding the clearing and handling of goods at customs was responsible for such huge delays that the SMN agent in Batavia complained that work that took hours elsewhere could be translated to days in Batavia.[30]

Prince Hendrik, younger brother of King Willem III, just happened to be the honorary chairman of the SMN. Hendrik, who was nicknamed 'The Navigator', was an important promoter of the interests of merchant shipping and he was also the inspiration behind the new 'national' steam shipping company which he had personally heavily subsidised. In June 1871 he wrote a pressing letter to the governor-general in the Dutch East Indies in which he drew attention to the deplorable state of Javanese harbours, not least the one in Batavia. To the prince's mind, the main Javanese harbours were insufficiently equipped to meet the requirements of modern shipping.

> *I must admit that I am overwhelmed by a feeling of sadness whenever I am reminded that at the anchorage in Batavia there is nowhere for private steam ships to dock or be repaired and that the loading and unloading is still so slow and laborious that much useful time is wasted.*
> *Apart from the fact that* Semarang *reflects the same kind of deficiencies and lack of facilities, there is only one foot of water at the entrance to this harbour.*
> *The anchoring place of the main colonial centre and the stockpiling harbour for Central Java are exposed to the high seas for much of the year as a consequence of the prevailing north-westerly winds of the rainy monsoon season, all of which leads to significant delays in trading whilst every year many people are killed in the heavy swell just outside the harbour entrance.*
> *The* Soerabaja *anchorage ground is naturally well protected and –apart from its dry dock – it has private workshops where the steam-powered machinery of the merchant fleets can be repaired. What must, however, be said of this harbour is that it is getting more and more inaccessible to large steam ships.*[31]

What was so striking about Hendrik's letter was its sharp tone and the fact that it ignored the conventions of etiquette usually observed by those in positions of authority. Hendrik suggested that the East Indian government should have put more effort into improving the harbour situation and stimulating the contributions made by private enterprise. He then turned to the East Indian governor-general and demanded swift action without any further delay. The prince indeed showed great faith in the power of technological achievement.

> *Through scientific innovation many things that would have seemed impossible in the past*

have recently been realised. This is a fact and something that we are confronted with on a daily basis and I am sure that Your Excellence will moreover recognise this as such because of the mechanical means driven by the steam power which we nowadays have at our disposal and which has indeed been deployed in various engineering works such as for the draining of the lakes of North Holland and for the excavating of sand wastes and granite mountains, all of which can become a major barrier to speedily designing and implementing the necessary improvements in our East Indian harbours.[32]

Priority had to be given to improving the Batavian harbour and Hendrik even put forward a plan of his own which involved creating a 2000-metre-long breakwater opposite the entrance to the harbour channel. Behind that dam it would be necessary to dredge the seabed to a depth of some five fathoms (i.e. approximately 8 m) so that large sea-going vessels could moor there safely. The excavated ground could be used to create a new terrain to the west of the harbour channel for a dockyard and wharf intended for the maintenance and repair of ships. A coaling-station could be situated on the western harbour jetty.

Hendrik was keen to receive a speedy reaction concerning the feasibility of his plan or to learn of any possible alternatives; three months later the prince once again inquired, by letter, how things were progressing.[33] The princely pressure was certainly effective. Hendrik's harbour letter was promptly furnished with comment and sent on to the Department of Public Works and to the Batavian Chamber of Commerce and Industry. Both organisations ultimately rejected the prince's plans and submitted instead their own proposals.[34]

A new harbour for Batavia

The governor-general sought the advice of a number of bodies and individuals. He had meanwhile become very conscious of the high degree of public sensitivity to the harbour issue. In the end, three new harbour plans were presented. The first plan came from the Batavian Chamber of Commerce and Industry and involved the creation of a new harbour basin to be situated right next to the existing harbour channel and close to the Kali Besar trading centre. The plan's ingenuity resided in its simplicity: by constructing a third pier at a certain distance from the existing harbour jetties a harbour basin would thus be created. Ships would be able to moor on the new pier, freight could be loaded and unloaded and passengers could embark and disembark. On the other side of the moored ships, the waterside, consignments could similarly be transferred, like in the past, via loading proas which would then pass down the harbour channel to the lower city area of Batavia.

Those who opposed the plan argued that any harbour in that location would rapidly become silted up so that within the space of a couple of years the basin would be too shallow to accommodate ocean-going steamships. Frequent dredging would thus be required and that would make the whole plan an expensive option. The other two plans that had been submitted were based on the similar principle of constructing a harbour in a different place within the bay of Batavia.

One of the locations was in the west of the bay, near to the island known as Onrust. By constructing a connecting dam between the coast and the islands of Onrust and Kuiper, a spacious harbour basin could be formed. The dam would provide shelter from the wind

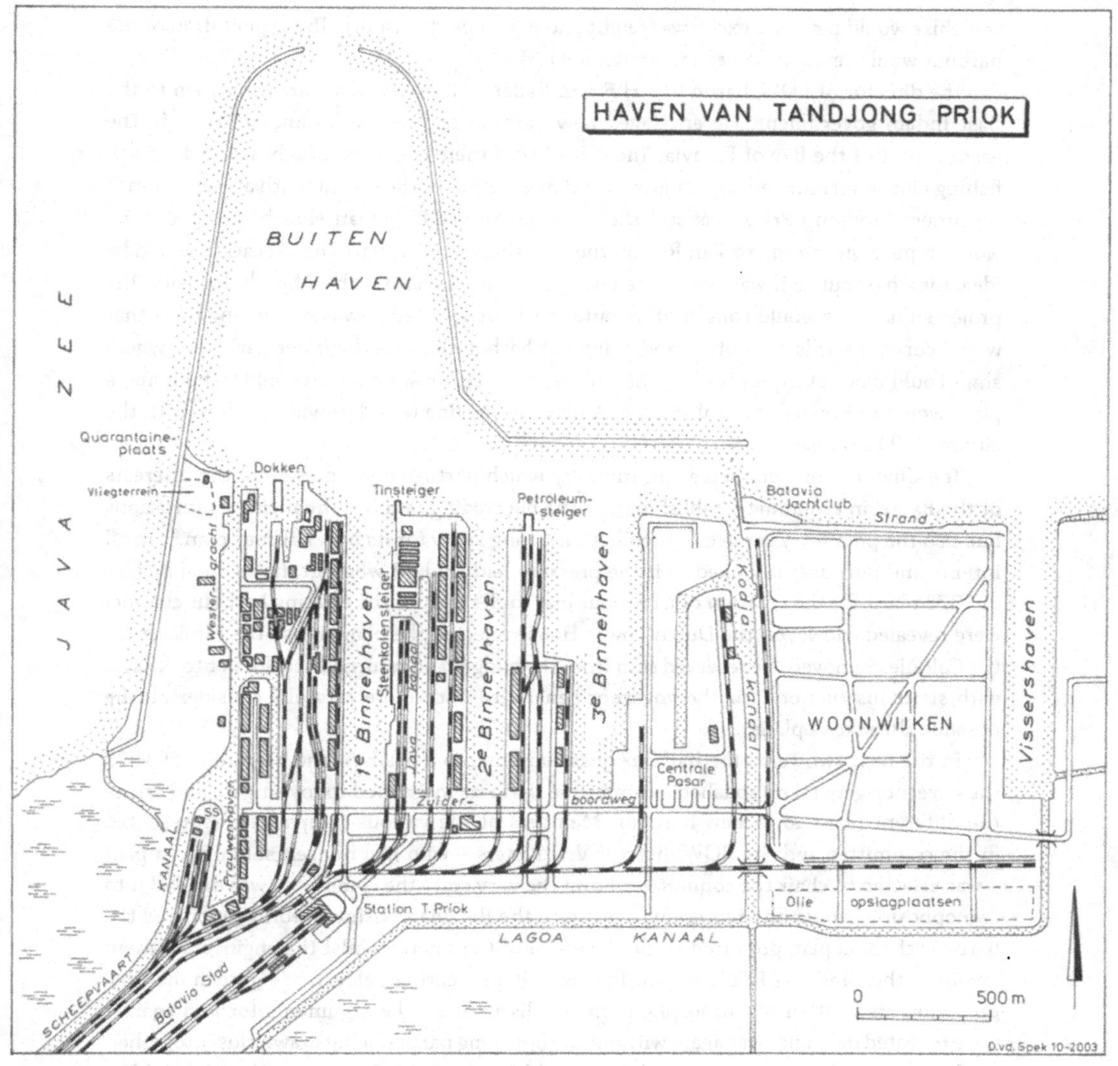

A map of the Batavia harbour, Tandjong Priok, around 1930.

and waves for boats anchoring there whilst at the same time serving as a loading and unloading quay. The distance from the new harbour to Batavia was 20 km. To guarantee good connections a railway line could be constructed. It was, of course, no coincidence that this was the plan dreamt up by the Netherlands East Indian Railway Company (Nederlandsch-Indische Spoorweg Maatschappij, NISM) which also held the franchise for the railway line between Batavia and Buitenzorg (the seat of the governor-general). The harbour line would be a welcome addition to the expanding railway network of the NISM, especially as the franchise would facilitate exclusive freight and passenger transport. The exploitation of the harbour would remain a government-steered task.[35]

The director of BOW, baron W.E.H.F. van Raders, put forward a plan of his own to the East Indian government. He envisaged new harbour facilities at Tandjong Priok, in the eastern part of the Bay of Batavia. The spit of land there consisted of a beach and a small fishing village surrounded by mangroves and marshland. In the seventeenth and eighteenth centuries, Tandjong Priok was still the place where the European elite based in Batavia would repose. According to Van Raders the conditions in that particular location would be ideal for a harbour as it was one of the few spots that was not troubled by silt deposits. The projected harbour would consist of an outer harbour situated between two long piers that would curve towards each other and a dug out harbour basin – the inner harbour – where ships could moor at a quay wall. In the surrounding terrain warehouses could be built and a place would be created for coal storage. A new railway line would provide the link with the city some 9 kms away.[36]

The Chamber of Commerce and Industry which particularly represented the interests of the Batavian economic establishment – the big trading retailers, financial organisations but also the proa ferry services – mobilised a strong lobby for its own harbour plan.[37] To all intents and purposes it looked as if the pressure had paid off when in the colonial budget of 1873 plans for the creation of a harbour just to the west of the existing harbour channel were revealed. However, the Dutch Lower House rejected the plan and so the Minister for the Colonies removed the reserved sum from the budget and returned the dossier to Batavia with strict instructions for the governor-general to once more carefully consider all the possible alternative plans.

In the next two years the harbour question came to a head. At the beginning of 1873 the governor-general established a committee that was instructed to investigate and assess the different plans for a new harbour. Members of the various camps were represented in the committee and the BOW director Van Raders – who had himself submitted a plan – was selected to chair the committee. From the very start the committee was divided into two opposing camps: the representatives from the Batavian business world supported the harbour channel plan produced by the Chamber of Commerce whilst the engineers were in favour of the Tandjong Priok option. The committee meetings, eleven in eighteen months, proceeded in a rather icy atmosphere. In the discussions, the arguments for and against were repeated over and over again without the opposing parties being drawn closer together. A whole range of economic, technological and hydrological information was amassed but there, too, the various visions and reports tended to contradict each other.[38] During the technological discussions it was hard to conceal the fact that it was really a conflict of interests that underlay the whole harbour choice debate. The Chamber of Commerce rightly feared that having a brand new harbour in a completely different place would constitute a direct threat to the proa ferry services. The other problem was that the prominent position of the commercial establishments and storage enterprises in the Kali Besar area would immediately be weakened as the construction of another seaport away from Batavia would open up all kinds of opportunities for newcomers, thus undermining the somewhat monopolistic position of the established concerns.

It was particularly the NISM that constituted a formidable rival for the proas. In view of the fact that operating profits had been rather disappointing the harbour franchise would be very welcome for that particular private railway company. The main problem, though, was the major difficulties encountered while laying a line in Central Java. An exclusive transport contract between Batavia and a new seaport would considerably strengthen the position of the NISM. Unfortunately the railway proposal gained little support from the committee.

Next to the old Batavia business establishment and the 'newcomer' NISM there was a third party, which can be roughly denoted as 'the engineers'. Their stakes were of a completely different nature. In late nineteenth-century East India the power of the engineer was growing. After having been educated in Delft, many young engineers then went on to seek employment with the railways or with what was then a modern department like BOW. With what amounted to a virtually unshakable belief in the blessings of technology they were fired up to develop the East Indies. By contrast, though, they were highly suspicious of the intentions of the European commercial sector in the East Indies because that was the very body that stood in the way of technological advancement and the elevation of the colony.[39] For the BOW engineers on the harbour committee, and in particular for chairman Van Raders, the creation of a brand new harbour would constitute the perfect opportunity for them to display their know-how. With a somewhat superior air they criticised the plans put forward by their opponents. Van Raders c.s. were convinced that instead of muddling on in the harbour channel area it would be much better to start afresh with a modern harbour and the circumstances at Tandjong Priok seemed ideal.

By chance one of the engineers in the committee used the term 'Old Batavia' at one of the meetings. This triggered a fierce discussion in which the hidden agenda of both parties clearly emerged. Representatives from the Batavian business community began to wonder in agitated terms whether Tandjong Priok would then become 'New Batavia'. Even though that notion was adamantly denied by the BOW people, the businessmen's suspicion was not exactly unfounded: the engineers were really keen to see a new commercial centre spring up around Tandjong Priok.

All in all, the harbour question reflected many parallels with discussions on large-scale infrastructure projects continuing elsewhere in the colony. In West Sumatra the bickering about a franchise for a railway line leading from the Ombilin coalmines continued for no less than 23 years. There, too, economic and political arguments had been packaged and presented as technological arguments and bureaucratic procedures and once again it had been the engineers who had finally come out on top.[40]

On 15th July 1875 the committee submitted its final report to the newly appointed governor-general, J.W. van Lansberge. The committee members had been unable to reach a consensus and so their recommendations remained divided. The majority opted for a new harbour near to the existing channel but a minority was still in favour of the Priok option.

As opinion in the East Indies had become permanently divided, a new committee with two hydraulic engineers: J. Waldorp and P. Caland[41] was established to determine matters in the Netherlands. In one of the assignments Waldorp had come up with a design for the harbour expansion of Amsterdam whilst Caland had been closely involved in the construction of the New Waterway of Rotterdam. After having scrutinised the plans they arrived at the surprising conclusion that if a seaport were to be created in Batavia 'no other place would be more suitable than Tandjong Priok'. They estimated that constructing a harbour in that spot would cost some 20 million guilders which amounted to almost double the cost of the original Priok plan.[42]

The decision led to disbelief and furious reactions in Batavia where the Chamber of Commerce and Industry even went as far as to lobby for the instant dropping of the entire harbour construction plan. It was surmised that it would be better if the money was invested in railway construction.

The reason why the Tandjong Priok choice came as such a surprise to everyone was because originally it had been the plan that had attracted least attention. During the decades when the discussion had, however, dragged on the maritime circumstances had changed considerably due to the speed of expansion of the whole steam-shipping branch. In their decision Waldorp and Caland had particularly taken into consideration precisely these changes, these new circumstances which required a harbour with considerable depth and capacity, modern means of loading and unloading and good connections with the hinterland.

Ultimately then, the notion that the decision had fallen with Tandjong Priok represented a great triumph for the engineers and the BOW director, Van Raders, accepted and celebrated the decision as a triumph for East Indian engineering in general and for BOW in particular.[43] The design was not further worked out in the East Indies, though, but rather in the Netherlands. The Minister for the Colonies who was thoroughly exasperated by all the quibbling in the East Indies asked Waldorp to take responsibility for the final harbour plan for Batavia and to also draw up the construction specifications and conditions. It would have to comprise an outer harbour and a water surface area of 140 hectares closed off by two piers curving towards each other. The inner harbour was to be 110 m long and 185 m wide. It would have to be possible for ships to moor on the western quay where steam cranes would be installed for loading and unloading purposes. There was to be capacity for freight storage in seven different depots that would be built right on the pier. On the eastern pier mooring landing stages would be constructed on screw piles for the loading and unloading of salt and tin. At that point there would also be a coal storage place and a bulk storage station. As a temporary solution, a 4000-ton floating dock would be borrowed from the naval base in Soerabaja. In the design, allowances were also made for the construction of a permanent dry dock.

The Batavian harbour works

In 1876 the Minister for the Colonies put out a call for tenders for the harbour construction works. Even though the government had hoped to reap a crop of 'solid contractors' from the advertisements placed in the official government publication paper *Staatscourant* (Gazette) and various Dutch and International technological journals, the whole tender turned out to be rather disastrous. Only three companies applied for the contract. The bids put in by the French firm Dussaud, a renowned harbour builder, and the British-Dutch consortium Lee & Van Hattum both far exceeded the planned budget of 19,000,000 guilders. Moreover, they were only prepared to take on the contract under conditions that would diminish any risks.[44] The British contractor, Le Fevre & Sons, was the only applicant to remain within the budget margins but after thorough screening by the ambassador in London they proved to be nothing more that a 'band of swindlers'.[45]

The work would therefore have to be done by the government, preferably under the supervision of Waldorp but the fee he demanded was considered too high. After Waldorp had declined the request to travel to the East Indies to supervise the work in person the Dutch government sent out J.A. de Gelder who, as the 'assigned engineer', had been very involved in the former design and preparation phases. According to Waldorp that was very important because:

> *he who is responsible for supervising this very difficult job in the East Indies must be perpetually aware of the overall programme and of the way in which even the smallest detail is to be executed. All of that must be clear at every single stage if the implementation is to be securely executed.*[46]

Waldorp's proposal that he could possibly function as a kind of overseer from The Hague was firmly rejected by the Minister for the Colonies:

> *From the moment that operations commence in the East Indies the Minister for the Colonies will only intervene to provide help if expressly invited to do so by the East Indian Government.*[47]

The responsibility for operations was thus clearly in the hands of the East Indian government. The construction of that particular harbour was to be one of the most extensive and expensive government projects in the East Indies in the nineteenth century. The pressure upon De Gelder was thus immense, as was later mentioned in the commemorative book marking the fifty-year jubilee of the Dutch Royal Institute of Engineers in 1894. De Gelder tackled the job in a way that ultimately commanded respect:

> *When, on behalf of the government, the engineer De Gelder was made responsible for the execution of the harbour work - after two of the biggest European contractors had made clear their objections to the job - the exceptional trust placed by the top administration in a mere public works engineer met with general surprise. That surprise only increased when, accompanied by personnel recruited in the Netherlands, including a great number of craftsmen, he arrived in Batavia and promptly proceeded to spend millions on ships and tools [and] even went about importing wood from America.*[48]

After his appointment De Gelder only had three months in which to prepare for his trip to the East Indies. In that period personnel had to be appointed, detailed designs had to be drawn up and materials and production means such as ships and machines had to be ordered. Two engineers from East Indian Public Works: L.G.B. Bouricius and J.H.H. d'Arnaud Gerkens who, at that time, just happened to be on leave in the Netherlands assisted with the preparations. Bouricius set about designing the required bridges whilst De Gelder travelled to Great Britain to purchase and have constructed a number of dredgers and steam hopper barges and to have a dry dock constructed. In the Netherlands he placed orders for smaller material. On top of that there was equipment such as: screw piles, concrete mixers, steam cranes, drilling machines, centrifugal pumps, traction engines and stone crushers that he wanted to have shipped from Europe to the East Indies.

Land surveying in connection with the Tandjong Priok harbour plans in about the year 1878.

De Gelder wanted to recruit half of the personnel in the Netherlands and the rest in the East Indies. In the Netherlands people were enthusiastic. For the 32 vacancies there were no less than 500 applications and that was not just for engineering work but also, for instance, for seamen wishing to join the fleet and bookkeepers.[49] In March 1877 the whole company, together with their families, set off for Batavia on the SMN steamship 'Koning der Nederlanden' stopping on the way in London to pick up material. Various of the jobs were started on board as the journey was to take some two months. Whilst the craftsmen occupied themselves with making all the equipment ready for use and maintaining it, the foremen were instructed on the East Indian approach to work and were told about the available local materials. The administration and the accompanying regulations were settled on board: the most important issue was to find a way of avoiding fraud.

A special organisation was set up for the implementation of the plan, for the Batavian harbour works (BH) that did not officially fall under BOW but was directly answerable to the governor-general. As the man in charge, De Gelder had overall responsibility for the BH, which consisted of three divisions. The first division was concerned with the execution of the actual harbour works in Tandjong Priok, the second division with the railway constructing, the channel and the road to Batavia and the third division was responsible for the stone quarrying in Merak in West Java which was to provide the filling material for the harbour dams.

The administration was 'commercially' set up so that the costs per division were carefully budgeted and recorded with a view to material expenses, the depreciation factor and the remnant value of certain saleable surplus materials. Comparisons were finally made between estimates and actual costs.[50] For each division, the costs would be broken down and checked. At the head office the various separate components would be combined to compile technical and administrative reports. Each month the head of the Batavian harbour works would submit progress reports to the government which would subsequently be published in the *Javasche Courant* (Javanese Newspaper). The monthly reports gave an insight into how the work was progressing, the technological difficulties, and the financial and social aspects such as the health of the labourers.[51]

Together with his assistants De Gelder had drawn up a work plan which, in two big folio volumes, provided detailed descriptions of how the work was to be executed. All the administrative directives and technical instructions were included in the work plan. This extensive documentation would prove useful during later harbour extensions and when creating other harbours. For instance, the engineer Johannes de Rijke was able to draw on the experience gained in Priok when creating harbour works in Japan.[52] During the actual construction work the BH engineers sometimes had to drastically adapt Waldorp's original plans either to meet the demands of local situations or else because a cheaper option had been found. In the end, the construction of the harbour piers and quays had been so thoroughly amended that according to De Gelder at least four million guilders had been saved.[53]

The construction of Tandjong Priok[54]

The harbour project certainly constituted heavy work, especially in the early phases. Before sunrise engineers and workmen would sail to Tandjong Priok in small boats and they did not return until after sunset. The main dangers were the tropical sun and malaria. The working days were ten hours long including the one-hour lunch break and on Sundays maintenance and repair work would be carried out on the machines and equipment. De Gelder was well known for his severe way of managing people and for being someone who demanded a great deal from his employees. Through his system of penalties and rewards he was able to exercise much control over his personnel. According to his assistant, M.J. van Bosse, punishment and dismissal were daily occurrences but at the same time the BH director would reward achievement with bonuses and gratuities.[55]

The dredger boats that had been ordered in England finally arrived in the East Indies in 1878 after a laborious voyage. The ships proved to be insufficiently seaworthy for such a long trip and one ship was lost after having broken anchor during a storm near Ceylon. The replacement ship, built in the Netherlands, did not arrive in Batavia until 1880. At a speed of 13 to 14 buckets per minute (around 4.5 m^3) the bucket dredgers dug up mud to a depth of 8.5 m. The hard coral reefs slowed down the work because they first had to be broken up with dynamite. In all, the three big dredgers managed to remove 2,841,328 m^3 of sand and coral from the inner and outer harbours within the space of four years. The smaller dredgers, two of which had been constructed in England and two in the Netherlands, worked at a speed of 2 to 3 m^3 per minute to a depth of 5.40 m. Those dredgers were deployed for the construction of the shipping channel, the inner harbour and other important canals and together they managed to remove 1,221,749 m^3 of spoil. People complained profusely about the dredgers that had been constructed in England because the constructors had not kept to the specifications and the equipment had not been tested beforehand which meant that when it came to the assembly phase in the East Indies many parts did not even fit.

Twelve steam-powered vessels were ordered for the transport both of the dredged up sand and of stones for the harbour jetties extracted from the quarries in Merak. These hopper barges, most of which had been built in England and Scotland did prove to be a good investment. Because of their great flexibility they could also be deployed for other ends. After all the construction work had been completed three of the ships remained in Tandjong Priok where they were incorporated into the harbour's fleet whilst the others went to the Government's Navy, the Beacons and Pilot Service and the Department of Financial Affairs where they were deployed to help in the combating of opium smuggling. For the maintenance of the ships De Gelder had designed and commissioned a cylindrical dock that was built in Newcastle before being shipped out to Batavia in parts where it was then assembled. An engine factory was built especially for the maintenance of the fleet and all the necessary machinery and it was equipped with a smithy, an iron foundry with ovens, a copper foundry, a boiler room, a work bench shed and turning room, a carpentry and joinery, and a big steam engine.

The harbour dams consisted of over half a million m^3 of trachyte, a volcanic type of rock mined in Merak on the coast of West Java.[56] Outside the quarry, a temporary harbour was created with two screw pile jetties, a coal jetty and a loading and unloading bay equipped with two steam cranes. The hopper barges were used to transport the quarried stone to Tandjong Priok but the material could not be directly dumped as the dams would then sink too easily into the soft seabed. The problem was resolved by first sinking heavy trachyte blocks into the seabed on either side of the harbour pier; clay was enclosed between the heavy blocks. Sand was then placed on top before another layer of trachyte was introduced. The harbour pier was to be built up from the low water line which would be protected from waves on the sea side by concrete blocks. The western pier was considerably better reinforced than the eastern dam which would have much less buffeting to endure from the powerful waves.

The quay walls, one of which was 1000 m long and ran along the western side of the basin while the other was 300 m long and skirted the projected coal harbour (which was later to become the location for the dry dock) consisted largely of concrete. As the basin was not yet ready, a gully was excavated along the line where the future quay was to be built and then cordoned off with wooden sheet piles. After that the concrete was cast and the portion above the imagined waterline was finished with trachyte masonry as that would be more resilient to impact and thus less vulnerable to damage. The sharp coral on the seabed hampered gully excavation progress. The local labourers were not even prepared to do that work for four times the normal wage because of the high risk of being seriously wounded. Ultimately forced labour had to be resorted to but that really went against the grain for the engineers who, in contrast to many civil servants in the East Indies, were opposed to forced labour. The engineers preferred to employ people who worked on a voluntary basis, especially in view of the fact that such individuals were always more intrinsically motivated.

By mid-1882 the quays were ready. Along the top of the quays there were two sets of rail tracks, one for the loading and unloading cranes and the other for the railway line. Seven hangars were built straight onto the harbour wall for the storage of all the goods that needed to be inspected by customs officers. Behind those sheds were seven storage warehouses that were two stories high. The quays would be lit up with 29 electric arc lamps which would be switched off on moonlit nights in order to save electricity.

There were certain other construction work jobs that were carried out by BH that were not strictly related to the official harbour construction activities, such as the building of houses to accommodate warehouse employees, policemen and the actual European and local inhabitant harbour personnel. For the Europeans semi-detached dwellings were built; for the locals the BH provided semi-permanent dwellings with a wooden framework, wooden floors, pantiled roofing and walls made of woven bamboo. The BH was also responsible for the building of the station and other public service buildings.

On 27th August 1883 the Batavian harbour work activities were hard hit by the eruption of the Krakatau volcano. The tidal waves created in the wake of the eruption completely wiped out the stone quarry at Merak. Only the bookkeeper who had left the site to send word to Batavia and a handful of workmen survived the catastrophe. The harbour works at Priok, some hundreds of kilometers away, were subjected to three successive tidal waves and De Gelder's assistants managed to save the ships that were moored there. A 300 m section of the eastern harbour jetty subsided but was soon rebuilt.

By 1883 the harbour was more or less completed. Though many trade and shipping facilities were still lacking, Tandjong Priok at least provided a safe mooring place where non-stop loading and unloading could continue. As shipping movements would slow down the final construction work, it was only in bad weather that ships were temporarily allowed to enter the actual harbour. Until 1885, ships were simply redirected to the old Batavia harbour when weather conditions were normal.

A technological triumph

No plans regarding harbour management had been made beforehand and the most important question was really whether the harbour should be privately run or left in government hands. After leaving BH as director De Gelder put in a bid for a harbour management franchise together with N.P. van den Berg, nota bene the former vice-chairman of the Chamber of Commerce and the individual most opposed to the Priok plans during the Batavia harbour debates.[57] The alliance between De Gelder and Van den Berg was rejected but their calculations were gratefully accepted as a basis for efficient harbour management. The government was wary of private ownership, its main consideration being that the harbour should first be allowed to create for itself a position in shipping and trading circles.

> *By temporarily suspending various duties it has been possible for the shipping and trade sector to become familiar with the new harbour; by offering much support the original objections of*

> *individuals who were initially put at a disadvantage by the new situation, have been eradicated. Now that a more established situation has been created the management is slowly becoming regulated.*[58]

Such a smooth transition could not really be expected of any private entrepreneur. It was only for the construction of a dry dock that a private franchise was offered. D. Croll, former director of the Rotterdam Fijenoord wharf, signed a contract with the East Indian government to set up the TP Dry Dock Company, the first dock to be specifically created for merchant shipping. A new 4000-ton dry dock was to be built on the western side of the harbour. Until such a time, Croll would be loaned the naval base's floating dry dock. The Dry Dock Company was established in 1891 and the wharf soon came to form an indispensable component of the Tandjong Priok harbour.[59]

The creation of the Tandjong Priok harbour had been of tremendous importance to East Indian Public Works because in that way the department had been able to prove that it was capable of tackling major projects. Public Works had a reputation for being slow and expensive but this time the works had been completed within the projected timeframe and within the stipulated budget. De Gelder who had led the project with a firm hand until 1883 was viewed as a true hero by East Indian engineers. After his departure as head of BH, where he was succeeded by M.J. van Bosse, he was to be appointed to the influential East Indian Council.

The emergence of a maritime network

After the millions that had been spent on Tandjong Priok no money was invested on harbour improvements for the trading towns of Semarang and Soerabaja and that was a bitter pill to swallow. In Semarang the local Chamber of Commerce had spent decades lobbying for better sea connections, also in collaboration with the Batavian Chamber of Commerce. In fact the circumstances in Semarang were not unlike the old situation in Batavia. Semarang had a coastal strip that offered little protection against wind and waves, especially during the west monsoon season. The harbour channel that did offer access to the city centre was shallow due to the perpetual build up of silt. The first plans for improvement had been drawn up in the thirties and in each subsequent decade a new proposal had been produced but nothing had ever materialised. As the harbour channel was eventually no longer even suitable for proas a new channel, the Kali Baroe, was excavated in 1872 that connected with the city river via a lock. It was not long before the new channel also silted up and, there too, the piers had to be regularly lengthened as the coast was encroaching at a rate of 14 m per annum.[60]

After complaints lodged by shippers and traders about the lack of mooring points and storage possibilities the government decided, in 1894, to have a harbour plan drawn up that was to incorporate ideas for a new and extended customs yard to the east of Kali Baroe. Local traders did not see that as an improvement but rather as something that was detrimental to their interests and so once again plans to do something about the harbour were shelved, even though the harbour had become vitally important to the sugar industry in Central Java.[61]

In Soerabaja, too, there were problems despite its sheltered coastal position. Ocean-going vessels would anchor in the bay and load and unload with the aid of lighters that were able to sail over the Kali Mas - a branch of the Brantas river - and into the city but in front of the quays, notably outside the customs offices, huge queues always developed. From 1875 onwards various plans were submitted on proposed ways of improving the harbour at Soerabaja but those plans were similarly rejected by the government on the grounds of cost.[62]

At this point it should be mentioned that Soerabaja did in fact have another harbour facility; the Marine Etablissement, i.e. the naval base. Unlike the trading port, the naval base in Soerabaja did provide moorings for large ships in its deeply dug out basin but the facilities were only intended for the navy's warship fleet. The harbour basin also had floating docks where navy and government vessels could be repaired. To make sure that access to the naval harbour remained possible and the basin itself sufficiently deep, it was necessary to maintain a perpetual dredging regime, an activity that would be unavoidable in the event of the creation of a possible trading harbour. The engineers at the East Indian Public Works department had little to do with the naval base as that fell under the authority of the Naval (or Marine) department. In view of its distinct status in relation to the other East Indian harbours the naval base will not be further discussed in this chapter.[63]

In the first years of its existence the new Tandjong Priok harbour's achievements were little more than mediocre. Due to ingrained habits, the closer proximity of the lower town area of Batavia and the stubborn resistance of the local trade sector, many ships still sailed on to the old bay and harbour channel instead of entering the brand new harbour, despite the better facilities provided there. What was crucial to the fate of Tandjong Priok, but also to the other harbours of the archipelago, was the creation of the KPM in 1888. The new steam ship company, established at the instigation of the Dutch government which had furnished it with heavy subsidies, was set up in order to deliver consignments to all parts of the East Indian archipelago. The packet service was intended for the government-related transportation of civil servants, military people and post and had to abide by fixed timetables. In addition to that ordinary passengers and freight could be transported. Packet transport was a shared public-private enterprise: the government initiated and controlled the service but the actual implementation was in the hands of a private shipping company.

Roughly speaking, after 1850 the packet services fell into the hands of two companies. In the first 15 years it was the Cores de Vries firm that was responsible for the main service. After 1866 the franchise passed to the Dutch East Indian Steam Shipping Company (Nederlands-Indische Stoomvaart Maatschappij). That was a daughter company of the British India Steam Company which had, over the course of time, created a virtually monopolistic position for itself in the otherwise poorly developed coastal steam shipping sector, a situation that was to lead to increasing discontent in the 1880s. The colonial government wanted to put an end to British dominance where East Indian coastal shipping was concerned, certainly in view of the fact that the Dutch East Indian Steam Shipping Company considered Singapore, and not Batavia, to be the pivot in its network.

That was the reason why the Dutch government actively stimulated the setting up of a new packet service company that could serve as a partner to boost expansionist aspirations within the archipelago. The new coastal shipping company was lined up to take over the official inter-insular shipping from the British enterprise and to – via the Javanese ports –

feed the large international SMN lines and the Rotterdam Lloyd company (RL). The KPM received the franchise for 13 lines, 6 of which started and ended in Batavia. Under such imperative shipping line politics, traffic was directed to Batavia and the expanding network of scheduled shipping services centred around Tandjong Priok and later, also, Soerabaja.

A symbiotic relationship developed between the KPM and the East Indian harbours. The KPM depended on good, efficient harbour facilities in order to build up its very diverse entirety of scheduled services. The KPM connections improved the disclosure of the ports' hinterlands and created a new 'foreland' which, via connections with international lines, would stretch to far beyond the local region. Within a relatively short space of time the KPM managed to link up big and small coastal points, in a way forming a thread drawing together all corners of the archipelago. Trading cities and islands which had previously only been linked together at regional level were suddenly brought together in one huge tapestry in which several large harbours and trading places became the main hubs. According to J. à Campo, it was the scheduled service network of the packet boats – and notably the KPM – which, after 1888, was to turn the Dutch East Indies into one big traffic and geographical whole so that the maritime infrastructure actually affirmed the geopolitical reality of the colonial state.[64] In reality the packet service coloured in the political land map of the Dutch East Indies.

Harbour improvements outside of Java

Towards the end of the nineteenth century it was not in Java but rather in Sumatra that the main harbour improvements were emerging. On the west coast a new harbour was built at Padang and on the east coast a new harbour was developed not far from Medan whilst in the far north of Sumatra the private coaling-station of Sabang was created that was later to become a full-blown export harbour. Harbour construction on Sumatra reflected the rapid economic rise of the Outer Islands and the integration of all those areas into one big Dutch East Indian political and economic whole. Previously Sumatran trading links, even in the western parts, had been very oriented to Singapore where products such as rubber and tobacco had been exported and textiles and foodstuffs had been imported.

The first big harbour project to get under way after the Tandjong Priok project was the construction of a harbour in West Sumatra where shipping services were sought to transport all the coal extracted from the Ombilin mines. After a fierce battle centred on the construction of a rail link – should it be a privately or a state run line – it was eventually the State Railway Company that laid a line from the mines to the coast between 1886 and 1893. Simultaneously, at a 7 km distance from the city of Padang, an entirely new harbour was constructed known as the Emma harbour.[65] The harbour came to reside in a natural bay that was even further protected by the construction of two breakwaters. The ships were able to moor on screw piles that connected directly with the railway line. After the construction work had been completed the State Railway Company handed over the brand new harbour, which had cost 3,400,000 guilders to construct, to the regional water board.[66] Emma Harbour thus became the second biggest harbour in the East Indies where ocean-going steam ships were able to moor on landing stages.

On the Sumatran east coast the plantation economy expanded rapidly in the last quarter

Street scene in the Tandjong Priok harbour area. To the left one can see the Dutch Steam Ship Company building, c. 1925.

of the nineteenth century. The fast rise in cultivation enterprises in that area, chiefly involving rubber and tobacco, also placed demands on the coastal harbour facilities. In 1890, partly at the instigation of the local trade sector, the Deli Railway Company created a harbour on the coast not far from the main town of Medan, the most important trading place in the area. The harbour was built on an island in the mouth of the rivers Belawan and Deli. Up until then ships had simply dropped anchor in the river so that goods had to be transported to and from the quayside by lighters. In the new harbour situation ships could be directly loaded and unloaded. Goods could be transported from the hinterland by rail because the Deli Railway Company had provided rail links between Medan and the districts of Serdang and Langkat. Within a couple of years the small jetties and the small warehouses proved counterproductive to speedy dispatchment and so in 1895 the Deli Railway Company, which still owned the harbour works, made the waterfront 350 m longer and constructed a number of new warehouses. Not even that expansion was enough to cope with the volume of freight because export activities on the Sumatran east coast were growing apace, at a rate of 17 percent per annum. On top of everything else, Belawan had to deal with a sandbank that hampered access to the harbour so that ocean-going steamers requiring a large draught were unable to enter.[67]

On the northern coast of Sumatra as well, on the island of Poeloeh Weh, the new harbour of Sabang was built at the end of the eighteen-nineties as part of a private initiative. It had originally been a coaling-station strategically situated in the Straits of Malacca exactly on some of the region's main shipping routes. Before very long it became clear that it could easily

The harbour and coaling-station of Sabang (Sumatra), c.1920.

do double duty as a dispatch point for products originating from Atjeh and the east coast of Sumatra. The harbour had a 1.35 km^2 sheltered deep bay and a 220 m-long trade quay. The 610 m-long coal quay was provided with cranes, transporters and a refuelling point, and with a grab crane. Though the harbour remained in private hands, the contribution made by the government was considerable. The government had substantial financial interests in the harbour but also understood its strategic value situated, as it was, so close to Singapore and Penang and so bestowed upon it the status of a free port.[68]

On the eastern side of the archipelago it was Makassar, in the southwest of the island of Celebes that profiled itself as an important transhipment harbour. Already in 1846 the government had made the strategically situated Makassar harbour a free port in the hope that as a transhipment harbour it might well be able to compete with Singapore. Rather than relying on a productive hinterland it was the transhipment market that was important to Makassar. It relied especially upon the goods brought in by 'traditional rigged sailing boats' from other coastal places such as Celebes, the Moluccas and other islands. In the latter half of the nineteenth century steam packet services contributed much to the shipping traffic in Makassar.

Makassar had a roadstead, sheltered by a number of islands, where large ships used to anchor. The harbour itself consisted of a 490 m-long quayside area where the loading proas could moor at six different landing stages. Situated immediately behind were the warehouses and the offices of the trading houses and the alleys connecting the quay with the trading district.

Towards the end of the nineteenth century, the lack of harbour facilities constituted

The Makassar harbour (Celebes), 1919.

a problem for the steam ship companies with ships that frequented the port of Makassar. Between 1896 and 1900 the number of ships using the harbour at Makassar had doubled.[69] As a free port Makassar levied no quay fees or other dues. The financial means to actually improve the harbour works were thus insufficient. The proposition that harbour dues should be imposed led to fierce protests from the local trading sector. One complicating factor was the fact that the government had come into conflict with the indigenous rulers in South Celebes. One of the immediate causes of the war had indeed been the matter of the levying of import and export rights. The rulers were subjugated after a military expedition mounted in 1906 and the harbours of Boni and Loewoe were closed to international traffic. At that point Makassar also lost its status as a free port. Not long afterwards the KPM expanded its coastal boat services, the underlying idea being to raise the export level in South Celebes while simultaneously perpetuating the subjugation.[70]

Despite the fact that Makassar had lost its status as a free port these developments led to considerable expansion in the volume of steam ship traffic. In 1908 another 500 m of quay space was built. The idea was to introduce a screw pile landing stage in front of the existing quay and to fill up the space in between but because the seabed was so soft the piles could not be inserted deep enough and there was a very real chance that the landing stage would capsize under the lateral pressure of the bolstering up operation. The problem was resolved by situating the filling up material 10 m away from the landing stage and closing it off with a stone wall. The open space between the landing stage and the quay was then bridged with a series of footbridges.

Specialisation and professionalisation

Soon after the turn of the century most East Indian harbours were in considerable difficulty because they were hardly able to cope with the rising trade levels and shipping flows. The shipping companies such as KPM, SMN and Rotterdam Lloyd pushed for improvements, just as did the commercial establishments. The KPM steadily expanded its number of shipping lines and made strict demands on the provisions in the harbours it frequented. If necessary, the company would even be prepared to organise the harbour facilities. The company endeavoured to control the whole business column – from agency to seaport – and, if need be, to even supply the lacking or inadequate facilities.[71]

A new harbour question

After the turn of the century it seemed as if history was repeating itself because a new harbour question arose and, once again, the business world and Public Works turned out to be each other's biggest opponents. Soerabaja was not just the most important trading place in East Java with an extremely productive hinterland that was moreover linked up by a railway line but it was also the archipelago's 'gate to the east'. In the first decade of the twentieth century freight transport increased by no less than 60 percent.[72]

In Soerabaja ships could drop anchor safely in the protected bay and just like in other places the proas were responsible for the transport of goods from the mooring point to the city. They sailed 3.5 km upstream up the Kali Mas, a branch of the Brantas river, to finally load or unload at a place near to the 'Red Bridge'. Because of all the silt, the river was difficult to navigate and it was certainly the case that when the tide was low the proas were hardly able to reach the quays which meant that the freight had to be carried over long footbridges to and from the quayside. The 350 large proas and the even bigger number of smaller lighter ships were responsible for the gigantic volumes of traffic on the river.[73]

Just to make matters worse, there was another problem and that was the matter of the location and facilities for customs provisions. The customs shed, the Kleine Boom, was located some 500 m from the river mouth and that was where all the imported goods were inspected. A few kilometres further there was the Groote Boom, a terrain with an area of some 3 ha equipped with sheds and warehouses. That was where all the taxable goods were unloaded to be checked by the customs officers but the facilities were just far too minimal for the fast-growing trade and shipping activities of Soerabaja which meant that long waits and endless queuing was required.[74]

By the end of the nineteenth century, after the 1875 plan to excavate a seaport fell through because the exploitation costs were too high, the harbour problem simply became insurmountable for the city. In 1897 the engineer, W. Jongh, who was affiliated to the State Railway Company suggested that a harbour should be created on the eastern side of the Kali Mas where the ships could then load directly from the quays. Once again, finance was an important limiting factor. In 1903, after various seabed composition investigations had been carried out, a special harbour committee recommended that that a screw pile landing stage should be built on the sea embankment to the east of the river. After many objections had been made to this plan, a new idea was promoted, a proa harbour scheme known as the Rahder plan that involved improving the navigability of the Kali Mas by deepening it. In 1906, W. van Goor, chief engineer of the Public Works department in Soerabaja presented his own pier plan which envisaged a mooring spot for ocean-going vessels to the west of the river. He wanted to either create a quay wall with sheds and warehouses behind it or – if necessary – build a pier that would project into the sea so that ships could berth on both sides. The pier would then be connected to the shore by a dam or bridge. He, too, saw it as essential to deepen the river. Once again many objections were made, both by the trade sector which saw much more potential in the proa harbour plan, and by some Navy officers who feared that the arrival of a pier would mean that the entrance to the naval base would certainly get silted up.[75]

The harbour at Soerabaja had turned into a proper East Indian relations bone of contention. Just as in the case of the Batavia harbour question, there were two irreconcilable stances: local interests that stood to benefit from a more modest solution – landing jetties, a proa harbour – and East Indian 'national' interests that supported an altogether more thorough and expensive approach. Inevitably the contrasts were magnified by the increasing political and economic interests that steam transport and harbours had got caught up in. Apart from anything else, the sugar industry and the trading firms were certainly not intending to relinquish their warehouses, storage enterprises and proas for a new harbour. It would be especially the KPM and the big shipping companies such as SMN and RL that would stand to gain from a new harbour. The question was whether Soerabaja would flourish as an import and export port or whether a new harbour would be able to secure a position as a transit port. When it came to transhipment it was not just Batavia but also Singapore that was hungry for a share of the market.[76]

In 1907 the East Indian Council had placed high stakes on Soerabaja's potential to become a significant port city. With all the products from the hinterland such as sugar, coffee and tobacco, trade could easily develop but also because of its favourable location Soerabaja had the potential to become a world harbour with a great future.

> *It is an indisputable fact that within our entire colonial state there is no other harbour that can overshadow Soerabaja and that is why, also with respect to its location in relation to Australia, it plays and will continue to play an important role among East Indian harbours.*[77]

In 1909, in order to put an end to all the quibbling – just like three decades before in Batavia – the question was brought before an independent committee of Dutch harbour experts. It was their responsibility to clinch the final decision. The two Dutch experts brought in to resolve the Soerabaja harbour dispute were professor Jacob Kraus and the engineer Gerrit de Jongh. Krauss saw the task as a choice between two points of view:

> *[...] whether it is to be agreed that everyone is going to remain satisfied with the existing loading and unloading system [by means of lighters] or whether opportunities must be created to allow sea-going vessels to make direct use of the permanent quay.*[78]

Kraus and De Jongh were both very conscious of the fact that there was more attached to settling the differences of opinion surrounding the harbour than simply making a number of technically sensible choices.

[...] we also pride ourselves in the fact that [our discussions] will contribute to bringing closer together the now very divergent opinions concerning the harbour issue so that the opposition which under the present circumstances all potential solutions will meet with will, at least to a large degree, be averted.[79]

They furthermore proposed in a letter addressed to the Minister for the Colonies that it should simply be a condition that they speak with all the various parties involved: representatives from the commercial sector, the engineers who had worked on the harbour plans and the customs authorities.[80]

The mere fact that Kraus and De Jongh had been picked for this function was indication enough that the East Indian harbour question was being given high priority in The Hague. Jacob Kraus was widely respected among engineers as an authority in the field. Kraus, former professor and chancellor of the Technical College in Delft, had been Minister of Public Works between 1905 and 1908. During his long career Krauss had been closely involved in the designing and constructing of various South American harbours such as in the Chilean port cities of Valparaiso and Talcahuano.[81]

The other expert involved in the East Indian harbour question was Gerrit de Jongh, director of the Rotterdam municipal works authority. In his term of office that had spanned nearly three decades De Jongh had worked on the Rotterdam buildings and infrastructure, and in particular the port. Towards the end of the nineteenth century De Jongh had been responsible for the creation of an extensive complex of harbour basins on the left bank of the Meuse river. In technological terms De Jongh was known as an 'utter realist' who, because of his frank and rather direct approach had earned the nickname 'Bold Gerrit' and who had managed to get the municipal council of Rotterdam to support his views.[82]

Kraus and De Jongh sailed off to the East Indies to examine the harbour situation in Soerabaja but it soon became obvious to them that they would also have to include various other harbours in their recommendations. Indeed, in Semarang the complaints were not unlike those lodged in Soerabaja, and even the modern harbour at Tandjong Priok had reached its limits capacity-wise. Krauss and De Jongh made a tour of the main East Indian harbours before producing three different reports: one on Soerabaja in line with the original task, one on Makassar and finally one on Tandjong Priok and the other East Indian harbours.[83]

The reports produced by Kraus and De Jongh

Kraus and De Jongh's research did reflect a standpoint which, for the first time, viewed the harbours as a comprehensive whole. Previously the harbours had only ever been considered in isolation but this time Kraus and De Jongh examined the functions fulfilled by the various harbours, what their relationship was to the hinterland, how they interlinked with other harbours and what requirements they should therefore satisfy. What Kraus and De Jongh finally put forward were plans for the drastic reconstruction and expansion of many East Indian harbours and those proposed changes were l, for the greater part, executed. What they emphasised in their recommendations for Soerabaja was that national interests should

The harbour of Soerabaja situated near to the estuary of the Kali Mas, 1920.

prevail. Local interests were of secondary importance.[84] In the case of Soerabaja that meant opting for a large-scale approach so that the harbour would be thoroughly suited to the quick throughput of goods both from the hinterland and from other parts of the archipelago.

In the reports, the situation in Soerabaja was compared with that of Rotterdam a couple of decades before – a situation that was obviously very familiar to De Jongh. Rotterdam, too, had possessed an excellent harbour approach from where freight could be transported on by lighters. Access to the city itself, via the river, was very bad until the time that the New Waterway opened up the harbour. It had cost 36 million guilders to construct the 'Nieuwe Waterweg', which was much more than the proposed improvements for Soerabaja would cost, whilst the number of shipping movements in the East Indian port city meanwhile far exceeded those of nineteenth-century Rotterdam. Unlike 'its European sister', Soerabaja would never becoming a first-class world port.[85]

The proposals put forward by Kraus and De Jongh were not unlike the pier plan drawn up by Van Goor in 1907. A 1200 m-long pier was planned along the Straits of Soerabaja and to the west of the Kali Mas. Along the whole length of the pier a quay wall would be built so that sea ships could berth there as the depth along the quay would be at least 13 m. On the 200 m-wide terrain behind there would be loading and unloading equipment such as steam cranes, open air and warehouse storage facilities, railway lines and connecting roads leading to the city. To the south of these new harbour facilities that were to be built out into the sea, a harbour basin would have to be dredged out that would be 950 by 1100 m with a maximum depth of 13 m. In the northeastern part of the basin a floating dock would be

The new Semarang harbour channel (Central Java), 1919.

created with a 14,000-ton capacity. To the west of the pier another harbour jetty would be built sheltering a basin for proa shipping. The Kali Mas would have to be widened but that plan was finally dropped because of the high value of the land along the riverbank. As an alternative, the quay walls were eventually replaced with more solid walls so that the river could be excavated right up to the quay. The mouth of the river was widened, however, and the western bank provided with a quay wall.[86]

Of all the three main Javanese trading cities, the harbour circumstances in Semarang were the worst by a long way. The build up of silt seriously hampered shipping movements in the harbour channels. From that point of view, the situation was much more acute in Semarang than in Batavia or Soerabaja. In 1894 a harbour plan was again rejected, leading to bitterness and discontent in the city. Meanwhile, though, trading and shipping traffic in Semarang was on the rise. In the space of ten years sugar exports alone had virtually doubled, having gone from 153,000 tons in 1899 to almost 265,000 tons in 1909 and having a value of 36 million guilders. That was half of the total amount of sugar being exported from Soerabaja.[87]

When drawing up the harbour plans for Semarang, Kraus and De Jongh had called in the help of the engineering expert A.G. Lamminga, a man who was very familiar with the local situation, not only geo-physically but also in terms of personal relations and commercial sensitivities. It was quite clear that Semarang really needed an improved harbour because its anchorage ground provided very little protection for ships and during the west monsoon season the transferring of goods was a perilous business. The seabed at Semarang shelved so gradually that it was only 4 km out from the coastline that a depth of 9 m was reached.

The western quay of the first inner harbour of Tandjong Priok as it was before 1918.

On top of that the seabed was very soft which meant that erecting any structures in the sea would always be a difficult and expensive operation. According to Lamminga's rough estimates it would cost around 35 million guilders just to create a deep harbour basin.[88] That was a sum that was out of all proportion to the interests of Semarang as a port city which was why Kraus and De Jongh recommended that in this particular case it would be wise to create a proa boat harbour to the east of the harbour channel. Originally a water surface area of 8.5 ha was planned but ultimately a larger harbour basin of 13 ha was created. The proa harbour with its 75 m-wide passage linked to the harbour channel was given a spacious entrance that branched off towards the pier into two small basins that were 55 and 65 m wide and a fishing harbour. Alongside of the three basins there were quays with a total combined length of 1400 m. Warehouses were built on the terrain surrounding the new proa harbours and the new harbour was equipped with two steam cranes and a small dry dock.[89]

In that way Semarang did not get the seaport that had been lobbied for so earnestly for so many years but the proa harbour did at least provide facilities for the safe and sheltered transfer of goods. In Semarang the interests of the proa ferries were thus protected, unlike in the bigger sister harbours of Batavia and Soerabaja.[90]

By about 1910 even the capacity of the Tandjong Priok harbour works, then in use for 25 years, had already become too restricted. In conjunction with the increased import and export levels and the rise in transhipment activities there was even evidence of major congestion and ships sometimes had to wait days before they could be loaded or unloaded.[91] In terms of tonnage, the volume of freight being handled had quadrupled since the harbour had opened which meant that not only all the quays were too confined but also the storage space in the warehouses and entrepôts had become too cramped. The other problem was that the harbour side railway connection hampered the speedy handling of goods. To the front of the sheds and warehouses there was just one railway line and the connection with other lines constituted a time-consuming bottleneck.[92] A problem of another kind was the bad health situation. Many of the harbour workers in Tandjong Priok were plagued with poor health. Around 80 percent of the dock workers suffered from malaria. And health risks were not confined to the dockers: in 1908, 14 of the 139 BOW employees stationed in Priok died.[93]

The proposals that Kraus and De Jongh made for Tandjong Priok were quite far-reaching. They suggested that to the east of the existing inner harbour a second harbour basin should be excavated (1000 m long and 120 m wide) that would be provided with quay walls on three sides. A third basin was also immediately planned. To make room for all these changes, the eastern harbour dam was lengthened and made to form a big curve shearing off towards the east. The western dam was also extended further up the coastline so that in between the two extremities new land area was freed up for harbour-related business. A section of the outer harbour was deepened to 13 m so that the most recent maritime norms could be met: in the Suez Canal the deepening efforts were fixed at depths of 12 m.

The railway network was reorganised and modernised. Kraus and De Jongh initially wanted to move the Priok station westwards but in the end the decision was taken to build a new station to the south of the inner harbour. All in all, the reconstruction work at Tandjong Priok amounted to the most extensive of all the harbour improvements which had been proposed by Kraus and De Jongh.

In the Outer Islands it was especially the situation at Makassar that Kraus and De Jongh focused on. Building-technically the expansions of 1908 had been a fiasco. For a whole year the banked up area had been subsiding and the screw pile landing stage had drifted 60 to 70 cms in a seaward direction. It was only round about the time when Kraus visited that the constructions had finally settled and stabilised, partly because the new warehouses were made of lighter material so that further subsidence was avoided. Shipping traffic had meanwhile further expanded so that the 500 m of quay was no longer sufficient for the annual freight flow of 350,000 tons, especially if all the goods had to be transported to the quayside over narrow walkways. While travelling around to review the situation Krauss had even discovered that sometimes ships were kept waiting outside the harbour for a place on the landing stage for so long that they occasionally just sailed off without having done the business that they had set out to do.[94]

In order to expand the harbour capacity Kraus and De Jongh proposed, in collaboration with the director of BOW, that a 550 m-long quay wall should be built to a depth of 11.50 m. Behind that wall a 130 m-wide harbour terrain would be created including a quay surface area 50 m wide, a street that would be 30 m wide and finally a strip with warehouses and sheds that would be some 40 m wide. The new quay would be situated at a 300 m distance from the screw pile landing stage and the space in between could be used to create landing stages for proas. Already during the construction of the new harbour facilities it became evident that 550 m quay would not be sufficient, and that it should be extended. By 1918 the new quay had grown to 1340 m.[95]

In Belawan, Kraus and De Jongh were also called upon to arbitrate in a harbour dispute.

The economic growth of Deli, the increased political grasp on North and East Sumatra and the increasing level of competition with British harbours in Singapore and Penang was all reason enough to set up a KPM line that would start from Sabang in the North and sail up and down the east coast of Sumatra. The main objective of that line was to ship out the products leaving from Atjeh and Deli via Sabang rather than via the British harbours. To make that line a political success, the harbour facilities in East Sumatra would have to be improved. After various studies had been carried out the choice fell with Belawan because of its close proximity to Medan. A solution had to be found for the shallow harbour mouth that was only navigable for small steam ships. Initial dredging tests were unsuccessful, but Kraus and De Jongh showed during their visit that a suction dredger would be capable of digging out a navigation channel and maintaining the gully at the required depth. The Belawan harbour was still in private hands but at the instigation of Kraus and De Jongh the government purchased the harbour works in 1913 from the Deli Railway Company for a sum of 500,000 guilders. The following expansion made Belawan one of the principal ocean ports in the western part of the archipelago.[96] The navigation channel was deepened, so that ships requiring a draught up to 9.85 m could enter the harbour.[97] New mooring places were created between 1913 and 1916, and storage space was built.[98]

New harbour management organisation

All the suggestions made by Krauss and De Jongh went further than just pertaining to technical improvements; they also extended to including harbour management. During their tour of the East Indies the somewhat haphazard management of the ports had not escaped their notice.

> *In Priok part of the responsibility for operational affairs lies with the harbour master, another part with the controller of customs and the rest with the Public Works engineer, all civil servants who fall under three different respective departments: the Navy, Financial Affairs and BOW. On top of that the Department of Government Companies, as overseer for the railways, and Inland Administration are also involved in harbour matters whilst ultimately the municipality of Batavia exercises its interests as the terrain beyond the customs barriers belongs to that particular body.*[99]

One of their most important recommendations, therefore, was that port management in its entirety should fall under the control of BOW. Responsibility for the harbour was to move to a port director who would in turn be answerable to the BOW director. An Assistance Committee composed of representatives from various departments, the city council and the commercial sector, would meet for monthly consultation with the port director. All the recommendations for rigorous change were taken on board and implemented by the East Indian government so that in 1912 this administrative system model was adopted, first in Tandjong Priok and then gradually in subsequent years in other harbours as well. Within BOW itself the port department was also reorganised. Finally, all the ports were hierarchically divided up into large, medium and small ports. Tandjong Priok was the biggest one followed by Soerabaja.[100]

Even though all the harbour responsibilities had been formally laid down, shipping companies such as the KPM, SMN and RL remained powerful partners, not least because they, too, had really professionalised their operational management and had their own offices, institutes (such as the Nautical Institute), proa transport services and pools of employees stationed at the various harbours.

Kraus and De Jongh also had opinions about the financial running of the harbours. They urged that harbours be run along the lines of the 'rentability' principle which meant that each and every harbour had to make sure it was a profitable going-concern. It was a principle that was adhered to in practical terms but not rigidly abided by at critical times; the colonial administration saw the ports as indispensable basic economic reserves.

Perhaps the true significance of the advisory work of Kraus and De Jongh was best formulated during a lecture given for the Dutch Royal Institute of Engineers some 13 years later by which time most of the proposals had been implemented.

> *If the harbours of the Dutch East Indies want to play their rightful part in international transport, something to which they are geographically entitled and which the country is thus obliged to achieve and maintain, then developments in the field of harbour matters may not in any way lag behind those of the best in Western Europe or modern America. The view that in regions below the Equator people should be satisfied with less, that lighter (proa) harbours are anything other than makeshift measures, that large ocean-going vessels will only remain in the Atlantic Ocean and will not set their compasses for the countries of the Pacific, that in many respects conditions there are so very different from our norms that all direct comparison is pointless, none of that holds water as far as the harbours and their administration is concerned. Even the varying labour conditions which have been appealed to many a time are changing with amazing rapidity just as, in recent times, they are changing in the Netherlands and elsewhere. Just like in the older countries, the harbours of the East Indies need to be perfectly mechanically equipped.*[101]

All the different technological advances, such as better dredging methods and the speedier production of qualitatively superior materials, made it possible for great strides to be taken in harbour improvement. One important innovation that certainly speeded up construction work was the use of concrete caissons. Caissons are large boxes that are sunk into the water and then filled with stone, sand or concrete. In 1901 Kraus had come up with a caisson design for the harbour of Valparaiso in Chile and so those were the first quay-walls to be built in that fashion. Subsequently the same method was implemented to do quay repair work in Rotterdam. Ultimately the caisson construction method came to be widely used in harbour construction in the East Indies. In Soerabaja the Holland Contracting Firm (Hollandsche Aannemings Maatschappij, HAM), a newly created daughter of the Reinforced Concrete Construction Company (Hollandsche Maatschappij tot het maken van werken in Gewapend Beton) built various piers but it was confronted with serious problems. After having been sunk, some of the caissons shifted from their position and capsized. By means of all kinds of stunts and emergency measures the concrete boxes were floated again and repositioned. One caisson even broke while being towed from the dock to its final destination. The problem with the caissons became the grounds for a protracted conflict between the contractor, HAM, and the customer, BOW, and finally had to be settled in court.

The SS Moena on the
Belawan quay complex
(Sumatra), 1928.

The caissons used in Tandjong Priok were more successful. There the harbour director, A.J. Dijkstra, had adapted the design so that they were more stable and easier to mass-produce. After that the engineer J.W. Maas of HAM devised a way of making caissons even faster and in series. In a relatively short space of time it thus became possible to extend the harbour at Batavia by giving it two extra basins. The caisson design that came to be known as the Priok-style was later used during harbour construction work in Rotterdam.[102]

The general advisor for port affairs

The specialisation in port affairs, already triggered during the construction of the harbour at Batavia, was given a new kind of impetus by the Kraus and De Jongh committee. Perhaps the most significant step taken was that of the appointment of a general harbour advisor at the Department of Public Works.

In 1914 an engineer by the name of Wouter Cool had formed part of the committee called upon to help resolve the difficulties that had arisen during the execution of the various harbour works in the East Indies, notably in Soerabaja and Makassar. Cool was also someone who had been drawn from the Rotterdam Municipal Works pool where he had served under De Jongh for many years and had witnessed Rotterdam's rise as a world port. In 1915 Cool was appointed general advisor for harbour affairs which was a newly created position. Cool, who was instantly put in charge of the Harbour Affairs bureau, had to completely reorganise the East Indian harbour system. The appointment did give rise to unrest in BOW ranks because Cool was not very familiar with the East Indies.[103] Yet again, the government had refused to appoint an East Indian engineer or civil servant and had chosen in preference a Dutch expert. Ironically it was precisely Cool who was to make an important contribution to the emergence of a specific 'East Indian type of harbour expertise'. Just as had been the case in Rotterdam, Cool was keen to learn about every detail of the specific local circumstances. From the very time of his appointment Cool travelled around to the different harbours in the archipelago where he not only spoke with harbour administrators or shipping representatives but also with railway authorities, trading agents and plantation owners in the hinterland.[104]

Cool was also the initiator of a series of publications on harbour matters in the East Indies including a collection of technical problems where the solutions found were of special relevance to the East Indies.[105] Cool commenced his *Technische lessen* (Technical lessons) by describing the general East Indian problem of shifting sandy coasts leading to the situation that the river estuaries, where most harbours have sprung up, tend to silt up or even slightly change their course during the west monsoon season. The first pages were therefore totally devoted to the matter of dredging and monitoring the depth of navigation channels in harbours such as Banjoewangi (East Java), Bandjermasin (Borneo) and, of course, the western sailing channel (Westgat) of Soerabaja. The next thing to be discussed was the construction of dams, such as in Tandjong Priok. After the issue of harbour access came the matter of the building of the harbour components such as landing stages and quays. All the experience gained with reinforced concrete construction and caisson construction, all of which had been widely applied for many years while extending harbours, was described extensively in Cool's book. In his Technical lessons Cool was not so much preoccupied with

Emma Harbour near Padang (Sumatra), 1919.

reporting all the successes as with examining in detail what had gone wrong and why. Cool emphasised that he was not doing that to criticise the work of fellow engineers but rather in order to learn from the mistakes made. What he was proud to note was the fact that precisely when unforeseen problems arose, engineers proved able to display their 'talents to improvise in amazing ways'.

The Technical lessons can be regarded as an accumulation of East Indian technological harbour know-how. It was no longer the case that only European standard works were consulted, specific East Indian situations could also be seen as exemplary for new construction projects. What was also very important was the manual for harbour management, as the directors and harbour masters in the entire archipelago would be basing their policy on similar instructions and notions.[106]

In 1920, the last year of his appointment as general advisor for harbour affairs, Cool published the book *Nederlandsch-Indische havens* (Dutch East Indian harbours) in which in two sections – a text section and a map section – he provided an overview of the harbour network in the archipelago. In his book Cool explained the hierarchical divisions as laid down by himself when reorganising the port network. He listed seven large harbours: Tandjong Priok (Batavia), Soerabaja, Semarang, Tjilatjap, Makassar, Emma Harbour (Padang) and

Belawan (Deli), eight medium-sized harbours: Cheribon, Banjoewangi, Amboina, Menado, Badjermasin, Pontianak, Palembang and Benkoelen and the remainder were all classified as smaller harbours. The privately owned harbour of Sabang was dealt with in a separate chapter.[107]

Of all the large harbours listed above it is only Tjilatjap that has not yet been dealt with in this chapter. That harbour, situated on the southern coast of Java, was really the one with the best natural anchorage of all on the entire island. The sheltered mouth of the Dona river was naturally at least 7 m deep but it was not on any main international or local coastal shipping routes. Apart from Tjilatjap and the more westerly situated Wijnkoops Bay (with its harbour of Pelaboehan Ratoe) there were no good anchoring places along the steep and rocky south coast of Java. In its hinterland Tjilatjap had to withstand sharp competition from the northern Javanese harbours of Tandjong Priok and Semarang which catered for much more frequent shipping services. The Preanger area was a hinterland with great potential for Tjilatjap. The mountainous region of West Java saw rapid economic growth in the post-1870 era, particularly in tea and coffee production but instead of taking those products to Tjilatjap, much of what was harvested was transported to Tandjong Priok: Batavia was the main trading centre, the port facilities there were superior to those provided on the south coast and there was the bonus of a rail connection. In the 1880s Tjilatjap was linked up to the railway network by the line that led to Djokjakarta. The rail link helped to boost the harbour's importance and so in 1886 screw pile landing stages were constructed and a quayside area with loading cranes, customs sheds, warehouses, coal storage bunkers and other modern harbour facilities was created.[108] It was especially for Central Java's sugar industry that Tjilatjap became a good alternative to the much-maligned harbour of Semarang. As trade traffic grew it became necessary, in 1910, to introduce a second jetty which would extend further out to sea so that ships requiring a draught of 8 m could berth there. At that stage the storage facilities were also expanded. As the freight level continued to grow – between 1910 and 1914 it rose by 45 percent [109]– it would once again be necessary, several years later, to increase the mooring and storage capacity but according to Cool before that could happen the harbour would have to be radically reorganised, not least because on the landing stage side there would be no more space for extra storage facilities. The only place for expansion was behind the landing stage. The users of the harbour, such as the shipping enterprises, were averse to the plans produced by BOW engineers and advocated that the screw pile landing stage should be widened. In 1919 the government did eventually agree to the plans for large-scale renovation so that in view of the existing harbour situation a new quayside wall could be constructed alongside the river.[110] In the early twenties, though, the government dropped all the harbour renovation plans because for several years there had been a decline in trade and shipping.

The publication of *Nederlandsch-Indische havens* may be viewed as the rounding off of a period in which East Indian harbours had seen radical change and had been modernised. In the space of a mere half century notions about the functions of harbours, about how they should be equipped and how they should interrelate had changed tremendously. They had gone from being simple transfer points dominated by customs services to being complete harbour industries that formed a solid link between the two growing sectors of trade and shipping.

By the 1920s a harbour was no longer seen as an autonomous entity but rather as part of an East Indian network. This was especially evident in the writings of Cool who felt that overall harbour management was the business of central government:

> *Harbours are not only objects of national importance but also of international significance and even more so in an archipelago than on continental land; clearly defined state intervention is thus essential. Such a situation can partly ensure that a harbour is not negligent when it comes to meeting its general duties, does not doze off because of lack of competition or become a financial risk by simply wasting money.*[111]

To conclude, it may be asserted that the 1910-1920 period was decisive for East Indian harbours. The rigorous research carried out by Kraus and De Jongh triggered a very active kind of port policy in which the actual harbours were modernised and their managing made more commercial. The execution of all the main large-scale projects was, to a large extent, the responsibility of Cool who, in his capacity as general director for port affairs, pursued the line first followed by Kraus and De Jongh when strengthening ties between the various ports. By about 1920 most of the bigger works had been completed, it was just the expansion of Belawan and the excavation of the third inner harbour for Tandjong Priok that that were not rounded off until the mid-twenties. The harbours were run like enterprises which meant that the income from port activities not only had to cover day-to-day expenses but that – as far as possible – the investments made had to be earned back. The harbour enterprises also rented out facilities such as cranes, warehouses and terrain and charged for the supplying of water, electricity, housing et cetera. Other activities carried out in and around the harbour were left, as much as possible, to private companies. In the 1910-1920 period the foundations were laid for a harbour network that would remain in tact in the archipelago for at least another fifty years.

In 1930 the *Handbook of the Netherlands East Indies* (published by the Division of Commerce of the Department of Agriculture, Industry and Commerce) gave an overview of the harbours of the archipelago. The editors established with undisguised pride that in a space of 20 years, though not quite all the goals had been achieved, the Dutch-East Indies had come to possess a series of modern, well equipped harbours easily able to compete with harbours overseas.[112] In the thirties no major physical changes were made to the harbours. Between 1874 and 1937 shipping activities in East Indian harbours reflected a steady growth curve with the exception of a slight recession in the 1914-1918 period when the First World War was in full swing and between 1930 and 1933 which marked a period of widespread economic depression. Touwen calculated that in twelve selected harbours in the Java Sea area, including Tandjong Priok, Soerabaja, Semarang and Makassar, the volume of freight shipped internationally had increased by four percent per annum going from one million m^3 in 1874 to almost 13 million m^3 by 1937. Inter-island shipping volumes accelerated even more rapidly to 70 million m3 in 1937 in the Java Sea ports and 140 million m^3 for the all the harbours in the archipelago.[113]

After 1942

In February 1942 Japanese bombers managed to incur considerable damage to the harbours of Java. First Soerabaja, where obviously the naval base constituted a strategic target, was attacked by the Japanese and that was followed a few weeks later by the bombing of Tandjong Priok. Before the Japanese proceeded to overrun the remainder of the archipelago the port managers sabotaged their own harbours, for instance by sinking ships in the harbour entrance. During the three years of occupation the Japanese did little on harbour maintenance work so that after 1945 the harbours were in a bad state of repair, especially in Java. The returning colonial power attempted to renovate the harbours so that trade and shipping activities could once again be restored. Also in connection with the bringing in of troops as an answer to the Indonesian struggle for independence, it was crucial for the colonial government to keep the harbours open. During that turbulent period rebuilding activities were slow and cumbersome. During the early years of Indonesian independence it became clear just how dilapidated the harbours had become after almost a decade of combat and how neglected the maintenance work had been. The harbours were, of course, plagued by the familiar silting up problem. According to a report published in 1951 by the American engineering bureau of J.G. White it would take five years to get the Javanese harbours back to normal using even the most modern of dredging equipment.[114]

In the Indonesian press there were weekly complaints about congestion in the harbours, notably in Tandjong Priok. Ships were unable to find space to moor, goods could not be stored and overland traffic was unable to access the harbours. Even the shipping lane between Tandjong Priok and the city, by then known as Djakarta rather than as Batavia, was silted up.[115] The Indonesian government developed plans for harbour improvements and also created new quays in Tandjong Priok but because of the political and economic troubles it was impossible to thoroughly tackle all the problems with infrastructure. By the end of the 1960s a number of quays in the harbours of Tandjong Priok, Soerabaja and Belawan had simply been rendered useless by the extreme seabed silting up processes. The rather sombre state of affairs in Indonesian harbours was reported in a survey conducted by a United Nations committee:

> *In the case of many valuable assets such as quay walls, port buildings, roads, rail tracks, dry docks and slipways repair was long overdue, some structures having fallen into disuse. Many small harbour craft were unserviceable and even more lay sunk alongside jetties. Many quay cranes and other items of cargo-handling equipment were out of commission, while port workshops were not fully operational. Port radio equipment was hard to use as most of the sets were out of date and did not have t spare parts.*[116]

The harbour network needed to be thoroughly reorganised and that need was only amplified by the emergence of container distribution. In the seventies, containerisation was to create a revolution in world shipping which meant that a new era had dawned for harbour activities everywhere, an era that will not be further explored in the context of this chapter.[117]

Conclusion

In the space of a century and a half, a maritime network had developed in the Dutch East Indies in which harbours were the essential nodes. Harbour development in the East Indies really falls into three phases. The first period, during which there was no specific harbour policy, continued up until 1870. During the second period, starting in 1870, a whole new concept emerged. For the first time harbours came to be seen as fulfilling a strategic function in the economic and political development of the East Indies. The creation of Tandjong Priok, a brand new deepwater port for Batavia constructed between 1877 and 1885, marked the beginning of a period in which the development of harbour works in the archipelago was to accelerate. In the third phase, commencing in 1910, this process was further perpetuated leading ultimately to the modernisation of harbours and to the perfection of harbour management and policy. The Kraus-De Jongh advisory committee and the general harbour advisor, Cool, were the ones to instigate harbour expansion throughout the East Indies and to turn it into a comprehensive, hierarchical system.

Just like all the other East Indian systems pertaining to infrastructure, the East Indian harbour system may be seen as a 'socio-technical system', that is to say as 'a social organisation developed in conjunction with a specific set of technological innovations'.[118] What characterises such systems is the fact that they are large-scale, develop in a phased way, have a certain own dynamics and generally enjoy the involvement of various social groups such as companies and government services. Different socio-technical systems can also interlink as is illustrated by, for instance, the combinations: harbours-steam ships or harbours-railways.[119]

Whilst shipping and the railways had developed at an amazing rate the harbours lagged behind during the 19th century. Shipping companies such as KPM, SMN and RL, and railway enterprises like the Netherlands East Indian Railway Company, the State Railways and the Deli Railway Company, were eager to be involved in harbour improvements, as was illustrated by Batavia, Emma Harbour and Belawan. But the developing or innovation of large-scale infrastructure networks, such as harbour works, was extremely capital-intensive.

In the first phase the colonial government hardly intervened at all in harbour matters, confining itself to upkeep but it was the creation of Tandjong Priok that was to mark the beginning of a period of active government harbour intervention. Lurking in the background were also various political-strategic considerations including a protectionist attitude. It was especially the relationship with the British colony of Singapore, which had manifested itself as a formidable competitor on the periphery of the archipelago that had prompted the East Indian government to adopt an interventionist political line in the shipping and harbour arena. As Van Doorn surmised, the harbours thus became an operational instrument in the whole colonial project. What also fitted nicely into that picture was the emergence of the engineer in the docks. Whereas in the past the harbour had been the domain of the harbour master, who was a naval officer, it was the East Indian engineer who was to take his place. In the battle for the new harbour for Batavia it had been the engineer who had triumphed with the choice of Tandjong Priok as a location, a project in which like with a tabula rasa a completely new harbour had risen from the drawing board and materialised.

The East Indian engineers moved up through the ranks of the BOW department to ultimately become decisive figures in the technological and project-oriented approach to harbour development and management. It was no longer natural forces or geography that were the determining conditions for a harbour but rather technological skill and expertise.

With the further growth of trade and shipping round the turn of the century, large-scale, technological solutions started to be sought to the capacity deficit in East Indian harbours. The recommendations of Kraus and De Jongh, implemented by Cool, were in actual fact a continuation of the process of growing intervention on the part of engineers that had been set in motion round 1870. The engineers could deploy huge budgets – far exceeding the restoration funds that had been requested earlier in the 19th century and had always been declined by the government. In the course of 50 years they had created an entire port system, which had even spurred on its own development of expertise instead of relying on standard models from the mother country. Although it took longer for harbours to become the domain of engineers than, for instance, railways or irrigation, the harbour sector did eventually become an important factor in the emancipation of BOW engineers.

The harbours of the East Indies, in close connection with conscious scheduled ocean-going and inter-insular shipping, became instruments in the policy of colonial administration. This policy was directed at integrating the archipelago surrounding Java both economically and politically, the Java Sea being the pivotal zone. Up until then the Outer Regions had been more oriented towards the remainder of South-East Asia than towards Java, but the colonial administration succeeded in integrating the East Indian archipelago more closely and the main Javan sea ports may be viewed as the nodes in this politically determined maritime network.[120]

In the late nineteenth century and the early twentieth century the development of the East Indian harbours converged with worldwide port developments. What we see in Batavia, Soerabaja, Belawan and Makassar is the perpetual search for space: deeper basins, more metres of quay and greater storage space. Physically the harbours came to be more and more divorced from the city whilst economically their positions became increasingly autonomous. The traditional bonds with the urban elite loosened because the harbours had become incorporated into a wider economic system.

Simultaneously the history of East Indian ports clearly reflects the difference between harbours in the colonies and harbours in the West. Tandjong Priok, Emma Harbour and Belawan were essentially bureaucratic and technological creations that had not emanated from a 'natural' evolutionary process. Their history was closely linked to the increasing political and economic hegemony of the Netherlands as coloniser in the archipelago. The ports described in this chapter were in the first place colonial harbours. Colonial harbours had different origins to their counterparts in Western Europe or America as they had been created for the one-sided export of goods from the hinterland. They thus passed through other developmental phases than those described in 'Anyport' or other related models. In contrast to, for instance, Rotterdam – as described by Meijer in his study – ports in the Dutch East Indies/Indonesia were there to fulfil a transito function and did not later develop into industrial or network ports. One may still question whether the end of the colonial era did give rise to change in the colonial nature of the ports. The paradox is that physically and technically Indonesia inherited a modern harbour system but it was a system that had been one-sidedly organised from a colonial economic and strategic perspective. This paradox was one of the factors – of course alongside other factors like internal political turmoil and economic instability – that hampered the development of the Indonesian port system in following decades.

A Javanese dam in the Brantas region (East Java), 1993.

CHAPTER 6 **LIVING WATER**

The development of irrigation technology and waterpower

Table 1. **Sawa acreage in the government-owned areas of Java and Madoera (in 1000 ha)**

Year	Total sawa acreage*	Acreage provided with modern irrigation**	Percentage
1895	2,365.4	50	
1896	-	60	
1897	-	70	
1898	-	80	
1899	-	90	
1900	2,450.1	100	4.1
1901	2,460.4	110	4.5
1902	2,456.8	120	4.9
1903	2,449.0	130	5.3
1904	2,452.5	140	5.7
1905	2,457.7	150	6.1
1906	2.463.3	160	6.5
1907	2,488.7	170	6.8
1908	2,502.7	180	7.2
1909	2,516.7	190	7.6
1910	2,530.7	200	7.9
1911	2,544.7	232	9.1
1912	2,558.7	264	10.3
1913	2,572.7	296	11.5
1914	2,586.7	328	12.7
1915	2,600.7	360	13.8
1916	2,614.7	391	15
1917	2,628.7	423	16.1
1918	2,642.7	455	17.3
1919	2,656.7	487	18.4
1920	2,670.7	519	19.5
1921	2,684.7	551	20.5
1922	2,699.6	583	21.6
1923	2,732.1	614	22.5
1924	2,747.7	646	23.5
1925	2,745.5	678	24.7
1926	2,767.1	704	25.5
1927	2,770.7	730	26.4
1928	2,756.0	757.7	27.4
1929	2,782.3	814.7	29.3
1930	2,840.2	883.4	31.1
1931	2,859.1	934.2	32.6
1932	2,864.5	944.3	33
1933	2,857.5	987.6	34.5
1934	2,881.0	993.1	34.5
1935	2,903.2	1054.2	36.4
1936	2,964.9	1154.7	39

* This information is based on data obtained from land surveys carried out between 1907 and 1920. From 1920 onwards the figures are correct whereas for previous years they have been estimated as closely as possible.

** The figures up until 1928 are estimates based on information extracted from the Civil Public Works (BOW) report. The years 1900, 1910 and 1919 serve as the reference points.

Source: Happé, 'Eenige beschouwingen over bevloeiingswerken op Java en Madoera', 1, II.26.

Graph 1. **Annual government expenditure on East Indian irrigation works (in millions of guilders)***

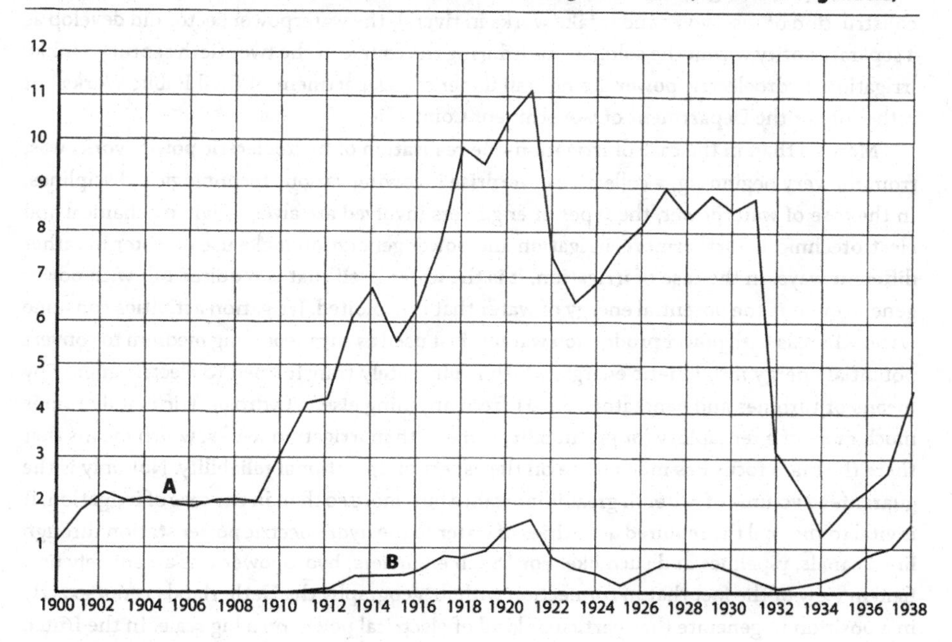

A - Spendings on irrigation works in the East Indies
B - Spendings on irrigation works in the Outer Regions
* Research and construction.

Source: Van der Meulen, 'Irrigation in the Netherlands Indies', 149.

Java has a large number of canals, dams, aqueducts, reservoirs and other irrigation works, all of which serve to support agricultural activities. At present there are more than 110 million people living on this main island of the Indonesian archipelago, all in close proximity to each other. About half of the population is employed in the agricultural sector and the staple diet is rice. It is therefore understandable that these works, which constitute components of networks or systems that are sometimes quite extensive, are of vital importance to the people of Java. Just like rice cultivation, irrigation – the providing of artificial water supplies for agricultural crops – has been familiar in Java for a long time but the basis for the present systems was laid at a time when the Netherlands still had considerable influence as a colonial power. In the space of over 100 years, the Dutch East Indian state in Java has made around 1.3 million hectares of arable land suitable for rice cultivation with the help of 'technical irrigation' (i.e. irrigation works designed and constructed by engineers). Towards the end of the colonial era that amounted to some 40 percent of the total acreage of irrigated rice fields or *sawas* (see Table 1). Elsewhere in the archipelago around 100,000 hectares of agricultural land was provided with modern irrigation facilities. (See for all the government expenditure on irrigation works Graph 1 and also Appendix 1). The hydraulic infrastructure was also, and still is, deployed for another purpose, namely for hydropower. Though in certain respects irrigation and hydroelectric ends have similar features (both require the construction of reservoirs and intake works in rivers), the waterpower sector did develop as a separate entity within the colonial state during the course of the twentieth century. Unlike irrigation, hydroelectric power did not fall under the Department of Civil Public Works but rather under the Department of Government Companies.

More so than in the case of irrigation, the realisation of hydroelectric power works was, from the very beginning, a collective enterprise involving various technological disciplines. In the case of waterpower, the types of engineers involved are always civil, mechanical and electrotechnical. Furthermore, irrigation and power generation make use of water in rather different ways. In the case of irrigation, it is the water itself that is required but with power generation it is the potential energy of water that is exploited. Irrigation activities consume water whereas with power production water is just used as a transporting medium to convert potential energy into kinetic energy, which is ultimately transformed to electric energy by means of turbines and generators. Apart from anything else, a hydroelectric installation is much more of a rentability (or profitability) object[2] than irrigation works, which means that there the main focus lies much more on the aspect of operational reliability. Not only is the guaranteed volume of water of great importance but, more so than in the case of irrigation, it is vital to channel the required quantities of water to the hydroelectric power station through lined canals, pipelines and such like. For Dutch engineers, hydropower was an entirely new field in view of the fact that in connection with its geography the Netherlands is not exactly in a position to generate that particular kind of electrical power on a big scale. In the Dutch East Indies, however, the conditions were right because of the huge level differences in the river basin areas where rivers and lakes provide the necessary discharges and volumes of water. Though tropical irrigation was a new field for Dutch engineers who, in their own country, had been more occupied with water drainage systems, it was not a field that was completely alien to them.

This chapter reveals how modern irrigation and waterpower developed in the Dutch East Indies while demonstrating what precisely the other developments were that went hand in hand with those particular areas of progress. In the first place there is the transformation process of the East Indian state to consider, in which exploitation and the maintaining of order on the basis of a simple administrative organisation made way for the support of entrepreneurs and welfare care backed up by a diversified and specialised government bureaucracy.[3] The irrigation development process corresponds to that process of state formation: important structural works were created during the early colonial state period (before 1870), the first irrigation systems were realised during the period of transition from traditional to modern administration (up until 1920) and all the other systems were developed during the time of the modern colonial state. The expanding role of engineers and the formation of a service-oriented and caring state may, in effect, be viewed as two different facets of the same process of modernisation within East Indian society. When it came to the development of waterpower, it may be asserted that as of 1885/90 there was evidence of a cautious switch from private initiative to state intervention. There, too, a turning point was to be seen around 1920. At that time large structural works were either being built or were in the pipeline and there were clear signs that a state-steered organisational structure was about to emerge.

The development of modern irrigation and waterpower in the Dutch East Indies was partly also made possible by advances in the fields of mathematics and physics, especially in the area of geometry (land surveying or geodetic surveying) and in the field of hydraulics (the theory and principles of flow) and, partly as a consequence of all of that, in the area of (East Indian) hydraulic engineering. Developments in hydrology (river gauging) and hydraulics (also through model-based research conducted in hydraulic laboratories) were important both to the field of irrigation and to the field of waterpower. Hydropower engineers made grateful use of all the experience gained from irrigation activities. When it came to the matter of designing and constructing hydroelectric power plants, much expertise and knowledge was gathered and imported from countries such as Switzerland, Scandinavian countries, France, Canada and the United States.

Developments in technological and political-administrative fields converged via the groups or organisations that were involved. In the case of irrigation and waterpower these include, for instance, engineers (working with the Dutch East Indian Public Works and the Government Companies departments), inland administration civil servants (Inland Administration) and later also agricultural experts (the Department of Agriculture). By the time the importance of hydropower and thus also of hydropower engineers had increased, the battles between the various services within the colonial state organisation had largely been fought. Much more than was the case with irrigation activities, hydroelectric power developments took place within an arena where international companies and financiers were active alongside the East Indian state.

Pioneers' works

In the first half of the nineteenth century Dutch engineers started creating small-scale irrigation works on the island of Java. The main concern was to build permanent weirs in rivers in order to back up the existing irrigation systems that served to support rice cultivation in the rainy period of the west monsoon season (November until April). Generally the works of engineers supplanted the older structures created by Javanese people or works that had been commissioned by the local colonial administration but had been built under forced labour conditions, what was commonly known as corvée. Later more and more engineering works started to spring up in new locations but everything still went hand in hand with traditional irrigation. The population, which had definitely been familiar with traditional rice cultivation since the beginning of the Christian Era, made weirs from wood (bamboo and bundles of branches) and stones (sometimes placed in conical baskets). Those were temporary measures: each year the dams would either be washed away in the rainy season or, just like the connecting canals, they would be seriously damaged.[4] The works constructed by the local colonial administration were larger than those of the local people but otherwise very similar. In effect the colonial administrators followed in the footsteps of the Javanese rulers before them by providing what was just a larger-scale version of the existing technology. Likewise, those structures were perpetually subjected to renovation. It was the engineers who constructed works that were expected to be more durable; in time their works were indeed to become permanent. At first they also made use of wood but as time passed they started to replace that with materials such as brick ('Waal bricks' transported from the Netherlands as ballast in ships) and concrete. The rivers, which had to cope with sudden large flows (i.e. *bandjirs*) in the rainy season and thus often burst their banks, were not so easy to control. The painstakingly slow process of erecting a fixed weir in the Sampean river well illustrates the complexity of the problem.

In 1832 the East Indian government in Batavia sent the engineer C. van Thiel to the Sampean area in East Java with orders to replace the temporary dam in the Sampean river with a permanent one. The irrigation area in question affected by that river was approximately 10,900 ha. The dam that Van Thiel went on to construct consisted of a wooden framework (made of *djati* wood or teak, a tropical hard wood) placed between masonry supporting walls that was filled with river stones. It came to be known as a framework dam and it was the first modern irrigation structure to be built in Java, thus making the Sampean delta the first area in the East Indies to be technologically irrigated. It was a success but it was also subject to wear and tear. Another engineer, S. Dik, therefore constructed a new dam that was completed in 1852. What was special about that dam was that it had been constructed so that the water flowed over it in a sidelong fashion: it was a proper overflow, Javanese-style. The framework dam and all the maintenance it required had turned *djati* wood into a scarce commodity in the area but the new dam was made of stone. Dik's structural work did not last very long because in 1857 the river destroyed a large section of one of the guiding walls and subsequently started flowing in a different direction. For almost 20 years the local authorities were then burdened with the task of renewing temporary dams at that spot on a yearly basis. Later that came to be described as a disaster for the area:

> *It may be asserted that these temporary dams ultimately depleted the whole region. Up until 1872 the work consumed an average of 25000 day shifts per annum (i.e. an average of 300 labourers per day, each and every day!), forced labourers were drawn from an eighteen pole region (approximately as far as Bondowoso and Besoeki [some 27 kms]). The dam devoured entire forests full of unplanted woodland wood and bamboo (djati had long since been virtually eradicated) so that in 5 of the 6 districts of the Panaroekan division there were only about 125000 bamboo shoots left. [...] The native civil servants had little choice but to work for and on the dam. If, in return, the irrigation had been satisfactory then the efforts would at least have been somewhat rewarded but unfortunately that was not the case and the prosperity of the region visibly deteriorated.*[5]

In the end, a weir was constructed that had been designed by J.C. Schumm, another engineer. That weir reached completion in 1876. Because of the tremendous force of the water considerable improvements needed to be made. Ultimately it proved impossible to control the situation and so it was decided that a canal should be dug around the weir. One of the engineers who was involved in the planning and excavation of the canal was H.H. van Kol, later an SDAP (i.e. a former Dutch socialist party) parliamentarian in the Lower House. The canal that was completed in 1887 was intended for *bandjirs* only, but in practice it provided a new course for the Sampean. A rocky shelf in the river functioned as a new fixed weir. In 1900 an ejection sluice was built into the old dam so that the structure was given the appearance that it has retained until the present day.

The building of the weir in the Sampean and other such structural works was all the result of colonial policy. At that time a policy known as the Cultivation System was in force. Two of the commercial crops that the Javanese were forced to cultivate under that system: sugar cane and indigo (which was less important) required irrigation. Those crops were grown on part of the irrigated area that was in use for rice cultivation. Most of the sugar cane was cultivated in East Java, including in the Sampean delta. It was precisely in 1832 at the time of the building of Van Thiel's dam that the Cultivation System came into effect throughout Java. The provision of irrigation water was an important aspect of colonial policy. In the eighteen-forties, in conjunction with all the other services that the Javanese were compelled to provide, the system ultimately led to mass starvation in Central Java. As a result, subsequent irrigation works were also created with a view to improving agricultural circumstances for the local inhabitants. The series of famines also led, in 1854, to the establishment of a civil public works service, the so-called Bureau of Public Works (in 1866 transformed into the Department of Civil Public Works or BOW) and the agency's main task was to construct irrigation works in Java. The irrigation engineers fell under the control of the civil servants, especially the heads of regional administration: the residents, who were affiliated to Inland Administration (Binnenlands Bestuur, BB). The person who usually took the initiative when it came to irrigation matters was the resident and only occasionally would he draw on the experience and expertise of engineers. In reality the establishment of East Indian Public Works made little difference to the situation as BB remained in charge of water-related affairs. Public Works also had to cope with limited financial means and a shortage of personnel and so it was not even always able to meet the requests for technical aid, as was seen in the years after 1857 in the Sampean area. There was also a problem of a different nature though. The engineers who had first been militarily educated but who later, after the establishment of the engineering college in Delft in 1842 had become civil engineers, knew little about irrigation engineering. They were highly critical of the works of the Javanese and of the local administration officials but despite boasting about how scientific their approach was there remained little, at first, to distinguish their efforts from those of the locals as was borne out by events in the Sampean region.

Meanwhile the irrigation engineers strived to gain more power and more influence and thus became embroiled in a power struggle with the Inland Administration people. One particular engineer who managed to do much to improve the position of engineers in general was H. de Bruyn who was twice director of Public Works in the 1861-1877 era. He not only linked the construction of modern irrigation works to financial gain but also to prosperity, progress and civilisation. Even though it was thanks to him that in many areas proper background research was undertaken, the number of irrigation works realised before the final decade of the nineteenth century remained minimal. The BB continued to occupy a dominant position. Over the course of time the quality of the works being constructed did improve and that was partly why the struggle for emancipation on the part of engineers started to be successful. One of the main keys to success for the engineers was their capacity to control the circumstances within which the works had to be realised or to at least predict certain events. One important activity was the building of weirs in Javanese rivers and that involved two main aspects: the creating of proper structural foundations and the insight into the river's flow regime, which was a hydrological matter.

Precipitation and river discharges

Regarding the hydrological issue, it was the engineer A.P. Melchior who developed a clever method. Melchior's method was to remain in use throughout the colonial age and to even extend beyond that era, despite the sometimes critical comments that were made. It was because of the dam problems in the Sampean river that Melchior was inspired to calculate how maximum discharges for Javanese rivers could be estimated in relation to rainfall.[6] In the Netherlands there was no such thing as precipitation-discharge formulas (hydrographs) because the big Dutch rivers had a flow pattern that did not really lend itself to such an approach. Since the engineers had no available national examples they looked across their borders to find appropriate examples. Round about 1890 the engineer J.E. Meyier (who will be referred to again later in this chapter) came up with a formula that had been developed in Austria by R. Lauterburg and had been published in 1877.[7] Lauterburg presumed that the maximum discharge was a function of the surface of a river's basin area, its so-called catchment area. The time dimension (for instance in the form of the duration and intensity of rain storms) is only roughly included in the calculations and the river's length is not considered at all. What was viewed as a great advantage of the method was the fact that only a few discharge measurements were necessary in a given area to be able to establish the value of the various factors. With those values it then became possible to estimate the discharges of other rivers in the area.

In the case of the European rivers to which the method was applied the results gained from the Lauterburg formula were viewed as fairly accurate (with deviations of 15 to 20 percent between measured and calculated discharges) but in the case of Javanese rivers the average differences were 65 percent![8] Especially in areas where the catchment areas were small (smaller than 300 km^2, a normal size for Java) the Lauterburg formula produced much larger peak values than had up until then been registered.[9] Melchior's method produced results that seemed to correspond more to the maximum discharges measured in the 1890/95 period, though for the bigger catchment areas the estimations appeared to be a little on the low side. Melchior not only based his formula on Lauterburg's calculations but also on the reasoning of another Austrian, C. Pascher, who had developed his method to calculate the peak discharges of the Wien river in Austria.[10]

The Melchior formula was as follows $Q = (\alpha * q * F)$, in which Q represents the maximum discharge that has to be calculated (in m^3/s), F stands for the surface area of the catchment area (in km^2), q stands for the predicted highest local rainfall level (expressed in terms of m^3/s per km^2) and α for a factor that indicates what proportion of the rain immediately superficially flows away, the so-called runoff factor. Melchior combined graphical and numerical steps in his calculations. The surface of the river area was determined by looking at a map and drawing an elliptical figure around the catchment area (the shorter axis of which had to be at least two-thirds the size of the longer axis). Alongside that the length and average gradient (longitudinal slope) of the river had to be known (with the exception of the highest ten percent of the area) and, of course, rainfall data had to be known. Melchior furthermore took a value of 0.52 for factor α. (See the calculation example using the Melchior method for the Pekalen river in East Java).

Maximum discharge calculation example for the Pekalen river (East Java) using Melchior's method

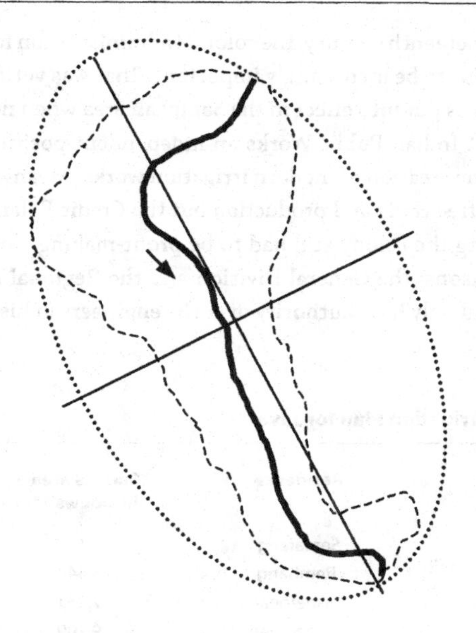

1. Acreage of the catchment area $F = 169$ km².
2. The longitudinal axis of the ellipse is 28.4 km, so the determined length of the short axis is $2/3 \times 28.4 = 18.9$ km. The resulting area of the ellipse is then $nF = 1/4\ \pi \times 28.4 \times 18.9 = 422$ km².
3. The length of the river is $L = 39.2$ km. By omitting 10% of the river alignment in the highest reaches of the catchment area the remaining 35.3 km has a free drop of about 1700 meters. The resultant average hydraulic slope is thus $i = 1700/35300 = 0.0480$.
4. The maximum rainfall of each of the 4 rain gauges - inside or outside the river basin - is 146, 165, 244 and 236 mm respectively, leading to a derived average maximum rainfall $h = 1/4\ (146 + 165 + 244 + 236) = 198$ mm, say 200 mm.
5. On the basis of $nF = 422$ km² and a special table a value of 3 is assigned to q. Then $Fq = 169 \times 3 = 507$ km². By maintaining the average hydraulic slope $i = 0.0480$ and by using a special graph an average flow velocity $v = 1.35$ m/s is obtained.
   The resulting flow time $T = 1000 \times L/ 60\ v = 39200 / (60 \times 1.35) = 484$ minutes.
   Subsequently, by using another graph, a more exact value of q is determined which is
   $q = 3.8$ m³/s per km².
6. By adopting this latter value of q as the basis the calculation is repeated, thus leading to the following results: $F \times q = 169 \times 3.8 = 642$ km², $v = 1.42$ m/s, $T = 460$ minutes, $q = 3.95$ m³/s per km². A following, repetitive, calculation renders no better value of q.
   A value of $q = 3.95$ m³/s per km² is thus accepted for the last computation which is an augmentation of q by 8% on the basis of $T = 460$ minutes and the use - once more - of another table, resulting in $q = 4.27$ m³/s per km².
7. Ultimately, the calculated maximum discharge is $Q = \alpha F q = 0.52 \times 169 \times 4.27 = 372$ m³/s.

Sources: Nijman, 'Bepaling van de maximale afvoer van rivieren volgens Melchior', 12; Van Maanen, *Irrigatie in Nederlandsch-Indie. Een handleiding bij het ontwerpen van irrigatiewerken ten dienste van studeerenden en practici.*

Irrigation and management systems

Towards the end of the nineteenth century, the colonial administration found research of the type conducted by Melchior to be increasingly important. That was verified in 1885, shortly before the final solution was put into effect in the Sampean area when new regulations were introduced that gave East Indian Public Works an independent position and, at the same time, a new impetus for the creation of modern irrigation works. Meanwhile the Cultivation System had given way to free trade and production but the Credit-Balance Policy continued to exist: for the time being the colony still had to be profit-making. The BOW department was divided into two divisons: the General Division and the Regional Division. From that time on the resident would only have authority over the engineers in his particular region in

Table 2. **The1890 General Irrigation Plan for Java**

Irrigation projects*	Residence	Surface area in 'bouws'**	Construction period
1. Demak	Semarang	47,554	1853-1894
2. Kening	Rembang	3,440	1889-1890
3. Pategoean	Pasoeroean	2,853	1887-1891
4. Pekalen	Pasoeroean	9,709	1883-1893
5. Manggis	Kedoe	6,110	1850-1908
6. Tjihea	Preanger	6,900	1891-1901
7. Babakan	Pekalongan	3,891	1890-1894
8. Kaboejoetan	Pekalongan	5,625	1890-1902
9. Pemali	Pekalongan	43,745	1893-1903
10. Tjomal-Tjatjaban	Pekalongan	37,402	1896-1906
11. Sampean	Besoeki	15,387	1875-1901
12. Indramajoe	Cheribon	48,671	1857-1917
13. Genteng-Sragi	Pekalongan	17,233	1892-1919
14. Magetan en Ngawi	Madioen	35,457	1891-1922
15. Zuid-Bagelen	Kedoe	35,540	1864-1906
16. Serajoe	Banjoemas and Kedoe	18,711	1880-1919
17. Solovallei	Rembang and Soerabaja	223,000	Stopped in 1898
18. Loesi	Rembang	21,000	Not executed
19. Sindangpitoe	Djokjakarta	36,000	Not executed
20. Tangsi	Kedoe	2,875	? -1909
21. Molek	Pasoeroean	5,669	1901-1904
22. Kedoengkandang	Pasoeroean	6,935	1904-1915
23. Tjioedjoen	Bantam	31,947	1905- ?
24. Djengkellok	Pekalongan	10,409	1904-1910
25. Kertosono en Waroedjajeng	Kediri	20,315	1903-1911
26. Zuid-Djember	Besoeki	70,000	1908- ?

* Projects 20 to 26 were conceived later.

** 1 bouw = 0,71 ha.

Source: Ravesteijn, *De zegenrijke heeren der wateren. Irrigatie en staat op Java, 1832-1942*, 162.

The fixed weir in the Pemali river (Central Java), a. in the rainy season, 1995; b. in the dry season, 1993.

The distribution sluice at Songgom that forms part of the Pemali irrigation system during the wet season of 1995; the water is coloured brown by the silt.

terms of the latter division, which was much smaller in its totality and less significant than the former. The General Division that covered the entire archipelago comprised two already well-known units: the Technical Office and the regional Public Works divisions.

It was really through the establishment of the third service unit: the Irrigation Brigade that any kind of new élan emerged. The division was requested to carry out preliminary investigations in preparation for the provision of irrigation services for all the government land that was being considered for rice cultivation. A number of years later the brigade was integrated into the Technical Office. The regulations furthermore introduced a new approach to irrigation questions. The approach was not only based on the notion that it was essential to collect data beforehand but also on the idea

> *that it was necessary to consider irrigation works in their entirety and as cohesive organisms unable to operate properly unless the efficient operation of all the various components is carefully considered and ensured.*[11]

The details on rainfall levels, water transport (flow rates), water requirements, and such like, would have to lead to the development of proper irrigation systems complete with head works and whole networks of canals created for the purposes of distributing and transporting water. The activities began to take shape when, in 1890, an overall General Irrigation Plan was produced for Java that was to comprise 19 projects (see Table 2). Attention was also given to the managing of irrigation works and it was decided that the concept of 'technical' water management should be introduced. As of 1888 there were Irrigation Divisions that had been created for the exploitation and maintenance of bigger works as well as for the completion of smaller works. These control units were originally supervised by BB but later that responsibility passed to Public Works (the department to which the divisions had belonged from the start). Methods were also developed for the fair distribution of water so that it was not only the way water was distributed amongst the rice fields that was taken into account but also the apportioning of water to sugar cane irrigation and to rice irrigation.[12] In general, sugar cane and rice were irrigated at different times. According to the generally observed day and night ruling, sugar cane would be watered in the daytime and rice at night. Later, round about 1920, other alternatives were developed, one of which was the *wadoek* system which involved collecting water at night in small *wadoeks* or reservoirs and then gradually distributing it during the daytime. When it came to the matter of rice irrigation, various experiments were carried out in the last years of the nineteenth century to ascertain the level of involvement of the local population. This is illustrated by the efforts made in the Pemali region.

In the dry east monsoon period, the Pemali Plains along the northern coast of Central Java constituted an arid area but in the west monsoon period the same area could be very wet. Those were situations that were not only unpleasant for the local population but also for the local sugar refiners. Efforts to improve the water situation commenced with the construction of a weir in the Pemali river designed by Van Kol. While that was under construction an engineer known as A.G. Lamminga (who had previously been active in the Pekalen area and who had adopted there a systematic approach to design[13]) came up with a follow-up design in 1896: a general plan for irrigation in the area including a preliminary design for a complete irrigation system with supply and drainage canals and specific smaller irrigation works. The whole plan was based on a great deal of research, including research into water flow rates in the Pemali and rainfall levels in the river's basin. The works were completed in 1903 and, as a result, some 31,100 ha of land could be continually irrigated. Lamminga also involved himself in the development of a management ruling, the main principle of which was based on a traditional method where the *oeloe-oeloe* (i.e. water master) was an important official. That water master was always one of the village elders. Under the Pemali ruling the *oeloe-oeloe* was reinstated and given an independent position.

The Pemali works were successful and became exemplary for later works. Much the same applied to the management ruling. Lamminga, who was also active elsewhere in the Pemali-Tjomal area, became known as the founding father of modern irrigation technology in the East Indies. For one year (1910-1911) he also held the position of professor at the Technical College in Delft where, after the turn of the century, a chair was created for East Indian hydraulic engineering. In 1930 Lamminga was posthumously honoured when, in Tegal, a monument was erected in his name. Unlike his works the monument has since disappeared.[14]

Economic aspects

When the first few projects were undertaken not all that much thought was given to the economic side of things. All of that was to change when, in 1897, the Rentability Committee was established for the purposes of examining the economic viability of projects. In other words, the committee had to assess whether all the benefits justified the expenses. This was to mark the beginning of the end of all the relative autonomy that Public Works engineers had enjoyed since 1885. Apart from the engineers, the Inland Administration civil servants were also represented in the committee and were thus again given a structurally important say in matters. The rentability criterion was given new special significance at the time of the Ethical Policy introduced in 1901, which had come about as a direct consequence of the 'reduced prosperity' of the indigenous population, something that had become only too apparent towards the end of the nineteenth century.[15]

The new policy was directed towards improving the situation of the local population, the main policy areas being: irrigation, emigration and education (as formulated by C.T. van Deventer). It went hand in hand with a 'modernisation mission' in which modern technology was central. Ultimately the irrigation engineers who, through their work, had paved the way for the Ethical Policy were once again forced to step down. Irrigation was to serve a higher goal, namely that of stimulating agricultural activity in the interests of the local population. That was to become the touchstone for the rentability of new projects. Around about this time a new group of experts appeared on the scene in the form of the agricultural experts who, as of 1905, became united in the Department of Agriculture, Industry and Trade. The agriculturalists also had a representative in the Rentability Committee. One of the ways in which they manifested their presence was, for example, by showing increasing concern for the possible ways in which irrigation water could be used to transport silt to the fields as silt was essential to the fertility of the soil. The whole economy of irrigation works then suddenly became a rather urgent issue because of the ambitious Solo Valley works project.

In the Solo Valley, the area situated in the lower reaches of the Solo river in East Java, the population led a pitiful existence because of all the floods and droughts they had to withstand. Yet despite all else the Solo, Java's longest river, was still known as the 'auspicious Lord of the Waters' (Bengawan Solo). In 1893 BOW began an irrigation project there which, because of the approximately 158,300 ha of land that needed to be irrigated, was to constitute the largest project of its kind to be undertaken in the East Indies in the colonial period. A main canal was planned that was to be no less than 165 km long. Part of the project involved diverting the mouth of the Solo to the Java Sea. The river deposited so much silt in the straits near Soerabaja that it was feared that ultimately the deposits would be hazardous to shipping. Together all these works, which formed part of the General Irrigation Plan of 1890, constituted a daring sample of engineering technology and something that was impressive even on a world scale. The Solo project could have made its designer and executor, the engineer J.L. Pierson, the big hero of colonial irrigation intervention but unfortunately history was to dictate differently. Early on in the construction phase it became apparent that the budget was going to be massively overstepped and that was followed, in 1898, by a declaration issued by the Dutch government to the effect that the project would have to be suspended. Not even the pressure of a weighty state committee and the support of the engineering world led by the Dutch Royal Institute of Engineers was enough to turn the tables and get the authorities to resume the project. In 1903 and 1905 the entire project was officially called to a halt. The alternative option (which was adopted) was to initiate a small-scale improvement programme including the building of several large reservoirs. The first large Solo Valley reservoir was completed in 1917 and the second in 1933 but debate on the works continued, even after the 'definite' decision of 1929 had been taken not to pursue the matter any further following an inquiry conducted by the engineer D.C.W. Snell. Even today, the Solo Valley as a whole is still awaiting a solution to the problem and the report published by the above-mentioned committee is still very much in people's minds. In the area in question one can still see the remains of the project in the form of half excavated canal sections.

What the Solo works clearly demonstrated was the fact that even modern approaches to irrigation could fail. In this case the problem was, in the first place, down to economic factors. The majority of the members on the special investigation committee viewed the rentability of the works in a favourable light but it was the BOW director, J.E. de Meyier, who had his reservations and thus publicly voiced what was actually a minority view. During the Ethical Policy era the matter of focusing on agriculture for the ends of the local population weighed heavily which thus made De Meyier's reservations very significant so that opposition to the works only mounted. The big question was whether the long canals that had been envisaged would be capable of transporting the silt to the distant fields. Another general concern was the matter of how agricultural activities could possibly be coordinated in such a wide area. De Meyier's point of view gained ground because in the new situation, the East Indian vote carried more weight in colonial decision-making processes than before. The technical facets of the works were also attacked, though that did not come out into the open until a later date. When, around 1910, the engineer and new BOW director, J. Homan van der Heide, endeavoured to breathe new life into the Solo project De Meyier who, from the technical angle, had previously been in agreement with the other Solo Committee members had the following to say:

We are absolutely convinced that the structure intended to back up the water of the Solo river [the weir] would have been in serious danger of collapsing during its very first bandjir had it been possible to dam up the abandoned branch of the river and allow the structure to enter into use. A number of risky structural works would have to be built so that the 165 km-long main canal, which would be 80 m wide at its starting point, could traverse numerous side rivers. Even if the head works were to hold out so that the canal could be well supplied with water, each of these individual works would, in turn, be in danger of being subjected to seepage, annual subsidence and breaches, none of which could be repaired during a single monsoon season. Ultimately the lower sections of the canal would be deprived of water practically annually causing dangerous crop failures.[16]

It was not only the discussion surrounding the Solo works that gave the engineering community cause to contemplate the approach taken. Meanwhile problems had also arisen concerning the Melchior method.

New measuring results

Some 20 years after Melchior's method had been published it emerged that the peak discharges calculated according to his formula were considerably lower than the discharges that actually occurred in reality. Engineers wrote about this in their specialist journal *De Waterstaats-Ingenieur* (The Public Works Engineer). In the Pekalen river, for example, heavy local downpours gave rise to a *bandjir* of around 900 m^3/s and that was exactly the same river where Melchior had predicted that the peak would lie at 372 m^3/s. One of the reasons for the measurement's inaccuracy was the fact that the weir itself also overflowed. Even in the Sampean river basin where the Melchior formula had been developed, the formula's outcome did not correspond to reality. When designing the *bandjir* canal it had been presumed that the Sampean's maximum discharge would be 1000 m^3/s which was slightly more than the maximum calculated by Melchior. Up until then the highest discharge that had ever been recorded had been 1788 m^3/s in 1878! But apart from that extreme amount, nothing higher than 700 m^3/s had been measured since 1876 which was why 1000 m^3/s was taken as the norm. In 1887, though, around 2075 m^3/s passed through the river and in 1916 it even rose to 2700 m^3/s.[17] All kinds of discussions took place and some wondered if perhaps the differences could be attributed in part at least to a changing flow regime. It was particularly the determining of the α coefficient that was subjected to intense debate. Melchior himself was perfectly conscious of the limitations of his method which he described as an 'in no way completely satisfactory way of arriving at a solution to the issue in question'.[18] Despite everything the Melchior method remained respected, just like its inventor, because of the highly scientific way in which the matter in question had been tackled. The discussions then started to focus more upon the circumstances under which the method was implemented.

Melchior's approach stands or falls with the availability of reliable data both on discharges and rainfall so that theoretical maximums can be calculated and then compared with observed data. Generally speaking such data was usually either lacking or incomplete. Melchior had thus been forced to replace much vital data with hypotheses. Rainfall measurements tended to be scarce and were mostly aggregated to daily intervals. However,

Table 3. **Annual government expenditure on large and small irrigation works in Java**

Year	Large works	Small works	Total*
1890	832,000	-	832,000
1891	1.270,000	-	1.270,000
1892	1.547,000	-	1.547,000
1893	1.496,000	-	1.496,000
1894	896,617	-	896,617
1895	965,260	146,973 233,381	965,260
1896	889,628	277,263 234,957	1.123,009
1897	782,528	213,838 214,188	1.059,791
1898	871,879	388,484 377,248	1.106,836
1899	908,934	521,189 494,552	1.122.772
1900	606,556	409,532	820,744
1901	1.136,378	351,72 405,805	1.524.862
1902	1.647,900	624,547 667,040	2.025,148
1903	1.353,447		1.874,636
1904	1.405,412		1.899,964
1905	1.331,715		1.741,247
1906	1.378,650		1.730,373
1907	1.285,702		1.691,507
1908	1.205,600		1.830,147
1909	1.046,425		1.713,465

* The amounts do not completely correspond with the amounts given in Appendix 1. The money spent on the terminated Solo Valley works is not included.

Source: Ravesteijn, *De zegenrijke heeren der wateren. Irrigatie en staat op Java, 1832-1942*, 371.

Moveable weir and inlet sluice that forms part of the Sadang works (Celebes), 1940.

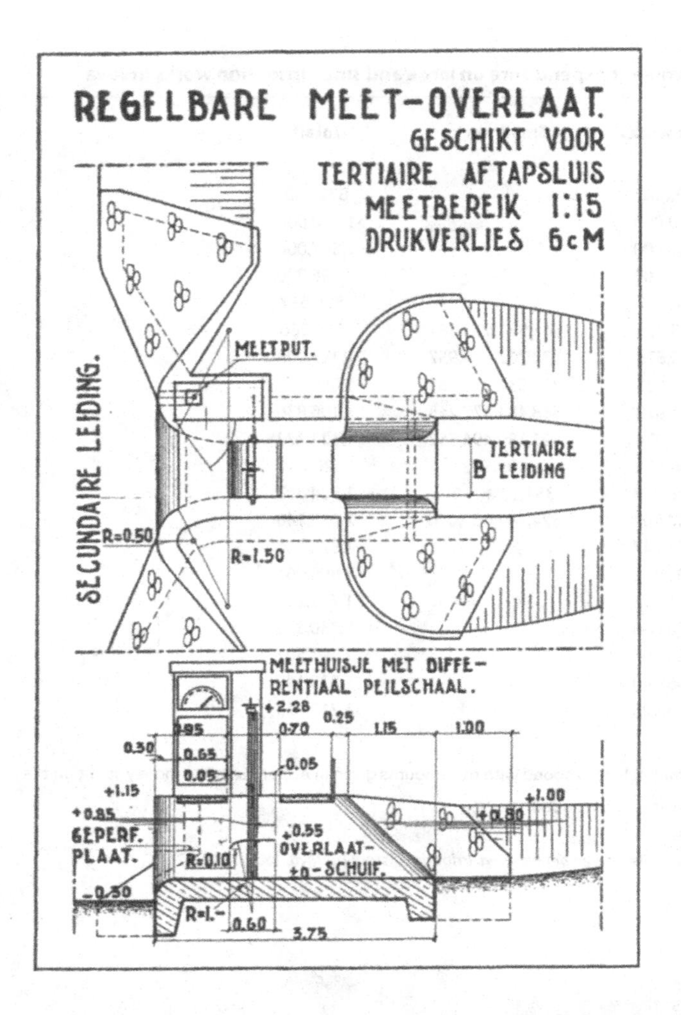

The adjustable broad-crested Romijn measuring weir: the plan, the longitudinal section and the test model (photo), 1932. A truly Dutch-East Indian invention: the sluice used to tap off water for the sawas can both control the flow and measure the amount of water that is allowed through.
(leiding = canal, meetput = stilling well for waterlevel gauge, differentiaal peilschaal = differential gauge, geperf. plaat = perforated plate, overlaat-schuif = broad-crested gate).

when trying to determine peak levels that is simply not sufficient as specific information is needed on the duration and intensity of individual storms. Rainfall intensity is variable and thus easiest to measure with a self-registering rainfall gauge. Since 1902 the Magnetic and Meteorological Observatory has been using such an instrument to measure rainfall levels. At first that only occurred at the three different weather stations of Batavia and Buitenzorg in West Java and Pasoeroean in East Java. It was only in Batavia that a self-registering gauge (which did not work all that well) had been in existence for a longer period of time, since 1879.[19] It was thus no coincidence that Melchior made use of the data collected in Batavia. He also used the data collected in Bagelen in Central Java. He even used European statistics to arrive at the conclusion that at 3 km from the centre of any storm, the intensity of the precipitation will be reduced by half. The only problem was that Javanese storms were not exactly comparable to those common in Europe.[20]

Optimisation

After 1900 a definite preference emerged for small-scale irrigation works designed to improve and help expand rice cultivation and that had partly to do with the engineering debacle in the Solo Valley but it was partly also related to the whole Ethical Policy ideology. Wherever possible, endeavours were made to link up with existing works constructed by the local population (see Table 3). In addition much attention was given to irrigation control and that was something which was, for instance, evidenced in the creation of Irrigation Divisions, also when modern irrigation works still had to be created. Management received priority over construction. In practice, another thing that emerged was a three-corned relationship between irrigation engineers, inland administrators (who had never really let go of the reins) and agricultural experts.

The abandoning of the Solo works project and all the scientific reassessing that was partly a direct result of all of that did not, in the long term, undermine the position of hydraulic engineers in the East Indies, even though many feared that that might prove to be the case. The construction of water works was much too important an issue to be so easily disrupted, also due to the fact that the development of hydropower was gaining popularity and that was somewhere where engineers were absolutely indispensable (see below). In the 1910 to 1920 period, construction activity in relation to irrigation started to pick up and the centre of attention switched to the Outer Regions (see Graph 1).[21] Round about 1920 most of the projects that had formed part of the General Irrigation Plan, including some seven that were introduced later, had been completed (with extra money flowing in for projects that were not remunerative) and the expenditure for new works reached a peak (of around ten million guilders per year). At that time there were more than 200 engineers working for BOW, rising to a maximum of 263 in 1930.[22] Because of the limited availability of Delft (and later Bandoeng) engineers, a number were recruited from abroad. For irrigation engineers it became possible – during this late-colonial state phase – to focus on the optimisation of various technological aspects of structural works and control. One important facet was the development of the whole laboratory world which was to contribute to a new generation of irrigation projects. Those projects were not simple imitations of what Lamminga had done but rather projects that had been extremely well thought through and executed and were even based on agricultural and economic feasibility studies. In that way the regulating and measuring of all the water flowing into the *sawas* could be carefully monitored and ultimately that was to culminate in the designing of something known as the Romijn measuring weir, a manually operated broad-crested weir fit for water level control and flow measurement, serving the distribution of irrigation discharges.

Right from the early days of state supported irrigation activities, engineers had made use of water measuring systems for the purposes of water distribution. General use was made of movable checks (measuring partitions of the Cipoletti or Thomson type) or of floats situated in straight canal sections. The problem with the floats was that the measuring was fairly inaccurate and with the checks there was always the risk of silt build up and water loss (level drop) over the check. It was useful to be able to combine measuring and flow control in one particular structure which was why, early on, engineers sought such a suitable solution. Lamminga, for instance, designed for his irrigation system in the Pekalen area a measuring and distribution structure that was not unlike the present-day American 'constant head orifice'. In retrospect Lamminga found his design too finely tuned and thus too difficult to operate.[23]

Another option that appeared in the mid-twenties was something that had been invented by the engineer P. de Gruyter. He took the British-Indian construction created by the engineer E.S. Crump and adapted the artefact to suit Dutch East Indian circumstances by making it adjustable.[24] The great advantage of that particular system was that the backwater level hardly affected the flow rate of the water that was allowed to pass through. The Javanese farmers were thus unable to influence the feed-through rate. The problem with the Crump-De Gruyter method was that the measuring system was complicated (two measuring scales had to be read) and the required head was considerable. In the flatter regions of Java where most of the engineering works were situated sufficient natural head was scarcely available. In 1923 an engineer known as S.H.A. Begemann introduced a much more successful method: the Venturi measuring system which then went on to be constructed on a big scale because it was very accurate and required little head. Unfortunately the measurement range was relatively small and that was sometimes resolved by building two Venturi meters alongside of each other.The other disadvantage was that if the downstream water levels were too high that distorted the measurement of the flow. All in all, though, the Venturi meters were a great success. In the Krawang area of West Java – the showcase area of the twenties from the point of view of irrigation achievements – Venturi meters became standard. In Demak, which was one of the oldest irrigation areas, Venturi meters also gradually replaced all the other existing artefacts.[25]

By means of experiments conducted in the Hydraulic Laboratory in Semarang, engineers tried to resolve the downstream water level problem. Provided that one was not too fussy about the requirements pertaining to the independence of the downstream water level, it seemed to be a good idea to place a metal flap behind the Venturi meter. Engineer A.L. Verwoerd placed a broad-crested weir behind the Venturi meter and that seemed to be a brilliant move. In fact the idea proved so good that before very long people started experimenting with the weir on its own because it really made the Venturi meter redundant! In 1932 engineer D.G. Romijn presented the results of the adjustable broad-crested weir that seemed to provide the answer to every engineer's questions: the weir could regulate the upstream water level and measure the flow, it required a small drop and was almost entirely independent of the downstream water level.[26] There remained one problem: by simply adjusting the sliding gate of the Romijn weir any interested party could influence the water distribution. The measuring weir that went down in history as the Romijn measuring weir (and not the Verwoerd weir!) was therefore encased in a measuring cabin.[27] In the irrigation works of the late thirties, such as the Tangerang system in Java (see below) and the Sadang system in Celebes, Romijn weirs became the standard water measuring and distribution structures. The Romijn measuring weir also met growing pressure to be economical with water. The international depression of the thirties had intensified the attention paid to frugality. At the same time the nationalistic feelings of the Javanese population that emerged provided solid grounds for the government to perpetuate the mission to modernise.[28]

What typified the irrigation activities of this period was the works constructed in the Tangerang Plain. The works of that region situated to the west of Batavia (Jakarta) were designed to irrigate around 52,500 ha of agricultural land with water drawn from the Tjisedane and the Tjidoerian rivers. All the construction work undertaken was based on thorough research and a well thought through general plan. The plain in question was dominated by private estates offering the population a paltry existence. As there was no Javanese administration like in other places, the population soon became restless. The general idea was to buy back the land and then 'elevate' the region by introducing both an irrigation plan and a road plan. The cutbacks of the nineteen-twenties and thirties slowed down progress on the works which were not in fact finished until after the country got its independence. The large mobile weir in the Tjisedane was provided with electrically operated gates so that at high water the town of Tangerang could be protected from flooding. One of the problems surrounding the works had to do with the Mookervaart shipping canal that had been created in the seventeenth century, which also contributed to the flushing clean of the canals in Batavia and which had its starting point just upstream of the Tjisedane weir. If nothing at all was done then much water would be lost via that canal as soon as the works became operational and instead of being beneficially employed the water would plague Batavia. However, there were also disadvantages attached to closing off the canal. In such a case the Tjisedane would have to deal with all the *bandjir* water and so at great expense it would also have to be widened. The idea that solved the problem was to provide the entry of the Mookervaart canal with a sluice for the conveyance of flushing water and surplus *bandjir* water and a separate lock for ships and bamboo rafts. Laboratory experiments were carried out to establish what might be the most suitable designs.

The final phase of the main works in the Tangerang Plain was completed under the supervision of provincial Public Works and was a consequence of the implementation of decentralisation which led in Java to the creation of three different provinces (West, Central and East Java). In the process, provincial councils partially replaced the previously created administrative councils. In the wake of all the administrative reforms, the General Water Ruling was inspired in 1936 by the double objective that had characterised the governmental approach during the difficult years: frugality and control. It stipulated that detailed culture plans (i.e. growing plans) should be created. These combined water and plant rulings aimed at improving water management which was something that had partly been made necessary by the declining sugar yields resulting from the depression. Because of that rice production levels rose but rice cultivation requires more water than sugar cultivation. The refining of all the water management rulings and activities made it possible for the government to tighten its grasp on the potentially restless population. The inclination to perfect technology and control was thus reinforced by the economic and political problems.[29]

Hydrological discussions

In the late-colonial period developments were also taking place within the scientific and less applied segment of the irrigation world. In 1933 an article first published in 1914 was republished in which Melchior's approach was described (without any background information).[30] It was particularly directed towards the engineers who neither knew about the article nor about Melchior's original theory. Without rejecting Melchior's approach as the standard way of determining maximum river discharges, a number of significant discussions were continued on hydrological matters. The three leading Dutch engineers in those discussions were: F.H. van Kooten, J.P. der Weduwen and S.H.A. Begemann. Van Kooten and Der Weduwen's contributions to the debate centred on establishing the best way of making use of topographical details such as surface area, the shape and length of the river and the river's basin, as well as utilising meteorological data such as the frequency, duration and intensity of precipitation. Begemann's interests lay in a slightly different direction: he endeavoured to apply mathematical statistics to the water movement rates within various river basins.[31]

In 1927 Van Kooten published a book (a German version of which appeared in 1932) in which he discussed all the factors underlying formulae such as those used by Lauterburg, Pascher and Melchior. In the process he made use of a large amount of information drawn from countries such as the United States and Germany as well as Dutch East Indian data. On the basis of all the data he was able to make much more accurate analyses than people had been able to make before him. Van Kooten examined a number of possible ways of calculating the discharges of rivers without giving preference to any one method. In that respect his book was more scientifically than practically oriented. By contrast Der Weduwen was interested in discovering a better calculation method. Originally drawn up for Batavia and the surrounding area, Der Weduwen's method – which was an adaptation of Melchior's – gradually began to replace Melchior's method, at least in areas where the river basins were small. In reality, though, his contribution came too late to be of much effect during the colonial period. There were also disadvantages to the new method; Der Weduwen had been forced to take a certain standard rain distribution level dispersed over the course of time. Der Weduwen was conscious of the shortcomings of the approach but due to the shortage of data there was really no alternative to making assumptions.[32]

Even in the late nineteen-thirties and early forties the lack of important data turned out to provide an important argument in favour of simplifying the approach. In a reaction to Der Weduwen's claims, engineer H.M. Verweij posited that it was not always the best thing to strive to achieve a method that would always work on the basis of simplification. In cases where precise details were available Verweij preferred Begemann's approach, especially where larger works were concerned.[33] For his analyses Begemann used measurements obtained from the Toentang river in Demak, a key area not only for irrigation activities but also for the development of hydroelectric power. Ultimately his statistical approach proved to be less significant. Even though many later referred to his brilliant work, Begemann's findings were not actually used in practice, not in the colonial era and not after the country gained its independence.

It was not only for irrigation works that reliable hydrological data was required but also for the generation of hydroelectricity. Not only was it important to consider maximum river discharges but it was just as important to consider minimum flows, particularly when it came to the matter of guaranteeing that expensive works would remain in operation so that electricity supplies could be sustained. Obviously a similar kind of guarantee had to be given for irrigation works but in the case of waterpower the stakes were, in a sense, higher because of the greater investment costs and the greater inherent business risks for the users of electricity (especially where industry and the railways were concerned).

Waterpower

In practice, then, one could add a fourth policy area to the earlier-mentioned Ethical Policy triad of Van Deventer's and that was electrification. After 1895, the private exploitation of hydroelectric power started to catch on, thanks to the franchises offered by the Dutch East Indian government. The 1900-1920 period reflected rapid development in waterpower provision as regards the dimensions, number and responsibility facets of the various works. The first implementation of waterpower in the East Indies had come as a result of private enterprise in conjunction with the need for power and light supplies, notably for mountainside crops (tea, rubber, and so on), light industry (including ice production) and mining industries (like the goldmines in Sumatra and Celebes). Those involved in private enterprise requested concessions from the colonial government so that hydroelectric power could be generated that was not actively exploited by the administration itself. Gradually, with the help of waterpower, electricity started being generated in a number of the bigger towns. Since 1906 the Bandoeng Electricity Company (Bandoengsche Electriciteits-Maatschappij, BEM) has been generating hydroelectric power for the municipality of Bandoeng. In 1911 the town of Semarang got its own hydroelectricity in the form of the Toentang plant that was exploited by the General Dutch East Indian Electricity Company. At that period the first director of the Government Companies department, H.J.E. Wenkebach, requested that the subject of waterpower be given serious consideration.[34] The State Railway Company became interested in using hydroelectric power so that the railway network could thus be partly electrified. To that end the State Railway Company established its own waterpower bureau in 1912. In 1914 building work commenced on the first state hydroelectric plant (of Tjatoer near Madioen) which was financed out of the railway budget. The plant's first electricity started being generated in 1917.[35] The water supply method was the same as that frequently used for irrigation works.

In 1917 the railway bureau was merged with the Government Companies' Electricity division to form a new organisation that came to be known as the Service for Waterpower and Electricity (Dienst voor Waterkracht en Electriciteit, DWE) which had its headquarters in Bandoeng[36], the Javanese waterpower centre. The former State Railway Company's chief engineer, P.A. Roelofsen, was made the director of DWE. The way in which the service was organised was proof enough of the fact that waterpower requires the cooperation of a number of disciplines: there was a hydroelectric division and an electrotechnical division. Civil engineers, mechanical engineers and electrotechnical engineers were employed there including a number who came from Switzerland where waterpower was a familiar phenomenon. Indeed, with their presence, they really highlighted how completely new that form of energy generation was to Dutch engineers.

Nonetheless, not everyone was in favour of foreign engineers specialised in hydroelectric power being appointed to key positions. The fact that engineers might travel abroad to learn how things were done was one matter, indeed similar work trips had been made in conjunction with irrigation technology but actually employing foreign engineers, in this case Swiss engineers, was something that certain Dutch engineers found hard to stomach. Two of the leading antagonists were the engineers R.A. van Sandick and C.W. Weijs, the latter claimed that:

> *between using water for irrigation and for waterpower there need be absolutely no friction and no disagreement and that in effect through studying irrigation the younger engineers become prepared to knowledgeably carry out all the research, the measurements, the reconnaissance work and to do all the calculations required to create an as clear as possible view of the waterpower that is made available.*[37]

Weijs, who was professor of irrigation in Delft, was lobbying here for his own graduates who did not always find it easy to find work in the East Indies (neither in the hydroelectric power sector nor in the irrigation branch for which, post-1913, foreign engineers were being recruited as well [38]). He himself was an example of an irrigation engineer who only late in his career had come into contact with the phenomenon of waterpower. That had occurred when, after 26 years of service with East Indian Public Works, on the Pamanoekan and Tjiassen estates (the biggest privately-owned estates in Java) he started working on the realisation of a water-driven power plant. As a matter of fact, Weijs himself had at that time appealed to the firm Mainz & Co. via the company's representative in the East Indies who was a Dutch engineer and whose superior was a Swiss chief engineer! However, the solution was not to be found in appointing experienced irrigation engineers to work in hydroelectric power plants and then filling up the vacancies with engineers who had recently graduated. Experienced engineers were usually difficult to find. An implicit defence of the resolution to take on foreign engineers (and all in all that did not amount to very many) is to be found in the many comments implying how alien waterpower still was to the East Indies:

> *Up until recently hydroelectric power had something of the unknown and mystical to Dutch technologists that powerfully attracted the average Dutchman to the mountains.*[39]

Although elsewhere in the archipelago hydroelectric power works were also either already in use or still under construction, especially in Sumatra (for various industrial ends like for instance the mining industry), in the late-colonial era attention was especially focused upon the Bandoeng upland plain. There it was not only the case that a number of new works were being constructed but it was also the case that from the organisational point

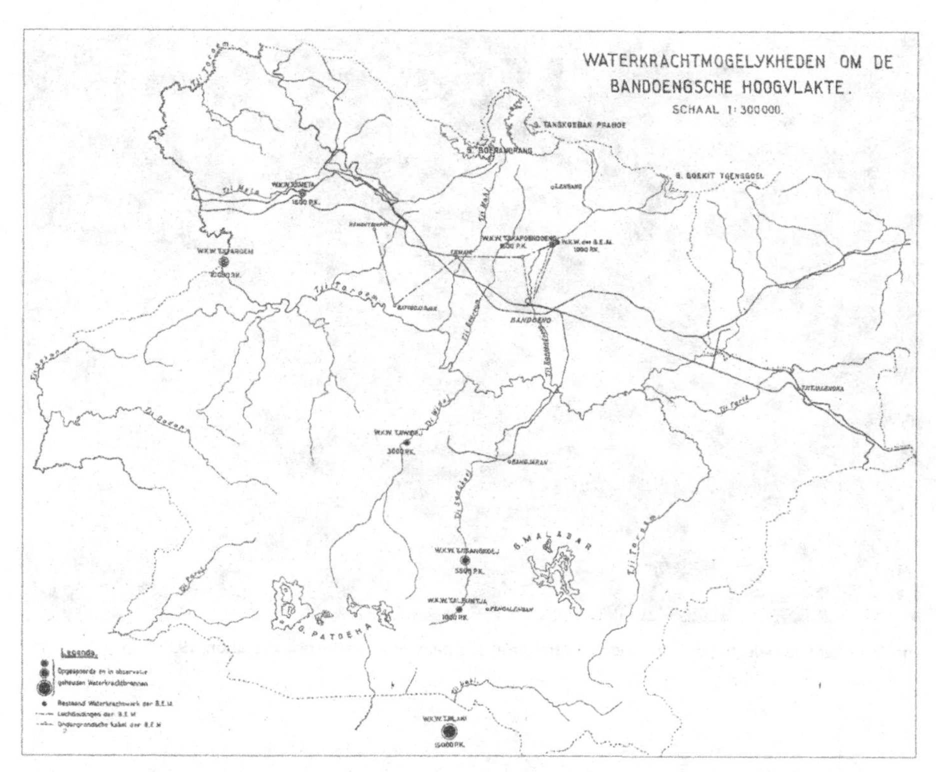

A map of the Bandoeng plateau (West Java), 1934. This was an area where the focus was very much on the development of colonial hydropower.

of view new coalitions were being established. In 1920 the BEM was amalgamated with the General Electricity Company for Bandoeng and Surrounding Areas (Gemeenschappelijk Electriciteitsbedrijf Bandoeng en Omstreken, GEBEO). It became the first occasion when the state, local councils and private companies worked together. The management of the BEM power plant was taken over by the state power station.[40]

Bandoeng

West Java was the part of the island with most water but what also made it interesting in terms of hydroelectric potential was the presence of the island's leading cities. The four biggest customers of the Bandoeng waterpower station's electricity supplies were: the Priok-Batavia-Buitenzorg electric railway line, the places Batavia and Buitenzorg, the city of Bandoeng (for public power supplies) and the Malabar radio station. In order to be able to serve all these places, state hydroelectric power stations were built in the Tji Anten (28,000 hp), the Tji Tjatih (7500 hp), the Tji Kapendoeng at Dago (3000 hp) and the Tji Saroewa at Pengalengan (4500 hp). In view of the fact that it was anticipated that the

The Ketenger hydroelectric power station (Central Java): the penstock and the power station, 1939.

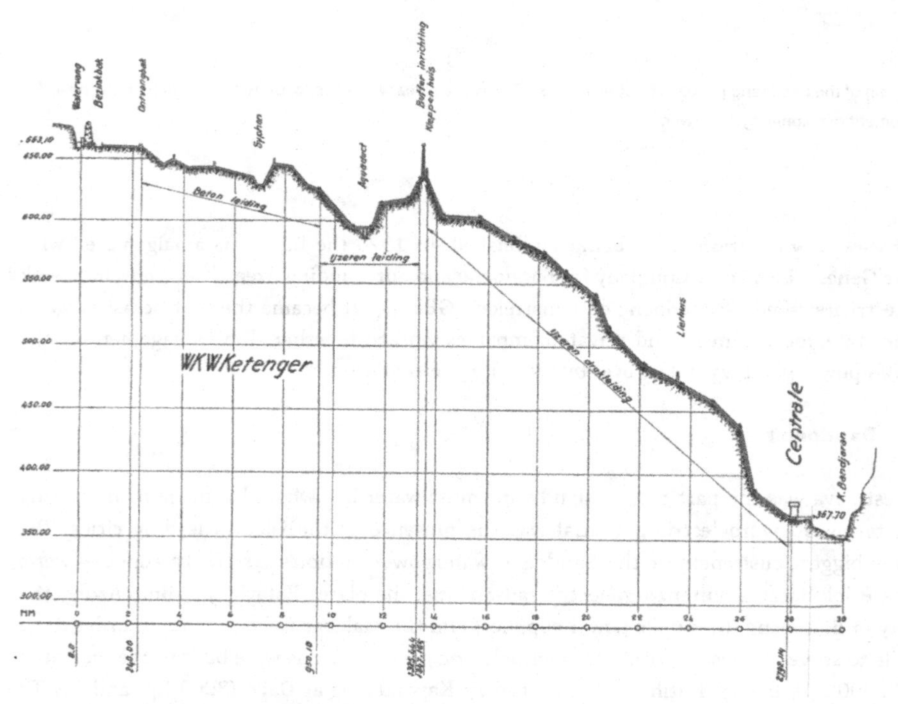

The hydroelectric power station at Ketenger: linear profile, 1939.

Table 4. **Overview of the suppliers and the components required for the Sengguruh power station**

Supplier	Component
Morgan Smith, York, Pennsylvania, USA	Turbine (turbine spiral casing made of reinforced concrete and produced by the Indonesian Water Power Exploitation Company – WEMI)
Woodward, Rockforf, Illinois, USA	Turbine regulator
General Electric Company, Schenectady, USA	Generator
De Bromo, Pasuruan	20 ton trolley
Ruhaak, Surabaya	Batteries case
Engine factory De Vulkaan, Surabaya	Water resistor, 70 kV power pylons
Westinghouse Electric Corporation, Sharon, Pennsylvania, USA	3300 kVA 6/70 V transformer
Design and structural supervision	WEMI

Source: Goemans, 'Waterkrachtwerk Sengguruh. De plaats van de centrale Sengguruh in het Oost-Java energieopwekkingsysteem'.

demand for electricity would only rise, new works were planned including reservoirs with a total capacity of 40 million m^3.[41] One of those works, Tjisankoei I, made use of an old and somewhat neglected local reservoir which was promptly renovated and enlarged.[42] Most of the works were constructed by the state itself with the exception of certain components such as pipelines.

What formed the backdrop to all these construction activities was the transition to a situation in which things would be more state-influenced. In the early twenties there was a turnaround in the world of waterpower:

> *Before the war and during the first years of the war, when the State was paving the way for the establishment of hydroelectric power stations in Java, the solution to this double-sided technological and commercial question appeared to be rather less complex than it actually is at present. [...] Gradually, influenced by the world's economic crisis, which was to be seriously felt later in the Dutch East Indies than elsewhere in the world, the whole situation changed quite dramatically during the course of 1921. Ultimately those changes were to be decisive for the plans originally made for power supplies in West, Central and East Java.*[43]

Various matters came to light in DWE's annual reports of the first half of the twentieth century like, for instance, in the 1923 report where it was established that because of the world economic situation expenditure and activities in the field of waterpower would have to remain limited. The East Indian government was directly involved in these matters in view of the fact that it had

Flooding in the Solo Valley (East Java), 1995. The Dutch engineers did not always succeed. The major Solo Valley works had to be halted around 1900. Today the Solo Valley is still an area where the water situation is not yet fully under control.

very different views from [those of] other [governments]. In fact everywhere else in the world, even in countries that are financially much worse off than the East Indies there has been, since the war, an exceptionally industrious approach to waterpower, not least because of railway electrification.[44]

It was especially the Javanese railway developments that proceeded slowly. By contrast, the works planned for the Bandoeng upland plains were reasonably on schedule. Upon completion an annual output of around 100 million kWh per year was achieved.[45]

From the experience gained from such works the engineers learned how to process a number of matters better. Indeed, dealing with water in an economical and controlled fashion was much more important in the case of hydroelectric power than with irrigation. They carefully studied, for instance, the way in which a lined canal behaved so that they were able to determine the relevant wall roughness. The much-used formula of Bazin did not suffice and the DWE preferred the Gauckler-Manning formula.[46] Also in the case of the sloping gutters, frequently adopted in waterpower plants, the formula proved very feasible. Experiments were furthermore carried out with different construction methods using, for instance, pressure pipes of reinforced concrete, and experiments were also done with models in the Bandoeng Hydrodynamic Laboratory that had been established in 1927. One of the model tests concerned the intake works for the later hydroelectric power stations at Senggoeroeh/Sengguruh in the Brantas. For this last project the maximum flow rate established using the Melchior theory was 920 m^3/s but what was actually measured was 1200.[47]

A number of waterpower initiatives were embarked upon just before the Second World War, one of which was to allow the Billiton company to create large works in the Asahan river in Sumatra for the purpose of fertiliser production and another was the Sengguruh works plan in the Brantas[48], which showed just how international the network of parties involved had become. Working on the Asahan project were Swiss advisors, Dutch engineers and civil servants, and financiers from France and the United States of America. After all, it was not just a case of developing a hydroelectric power plant but it was also a case of organising large-scale nitrogen production for fertiliser. From the overview of the parts suppliers for the Sengguruh power plant (the construction of which was resumed in 1949 after work had stood still for six years) given in Table 4 one can see just how many international companies were involved in supplying the required goods.

After 1949

Despite the political rift with the Netherlands, the Indonesian state retained many of its colonial state characteristics. A high degree of continuity is apparent both in the fields of irrigation and waterpower if one looks at the colonial water works that are still in use today but also if one studies the extensive programme for rehabilitation and the expansion of modern irrigation and hydroelectric power facilities that has been completed in independent Indonesia. In this last case, it could be said that with irrigation the wheel was, up to a point, being reinvented from the point of view that engineers always embarked on head works and only considered the entire system and control matters when problems arose.

There are more similarities, though: there is still a degree of competition between irrigation engineers and (non-technical) civil servants, though the dominant position of the former has been consolidated. Present-day discussions in Indonesia on large-scale versus small-scale irrigation, absolute versus proportional water distribution, water management based on entire river basins and environmental issues are all very reminiscent of similar kinds of discussions that took place during the colonial period. The water situation in the Solo Valley is still insufficiently regulated. Even in the case of new developments there is evidence of continuity.

Apart from working on expansion the engineers also carried on working on controlled water systems which included creating 'multipurpose' reservoirs designed both for irrigation and waterpower generation, linking up irrigation systems and introducing new management structures such as water users' associations and river basin management.

Various such matters came to the fore in the plans unfolded by the engineer W.J. van Blommestein who, as head of the section of Irrigation and Drainage for East Indian Public Works, developed a welfare plan for West Java that focused on increasing rice production and strengthening the economy.[49] This innovative plan involved works in various river basins that could be implemented for a whole range of different purposes such as: irrigation, drainage, domestic water provision, electricity generation, shipping and industry. The basis of the plan was to have three reservoirs situated in the Tjitaroem area (Djatiloehoer, Tji Rata and Sanggoeling). The Indonesian government did execute the plan. The first reservoir, the one at Jatiluhur, was completed in 1965 with the aid of French funding. The dam is a rockfill dam that is about 80 m high and the reservoir has a useful capacity of three billion m^3. Reservoir building continued in the 1983-1988 period and various other activities linked to the entire network continued until 1993.

Thanks to such projects and thanks to previous efforts of the past, large-scale hydroelectric generation only really began to take off in the time of the republic. The Asahan river project referred to above is also a project that did not get going until after the country gained its independence. The Asahan river with its virtually constant discharge and a couple of natural 'steps' in its path (i.e. the Wilhelmina, now the Siguragura waterfall and the Tangga waterfall) constitutes a perfect source of waterpower. Already in 1939 a start had been made on felling jungle growth so that an access road could be created and in 1941 the first preparatory steps were taken at the Wilhelmina waterfall. Because of the Japanese invasion the whole project was abandoned. It was not until 1982 that two power stations were completed in the river, the Siguragura and the Tangga plants and both were backed by Japanese(!) funding.[50]

Three years before, in 1979, Van Blommenstein had proposed that his welfare plan first launched in 1949 be pursued in the form of a development plan for the whole of Java and Madura and, once again, the overall plan had various objectives.[51] The plan aimed to connect up irrigation systems throughout the whole of Java and it would involve transferring water from the wetter western part to the drier eastern part of the island; a siphon would be used to transfer the water to Madura, which was also a very dry place. The Indonesian government was enthusiastic about Van Blommestein's plans but the same could not be said of the evaluation mission set up by the Dutch Ministry of Foreign Affairs which asserted that in view of the results of the Green Revolution the plan would not be necessary. What in fact emerged later was that the population level in West Java had grown so rapidly that there was, meanwhile, no longer evidence of a water surplus. As will be explained in Chapter 9, Dutch engineers became further involved in irrigation development in the former colony in all kinds of possible ways.

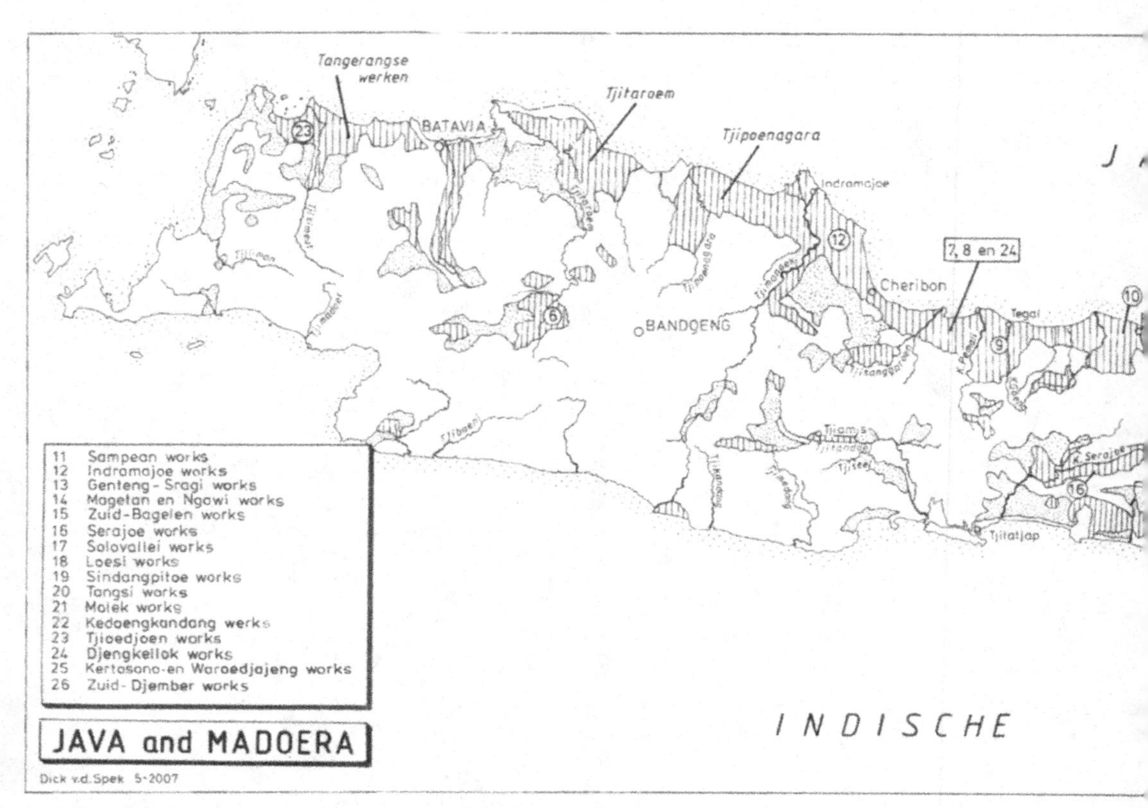

Map of the irrigation areas in Java and Madoera around 1940.

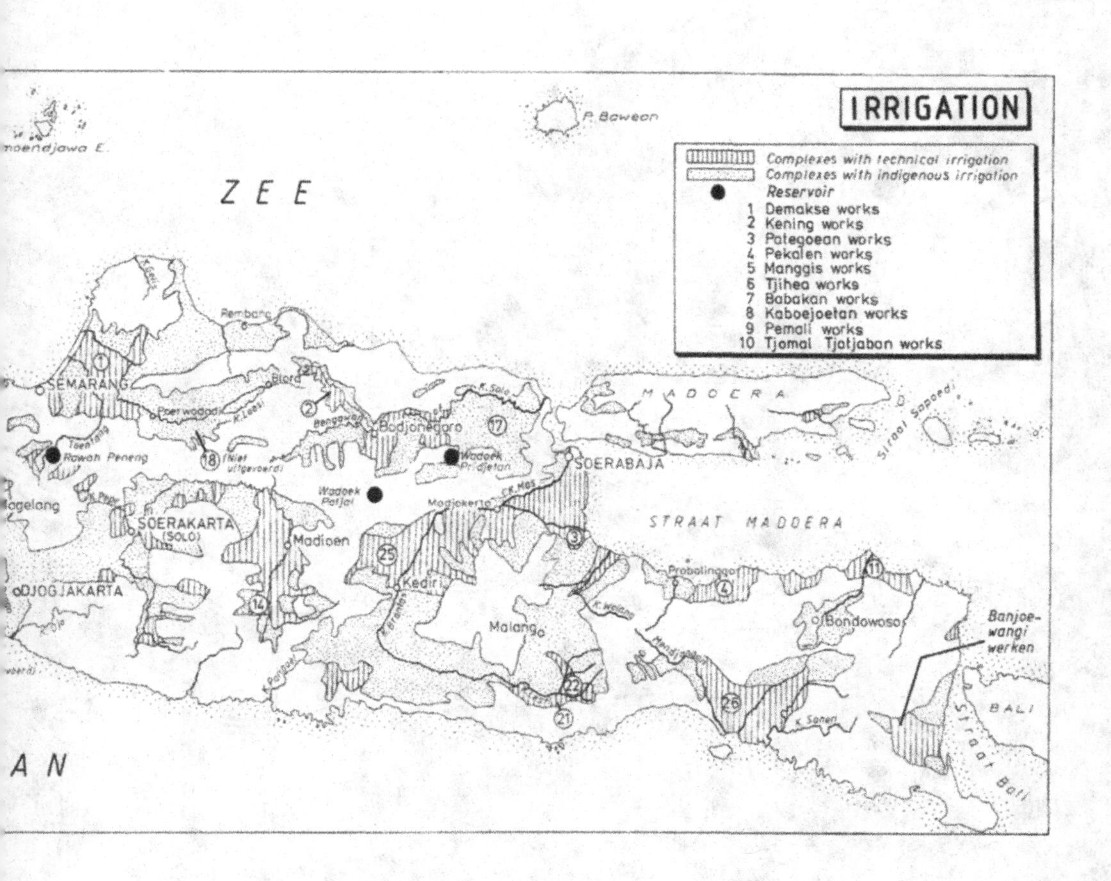
IRRIGATION
Complexes with technical irrigation
Complexes with indigenous irrigation
Reservoir
1 Demakse works
2 Kening works
3 Pategoean works
4 Pekalen works
5 Manggis works
6 Tjihea works
7 Babakan works
8 Kaboejoetan works
9 Pemali works
10 Tjomal Tjatjaban works
ZEE
P. Bawean
MADOERA
STRAAT MADOERA
SEMARANG
SOERAKARTA
(SOLO)
DJOGJAKARTA
Madioen
Bodjonegoro
Blora
Poerwodadi
Rawah Peneng
Wadoek Patjal
Wadoek Pridjetan
SOERABAJA
Modjokerto
Kediri
Malang
Probolinggo
Bondowoso
Banjoe-wangi werken
BALI
Straat Bali
Straat Sapoedi
AN

Privately designed attached dwellings situated on the Karre Road in Semarang. Photographed by H.G. Tillema around 1910.

CHAPTER 7 **FOR KOTA AND KAMPONG**

The emergence of town planning as a discipline

Obviously it cannot be denied that in the past many a regional administrative head or even the B.O.W. made important and sometimes very accurate decisions concerning town expansion – but that can hardly be said to constitute town planning.
(H.T. Karsten 1920).[1]

At the time of writing *Indiese stedebouw* (Indian town planning) the author of this quotation had been living and working in the Dutch East Indies for some six years. Following his studies in architectural engineering at the Technical College in Delft (1904-1909) Thomas Karsten travelled to the East Indies to take up the position of 'chef de bureau' at his friend's architectural bureau in Semarang. The friend in question was the engineer Henri Maclaine Pont. It was not long before Karsten made a name for himself as an enthusiastic and gifted creator both of individual building designs and of urban development plans. With his book *Indiese stedebouw,* the first East Indian publication to deal with the objectives and the essence of modern town planning in the Dutch East Indies, Karsten's name became permanently linked to town planning in the Dutch East Indies during the first half of the twentieth century. He wrote the following about the basic conditions required for any town planning practice:

What is essential is a perpetual interest in and knowledge of local issues [...]. Town planning can only be good if planners are well-informed about local conditions and constantly in touch with people's needs and the needs of industry.[2]

At the beginning of the twentieth century town planning was a relatively young discipline. Even though people had contemplated city layouts for centuries, it was not until the end of the nineteenth century that the field was given renewed impetus through publications such as *Bijdrage tot de kennis van den stedenbouw* (Contribution to the knowledge of town planning) (1880) by H.W. Nachenius, *Der Städtebau nach seinen Künstlerischen Grundsätzen* (The artistic foundations of town planning) (1889) by Camillo Sitte and *Garden Cities of Tomorrow. A Peaceful Path to Real Reform* (1898) by Ebenezer Howard. With a view to economic and social development these books described the demands, possibilities and problems surrounding future town and country development, and design. In many respects the problems encountered in the Dutch East Indies were so divergent though that European and American publications on the subject offered little solace.

Karsten's reference (see opening quotation) to the lack of a town planning discipline and the negligible intervention of the Department of Civil Public Works (Departement van Burgerlijke Openbare Werken, BOW) in the systematic development of East Indian towns did not imply that up until the 1920s every form of planning was totally lacking in the East Indies. Over the centuries various native and foreign powers had established settlements that adhered to a more or less systematic planning regime. Karsten did not wish to deny this, he merely wanted to point out that on the whole during the period of Dutch rule town planning had not existed because predetermined rational and practical considerations, a methodology, institutions, regulations and directives had been lacking.

The emphasis that Karsten placed upon the need for a systematic town planning approach and for systematic practice was to culminate, in 1938, in the presentation of a draft for an ordinance in the field of town planning. The text of this draft ordinance was the result of the findings and recommendations made by the Town Planning Committee that had been appointed by the government in 1934. Among the most prominent members of the committee were J.H.A. Logemann (chairman), professor at the Law College in Batavia, M. Soesilo, a practising engineer and architectural supervisor employed at Karsten's bureau in Bandoeng, the engineer J.P. Thijsse, at the time employed by the Municipal Works Service in Bandoeng, and Karsten.

By describing all aspects of the town planning process in a lucid and coherent fashion the ordinance provided architects, administrators and the general public with the necessary guidelines and insight. Despite extensive debate over the text of the ordinance and its elucidation during the first planning workshop in 1939, the war situation in Europe and Asia delayed its approval and decree until after the war. Though the delay was undoubtedly hard to accept for the committee members involved, it detracted little from the legal and practical significance of the ordinance being the first Dutch legal regulation to deal with all aspects of (tropical) town planning ranging from design to construction.

Another effective aspect of the ordinance was the fact that it did not confine itself to the discipline of town planning only. From time to time the authors stressed the importance of and the need to maintain a broad planological perspective and the need for cooperation between town councils and regencies in order to create regional plans and a national plan. Because of this broad and forward-looking view it is hardly surprising that when in 1948 the government decided to establish a spatial planning committee the ordinance was taken as the basis for the committee's activities.

Changes in the nineteenth century

Far-reaching developments

Nineteenth century colonial administrators and engineers abided by the instructions of the Dutch government and operated in its interests. Given the colonial character of the Dutch East Indies this meant that its administration and its development were organised in such a way as to serve the colonial power and the economy of the European part of the kingdom. Where architecture was concerned that meant that infrastructure, irrigation works, staff residences, offices, warehouses and utilitarian works constituted the most important building assignments. This situation changed slightly when in 1870 the Cultivation System was abolished and the Agrarian Act was adopted. Even though the colony continued to be centrally governed from Batavia and mandated by The Hague it was from that time on, no longer the sole prerogative of the government to establish trade relations and exploit the colony: private persons were also free to trade with and within the colony.

The result was an instant increase in the number of Europeans who settled in the East Indies. Between 1870 and 1900 the number of European civilians and military people in the colony increased from 44,200 to 90,800.[3] Both in the countryside and in the city their arrival created employment possibilities that appealed to the many indigenous inhabitants. The arrival of the Europeans thus not only altered the composition of the population and created new employment opportunities, it also generated a considerable migration of the rural population to the towns as well as from the Outer Regions to Java. Between 1870 and 1905 the indigenous population of Java grew from 12,100,000 to 28,300,000.

As the existing towns were not equipped to cope with such a population surge it was not long before an acute large-scale housing shortage arose. Endeavours to alter the situation by providing new housing areas and shopping facilities for the Europeans had an immediate negative effect on the housing situation among Indonesians as it were their houses that had to make way for such expansion plans. Consequently the housing shortage among Indonesians increased and led to a deterioration in their living conditions. This situation, together with the regular outbreaks of malaria, typhus and an occasional outbreak of the plague epidemic made it plain that drastic measures would have to be taken to put an end to this situation.

Standing up for architecture

More or less simultaneously and as a result of the new spirit of enterprise that emerged at the end of the nineteenth century and the prospect of the potential role that architecture might play in this new era, a debate commenced on the importance of good architecture and the emancipation of architects. What triggered the debate was the characterless and unoriginal architecture of many late nineteenth century buildings and structures. Many engineers who had been educated as architects were of the opinion that the cause of this problem was to be sought in the inadequate artistic training of the civil engineers who had been responsible for the design of many of those buildings. Lobbying for the rightful status of architects in the building industry they pointed out that civil engineers were primarily trained to design utilitarian works. Which meant that they were only partially conversant with the principles of good architecture. By contrast, architectural engineers were thoroughly familiar with the principles of architecture.[4] Because of his knowledge of the theory of form and his artistic training an architect knew how to make form and colour correspond to the function of a building and how to create, both in the interior and in the relationship between a building and its environment, harmony, rhythm and rest while carefully considering good spatial planning – and should thus be favoured as a designer of buildings.

Since BOW was responsible for the design and realisation of the majority of the buildings in the colony it was especially towards that department that the architects directed their frustration. It was not only the efficiency and sobriety towards which BOW aspired that elicited a great deal of criticism from them. It was also, indeed particularly, the application of 'standard designs' that evoked their criticism because they so obviously bit displayed a preference for productivity and a disregard for architecture. Standard designs were basic designs that adhered to instructions and fixed directives concerning ground plans and elevations while forming the basis for new designs. In the post-1870 period BOW had turned to standardisation in order to keep abreast of the rising demand for buildings. With the exception of the department's buildings and the offices where administration was housed, BOW engineers made use of standard design for all remaining buildings. Among them houses, schools, hospitals, post offices, prisons, warehouses, and watch houses.

The idea underlying the principle of standard design was that it facilitated a fast, efficient and relatively inexpensive way of working because it enabled an engineer to realise a technically and financially satisfactory design with the minimum of effort. Because standard designs were used all a designer had to do was to concentrate on the design details. Thanks to the expertise of those involved, only written instructions needed to be provided during construction. Where necessary adjustments could be made, for example because of divergent soil conditions or available funds, particular representative demands or specific locations. According to the architects, however, the problem with this approach was that it led to solutions that were architecturally completely unsatisfactory. Remarks similar to the following became gradually more frequent and more insistent: 'Our government does not employ great builders' and 'Uniform through their insignificance, insignificant through their uniformity, is the best way of qualifying public buildings of the last few decades'.[5]

In 1898, in order to boost their plea for good architecture, architects of the colony established the Association of Dutch East Indian Architects (Vereeniging van Bouwkundigen in Nederlandsch-Indië).[6] It also published a periodical, the *Indisch Bouwkundig Tijdschrift* (The East Indian Architectural Journal). Subsequently, in 1850 two Indian branches of the Royal Institute of Engineers (Koninklijk Instituut van Ingenieurs, KIVI) were established. The KIVI initially published a special section on the Dutch East Indies in its periodical *De Ingenieur* (The Engineer). Later on a separate issue was published, called *De Ingenieur in Nederlandsch-Indië* (The Engineer in the Dutch East Indies).

In the same year that the Association of Dutch East Indian Architects was set up, the East Indian architects gained the support of a prominent Dutch architect, H.P. Berlage. Despite the fact that he did not have any personal or design-technical experience with the East Indies or its architecture he wrote the directorate of the Life and Life Insurance Company 'De Algemeene' (Levens- en Lijfrenteverzekeringsmaatschappij De Algemeene) the following regarding a design for their new office in Soerabaja:

> *What seems to me to be generally most desirable is that in broad outline the East Indian building style is retained but that to that end the architecture is made slightly richer. The designer was, however, of a different opinion and applied a purely European architectural style. Presumably there was good reason for doing this, which I can accept, but I cannot but disapprove of this kind of European architecture because the façade resembles that of a villa-like shop in a small community designed by a small architect. It is this architecture that has made all our lovely cities and towns ugly.*[7]

Bolstered by the ideological support of the prominent architect Berlage, the East Indian architects continued to plea for good architecture. In view of the tense relations between architects and their colleague engineers at BOW the stance taken by the architectural engineer S. Snuyf was remarkable. Appointed as junior engineer of the accursed department he had already pointed out in 1908 that the then recent political and economic changes had led to a situation in which people 'were preoccupied with creating all the kinds of organisations that would make it possible for people to have a pleasant stay in the tropics' and 'everyone is doing his best to make our Indian years more pleasant'.[8] The conclusion he drew from this was that architects would have to follow suit and contribute by designing 'pleasant public buildings', 'cheerful schools', 'nice fresh offices' and 'good hospitals'.[9] Snuyf contended that the custom of 'always building white houses with their bad, or even worse applied Renaissance styles' needed to be challenged and avoided.[10]

Apart from the works of Berlage what Snuyf saw as exemplary architecture were the designs of the Dutch architects P.J.H. Cuypers and W. Kromhout. He furthermore concluded that the choice of a certain architectural style did, to a high degree, also depend on the environment in which a building was situated. He wrote:

> *When placed against a clear sky the silhouette of a building must be purer and more regular than when placed against the background of a wood or hill. In the latter case a fantastic silhouette can be very successful whereas in the former case clear-cut lines are a prerequisite.*[11]

Clearly BOW was not insensitive to the persistent criticism of its architectural designs. Just one year after his first remarks Snuyf was appointed head of the department's newly created Architectural Bureau, a position that he was to fulfil until 1912. In his capacity as head of the bureau he immediately put his ideas into practice: his 1909 design for the post and telegraph office in Medan was widely acclaimed.

A new century, a new policy

Administrative reorganisation

The debate on architecture and the standardization adopted by BOW towards the end of the nineteenth century was indicative of the lack of sufficiently qualified architects and the need for change. The appearance of buildings was not the only thing that needed to be changed. The new character of the colony, the rapidly changing composition of the population and the accompanying social, cultural and political changes forced the country's administration to contemplate its approach to the task in hand. In order to adapt to the new situation and to the related changing social relations the liberal policy that had been adhered to for some time was further considered and elaborated. It was to result in a new political line that came to be known as the Ethical Policy. The revised political line was to pave the way for administrative reform. One of the most far-reaching moves was the decision to decentralise the central administration. The legal ruling required to that effect was adopted in 1903. Both from historical and political perspectives the Decentralisation Act was an important document as it officially put an end to having a central administration operating from Batavia – a system customary almost from the time the Dutch had set foot in the archipelago.

The Decentralisation Act marked the beginning of gradual administrative reform and the realignment of the power configuration. A government commissioner and an assistant government commissioner were appointed in order to implement the objectives laid down in the Decentralisation Act, to advise the governor-general and to inform all the authorities involved. One of the first suggestions the government commissioner made was that local councils should be established: administratively local, autonomous authorities which, taking into account the various hierarchical relations, would be individually responsible for local administrative and financial policies. The proposition was well received and resulted, in 1905, in the approval and establishment of the ordinance required for the establishment of local councils.

This so-called Local Council Ordinance was an important decentralisation instrument. It made it possible to create local and regional councils steered by a European administration, and regions and departments governed by local-based administrations.[12] In accordance with their founding ordinance, the local councils had to provide:

> *maintenance, repair, renewal and construction of public roads including accompanying attributes such as planting, banks, dykes, verges, etc., as well as other works of local importance such as: squares, gardens, gutters for general use, sewers, flushing conduits, works for the acquisition or distribution of drinking, washing and rinsing water, general slaughter houses, markets and pertaining sheds.*

In addition they were responsible for the collecting of roadside refuse, the spraying of public roads, streets and squares, the fire brigade, cemeteries and public ferries. In connection with his responsibility for the implementation of the Decentralisation Act, the government commissioner – who was made decentralisation advisor as of 1912 – was involved in local authority affairs.[13]

The tasks confronting the local authorities were numerous and varied from sanitation works to public housing and *kampong* improvement, and from land issues to town expansion and improvement from the very start. Besides the municipalities had to deal with the presence of the many ethnic population groups, each with their divergent individual and regional habits and customs. The deficient or even totally lacking administrative experience of most of the local council members, the social and economic circumstances that deviated from those in the 'motherland' and the virtual lack of support and assistance from the Batavian government initially did not help to make administrators particularly enthusiastic about decentralisation politics. In the field of urban development it was no different: the approach required to achieve optimum results was more or less thwarted by the chronic lack of well educated and capable architects and engineers and the regional and even local variations in assignments, building types and town plans.

Unable to autonomously realise the town planning tasks, the local authorities regularly appealed to the East Indian government for understanding and support through the decentralisation advisor. One of the first hurdles facing the municipalities, which were primarily composed of Dutch citizens, was the land disposal right. In virtually every respect the traditional, indigenous rights and customs relating to land were different from Dutch rights and customs. In order to anticipate and elaborate on economic, social and cultural changes and organize and cultivate the land according to its own intentions and desires, the government in the nineteenth century had already started the recording of the various forms of indigenous landownership within the framework of the Dutch legal system. As the numerous variations and nuances of the indigenous system were very difficult to reconcile with the considerably simpler Dutch system this operation right up until the 1930s frequently lead to mutual irritation, incomprehension, conflict and amendments. Social objectives were secondary to state objectives.[14] Therefore considerations concerning relationships between indigenous land rights and the preserving of the community rarely constituted a significant basis for the Dutch authorities.

Another problem was that when local authorities had been established it had been

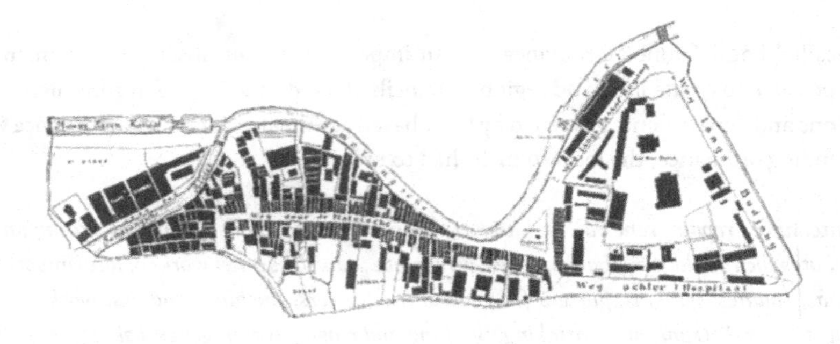

Map of a district in Semarang (Central Java) inhabited by native people. The situation as it was around 1880.

decided that the indigenous authorities (known as the *desa* administrations) that were situated within the municipal borders would be autonomous at managerial level. It soon became clear that such a juxtaposition of two administrative powers created an awkward situation that often lead to complications. The main problem arose from the fact that due to the separate status of indigenous municipalities and the lack of administrative powers of the European local authority in the *desa*, the indigenous authorities could not be forced to cooperate in the implementation of a municipal plan. They were thus able to frustrate the desired integral municipal approach and consequently emphasised not only technical but also the social differences.

The importance of hygiene

One of the reasons an integral approach was so desirable had to do with the improvement of hygiene, notably in the more densely populated regions of the larger cities. The first such concrete proposition to be made came from two Semarang town council members, W.T. de Vogel, a doctor, and H.F. Tillema, the pharmacist. In 1909 they lobbied for a new housing area to be situated in the hilly region a short distance south of the existing town. Drawing on their knowledge of medical matters De Vogel and Tillema argued that because of its hilly character and higher location the new neighbourhood would not only be visually attractive but, more importantly, more healthy. It would, in other words, constitute a good way of resolving a number of Semarang's major problems: insalubrious living areas.

From a professional point of view De Vogel and Tillema were both interested in changing the unhealthy conditions under which a substantial portion of the largely indigenous population lived. The fact that they then went on to try to present resolutions by drawing on town planning and architecture was something which, up until then, had been unheard of in the Dutch East Indies. Nonetheless the plea by De Vogel and Tillema was endorsed by the Semarang town council because its members recognised the need for a healthy city and an integral approach. A decision that underlines that fact that the municipal board in Semarang was forward-looking. Unlike many other boards it realised in an early stage that all problems were interlinked and that in view of the extent and urgency of the individual problems a coordinated approach was essential if optimum results were to be achieved.

Kampong dwellings in the Karang Bidaro district in Semarang. Photographed by H.F. Tillema around 1910.

Attached European houses in the Karang Bidaro district in Semarang, around the year 1910.

From a Dutch and European perspective this was not unique though. In the Netherlands and Europe the attention devoted to town planning, to housing and to the relationship between the two had emerged during the late nineteenth century from concerns about the abominable hygienic circumstances in which many people lived. Here, too, it was often those who were not engineers who sounded the alarm to contributing to a better living environment. The situation in the East Indies was similar. Up until the early twentieth century civil engineers and architects had hardly occupied themselves with questions such as town expansion and public housing. It was the demand for a healthy living environment free of illness and epidemics that was to bring about a change in all of that and to force civil engineers and architects to devote more attention to these issues.

The ideas and principles upon which De Vogel and Tillema had based their appeal were later described in a book published by Tilllema entitled *Van wonen and bewonen. Van bouwen, huis en erf* (On living and inhabiting. On building, house and property). In this publication which was presented during the tenth international housing congress in Scheveningen (the Netherlands) in 1913, Tillema described the housing issue in general and the deplorable housing situation in Semarang in particular. In the process, he not only drew attention to the issue of the layout of housing areas but also to the importance of water management and the provision of sufficient good quality drinking water. Tillema's arguments in favour of the hygienic and social changes required were substantiated with international examples, extensive figures, numerous illustrations and photos. It was particularly the photos that explicitly illustrated the architectural and hygienic misdemeanours varying from unsuitable building material such as corrugated iron roofing and walls fabricated from petroleum tins, to poor or unutilised building components. It was especially in that latter category that virtually everything was amiss including insufficient or no protection against the sun, little or no indoor ventilation and ventilation between houses, insufficient sun resulting from poorly selected and positioned trees, undivided water flows, blocked drains, stagnant water with breeding malaria mosquitoes, et cetera.

What Tillema was most critical about was the fact that uniform solutions were adopted and administrators and designers lacked knowledge. As a result his campaigns and publications were principally aimed at European politicians and citizens who took decisions but who, due to their often privileged social status were hardly ever confronted with the true

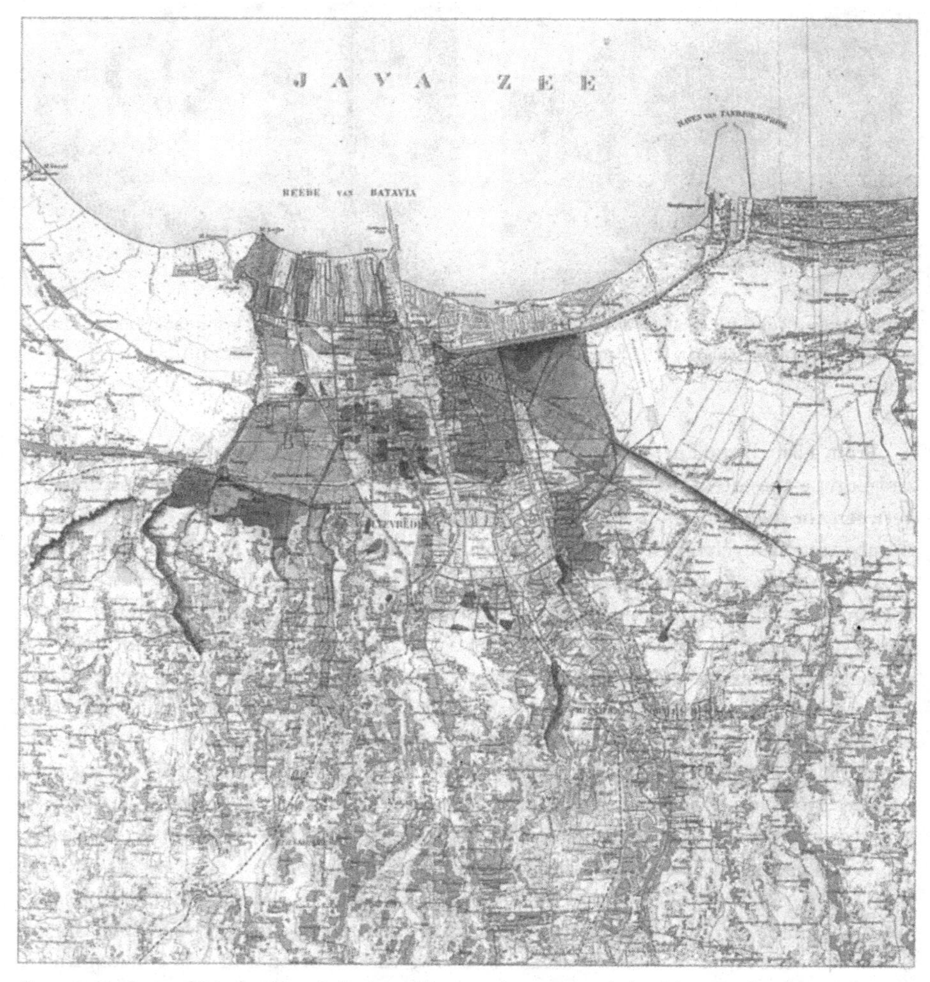

Topographical map of Batavia with an indication of the areas that were regularly subjected to flooding, 1913.

nature and extent of the problems. In his proposals for improvement Tillema stressed that in the designing and orientation of houses, parcels of land, public planting and town planning the specific hygienic demands of a tropical climate would have to be taken into consideration. More than in Europe it was crucial in the East Indies to make sure that air circulation was sufficient, sunlight was tempered (though explicitly not 'blocked'), waste materials removed, and water management under control. He also pointed out that both inhabitants and the town council were responsible for the creation of a good and pleasant living environment: the council because, apart from being responsible for good leadership, it was also responsible for drawing up a comprehensive urban development plan, and inhabitants because they needed to utilize their houses in accordance with the tropical climate.

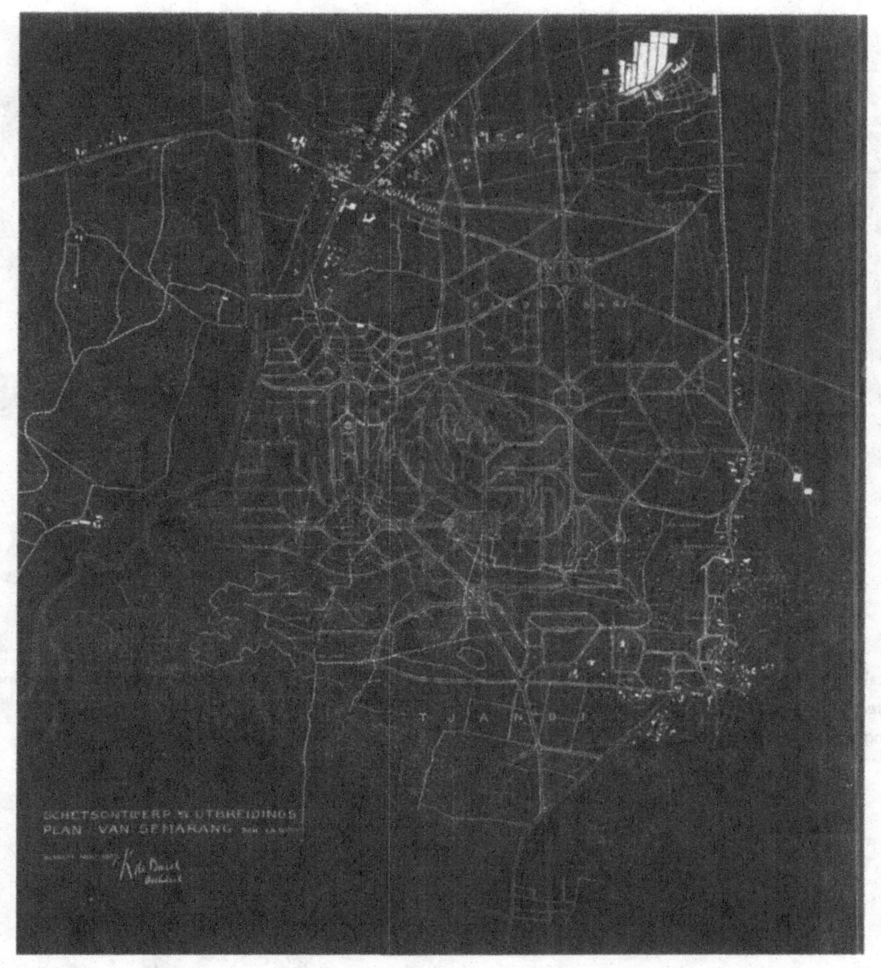

Blueprint of a design sketch drawn up by the engineer K.P.C. de Bazel in conjunction with the Semarang city expansion plans, 1907.

One example that serves to support the plea for an integral approach in the field of urban development was the state of affairs in relation to water management. The nature of the problems surrounding the supplying and discharging of water were so extensive in most East Indian towns that efforts to improve matters in other areas would be ineffective if the water management was not improved first. Problems with the supply and disposal of water was something that had been a constant source of concern from the moment the Dutch had first set foot in the colony. The floods during the monsoon, the lack of water during the dry season, the laborious provision of clean drinking water and the perpetual problems with the disposal of waste water were just some of the issues that had to be contended with. It was clear that no improvements would have a lasting effect if nothing was done to improve water management.

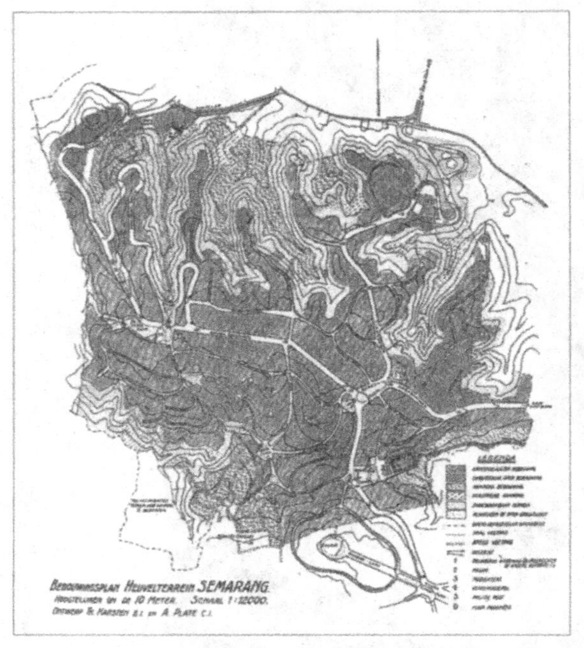

Building plans (scale 1:12,000) for the urban development of New Tjandi, the hilly area to the south of Semarang, according to the design of H.T. Karsten (engineer) and A. Plate, 1916.

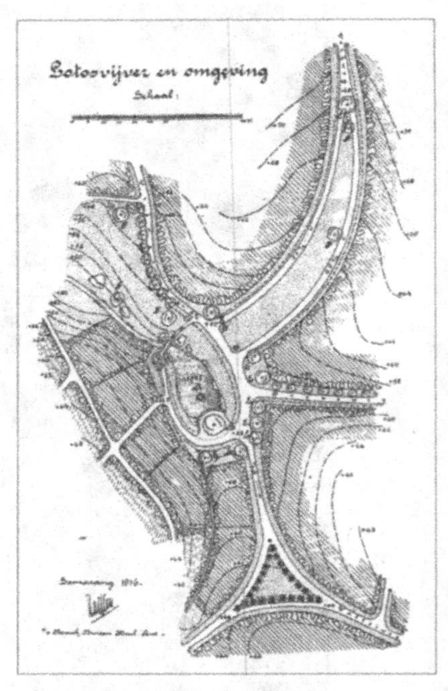

Map of the Lotus pool and its surroundings in New Tjandi, Semarang.

Artist's impression of the Lotus pool in New Tjandi looking in a northerly direction, Semarang.

One of the first towns to introduce an integral improvement plan for the water management situation was Batavia. After nineteenth century efforts to abolish the problems by creating new housing areas and after a systematic but incomplete improvement plan had not led to the desired result, governor-general A.F.W. Idenburg in 1911 gave Public Works and the Batavia town council instructions to draw up a general plan in consultation that would ultimately lead to the sanitation of the town. In 1913, as a result of these instructions H. van Breen, who was an engineer and a member of the Batavia town council, presented an overview of works he proposed for the improvement of the drainage and supply of water in the capital Batavia. To illustrate just how unhygienic the conditions were, especially in the densely populated *kampongs, he wrote:*

> *The drainage system in the kampongs is much worse than in the neighbourhoods inhabited by Europeans and prosperous Chinese. In many of the very densely populated parts there is absolutely no evidence of drainage facilities. Just here and there between the houses one finds a pit in the ground that is less suited to water transport than to pollution where the water collects.*[15]

Apart from pointing out that stagnant water constituted a major conveyor of stench, filth and disease, Van Breen also drew attention to the economic importance of good water management. He demonstrated that poorly maintained waterways obstructed intensive shipping movements on the major waterways and could in extreme cases even become dangerous.

The plan's main objective was to prevent flooding, to regulate local drainage, to put an end to long-term pollution, to end sustained pollution, and to remove and prevent any hazards and dangers posed to shipping. Because the municipality had already developed plans for the provision of drinking water Van Breen disregarded that particular issue in his memo. As Van Breen anticipated the town planning developments that were to materialise in the long term his memo and propositions concerned the area within the Batavian city boundaries and the various land areas that were soon to be developed for town expansion. Almost in passing Van Breen alluded to the advantages to be gained from various measures to be taken in connection with the existing traffic situation. He was of the opinion that the partial filling in of a number of large waterways would constitute an important contribution to the resolving of the traffic problems in busy parts of the town and that the presence of flushing conduits would make the hosing down of roads a great deal easier.

Municipalities unite

The proposals that were being put forward for Semarang and Batavia illustrate how municipalities gradually realized of the importance of treating the city as an organic entity. It was evident that in view of the recent economic and social developments and previous experience in the field of town planning it would no longer be sufficient to simply tackle problems in isolation. The main problem however, was that the extent and complexity of the requirements was completely disproportionate to the means available to deal with those requirements. The budget annually paid out to local councils by the government, known as the 'allocated sum', was certainly not sufficient. The councils were also having to cope with a permanent lack of sufficiently well-trained designers and administrators, and inadequate legal and organisational means. The range of these problems was magnified by the vastness of the areas that had to be administered and the geographical distances between the different administrative areas, all of which meant that communication was kept to the minimum.

From the contacts the municipal authorities maintained between themselves, it was almost immediately evident that the problems facing the individual municipalities were so general that the most sensible and effective thing to do would be to work together. In addition to that the municipalities fairly quickly realised that it would be imperative to receive practical and financial support from the central government if the existing situation were to be improved. In line with the decentralisation notion though, the government in Batavia originally ignored any request for support and assistance from the municipalities. The local authorities had, after all, been established on the basis of the idea that they would contribute to improvements in the execution of various government duties.[16] Involvement in local affairs and intervention from higher up was something which, in the eyes of the government, conflicted with the concept of decentralisation and the idea that municipalities as local representatives of the central government were responsible for the way in which affairs were run in their own territory.

At first hesitant but later more self-conscious, the municipalities endeavoured from the very beginning to convince the East Indian government that under the existing regulations they were hardly able to carry out their duties in a satisfactory manner. In connection with their direct responsibility for and the daily confrontation with local issues, they badly needed an organisation that facilitated the sharing of ideas and the sharing of experience. Preferably in collaboration with the central government. During the first conference on decentralisation organised by the municipalities the need establishment of an association was therefore discussed. One year later, in 1912, the Association for Local Interests (Vereeniging voor Locale Belangen, VLB) was set up during the second conference. The VLB operated as a platform where administrators, designers and other professionals involved could discuss various matters relating to administration and the way in which local authorities were organised. Its periodicals *Locale Belangen* (Local Interests) and *Locale Techniek* (Local Techniques) and the annual decentralisation conferences were important channels for presenting, studying, sharing and commenting upon various ideas and views involving all kinds of administrative and legal matters but also practical issues like water management, building regulations and other aspects of urban development.[17]

It is notably Local Interests and minutes taken during various council meetings that give an idea of the way in which economic and social developments forced local authorities to pay attention to general administrative matters and to consider the physical development of the town. Apart from dealing with the local administration many issues were discussed, such as: budgets, taxes and bye-laws, public health, public works, development companies, expropriation, social housing, building and housing inspection, gutters and sewers, water supply policies, premises managed by the local authority, roads, traffic, railways and tramways and sometimes also harbour affairs.

The decentralisation conferences that were variously convened in Bandoeng, Batavia, Malang, Semarang, and Soerabaja were attended by civil servants and a wide range of experts.

It is especially the preliminary advices and the conference proceedings that create a picture of the nature of those events, the subjects that arose and the views that predominated. They not only underlined the urgent nature of many questions but also illustrated how much more efficient it would be to no longer leave certain matters to the individual attention of separate municipalities but to regard such matters as national policy questions and thus deal with them in that way. The media used to discuss issues related to local policies were written preliminary advices, lectures or debates.

On a number of occasions it was in fact town planning or aspects thereof that constituted the conference theme: housing and social housing (1918, 1922, 1923, 1940), town planning in the Dutch East Indies (1922), expenses related to town planning (1933), and regulations concerning building alignment (1929). In that connection, D. de Iongh, mayor of Semarang and later director of the Department for Government Companies prepared a preliminary advice on housing in 1918. H.J. Bussemaker, mayor of Malang and later of Soerabaja, and R. Slamet, a Semarang council member, questioned the matter of the abolition of desas within the municipal boundaries and the subsequent reorganisation of municipal departments. A. Poldervaart, chief engineer and director of the Bandoeng Development and Housing Company and later director of Bandoeng's Urban Development Service, together with F.C. Frumau, director of the Public Works Service in Soerabaja wrote a preliminary report on the costs involved in urban development. One of the most notable contributions though, was the preliminary report entitled 'Indian town planning' presented by Karsten during the 1920 conference.

'Indian town planning'

'Indian town planning' – which despite its title mainly pertained to the island of Java – was a description of the objectives and essence of modern town planning in general and in the Dutch East Indies in particular. It described the various components of town planning and the means used to realise its principles. After establishing the scope of the discipline describing the various components and responsibilities, and identifying the organisational and dynamic model required for an urban development plan, Karsten listed the basic elements and the means used to realise any town plan: buildings and roads, the various building types, the way in which buildings are mutually situated, road type differentiation, length and cross section profiles, ordinances, and different official bodies.[18] Alongside the technological aspects and regulatory measures, he described the aesthetic side of town planning at length.

On the aesthetics of town planning and architecture, the relationship between aesthetics and society, and the role and duty of the government, Karsten wrote the following:

> *The building of any town or village is a government task accompanied by the obligation to give form to that as consciously and as well as possible; the good quality then becomes an expression of the community as such [...]. Just as in our times it would not make sense for all kinds of aesthetic requirements to be demanded of local town planners, likewise no sustainable effects can be achieved by having local communities make demands of the inhabitants with regard to architecture. One cannot extract from a community any more aspects of good form than*

actually reside in its mental level which is determined by fundamental social factors that are much deeper rooted than any ordinance.[19]

On the basis of the notion that 'good form and character' derived from the organic unity between external and internal matters, and thus constituted a criterion for a satisfactory design, Karsten argued in favour of organic plans. Plans in which the character of any one plan would largely be based upon local topographical, social and historical circumstances. With regard to the triple town planning task of creating space, facilitating spatial development and encouraging spatial experiences he noted that it was not simply possible to apply Dutch or European solutions to the East Indian situation. In order to reinforce the character of an urban development plan Karsten, in line with contemporary views on town planning elsewhere, was very much in favour of drawing up town plans that would be adapted to the local landscape. Hence in the Dutch East Indies the tropical climate, the vastness of the archipelago and the abundance of its flora were, to his mind, the salient elements that needed to be borne in mind and which in an almost natural fashion would lead to open, horizontal sections ('walls') of what would be largely open structures with much 'green'. Regarding design Karsten was of the opinion that the nineteenth century Western tendency to view what was 'beautiful as a more or less unnecessary externally imposed veneer', was quite remote from the Eastern view that good form is central to and inseparable from the entire design process.[20] In connection with this discrepancy Karsten – who was greatly interested in and had much appreciation of Eastern art philosophy – warned Western designers working in the East or in Eastern environments to be careful not to let their 'sober Western indifference' overshadow or eliminate the Eastern approach.[21]

According to Karsten, town planning in the East Indies of 1920 was transforming from a relatively unsystematic to a fully planned town planning design practice. As is evident from the opening quotation, to Karsten's way of thinking played an important part in the systemising of such a design practice. After all, it was the local councils that offered the administrative and technical continuity that he believed was necessary for good town planning practice and for the realisation of good town plans.

Apart from being attributable to a chronic shortage of craftsmen, the lack of continuity was largely a consequence of the physical and administrative circumstances that characterised the archipelago until the twentieth century. Firstly, the geographical distances between the central government in Batavia and the various local authorities elsewhere in the archipelago made consultation and the exchange of information between different parties sporadic. It also explained why the government in Batavia was hardly aware of specific local circumstances, problems and requests. Secondly, gradually acquired knowledge, insight, views and ideas periodically ebbed away because of the frequent changing and transferring of civil servants. Thirdly, the level of expertise of most technicians was so inadequate that also in that respect there was little continuity. Karsten maintained that this situation had changed for the better when the local councils had been installed. Through their direct presence and responsibility, administrators were immediately confronted with and answerable for local problems and circumstances. It was a condition Karsten deemed necessary for the emergence of proper town planning practice.

The clear and concise way in which Karsten brought together and described divergent

Table 1. **Dutch East Indian urban planning 1907-1950**

Places	Urban planning project	Designer / responsible organisations	Year of initiation
Semarang	Tjandi Baroe	Ir. K.P.C. de Bazel	1907
Batavia	New Gondangdia	P.A.J. Mooijen	1911
Soerabaja	Darmo	Ir. H. Maclaine Pont	1914
Soerabaja	Goebeng, Ketabang, Koepang, Ngagel	Municipal Public Works Service Soerabaja	
Semarang	Tjandi Baroe	Ir. H.T. Karsten	1916
Semarang	Pekoenden-Peterongan, Sompok	Ir. H. Maclaine Pont, Ir. H.T. Karsten	1917
Malang	Oranjebuurt	Municipal Public Works Service Malang	1917
Bandoeng	Extension Bandoeng North	General Bureau of Consulting Engineers and Architects	1917
Buitenzorg/ Bogor	Kedong Halang	Ir. H.T. Karsten	1917
Medan	Polonia	Municipal Public Works Service Medan	1919/1920
Malang	G.G.-buurt, Bergenbuurt	Municipal Public Works Service Malang	1920
Bandoeng	Extension Bandoeng South and revision of Extension Bandoeng North	Municipal Public Works Service Bandoeng	1927
Malang	Extension Malang West	Ir. H.T. Karsten	1931
Bandoeng	Sorghvliet	Ir. A. Poldervaart	1938
Batavia/ Djakarta	Satellite Town Kebajoran	Central Planning Bureau/ M.Soesilo	1949
Balikpapan	Recovery	Ir. H.Lüning	1949
Samarinda	Recovery	Ir. H.Lüning	1950
Palembang	Recovery	Ir. H.Lüning	1950

Source: Akihary, *Architectuur & stedebouw in Indonesië 1870/1970.*

town planning factors won him much support and contributed significantly to his position and influence in the field of town planning in the Dutch East Indies in the period before the Second World War. One of his earliest supporters was Dr. M.J. Granpré Molière, professor of architecture at the Technical College in Delft. In 1922 he devoted considerable attention to 'Indian town planning' in the journal *Tijdschrift voor Volkshuisvesting* (Journal for Public Housing). He commenced thus:

> *This is a publication that definitely needs to be acknowledged; it is in itself remarkable that what may truly be termed a complete work has appeared solely on the East Indies, complete in the sense that the issue is dealt with to its fullest extent; and that is quite an achievement if one considers the different races, the striking transition from primitive to cultured town forms, and so on that exist in the Dutch East Indies.*[22]

Granpré Molière's admiration was not only prompted by Karsten's analysis and description of town planning in the complex colonial society of the Dutch East Indies. He also considered 'Indian town planning' to be a refreshing change from what he perceived to be the prevalent tendentious, exaggerated or superficial approach to the town planning discipline. From his remark to the effect that in Dutch planning main roads were 'often so blatantly' at odds with Karsten's insights it may be concluded that Granpré Molière was of the opinion that the principles and views voiced by Karsten lent themselves perfectly to examination and application outside the East Indies.[23]

Apart from being founded on theories and ideals 'Indian town planning' was also based on experiences in the colony. In that respect the first example was Semarang. It was for Semarang that De Vogel, even before he and Tillema had presented their ideas for an expansion plan south of the existing town, in 1907 and of his own volition, asked the Dutch architect K.P.C. Bazel to design a preliminary plan for the future development of this area. After having become acquainted with the proposals by Tillema and De Vogel, the municipality decided some time later to ask Karsten to come up with a design for Nieuw Tjandi (New Tjandi). Karsten accepted the offer and presented his design in 1917.

Other plans followed. In Soerabaja the municipality purchased the private land area known as Goebeng in 1908, the idea being that it could serve as a new European cemetery. The cemetery plans never materialised. Instead a new European housing area was created. With a view to the need for even more European dwellings the authority then went on to purchase the estates of Ketabang and Ngagel in 1916. In 1917 expansion plans for Batavia and Medan were finalised. One year later, in Batavia, the engineers J.F. van Hoytema, F.J.L. Ghijsels and H. von Essen presented their designs for the predominantly European residential areas of Menteng and Nieuw Gondangdia.

The development of a town

One municipality whose development was closely related to new policy and administrative organisation was Bandoeng. The proposal put forward by governor-general, J.P. graaf van Limburg Stirum in 1916 to transfer a number of departments from Batavia to Bandoeng. As a result and based on the 1917 design by the General Engineering and Architects' Bureau (Algemeen Ingenieurs- en Architectenbureau) for the northern part of the town Bandoeng was to be rapidly transformed from an insignificant municipality to a town of worldly allure. What was envisaged for the northern Bandoeng expansion plan was mainly European institutes and residential areas. The southern part of the city remained the living and working area for the indigenous inhabitants and the Chinese population. As a rule, the indigenous and Chinese neighbourhoods tended to have intricate and very built up street plans. Unlike in the northern part of the city, many houses were not linked up to water mains or the sewerage system. The northern neighbourhoods were spaciously laid out: the through roads had a wide profile with plenty of green strokes were flanked by villas situated on spacious grounds. Admittedly the minor roads possessed more modest profiles and buildings but they, too, featured plenty of green areas.

One of the first institutes to be established in Bandoeng, partly in conjunction with the plan launched by Van Limburg Stirum, was the Technical College. After the college had opened its doors in 1920, some of the institutes to appear there a year later were the Government Companies department, the Post, Telegraph and Telephone Services, the Pasteur Institute, the National (Cowpox) Vaccine Institute (Landskoepokinrichting), and the Meteorological Institute.[24]

The way in which the plots were divided up in the expansion plan drawn up for North Bandoeng (West Java) by the General Engineering and Architectural Bureau (AIA) in 1917. Map dating from 1931.

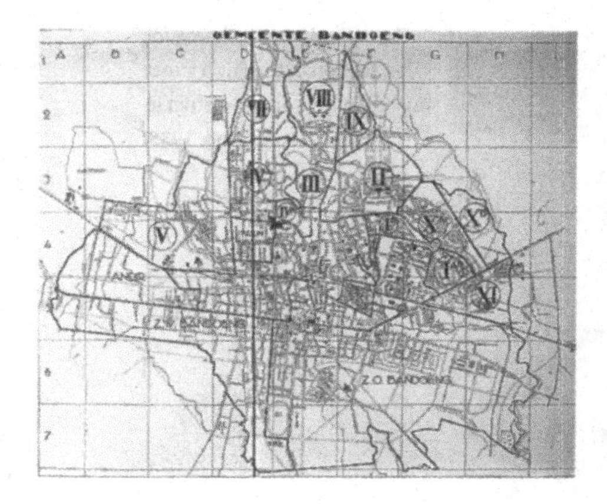

The ground use divisions for the Bandoeng municipality as revealed in a map made by the Urban Development Service, 1933.

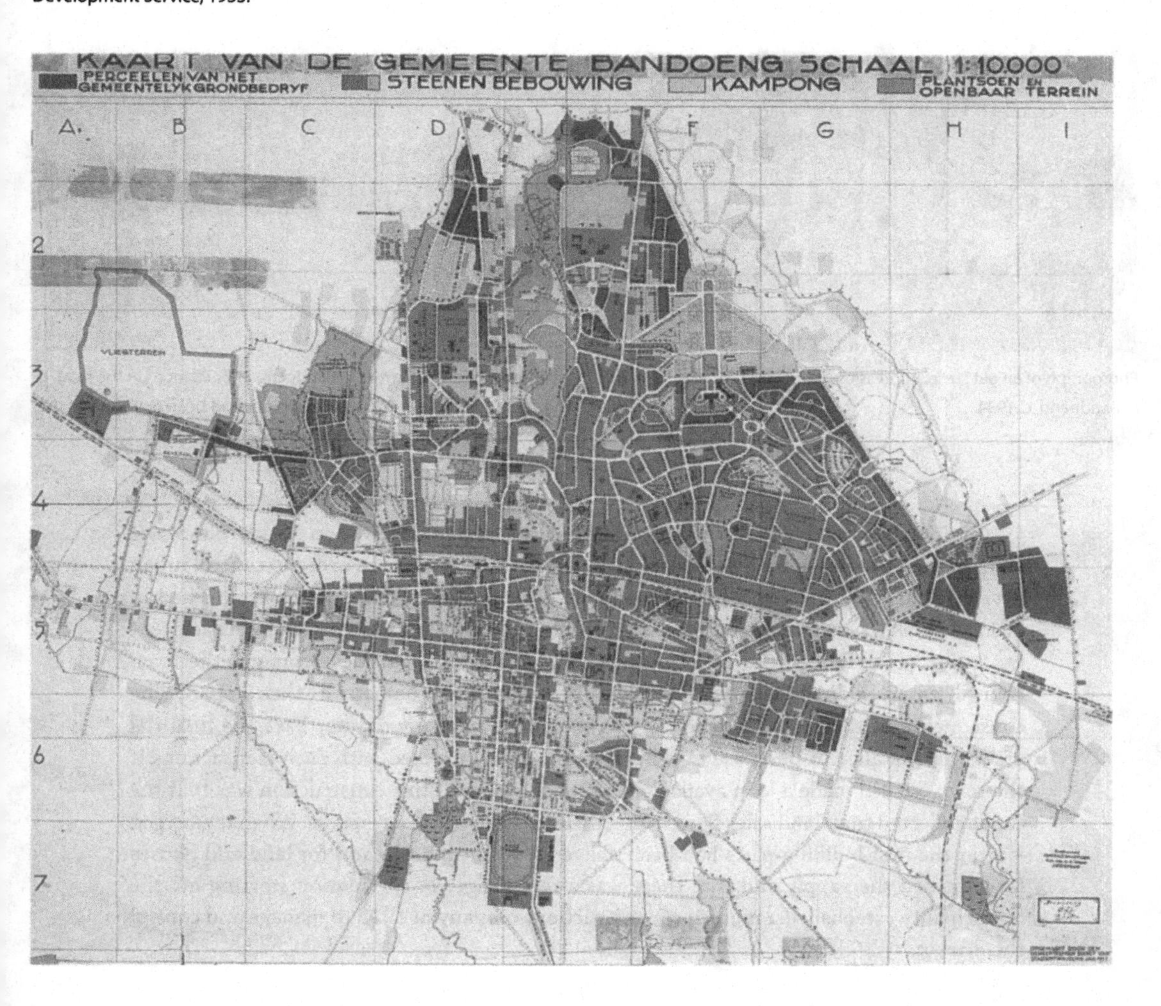

The new planting scheme for a section of the Frisia street situated in a European district in Bandoeng, 1935.

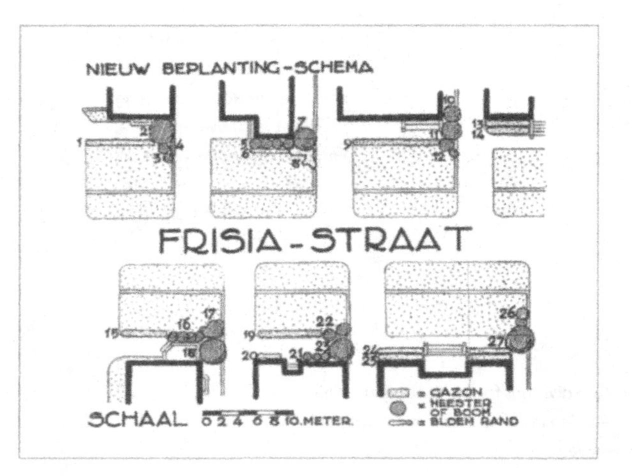

Photograph of an old street impression of the Frisia street in Bandoeng, c. 1934.

Photo of the Frisia street in Bandoeng after changes in the road profile and the individual plot divisions had been realised and the planting plan had been implemented, 1935.

The departmental transfer plan reinforced the systematic way in which Bandoeng worked on the expansion of its territory in a northerly direction. In order to control urban development as much as possible a municipal development company was established in 1917. Its task was to manage both the land within the municipal borders and in the adjacent expansion area. Though the company was responsible for urban development, the town council remained directly involved because of the municipal character of the company. The consequence of this was that when the development company did not have the financial assets for instance to purchase new land, it could only proceed with such transactions if the town council made a loan available. The consequence of this construction was that the funds collected from land sales flowed directly into the municipal reserves. When it emerged – after the first building plots had been delivered – that the demand for land and houses far exceeded the supply and that there was a real danger of speculation starting up, the municipality established a municipal construction company in 1918 to manage and control the market.

Another effective instrument that the municipality had at its disposal for the realisation of its plans was that of promulgating municipal ordinances in order to cultivate an orderly street image. Such things as 'public order, tidiness, cleanliness and health' were promoted. The advantage of this approach was that through its municipal companies the municipality was largely able to steer town planning developments so that, in time, it could apply for government subsidies.

By 1923 the Bandoeng development company had prepared and sold a total area of some 200 ha of ground for building purposes. When, as a result of amendments to Van Limburg Stirum's plan, the execution and the elaboration of the original plans began to stagnate after the mid 1920s and ground sales dropped, the expenses involved in the acquisition of the remaining 500 ha of ground started to weigh more heavily on the town's budget. It was a situation that regularly led to the accusation that the company was putting the municipality in financial jeopardy and which in 1932 forced the company to reduce its total amount of land to 350 ha. Notwithstanding these and other financial ups and downs – but thanks to its unyielding approach – the municipality of Bandoeng succeeded in realising a large proportion of the expansion of the northern area of its territory, creating an urban drainage and sewer system and fulfilling a number of town improvement plans, including *kampong improvements.*

Sustained housing shortage

The expansion plans developed in many municipalities did little to alleviate the ever-more acute housing problem. It was particularly in the non-western neighbourhoods that the problems remained urgent. The sustained housing shortage that was very much in the foreground during the first decade of the century made a systematic approach to the problem unavoidable, not just on humanitarian grounds but also because of medical and political considerations.

Round about 1920 it was possible to clearly distinguish three quite different types of housing areas on the grounds of ethnic, architectural, town planning and economic characteristics: European, indigenous and Eastern. The initial European districts were often densely built up and were a combination of residential and commercial buildings. They had often been built according to some kind of preconceived plan, the houses were spacious, stone was the predominant building material and European architecture and constructions prevailed. As the nineteenth century progressed the differences became more pronounced when the Europeans started building new housing areas. Situated some distance away from the old trading centres and very spaciously laid out with predominantly large detached houses and plentiful splashes of green, these neighbourhoods were the embodiment of the hierarchical relations within the colony. With just a few exceptions, places of work and shopping facilities were situated some distance away from the residential area. The new constellation did not, however, completely do away with the old situation. The predominantly indigenous population in domestic service in the European neighbourhoods perpetuated their customs by constructing semi-permanent houses in the direct vicinity of the European neighbourhoods. It was almost impossible to conceive of a greater contrast than that which existed between these two environments.

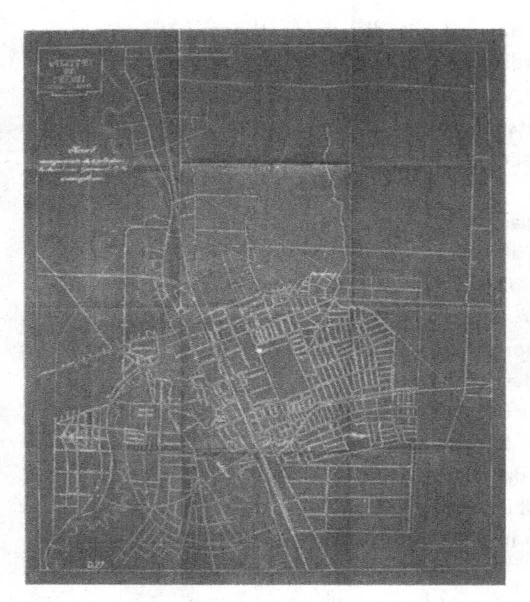

Blueprint of the layout for Medan, Sumatra (scale 1:10,000) including the five locations reserved for municipal housing, June 1920.

Blueprint of the building plans for 230 dwellings in the Petissah district (scale 1:1,000) according to a Medan Municipal Works Service design, 6-7-1920.

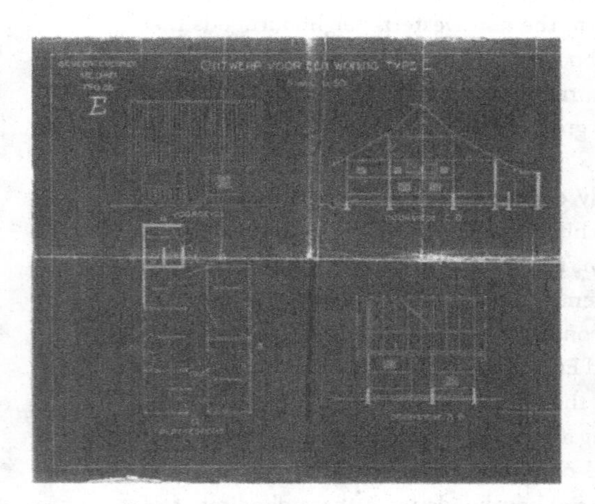

Blueprint of housing type E (scale 1:50) in the Medan municipal housing project, 30-6-1920.

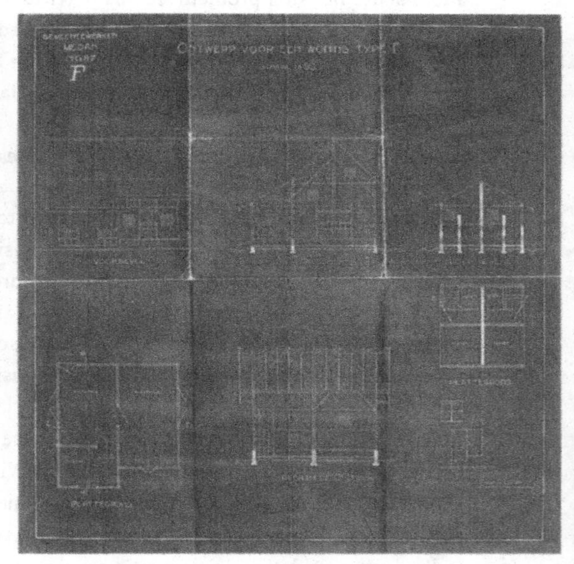

Blueprint of housing type F (scale 1:50) in the Medan municipal housing project, 1-7-1920.

Contrary to the European minority the indigenous majority lived in neighbourhoods (*kampongs*) where planned and unplanned construction seamlessly merged. It was notably in the unplanned areas of these neighbourhoods that hygiene and other such matters left much to be desired. Even though various homes had been built along traditional lines, most of the houses in these densely populated neighbourhoods were fabricated from non-permanent material and did not adhere to any stylistic or hygienic guidelines. The generally dark and damp dwellings and the narrow, uneven and unpaved streets created an environment that was far from desirable, both hygienically and socially. Sanitation facilities such as washing provisions and toilets as well as water supplies and waste water systems were often lacking or else were inadequate. There were small shops situated in the neighbourhoods, other employment was usually a short walking distance away.

The third population group, referred to as the 'Foreign Easterners' (Chinese, Arabs, Indians) generally lived in separate neighbourhoods. This was partly a consequence of a European law which ruled that up until 1919 in Java and Madoera and 1926 in the Outer Regions the Foreign Easterners were obliged to live in separate camps. In the areas populated by Eastern people ground occupation was intensive: dwellings were small and built closely together and the streets were narrow. Living and working was often combined in one building with business on the ground floor and the living quarters on the first floor. The main differences between the indigenous and the Eastern neighbourhoods was that the building material used in the Eastern sectors was often of a more permanent nature than in the indigenous parts and that the architecture and construction often evoked those in the country of origin.

Municipal proposals aimed at improving the often unhygienic living conditions, especially in the densely populated areas, found little support and stimulation from the central government. Already in 1907 it had rejected a plan for *kampong* improvement in Batavia. Other municipalities were also soon made to understand that they were to refrain from any intervention in *kampong* affairs.[25] The uncooperative attitude of the government did not prevent town councils and interested parties from occasionally taking steps to alleviate the housing shortage and perpetually examining and drawing attention to housing shortage and hygiene issues.[26] The result was that from 1915 onwards each time a governor-general planned a working visit, a tour around a poor urban *kampong* would be part of the programme. In 1920 the request of the municipality of Medan for a subsidy for a municipal housing project even met with a positive reaction. The subsidy enabled the municipality to construct 34 European dwellings and 238 local inhabitant dwellings in three different *kampongs* in the space of four years.[27]

Nevertheless, one-off solutions did nothing to structurally alleviate the housing problem for the municipalities. It was an utter waste of time: the problems mounted in proportion to the numbers of inhabitants. By 1920 the housing problem was far from resolved. At that time a mere 4.52 percent of the houses in Java and Madoera were fabricated from stone. All the remaining houses were fabricated from temporary materials and often just consisted of wooden walls and tiled or corrugated iron roofing. It was evident that especially where housing was concerned the municipalities would not be able to surmount the problems without the support of the government or other parties.

In an umpteenth attempt to convince the government that structural intervention was

Traditional dwelling according to a design created by the engineer H. Maclaine Pont in order to illustrate his lecture 'How can the endeavour to improve the housing situation in Java be turned into a national movement?' (1925)

unavoidable the Social-Technological Association (Sociaal-Technische Vereeniging, STV), established in 1918 as a sub-committee of the VLB, organised a conference on social housing in Semarang in the year 1922. Because of the scope of the subject and the relative lack of knowledge about the matter, the conference remained predominantly exploratory. The issues dealt with had to do with hygiene and with the social-political and town planning aspects of public housing. They were all discussed on the basis of preliminary advices questionnaire results and the designs of private architects. J.J. van Lonkhuijzen, chief inspector of the Civil Medical Service (Burgerlijke Geneeskundige Dienst, BGD) and the engineer J.T. Bethe, director of Public Works in Soerabaja, were of the opinion that improvements in the housing situation could not be achieved unless a healthy and hygienic environment was first created. This was not just a matter of cleaning up the actual houses and their immediate surroundings. The constructing of roads, the providing of sewers, the supplying of water and sanitation all had to be tackled first. The engineer H. Heetjans, who was the director of Municipal Works in Bandoeng, emphasised in his recommendations the importance of the link between land politics, extension plans, building and its corresponding regulations, and financing underlined how vital it was to possess reliable statistical material. To ensure that housing remained profitable at all social levels Heetjans proposed that private contractors should provide accommodation for the more well to do citizens so that town councils could focus exclusively on the building of cheaper dwellings and housing complexes.

During the conference the architects Karsten, Maclaine Pont and Professor C.P. Wolff Schoemaker focused on the technological-architectural aspects of the public housing issue. Their points of view were somewhat divergent. Even though they were all in agreement on the importance of generating good architecture and correctly situated buildings, Wolff Schoemaker and Maclaine Pont differed fundamentally when it came to the kinds of examples that should be followed and how plans should be drawn up. Wolff Schoemaker also contended that in the Dutch East Indies traditional 'architecture' was absent and that Hindu monuments, because or their lack of spaciousness, constituted unsuitable examples for housing. He was furthermore of the opinion that European materials and social conventions automatically resulted in different buildings to those dictated by Indonesian conventions.

By contrast Maclaine Pont, architect and inspector-engineer at the BGD was of the opinion that it was precisely in tradition local building styles that the solutions had to be sought. He studied various regional houses, experimented with the applied building materials and developed a type of house which, by making use of locally available materials and forms, would conform to Western norms of hygiene and comfort.

As there was no time for an in-depth or more specific exploration of the public housing not in indigenous housing problem during that first public housing conference the STV in 1925 organised a second conference exclusively devoted to the matter of indigenous housing and the sustainable improvement of poor *kampongs*.[28] Abdoel Rachman, assistant inspector of Building and Housing Inspection in Batavia, Frumau and the engineer J.J.G.E. Rückert produced preliminary advices on measures that would be required to improve substandard existing *kampongs*. Rachman and Frumau believed that it was utterly important to maintain the number of houses and the land available for public housing at acceptable levels. In order to ensure that local inhabitants were not expelled from their land they advised that a prohibition be placed on the selling of indigenous ownership rights to Europeans. They furthermore emphasised that there could be no systematic improvement of *kampongs* as long as autonomous desas existed within municipal boundaries because the administrative autonomy of the desas barred municipalities from having any kind of a say in matters.

Rückert endorsed the views of Rachman and Frumau but added a few supplementary remarks concerning aspects that stood in the way of improvement such as the lack of housing statistics, proper maps, legislation or financial possibilities. He also stressed the importance of maintenance after improvements had been carried out. To ensure the government did not confront the local population with a stream of changing ideas Rückert argued that during all phases of activity administrators and planners should engage themselves in direct and close contact with the population. Only in that way would the government be able to obtain the vital information required on various facets of the matter and to keep the population continually informed of its views and considerations. Where the duty of the government was concerned, he believed that if private companies were not able to provide enough accommodation it would then be the responsibility of the government to do so. The regent for Batavia, R.A.A.A. Djajadiningrat, supported the views of Rückert. He emphasized that the housing problem was indeed a material issue but also an important moral and hygienic, and thus ultimately, communal problem. The government could therefore no longer afford to leave the matter entirely to third parties. If it really wanted to deal with the problem the government would have to draw up and execute a proper relevant policy. Djajadiningrat and the conference participants maintained that the participation of the local population would be essential. Not just because it was the only way in which vital missing details could be obtained but also because it would involve the population when it came to realising improvements in their own environment and hence would improve relations between Indonesians and the local authority.

The housing conferences of 1922 and 1925 constituted the first communal occasions where experts and local authorities expressed views on the need to think systematically, ambitiously and collaboratively about public housing and *kampong* improvement. They were also the first events that prompted the central government to make advances in finding solutions to the problems. Not long after the first conference the government met the request to set up an institute exclusively designed to meet public housing interests by agreeing to the establishment of a separate Division for Housing within the Public Health Service. To alleviate the housing shortage the government also agreed to the creation of and participation in commercial entities, such as public limited companies and private companies.[29] After the second conference the government agreed to a directive which would allow municipalities to appeal for subsidies amounting to up to 50 percent of the costs involved in public housing. What really constituted an important decision, though, was when the government finally decided to draw up and introduce a simplified compulsory purchase procedure and to grant municipalities priority rights to land that was earmarked to become part of future expansion plans.

Far-reaching breakthroughs

Revision

The changes in the national policy regarding urban development that began to take shape in the mid-1920s enabled municipalities to contemplate the designing of plans that incorporated more than one issue and could be worked on for a substantial length of time. In concrete terms, it gradually became possible to embark upon public housing, extension plans and kampong improvement in mutual cooperation. In the second half of the twenties a number of important decisions were taken, notably in the field of acquisition and land use. In 1926 in what became known as *Bijblad* (Supplement) 11272, some guidelines for town extension and public housing were drawn up whilst the existing ruling on municipal priority rights to land was expanded.

The priority right was a ruling that had been adhered to since 1911. It gave municipalities the chance to acquire government land for future expansion and improvement plans. *Bijblad* 11272 extended this ruling by stipulating that municipalities might acquire priority rights if they possessed an urban development plan that had been approved and adopted by the town council and the government. Apart from giving an overview of improvements and expansions such a plan also needed to indicate the areas outside the municipal boundaries where the city wished to claim priority rights. In linking together priority rights and town plans *Bijblad* 11272 killed two birds with one stone: the municipalities were ensured, for some time, of the chance to expand their towns in a specific direction while at the same time methodical urban development was guaranteed.[30]

In the same year that *Bijblad* 11272 was adopted the government also agreed to what was termed the Disruption Ordinance, a ruling designed to stimulate the realisation of home, work and recreational zones. Furthermore, in 1927 and 1928 the government approved 'Regulations regarding the obtaining of free access to land on behalf of the State' and a regulation concerning structural financial support for *kampong* improvement plans.

The increasing willingness with which the government in Batavia agreed to the wishes and demands of the municipalities is striking. What remains to be seen though is whether the change in the originally aloof stance of the government is to be exclusively attributed to hygienic and humanitarian considerations or whether there were also underlying political-economic considerations. It is an undeniable fact that both on the European and on the

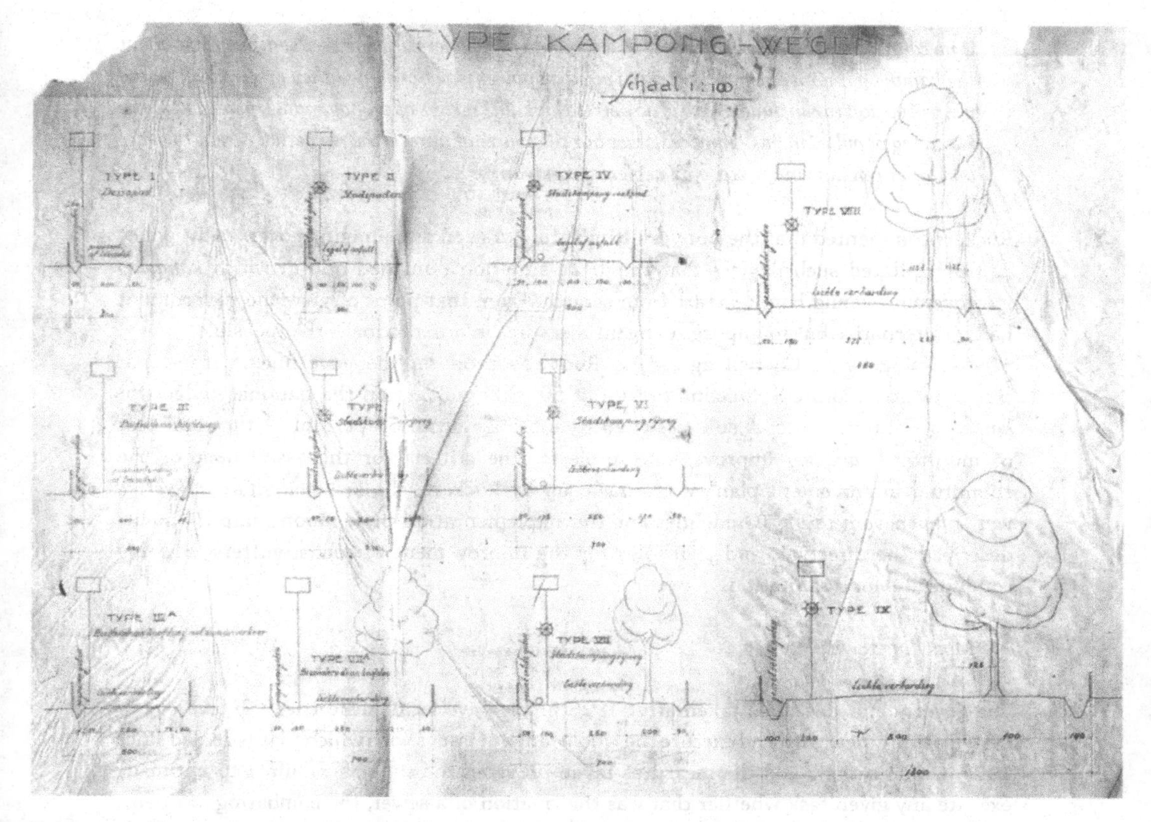

Various types of kampong roads according to the 'Guidelines for the implementation of kampong improvement' laid down by the engineer J.J.G.E. Rückert (1928).

Indonesian side political movements arose that opposed colonial policy. Their protests became ever more frequent and vehement during the 1920s. Moderate organisations like Boedi Oetomo and Sarekat Islam and the more extreme Perhimpoenan Indonesia were good examples of this trend. It is not unthinkable that the altered attitude of the government was to no small degree influenced by considerations pertaining to these very developments.

One indication in this direction is to be found in the plea that Rückert made before the People's Council (Volksraad) in 1928 for the granting of a government subsidy for municipal *kampong* improvement schemes. He contended that any financial contributions to *kampong* improvement would serve two ends:

> *May that segment of the population which inhabits the town kampongs still live in muffled resignation, unconsciously convinced that it cannot be any other way, one should not forget what a rewarding field it is to work with such a population for those who set themselves the goal of overthrowing the authority which allows such situations to continue without intervening. It*

is no coincidence, Mister Chairman, that a number of leaders are not at all enthusiastic about the kampong improvements. As long as political power is not obtained they prefer to allow the poor situation to continue; there is no better thinkable means of propaganda. Up until now the kampong population has remained quiet but the source of unrest, the perpetual threat to peace and order remains in the so badly neglected kampongs.[31]

Rückert commented that the not very stimulating or even discouraging government policy had precipitated such neglect that the 1928 situation remained one in which *kampong* improvement would have to start from scratch. From that point of view the government had no alternative but to impose stringent measures in order to make that possible

As the People's Council agreed to Rückert's proposal the government from that time onwards allocated a maximum sum of 500,000 guilders on the national budget for *kampong* improvements. A sum assigned to subsidize up to 50 percent of the total cost of municipal *kampong* improvements projects. The criteria for the assessment of the submitted improvement plans were drawn up by Rückert. His *Leidraad bij de uitvoering van kampongverbetering* (Guidelines for the implementation of kampong improvement) described the directives and principles for the improvement of roads, gutters, flushing conduits and local sanitation.

Modelling town planning

The government decisions taken after 1926 finally gave local authorities the leverage to systematically deal with and execute the whole range of issues with which they wrestled since 1905. Organisations and the measures taken provided the support required to optimally execute any given task whether that was the creation of a sewer, the combating of floods, the building of houses, the improvement of a kampong or the drawing up of a extension plan. This process had hardly been set in motion when, in the early 1930s, external factors temporarily upset the applecart. In the wake of the economic depression ensuing from the 1929 Wall Street crash, the government was forced to revise financial promises between 1931 and 1934. Especially in the field of *kampong* improvement this was a great disappointment because it implied that the modifications that were about to be implemented had to be postponed yet again.

In line with new policy regarding the coordination and steering of urban development the government decided in 1930 to set up two investigation committees: one to examine building restrictions and the other to look at building alignment. The research conducted by the Building Restrictions Committee was aimed at documenting all building restrictions issuing from general and local regulations. The committee was also responsible for advising the government on whether it was desirable to replace existing restrictions with legal regulations. If the answer to this issue was affirmative, the committee was invited to present proposals to that extent. The Building Alignment Committee consisted of members who also had a seat in the more extensive Building Restrictions Committee.[32] Its task was to investigated the way in which various municipalities assessed building lines.

Following the activities of these committees the government decided to establish a new committee in 1934. This new committee, the Town Planning Committee, was given the task of studying the issue of town planning. In other words, it was required to examine and formulate principles and rudiments for town planning in the urban areas of Java in order to replace *Bijblad* 11272 and all sorts of other prevailing measures in the field of town planning.[33]

The decision to establish a Town Planning Committee marked the start of the coherent approach to town planning that was to become customary from that time onwards in which attention was paid to the various design-technical, methodological and legal aspects. In view of the fact that many members of the Town Planning Committee had previously been members of the Building Restrictions Committee this almost automatically formed an important link between legislative and other issues related to town planning. Thus, through its individual members the committee pooled more than 25 years of experience and knowledge in the field of town planning.

Because of its high degree of expertise the committee was able to formulate a draft ordinance regarding town planning in a relatively short space of time. Already by 1938 a draft and elucidation for a Stadsvormingsordonnantie (Town Planning Ordinance) had been presented. One section in the draft text explained that the ordinance aimed to incorporate social, economic, political and field-intrinsic aspects:

> *The town plan and the town planning regulations order layout and construction, both on the part of the municipality and third parties, so that in this manner the development of a town is made to correspond to its social and geographical character as well as to its anticipated growth. The aim is also to strive for a fulfilment of the needs of all population groups in accordance with the nature of those needs and to achieve a harmonious functioning of the town as a whole, all of which must fit in well with the environment while observing its function in a general respect.*[34]

After the customary defining of the terminology used, the Town Planning Ordinance described the various components of an urban development plan: the general plan, the detailed plan, the zones to be opened up, the components and the building regulations where necessary. Further directions and indications for the elaboration and detailing of town plans could be issued by the government. This account was followed by a description of how existing works and parcels of land would be handled and what procedures were to be adhered to when it came to determining the various sub-plans, including measures concerning tolerance obligations, types of licences, mandates, compulsory rulings and building inspection. The ordinance closed with a description of the financial side of town planning: the determining of rights to compensation, damages and values, and the different ways in which contributions could be made to costs.

The *Toelichting op de Stadsvormingsordonnantie stadsgemeenten Java* (Elucidation of the Town Planning Ordinance for the Municipalities of Java) which, with its 204 pages was almost four times as long as the ordinance, by means of statistics and descriptions elaborately conveyed the actual situations in East Indian towns and the notions.[35] Regarding the extent of the descriptions, the authors observed that it was the versatility and newness of town planning together with the lack of overview and the need for overview, structure and policy that made a shorter description both impossible and undesirable. In order to enforce

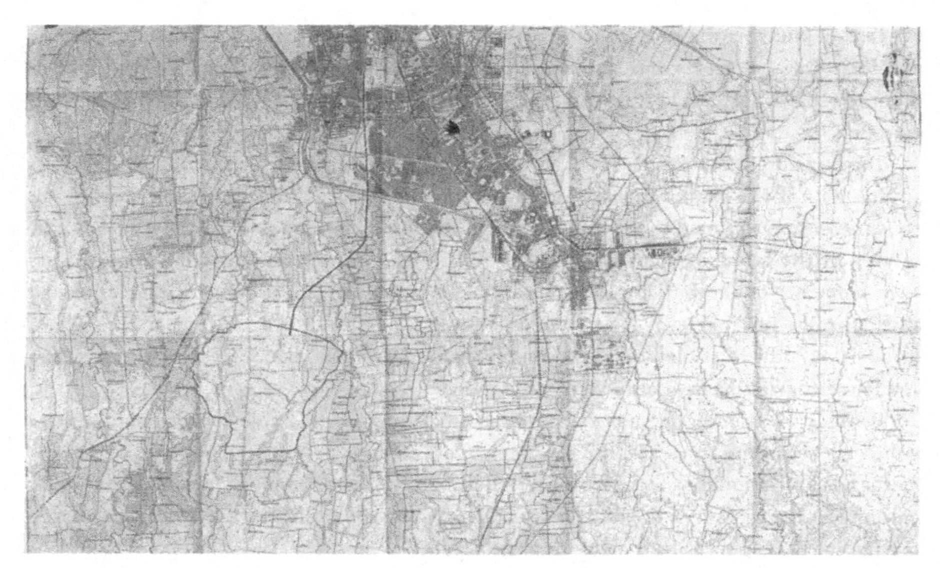

Map of Batavia indicating the area earmarked for the future satellite city of Kebajoran Baroe to the south of the existing municipal border, 1948.

these measures and because of the social consequences, the compilers maintained that the necessary 'supervision' would initially have to focus on social and administrative aspects and only at a later stage on matters of a technical nature. Excluding defence structures and other works such as railways and harbours together with the relevant land, the Town Planning Ordinance would apply to all municipalities in Java and would be enforced as soon as the governor-general gave his approval to the proposition.[36]

Though it was by no means a perfect end product, the Town Planning Ordinance was from many points of view (architectural, administrative, legal, economic, sociological, hygienic, defence-wise, cultural-historical, financial and fiscal) a document of great importance. It constituted a first effort to lay down the principles, methods and legal basis for town planning in the Dutch East Indies in one all-embracing ruling.[37] Apart from the various specific demands that the climate, the nature of the tasks, the vastness of the archipelago and the abundance of its flora placed upon town planning, the ordinance was a fairly general ideological and methodological description of town planning practice. As such, it might have been suitable for application outside the Dutch East Indies.

Untimely end

The draft version of the Town Planning Ordinance was discussed a year after its presentation during a Planning Workshop organised by the Planological Study Group in 1939.[38] What was pointed out a number of times during the debate was the fact that the ordinance aimed to achieve a degree of perfection which definitely did not correspond to reality. Those present therefore doubted whether 'the ordinance would speedily find its way into the Law Gazette'.[39] Another thing that was mentioned was the fact that because of the unfamiliarity of the many new concepts included in the ordinance, the observed transition period between the present and the future status quo would be too brief.

Other points of criticism asserted that the town planning discipline was approached too much from the housing angle and paid too little attention to traffic aspects – which meant that it leant too heavily on the Dutch Housing Act. Furthermore there was criticism about the fact that much was said about public housing while precious little had been achieved in that particular area and it was fair to assert 'that a nation was in need'.[40] The conference participants also objected to the fact that the ordinance used zoning schemes and zoning plans as town planning instruments even though experienced had shown that such instruments were ineffective. They were also of the opinion that the proposed procedures would cause friction because compared to the decentralisation advisor town councils had limited authority. Finally they stated that preliminary plans and planning elements required more legal protection and vigour than the draft text suggested as it was 'doubtful whether the Government would ever adopt a city plan'.[41]

In reaction to the criticism, Logemann and Karsten elucidated a number of points during the Planning Workshop. In general, Logemann and Karsten promised that any criticism or comments on the draft text deriving from conference participants or, for example, from the departments corresponding with those of the committee, would be incorporated in the text as accurately as possible. Due to the outbreak of war in Europe in 1939 and the Japanese occupation of the Dutch East Indies in 1942 this commitment never materialised.

Building up and rounding off

Resetting the course

When, after the Second World War and the Japanese occupation, Indonesia proclaimed its independence on 17th August 1945 it became clear that a reviewing of the political relations was necessary and unavoidable. Despite intensive inland armed combat and international pressure to acknowledge Indonesia's independence, it was not until 27th December 1949 that the Netherlands officially transferred power to the new federal government of the republic. In the intervening period administrative adaptations had been made in order to deal with current businesses. One of the adaptations on an administrative level was that the colony was no longer to be ruled by a governor-general and run by directors of departments but by a lieutenant governor-general and secretaries of state. The pre-war lower administrative units of residences, municipalities and regencies were retained. The position of advisor for decentralisation was abandoned altogether. The Provincial Ordinance of 1924 was further elaborated. Alongside the existing provinces of Central, East and West Java, the places and the islands of Sumatra, Borneo, Celebes (Sulawesi), the Moluccas and the Soenda islands were also given the status of province.

Because of the war and the ensuing hostilities, much damage had been incurred in many parts of the country, especially to harbours. In order to restore the 'normal' pre-war situation it was essential to speedily embark on repair and reconstruction activities. The administrators as well as the designers soon realised that the reconstruction work that needed to be done was tremendous but that the means available for doing that were restricted. They also realised that the only way to cope with the work ahead would be by centrally organising activities. It was therefore decided that a Central Planning Bureau (Centraal Planologisch Bureau, CPB) would be established, a centralised service that would fall under the Department of Public Works and Reconstruction that was, in effect, the successor to the Department of Transport and Public Works. All the designers who had any experience with town planning issues were employed by this bureau. They included J. P. Thijsse who, before the war, had worked for the Municipal Works Service in Bandoeng and M. Soesilo who had worked with Karsten in the thirties. Besides them only six more people (two urban developers, two architects, a sociologist and a lawyer) and administrative personnel were employed. Because he died during the war, Karsten was no longer able to contribute to the spatial developments in the archipelago in person.

From Town Planning Ordinance to Spatial Planning Act

In its activities the CPB worked on the principle that local town plans were by definition subordinate to reconstruction projects and that because of a lack of factual data they were temporary and would be subject to review as soon as the situation stabilised and more information was available. Just as had been the case before the war, the CPB staff were soon confronted with obstacles that were the consequence of the absence of a solid methodological and legal basis for town planning. One of the problems was that according to the still prevailing pre-war ruling, areas without a municipal status were not empowered to draw up or execute town plans. In order to put an end to that obstructive situation it was important to draw up an emergency ordinance as soon as possible.

As no proper legal foundation for such an ordinance existed, the draft text of the Town Planning Ordinance of 1938 was used. Some minor changes were made in order to adjust the text to post-war circumstances and it was then approved by the lieutenant governor-general in 1948. Shortly after he designated 15 cities and places where, because of recent and future developments, the ordinance could well be enforced. The first place where the Town Planning Ordinance was adopted and entered into force as of 1949 was Bandjermassin in East Indonesia. Later that year the following towns were to follow suit: Padang in Sumatra and in Java Batavia, Tegal, Pekelongan, Semarang, Salatiga, Soerabaja, Malang, Tjilatjap, Tangerang, Bekassi and the areas around Kebajoran and Pasar Minggoe .[42]

With the coming into force of the Town Planning Ordinance urban development and reconstruction plans, detailed plans and building regulations were, for the first time, given a solid legal and methodological basis. In order to be applicable to development outside Java and to reconstruction plans, a few adjustments were made to the contents of the ordinance. The lemma of the legal text in the *Staatsblad van Nederlandsch-Indië* (Law Gazette for the Dutch East Indies) mentioned this clearly:

> *Town planning. Rules in order to ensure town planning is well considered, in particular in the interests of a rapid and effective reconstruction of all areas hit by the turmoil of war.*

The broadening of the field to which the ordinance was applied was achieved by appending two articles to the original ordinance text. Article 51 stipulated that the ordinance should apply in cities and in other places where the drawing up of an urban development plan was necessary because of recent or expected urban planning developments. Article 52 determined that in cases where towns had no municipal status or had their own appropriate municipal departments, the resident would be empowered for a certain period of time to appoint other authorities to execute the powers of mayor and aldermen.

Whether the decreed Town Planning Ordinance met with the intention of the director of the CPB, Thijsse, that it should exclusively serve to meet the 'present abnormal circumstances' and should therefore just have the temporary character of an emergency ordinance is difficult to determine.[43] What is clear, however, is that the Town Planning Ordinance was an important point of departure for the tasks of the Interdepartmental Government Committee for Spatial Planning in Non-Urban Regions (Interdepartementaire Regeringscommissie voor de Ruimtelijke Ordening in Niet-Stedelijke Gebieden) that was established in 1948.[44]

In establishing this committee the government fulfilled the ambition of Thijsse to broaden the town planning discipline in Indonesia to include spatial planning and general planning in line with developments in Western Europe and North America. Thijsse was of the opinion that the Town Planning Ordinance was to be perceived as the forerunner to spatial planning. Political developments and developments in the field of planning would irrevocably lead to a planning practice that was no longer confined to an area within the municipal boundaries but which would also come to be applied to areas outside those boundaries. In the light of the ongoing building up of a federal state, Thijsse was of the opinion that the work should be organised in such a way that individual federal states would be made responsible for the organisation and the realisation of coherence between the various regional plans. In order to make sure that everything went smoothly the committee decided that there was a need for a law on the procedure and the decreeing of plans.

In spite of a number of drastic personnel changes that took place after 1949 as a consequence of the transfer of sovereignty from the Netherlands to Indonesia, the government committee continued its activities.[45] The draft text for the Spatial Planning Act that was subsequently handed over to engineer Laoh, the Minister for Public Works and Energy (Pekerdjaan Umum dan Tenaga Listrik) was largely based on the Town Planning Ordinance and on the Dutch National Plan and Regional Plans Act. It contained guidelines for a national plan, regional plans, the execution of zoning, approval and assessment procedures, building regulations, compensation and indemnification. Just as with the Dutch equivalent, the Indonesian act provided a national plan for the entire country or at least part of the country. In order to enable politicians and designers to make decisions based on the needs and demands of various groups and sectors, the national plan was divided up into what were termed 'facet' plans.

It looks like the Spatial Planning Act was never decreed nor implemented. J.W. Keiser, the lawyer involved in the legal sides of the Town Planning Ordinance and the Spatial Planning Act noted in 1951 that because an Indonesian translation of the text was not available the Indonesian government was unable to assess the law proposal. Looking for an explanation why the act was still not decreed in 1954, Thijsse wrote:

This regional and national planning law has not yet been approved to date although the draft was presented more than 3 years ago. One of the reasons of this postponement is certainly the expectation that the execution of this law will be very difficult owing to the lack of competent personnel.[46]

'Contribution to the morphology of town planning, particularly in Indonesia'

The lack of capable personnel to which Thijsse alluded was not a problem that only manifested itself in Indonesia after the transfer of power. Also immediately after the war he frequently alluded to it. He repeatedly emphasised that the civil engineers who had been educated at the Bandoeng Technical College were insufficiently equipped to take up leading positions in the field of town planning or general planning because of deficiencies in the educational system. In order to improve this situation, Thijsse proposed adapting the curriculum and introducing refresher courses for civil servants in the areas of planning and town planning as well as in sanitation and technical hygiene.

The inadequate education was directly related to the lack of good publications and manuals on town and other planning matters that were applicable to Indonesia. The first book that was to bring about a change was Thomas Nix' dissertation *Bijdrage tot de vormleer van de stedebouw in het bijzonder voor Indonesië* (Contribution to the morphology of town planning, particularly in Indonesia) that was published in 1949. Nix was an Indo-European architect who before the war had been employed by the architectural and engineering bureau Hulswit and Fermont in Weltevreden and by Ed. Cuypers in Amsterdam. In his book, Nix provided an accurate and orderly overview of the elements and components of an urban development plan. As the content was clearly based on pre-war activities in the field of town planning and broadly corresponded to the procedures and stipulations of the Town Planning Ordinance, that is to say to Karsten's ideas and theories, colleagues criticised the book for its unoriginal and rather handbook-like character. Although the criticism was not unfounded, the publication, precisely because of its content and appearance, demonstrated the urgency of the need for study material and the lack of expertise in the field of town planning: barely half of the 16 titles in the bibliography related to town planning in Indonesia.

Definite turning-point

At first the political and social changes confronting Indonesia from 1950 onwards had few repercussions as far as cooperation between the Dutch and the Indonesians was concerned. In retrospect though it became clear that the reconstruction era was to be the last period when the Dutch would prominently intervene in urban development and planning activities in the archipelago. Though they did not disappear from the scene immediately, their prominence gradually diminished after the transfer of sovereignty. The fact that Indonesia determined things for itself became evident when, for instance, President Soekarno decided to abolish the federal state in 1950 and once again centralise power under a central national government in the capital city which, by then, had been given the name Djakarta.

In the area of planning, apart from the different working atmosphere, the administrative alterations were not very influential. In the case of education, on the other hand, some sweeping changes were made. In 1950 the decision to unite all existing colleges in one national university situated in the country's capital, the University of Indonesia (Universitas Indonesia, UI) led to the conversion and incorporation of the civil engineering department of the Technical College in Bandung to the UI faculty of architecture.[47] The curriculum largely continued to be based on that of the Technical College in Delft. In order to supplement the civil engineering courses Soekarno instigated the curriculum for architecture students was to include lectures on art and on the history of Hindu-Javanese building construction and design.[48] Attention for the theory and design side of contemporary architecture and to town planning remained limited to a minimum.

Finally the Indonesian-Dutch society which had in many respects been sustained came to an abrupt end when in 1957 the Indonesian government refused to accept the inflexible attitude of the Dutch government regarding the political status of New Guinea. Because of the subsequent deterioration in relations, 50,000 Dutch nationals were forced to leave the country. That was this political upheaval that really put an end to Dutch involvement with Indonesia.

The temporary vacuum created by the speedy departure of Dutch experts in the field of town planning and planning was soon filled by colleagues from Germany, Austria and notably the United States of America. Not long after their arrival it was decided that the two UI faculties which had been situated in Bandung since 1951 should once again be made independent. Thus the former Bandoeng Technical College once more regained its independence and was renamed the Bandung Institute for Technology (Institut Teknologi Bandung, ITB). Another consequence of the breaking of ties with the Netherlands was a gradual adaptation of the curriculum at the faculty of architecture. In accordance with education in the United States the subject of architecture became more integrated into the curriculum while at the same time town planning and landscape architecture were added. Almost simultaneously a plan was presented to establish a complete and independent school for town and regional planning in Bandung. The underlying idea was that it would in time evolve into an independent institute for urban studies and research for South and South-East Asia. Although this idea never materialised ITB organised courses on town and regional planning in conjunction with the United Nations aid programme for Indonesia from September 1959 until April 1965.[49]

The result of the increasing influence of the United States was that town development and design practice which up until 1957 had been mainly modelled on Dutch practice gradually became more oriented towards other, mainly American, methods and approaches. It was a trend that was reinforced in the 1960s when the first Indonesian designers who had graduated in the United States returned to Indonesia and when opportunities to be financially supported to study in the United States expanded. The direction subsequently followed by Indonesian town planning gradually led to a situation in which the Dutch model, ideas and knowledge that had laid the foundations for East Indian town planning gradually made way for an all American approach.

The Karet Weir in the Bandjir Canal
in Batavia, the situation in 1970.

CHAPTER 8 **WATER IN THE CITY**

Drinking water, sanitation and flood control in urban areas

JAN KOP

It was not until the nineteenth century that the massive archipelago became so densely populated that sizable urban agglomerations beyond Batavia and Djokjakarta started to develop on such a scale that matters such as the public supplying of drinking water (hereafter often termed public water supplies), sanitation and the protecting of urban areas against floods from the sea and/or rivers (*bandjirs*) began to require a structured approach. What fell under sanitation in urban areas was subsurface drainage (groundwater level control), land reclamation for reasons of hygiene (for instance to combat malaria), urban drainage (surface drainage, getting rid of excess precipitation), the collecting, removing and treatment of waste water and the collecting, storing and treatment of solid waste. In the centrally bureaucratically governed colony it was to be expected that the challenge for that structured approach would be taken on by, and would be the responsibility of, the central civil authorities and, specifically, of the civil public works authorities. That meant that as of then and with respect to such public works, decision-making and other activities such as reconnnaissance, mapping, investigation, research, reporting, specifications preparation, contract allocation, execution and control would all in the future be centrally administered.

Centrally administered civil public works was something that would in fact have been impossible before the beginning of the nineteenth century because of the presence and utterly dominating position of the VOC (the United East India Company) in the seventeenth and eighteenth centuries and because of the fact that before 1800 there could not, as has already been indicated, possibly have been mention of 'de jure' civil public works because there simply was no public Dutch government that exercised any kind of authority there. Apart from anything else, it should be remembered that the VOC settlements in the archipelago were of a modest size. Up until 1800 even Batavia was a small city with no more than 50,000 inhabitants.[1]

In the rural areas it was generally possible to cope with primitive and small-scale drinking water supplies and sanitation and the protection mechanisms against flooding often constituted part of the irrigation projects; a situation which in many places persisted until the end of the colonial era (1942). One exception to this was the various pioneering projects in the archipelago that came as a direct result of private enterprise, such as the boring and refining sites of the Batavian Petroleum Company, the cultures (estates) on the eastern coasts of Sumatra[2] and in Java[3] and, for instance, the timber processing industry on the island of Simaloer, then known as Poeloe Simeuloë. There the relevant companies took full responsibility for the drinking water supplies, sanitation and medical supplies that they needed (coolie hospitals).

All in all, the Dutch East Indies did not actually lag that far behind the mother country when it came to tackling these kinds of government matters in the second half of the nineteenth century. It was not until 1853 – forced by the knowledge that cholera was caused by the drinking of infected water – that Amsterdam became the first city in the Netherlands to establish a company for the provision of piped public drinking water supplies to the people of the city. The company collected water from the dunes and, after having purified it, channelled it via a piped system to the inhabitants of the city.[4] Amsterdam was quickly followed in 1856 by Den Helder, which had basically resolved to do just that as long ago as in 1852. In the Netherlands it was not until 1963 that all the so-called uneconomical terminals – chiefly located in rural parts – had been connected up to the mains of the water supply network so that finally the whole country had access to running water.[5]

Due to the rapid population growth, a consequence of increased medical knowledge and medical care – both preventive and curative – and economic prosperity, especially during the course of the twentieth century, the era saw a speeded up growth both of small and large cities which was something that went hand in hand with a need for government care of public water supplies (public drinking water provision), sanitation and *bandjir* protection. The administrative decentralisation system that was introduced in 1903 made it possible to largely transfer those responsibilities from the central government to the local administrations – regions, municipalities – thus moving such responsibility closer and closer to the urban population.

The developing of a legal basis and an administrative framework

Despite the appointment of a chief Public Works inspector (1818), up until 1854 – the year of the introduction of the 'Government Ruling' – the division between militarily earmarked and executed works and civil public works was not always all that clear. *Bandjir* protection in urban areas was something that only became concrete after 1854 and was the direct responsibility of Civil Public Works. Public health – hygiene – was and remained the constitutional element for works related to public water supplying and sanitation. One can more or less picture the problems regarding management, legislation and the sharing of responsibilities as far as this particular civil public works sector was concerned if one realises that:

- in 1827 a Military Medical Service (MGD) was created;
- it was not until 1854 that by means of a Government Ruling a director was made responsible for public works 'upon the orders of the governor-general';
- in 1866 the construction of civil public works fell under the Department of Civil Public Works (BOW);
- it was not until 1882 that the Civil Medical Service (BGD), which was separated from the MGD, was established, at which time it was clearly stipulated that planning, design, and the creation and management of public drinking water facilities were in the hands of BOW whilst the BGD merely monitored the quality of the drinking water. The friction between BOW and the BGD – later to be known as the Public Health Service (DVG) – issued from the 'implanting' of the basic element known as public health into civil public works activities, especially the public drinking water supplying side of matters and the friction never actually disappeared. In reality the problems just continued to drag on until1942.

Despite the introduction of civil decentralisation in 1903, the central hand of power remained very apparent and present within the various regions, provinces and municipalities because the demand for recommendations by BOW remained linked to all the different projects.

What was finally very important for drinking water provision and for hygiene was the emergence of a public works ruling in 1910 in which the streamlining of water provision was organised throughout the country in such a way that the supplying of 'raw water'[6] for the purposes of drinking water provision was given priority and the flushing of waterways, in conjunction with urban water regulatory measures (urban water management), was something that came in second place: both those aspects took priority over water provision for irrigation and hydroelectric power purposes.

Table 1 gives an overview and a summary of the relevant legislative framework, the administrative rulings and the official bodies concerned, insofar as that related to public drinking water facilities, sanitation and *bandjir* protection in urban areas.

Table 1. **Legal framework, administrative rulings and organisations**

Date	Description
1818	Appointment of the chief inspector of Public Works.
1827	Establishment of a Military Medical Service (MGD).
1850	The first Mining Affairs ruling in the Dutch East Indies: 24th October 1850, (by Royal Decree).
1852	Establishment of the Mining Affairs Service (by Governmental Decree), 3rd June 1852. The service fell directly under the command of the governor-general.
1854	Introduction of the Government Ruling which stated that seven directors would be made responsible for controlling the general civil administration tasks under the command of the governor-general, including 'education and religious affairs' and 'public works'.
1863	The Mining Affairs Service became affiliated to Public Works on 13th May 1863 by Governmental Decree.
1866	Establishment of the Department of Civil Public Works (BOW).
1871	The Ground Level Affairs Service became attached to the Mining Affairs Service.
1873	Introduction of a new ruling for the Mining Affairs Service on 31st December 1873, by Governmental Decree, which declared that the Ground Level Affairs Service was to be given the following tasks: to conduct general research in conjunction with artesian drilling projects for sources of water so that towns and populated areas could be provided with drinking water once engineers have discovered and indicated where suitable natural sources are to be found or where good river water is available. It was a ruling that was to remain in force for approximately 40 years.
1882	The creation of the Civil Medical Service (BGD), which was to operate independently of the MGD. The planning, design, creation and maintenance of public drinking water facilities remained in the hands of BOW. The BGD was responsible for checking the quality of the drinking water. The BGD was to be furthermore chiefly devoted to the curative combating of epidemics.
1903	Decentralisation law no. 329 of 1903 followed by the insertion of articles into the then prevailing Government Ruling in conjunction with the implementation of the law relating to the Decentralisation Resolution of 1905 (determined by Royal Decree) which stated that regions or parts of regions should be given the opportunity to become self-supporting. The intention behind these rulings was that BOW should thus gain the opportunity to transfer the management of public drinking water facilities based on natural sources and artesian wells to the respective local authorities.
1907	The Mining Affairs Service was merged with the Government Companies department that had been established in 1907.
1910	Determining of public works regulations (East Indian Gazette no.177), supplemented by the BOW director (17-01-1912 no. 904/E), in which the priority regarding the supplying of water was allocated to the following areas: 1) water supply companies (drinking water), 2) flushing (mainly of urban waterways), 3) industry, 4) irrigation, 5) hydropower, 6) navigation (ships and rafts).
1915	At the instigation of the BOW director, on 25th June 1915, the G division was created within BOW for the purposes of 'occupying itself with matters relating to Irrigation and Sanitation'. In reality, the responsibility

for public drinking water supplies, sanitation and urban bandjir (i.e. flood) protection therefore fell under the following two departments: BOW and the Department for Education, Religious Affairs and Industry (i.e. the BGD). Problems arose from the fact that the BGD saw itself as responsible for the research and inspection and BOW for technical planning, design, execution, operations and control. See further 1924.

1922 The Ground Level Service became a separate service though it still fell under the authority of the Mining Affairs Service.

1924 The Sanitation division within BOW was moved to the Department of Education and Religious Affairs upon the understanding that only planning and research would be handed over to that department but that design, execution, operations and control would remain with BOW. The BGD was transformed; it was turned into the Public Health Service (DVG).

1933 The establishment of the Department of Transport and Public Works which incorporated the various Government Companies and BOW. In conjunction with public drinking water facilities, sanitation and urban bandjir protection there were two important duties that were transferred to the department (notably to the Irrigation, Hydro Power and Sanitation divisions): the creation and exploitation of drainage and flood protection works and all other works of a hydraulic nature, insofar as the responsibility was not specifically assigned to or left to other parties; the creation and exploitation of drinking water provisions, sewerage systems and other works linked to public health, insofar as such responsibility had not been specifically assigned to or left to other parties.

The positioning of the DGV under the Education and Religious Affairs department. The service keeps as its main tasks: soil sanitation, drinking water provision and the removal of disposed of waste. In view of all the economy drives taking place during the period of economic crisis, the provision of drinking water was to be given priority above sanitation. Indeed, in connection with the important part played by municipalities in the field of sanitation (sewerage works, waste water treatment) the friction between Transport and Public Works and DGV was most apparent in the area of public drinking water supplies, see further 1936.

1936 The establishment of a Central Commission for Drinking Water Facilities which could then advise the departments of Education and Religious Affairs, and Transport and Public Works.

Public water supply

Before public drinking water facilities were properly organised the people in the villages (*desas*) and cities obtained water by tapping it from rivers or natural springs. Alternatively they created shallow hand-dug pits where the water welled up in an artesian fashion or was easy to tap because the water table was simply high. Alternatively they collected rainwater. Even the *kratons* in Djokjakarta and Soerakarta had to organise their own water resources . This was done by creating an ingenious network of wells and by storing rainwater.

Thanks to the colonial reports that started being written in 1850 and the excellent overview provided by the Experimental Station for Water Purification it is easy to reconstruct the post-1850 situation and to establish how and when the transition from dispersed water collection and a distribution system without networks of water pipes switched to a centralised piped public water supply system. Quite often private enterprise had an important part to play in such matters. In 1890 the government gave the two gentlemen, F.H. Rijdman and G.H. Birnie, priority for a one-year franchise to provide Soerabaja with drinking water derived from the Demboelan spring in Pasoeroean. In 1900 the two men were compensated because by then, in conjunction with decentralisation, the city of Soerabaja had taken it upon itself to provide public drinking water facilities and had constructed its own natural source water connection fed by the Kasri springs. Later, Soerabaja also started to draw surface water from the Kali Mas. What is interesting is that as early as around 1840 Soerabaja had already considered switching to a public drinking water system.[7] The waterworks of Medan have the Deli Maatschappij (Company) to thank for their origins. The company received a franchise in 1905, valid until 1956 (no less), for the construction and exploitation of a natural source water system and so, to that end, initiated the Ajer Berisih Waterworks Company, established in Amsterdam.

The creation of municipalities led to the accelerated introduction of public drinking water facilities. In the nineteenth century the first nine waterworks companies were set up and in the 1900-1942 era that number increased to 140.[8]

If we return to the period before 1890 we see that in Batavia, up until 1628, people tapped the water required for the castle from the river. During the time of the Javanese siege (1628-1629) the population switched to artesian wells. The water that infiltrated into the mountains found its way through coarse underground sand layers that extended to Batavia where, when bored for, it could rise to above surface level. Until 1830 water was tapped from the Tji Liwoeng slightly upstream from Batavia. In 1843 mention is made of the first artesian boring project (which proved unsuccessful) and of the using of water drawn from canals which had been successfully filtered through limestone layers. Already in 1834 borings were made for the military corps, predominantly in order to supply drinking water for the various military encampments in Ambarawa and Semarang, amongst other places. In 1843 another borehole was sunk for the purposes of Fort Frederick in Weltevreden and in 1854 one was sunk on the island of Onrust. Any hole drilling carried out for civil purposes was at that time done by BOW. Where possible, water originally required for military purposes would also be made available to the civilian population (i.e. chiefly for the Europeans), for instance in the case of the artesian borings in Semarang in 1840 and 1858 and in the case of the boring in Grissee (near Soerabaja) in 1865. After 1871 all the boring work was concentrated within the Mining Service.

It was especially in the first place the military encampments that succeeded in supplying people in the immediate vicinity of the respective settlements with reasonably priced drinking water by means of the Norton wells which were also suitable for drawing up groundwater in artesian and non-artesian fashions. The Norton well consisted of a suction tube with a filter tube section attached to the lower end that had a wrought-iron point. By driving piles through the suction and filter tubes with the help of a ram that came down directly on the point it proved possible – depending on the composition of the soil – to sink wells to depths of 20 to 30 m. Normally, when necessary, the water would be pumped up with a hand pump.

In 1884 and again in 1912, the colonial government prohibited private companies from boring to depths of more than 15 m with the aid of Norton wells, simply to safeguard production from its own wells.

In 1901 Batavia switched entirely to artesian water. A report dating from 1901 describes how water was being obtained from eleven artesian wells, each with its own distribution network. Artesian well water was not, however, always completely favourable as regards water table, flow, taste and temperature.[9] The preference lay with cool spring water. In 1892 Cheribon switched to drawing off spring water by means of a 1000 m-long draining

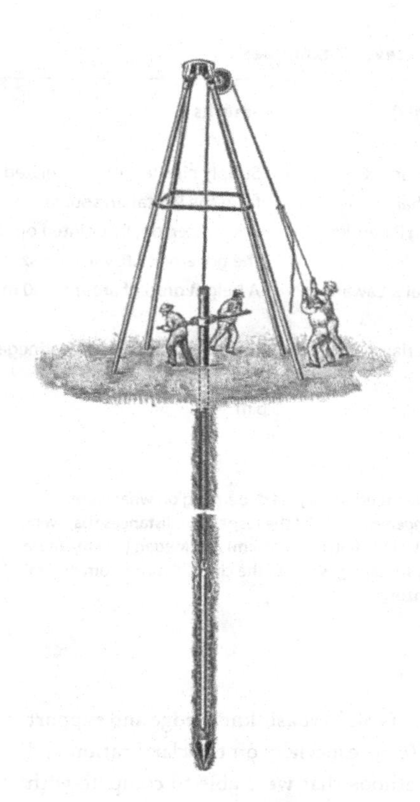

The Norton well: the sinking of the filter pipe with the help of an arrowhead and a rammer, nineteenth century.

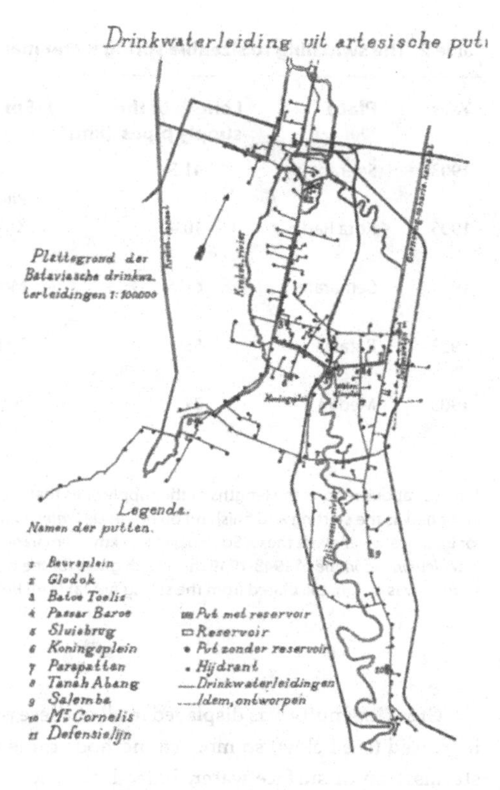

The Batavian piped water network as it was at the end of the nineteenth century.

trench. The availability of cast-iron piping was to herald a real breakthrough in the using of spring water. The cast-iron pipes also had the great advantage – with a view to the frequent earthquakes on the volcanic islands of the archipelago (Java, Sumatra, the Moluccas) – that they could withstand a degree of movement[10] (see Table 2).

Where there was a lack of fresh groundwater – in the form of springs, artesian wells or wells that had been dug by hand or sunk – people made do with surface water or the collecting of rain water. The bigger rivers which had sufficient water flows in the dry season (Moesi-Palembang, Batang Hari-Djambi, Kapoeas-Pontianak) served to supply surface water for the purposes of the large central public waterworks; the smaller rivers supplied the smaller agglomerations. Where necessary, reservoirs (*wadoeks*) were created so that water could be stored for the dry seasons. The *wadoeks* were either constructed as separate reservoirs or as storage reservoirs in the relevant dammed off river sections formed by means of what were termed valley closure dams. Where necessary – and either supplemented or not with plants for the purification of drinking water[11] – irrigation canals were also utilised for domestic supplies.

Table 2. **The switching to sizeable spring water mains – several examples**

Year	Place	Length of the supply pipes (km)*	Spring(s)	Remarks
1903	Soerabaja	41.5	Tojo Arang & Plintahan	Supply pipe lengths calculated from the Plintahan source
1905	Kota Radja	10.4	Kroëng Pineung	Source personally pointed out by the governor, J.B. van Heutsz
19...	Semarang	c. 15	Moedal & Lawang	A height drop of around 300 m
1923	Batavia	55	Tji Boerial	Including Mr. Cornelis (Jatinegara)
1905	Medan	37	p.m.	p.m.

* In the various reports the lengths of the supply lines that are given tend to vary, all depending on what were identified as the starting and finishing points. In *De Waterstaats-Ingenieur* of 1924 the respective distances that were, for instance, given were these: Soerabaja (43.5 km), Semarang (12.9 km), Batavia (50.5 km) and Medan (37 km). In *De Ingenieur in Indonesië* of 1948-1949 the length given for the pipes supplying water to the city of Batavia from natural sources was 67 km, calculated from the spring not far from Buitenzorg.

Great ingenuity was displayed in all of these projects and as basic knowledge and support increased (see below) so modern methods came to focus especially on the clarification and sterilisation of surface water. Indeed, they were methods that were able to compete with those of the day being implemented in the 'mother land' (the Netherlands) and Europe as a whole. The following example may serve to illustrate how the region was even in some ways ahead of its time. Already in 1916 an article appeared in the *De Waterstaats-Ingenieur* (The Public Works Engineer) journal which included a detailed design for the Soekamboemi facilities where surface water was to be drawn from an irrigation canal and sterilised by means of ozonisation with the help of locally produced electricity generated by hydropower (the Pelton wheel).

The mountainous character of the terrain was something that furthermore gave rise to the applying of surface water and groundwater acquisition methods that were not yet known in the Netherlands. Already around 1895, a hydraulic ram[12] was installed in Poeloe Weh for a 'civil establishment' designed to pump the water up 41 m from an available drop height of 6 m. When the spring abstraction centres at Karangan (from 1916) and Soembersari (from 1929) on the slopes of the Ardjoeno volcano no longer met the Malang drinking water demands, a 350 meter long tunnel was constructed (in 1934) into a water-bearing lava layer (andesite) so that groundwater could be collected. That only occurred after a survey had been carried out by the Research Service's Geological Technical Inspection department and permission had subsequently been obtained from the BOW director, after DVG recommendations had been made and once the East Javan provincial Public Works Service had given the go-ahead. In the case of Batavia, it had already been predicted before 1942 that in conjunction with the expansion of the city[13] and the existing suburbs of Mr. Cornelis and Tandjong Priok the spring water supplies, either supplemented with artesian water or not, would be insufficient for the city. Already the authorities had been entertaining notions of directly tapping and purifying water from Tji Liwoeng river water transported in the Bandjir Canal so that people could also profit from the water impounded in that canal upstream of the Karet Weir since 1918 and from the suppletion of surface water from the Tji Sedane via the Mookervaart canal which also profited from the movable weir in the Tji Sedane at Tangerang that had reached completion in 1929.[14]

In the low-lying coastal regions where the groundwater is brackish, where the estuaries are brackish to salty and where there is acidic bog water in the swamps (South Sumatra, Borneo), the population has to depend for its drinking water on the direct collection of rainwater. The minimum requirement is roughly 20 litres per head, per day for drinking purposes and for the cooking of meals. In the tropics, drinking water requirements can vary greatly. Just to quench the thirst alone people need about 3 litres per head, per day and that requirement can rise to as much as 15 litres per person, per day if the individuals involved are doing hard physical work. The bridging of the gap in dry periods creates such a problem that sometimes primitive methods have to be resorted to resolve the water shortage problem, such as storing 'roof water' in rusty oil drums. For washing and bathing purposes people then use brackish or acidic surface or groundwater.

What was remarkable was the way in which in the nineteenth century, despite the primitive communication means, the authorities in general and civil engineers in particular were able to keep abreast of technological and social developments in the field of public drinking water facilities in Europe and especially in Germany, the United States and the other countries surrounding the Dutch East Indies. Much the same applied to the necessary changes being made in the field of sanitation. Just as in the case of irrigation, careful note was taken of the activities and developments in British-ruled India. In the case of public water supplying some of the aspects looked at were reservoir construction, the determination of water requirements and the way in which water was distributed throughout the day, different water distribution methods – with or without flow rate limitations – the way in which drinking water was supplied by means of hydrants for the native population, the combating of leakage loss, and the way financing and retribution was applied to the various drinking water systems.[15]

After reinforced concrete had appeared on the scene in the twentieth century people ventured to erect water towers fabricated from that new material. A certain lack of knowledge about the basic principles of soil mechanical know-how did, however, came to the fore as evidenced by the water tower in Pasoeroean. The tower was built in 1917 but could not be put into use until 1924 because the consequences of soil subsidence first had to be rectified.

A huge leap forwards was suddenly made when the fundamental chemical-technological and biological knowledge required for the designing of processes started being made available locally, 'on own soil' so to speak. The same applied to the basic hydrodynamic, hydrological, geo-technical and constructional knowledge needed for structural works. On the one hand, the advantage of this was that all that knowledge was directly available on site whilst, on the other hand, the development of such knowledge could be perfectly adjusted both to current and local requirements. Exponents of such knowledge development were the Technical

A water tower in Pasoeroean (East Java), made of reinforced concrete,1917.

College in Bandoeng, established in 1920, the Water Purification Experimental Station (of the Medical Laboratory) at Manggarai, established in 1922 which continued, in 1935, as the Laboratory for Technological Hygiene and Sanitation in Bandoeng, the Hydrodynamic Laboratory of the Technical College in Bandoeng in 1928, the Hydrodynamic Laboratory in Semarang in 1930 and the Soil Mechanics Laboratory affiliated to the Bandoeng College of Technology in 1930 (see Table 3).

The level of education in the civil engineering department of the Technical College in Bandoeng was, from the very start, comparable with that of the Technical College in Delft. There was also very concrete interaction between the Netherlands and the East Indies. C.W. Weijs, for instance, a Delft professor in the field of hydraulic engineering, sewerage systems and water supplies (1913-1919) gained his professional experienced and won his spurs in Batavia and Soerabaja. Though the education at the Bandoeng College of Technology was particularly oriented towards irrigation, other related areas were well represented, like for instance: fluid mechanics, sediment transportation theory, hydropower, mechanics, statics and construction theory, not to forget education in the fields of drinking water facilities and sanitation. What drew particular attention were the highly academic articles in the fields of fluid mechanics, hydrology, soil mechanics and statics published by Professor C.G.J. Vreedenburgh. Standard educative publications also appeared in the fields of irrigation

Table 3. **Supporting institutes in the field of education and research**

Year of initiation	Description
1847	Establishment of the Delft Polytechnic (1908-Technical College, 1986-University of Technology). What was important was the Road and Hydraulic Engineering department (currently known as the Faculty of Civil Engineering and Geosciences) which trained civil engineers to deal with hydraulic engineering, irrigation, hydro power, drinking water provision and sanitation situations in the tropics.
1920	The establishment of the Bandoeng Technical College (THB; 1950-Institut Teknologi Bandung); where, up until 1950, it was only possible to train as a civil engineer and to gain a diploma equivalent to that issued by the Technical College in Delft.
1922	The setting up of the Water Purification Experimental Station (at the medical Laborartory) in Manggarai.
1927	The opening of the Medical High School in Batavia; significant in conjunction with the prevention and combating of 'water related diseases'.
	Establishment of the Hydraulic Engineering Demonstration Laboratory within the THB (headed by the engineer J.W.F.C. Proper, who later became a professor at the THB); referred to by Proper as the 'Hydrodynamic Laboratory'.
1930	The starting up of the Provincial Hydrodynamic Laboratory in Semarang (led by the engineer H. Vlugter, later a THB professor).
	The establishment of a Soil Mechanics Laboratory for the THB (4 years before the Soil Mechanics laboratory in Delft was created!)
1935	By decree of the East Indian government on 24-11-1934, the Laboratory for Technical Hygiene and Sanitation was transferred to the in 1935 established Association for the Promotion of Hygiene in the Dutch East Indies and the laboratory was situated in the grounds of the THB. At the same time the Water Purification Experimental Station in Manggarai was closed down. The laboratory was created for the benefit of 'education, science (research) and practice'. It did not belong to the TH but rather to the Public Health Service. At first the laboratory was headed by the engineer Dr. C.P. Mom (who in 1931 was inaugurated as a THB professor). Mom called the laboratory 'The Water Purification Experimental Station of the Public Health Service and the Laboratory for Hygiene and Sanitation'.

and sediment transportation (Professor H.C.P. de Vos) and applied mechanics (Professor J. Klopper). Professor C.P. Mom, also head of the Laboratory for Technological Hygiene and Sanitation, was the person responsible for creating the curriculum on public water supply and sanitation. Professor H. van Breen, the man of 'the small water management works in Batavia', was the person who was able to guarantee education on and the development of urban drainage and flood control (protection against *bandjirs*).

Thanks to all the experiments carried out in the Hydrodynamic Laboratory in Semarang the designing of hydraulic structures (fixed weirs, moveable weirs, measuring weirs, gates, flaps, stilling basins) saw major development. When it came to the matter of drinking water provision and sanitation, the Laboratory for Technological Hygiene and Water Regulation had an important part to play. The research conducted by that laboratory was on the one hand fundamental and on the other hand very clearly applied. The best way of illustrating this is perhaps by citing two random annual reports dating from 1928 and 1934 (see the respective appendices 3 and 4).[16]

General support for work carried out in the field was to be found in the manuals entitled *Indische Bouwhygiëne* (East Indian Building Hygiene) by G.W.F de Vos (captain of the Engineering Corps in the Dutch East Indian Army) that appeared in 1891 and the manual published by C.J. de Bruijn (major-general of the Engineering Corps) in 1927. The subjects dealt with were, for instance: water provision, sewerage works, the conveyance of faeces, solid waste removal, the disposal of the dead, private dwellings, encampments, hospital services, sports buildings and terrains, prisons, stables, workplaces, *pasars* (market places), *kampongs* and urban development.

In the case of the first public urban drinking water projects, the estimated daily water requirements were partly based on experiences in British-ruled India. For the further development of the public drinking water facilities, use was made of the experience gained in the various cities and regions of the archipelago. When assessing the requirements and supplies, a differentiation was generally made between the European, Chinese and Native (Inlander) population components or European, Chinese, Arabian and other Foreign Easterners, and the Native population. As far as the native population was concerned, a differentiation was made in the larger cities between city *kampongs* and the *desas* near to the city. The consumption of 'servants living-in with non-native masters' (Malang, 1917) was generally estimated 'p.m.' (as a memo).[17] In Table 4 several numerical examples are given of the estimated drinking water requirements in various cities around 1920. Alongside these water supply columns there was also the matter of the amount of water that had to be made available for specific industrial purposes and for other companies (ice factories, soft drinks producers), hospitals, military camps, fire-fighting, street spraying, and water losses due to leakage and industrial wastage.

Table 4. **Estimated drinking water requirements (c. 1920) in liters/per day/per capita***

Target group	**Batavia**	**Semarang**	**Soerabaja**	**Malang**
Europeans	150	150	150	150
Chinese and Japanese	100	90	90	90
Arabs and other Foreign Easterners	50	90	70	70
Indigenous groups:				
- city kampongs	50	50	50	50
- desas near the city	?	15	15	15

* The average drinking water consumption in the Netherlands (in 2000) was around 130 litres per day, per capita. In 1930 the estimate for the northwest of the province of North Brabant was 40 litres per day, per capita and in 1950 per mains connected inhabitant in Groningen (70% of the population had mains water supplies at that time) 50 litres per day. Source: VEWIN.

One problem was the way in which water was distributed and paid for in municipalities (cities) which had some kind of public water supply system. The following four kinds of distribution methods were implemented:

1. unlimited provision per property or directly to homes via an ordinary water metre;
2. unlimited provision per property or directly to homes using a coin metre system;
3. limited provision to properties by means of a perpetual and constant supply via calibrated fixed tap ends;
4. public hydrants, charged for or free of charge.

Where possible Europeans, affluent Natives, Chinese, Japanese, Arabian and other Foreign Easterners made use of method 1 involving monthly payments to the water company. Method 2 came into play in situations where people objected to the monthly payment system. In 1937 in Soerabaja, for instance, there were 14,000 households that made use of the coin metre system. Method 3 was implemented in cases where a maximum required supply per connection of 2 to 5 m^3 per month was presumed to be required and where it was possible for monthly payments to be made to the water company. The perpetual supply system required that the consumer should have a water reservoir in his own home. It was often the *mandibak* (fresh water reservoir) in the bathroom that served that particular purpose. This final supply system was particularly prone to fraud. By clandestinely drilling it was quite easy for people to widen the diameter of the tip of the tap and overflowing *mandibaks* amounted to losses.

The authorities tended to presume, as a foregone conclusion, that the majority of the population (the natives) lacked the means to obtain water via individual (fixed) connections – certainly during the great economic malaise of the thirties – in conjunction with the monthly costs being too high or the coin metre system being regarded as too expensive. By purchasing water from water vendors people were in that way able to adjust their consumption level to their immediate needs and means, even though that meant that ultimately they were paying more per litre. In order to stimulate public health among the urban population in particular but also generally, the water provision company and the government therefore searched for ways to provide safe drinking water at the lowest possible cost. The method that was opted for was number 4: distribution via hydrants (public stand posts) and many different types of hydrants were chosen. Generally, in order to prevent too much wastage, such water provision was not made entirely free of charge. For half a cent people in Cheribon could purchase a petroleum can full of water (18 litres). It was a system that fitted in well with the traditional

Table 5. **Water users in the areas provided with piped water**

	Initial situation	**After 5 years**	**After 25 years**
Europeans (directly connected)	40%	75%	95%
Chinese (directly connected)	15%	40%	70%
Chinese (via public hydrants)	20%	20%	25%
Native population (directly connected)	2%	5%	10%
Native population (via public hydrants)	20%	30%	75%

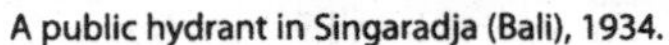

A public hydrant in Singaradja (Bali), 1934.

Public bathing and washing place, and WC in Bandoeng (West Java), 1934.

way of purchasing drinking water from Chinese water vendors. Despite slight reservations about the downsides to the hydrants themselves (not utterly tamper-proof: coins on pieces of string et cetera) hundreds such hydrants were installed in bigger cities (such as: Batavia, Soerabaja and Bandoeng). In 1916 the engineer S.J. Gompert recommended one hydrant per 100 'impecunious natives'.[18]

Other facilities that were provided in the bigger cities, especially for the indigenous population, were free or otherwise very cheap public baths and washing places. By 1918 there were, for instance, 100 public bathing places in Batavia and 15 public washing points.

It is also interesting to contemplate the speed at which the planned water supply actually materialised. In his article published in *De Waterstaats-Ingenieur* of 1924, the engineer W.H. Brandenburg (Public Works engineer in Probolinggo) provided the list given in Table 5. Nevertheless, despite all the efforts, it was established in 1940 that in Java only 1.5 percent, at the very most, of the needy indigenous population was directly connected to central drinking water facilities, even though progress had been witnessed in the *kampong* improvement projects involving either home or property connections to the supply system.

What is perfectly evident is the fact that publicly organised drinking water networks is something that instantly leads to improvements in the general public health situation, notably where water-borne diseases are concerned (see Table 6). In his study entitled 'Ongezond Batavia' (Unhealthy Batavia) published in the *Tijdschrift van het Koninklijk Instituut van Ingenieurs. Afdeeling Nederlandsch-Indië* (The Journal of the Royal Institute of Engineers. The Dutch-East Indies section) in 1913 W.J. van Gorkom reports the following facts. In 1813 and 1814 the respective mortality rates of 49 and 59 percent were entirely attributable to intestinal disorders.[19] Within the 1813-1864 period the death rate among Europeans dropped drastically from 22.7 percent (!) to 5.41 percent[20], chiefly because of the creation - instigated by the governor general Daendels - of the higher altitude and drier city

Plot linking in a kampong improvement project in Bandoeng, 1934.

region known as Weltevreden (malaria avoidance). As far as the European segment of the population was concerned the situation in 1913 was, according to Van Gorkom, as follows:

- Batavia, mortality rate of 2.91 percent, 1.79 percent of which was attributable to poor quality drinking water;
- Soerabaja, mortality rate of 3.29 percent, 2.51 percent of which was attributable to poor quality drinking water though the laying of spring water pipes had led to improvements;
- Semarang, mortality rate 2.98 percent, some 1.0 percent of whom died from drinking poor quality water. Indeed, vast improvements were seen after the introduction of spring water supply systems in the 1901 to 1903 period.

The mortality rate among members of the indigenous part of the population remained much higher right across the board. The death rate given by Van Gorkom for the 1911-1913 period is 6.89 percent. By 1930 the mortality rate among Europeans living in Batavia had dropped even further to 1.1 percent (with a reduction in the number of cases of typhus and dysentery to one in two hundred and an accompanying death rate of 6.5 percent).[21]

Sanitation in urban areas

Before 1873 – before the introduction of a new ruling for the Mining Service – it was not really possible to claim that sanitation measures were clearly centrally regulated by the government.

At that time the Dutch East Indies was also a relatively empty and thinly populated country, Java included. Towards the end of the nineteenth century it was only Batavia and Soerabaja that had population levels that exceeded the 100,000 mark and could thus be viewed as substantial cities (see Table 7). It was not until in the twentieth century that

Table 6. **Water-related diseases**

		Trans-mission	Cause of illness	Extent to which clean drinking water led to a reduction in the number of cases (%)
FAECAL-ORAL MICRO-BIOLOGICAL INFECTIONS	A. INTESTINAL CONDITIONS			
	* cholera	I, II	B	90
	* bacillary dysentery	I, II	B	50
	* amoebic dysentery	I, II	B	50
	* gastro-enteritis	I, II	B,V	50
	* entero-viral infections	I, II	V	10
	* diarrhoea-linked illness	I, II	B,V	50
	B. FEVER-LINKED DISEASES			
	* typhoid	I, II	B	80
	* para-typhoid	I, II	B	40
	* Weil's disease	I, II	V	10
	* yellow fever	I	B	80
SKIN AND EYE INFECTIONS AND ILLNESSES	A. INFECTIONS			
	* trachoma	II	B	60
	* conjunctivitis	II	B,V	70
	* scabies	II	A	80
	* septic skin ulcerations	II	B	50
	B. ECOPARASITIC INSECTS			
	* 'louse-borne fever'	II	A,B,R	40
PARASITIC WORM ILLNESSES	* hookworm (anchylostomiasis)	I, II	W	?
	* round-worm (ascariasis)	I, II	W	40
	* bilharzia (schistosomiasis)	III	W	60
	* guinea worm (dracunculiasis)	III	W	100
INSECT VECTOR ILLNESSES	* malaria	IV	A ? P	?
	* filariasis	IV	A ? W	20
	* yellow fever	IV	A ? V	10
	* sleeping sickness	IV	A ? V	80
	* dandy fever (dengue)	IV	A ? V	?

I = water-borne
II = water-washed
III = water-based
IV = insect vector

V = virus
B = bacteria
R = rickettsia

W = worm
A = arthropod
P = protozoa

Source: the 'Epidemiology' lectures reader compiled by Dr. J. Huisman, emeritus professor of epidemiology, Faculty of Civil Engineering and Geosciences, Delft University of Technology.

Table 7. **The population composition of the larger Javanese cities in 1890**

Place	Total	Native inhabitants	Europeans	Chinese	Foreign Easterners and Arabs
Batavia	104,590	67,659	7,891	26,932	2,108
Cheribon	19,342	15,561	398	2,537	846
Semarang	71,186	53,974	3,565	12,104	1,543
Djocjakarta	57,545	52,083	1,557	3,675	230
Soerabaja	117,986	100,482	6,575	8,775	2,154

Source: Boomgaard and Gooszen, *Population trends 1795 – 1942.*

those Javanese cities started to reflect explosive growth, closely followed by the rapidly expanding cities of Bandoeng and Malang with their large European districts because of the stated preference of Dutch people to reside in and work in the cooler and more mountainous regions.

Another matter that should not be overlooked is the fact that, certainly in the early nineteenth century, there was general public ignorance about cleanliness and how serious illness could be caused by lack of hygiene. Before people realised that there was a correlation between cholera and polluted drinking water it was the custom to burn down the 'infected dwellings' in an endeavour to combat the disease. With a view to public health, the battle against the plague was particularly sought in building better houses so that rats could not nestle in them. From the *De Waterstaats-Ingenieur* of 1918 the following may be quoted in relation to 'Public Health in Cheribon'.

> *When dealing with the plague it became evident that it is much better for natives to live in good, light accommodation than to dwell in dark hovels. The regional director of public works is therefore busy compiling new building regulations in which a separate service for building and accommodation supervision is envisaged. In the process, special attention will be paid to making the homes cheap and easy to keep clean, light and well ventilated and to using pan tiles for the roofing and wood for the framework. An atap roof[22] always becomes a nesting place for all kinds of vermin and it is impossible to keep bamboo struts or posts free of rats. If the government makes available cheap wood from the djati (teak) forests then the houses will not need to be expensive and it will be possible to keep them free of "the destroying attacks by white ants". The dwellings will be numbered and it will no longer be possible to erect new buildings without permission. People will have to keep kitchens and toilets separate from the main home area and it will be necessary to check whether the proposed site is actually suitable for building. Beyond that the natives will be free to do as they please.*[23]

At that time such resolutions and measures were being made and taken in all big cities like Semarang but also in smaller places like Probolinggo.

With the establishment of a G section within BOW in 1915 a definite incentive was given

Table 8. **Causes of death in Semarang in 1920**

Cause of death	Number of people	(%)	Perceived remedy
malaria	2,268	26.9	sanitation
amoebic dysentery	532	6.3	and
bacillary dysentery	72	0.9	provision of public
typhoid	191	2.3	drinking water facilities
cholera	0 (!)	0	
tuberculosis	348	4.1	structural
pneumonia	811	9.6	and housing situation
other lung diseases	99	1.2	improvements
syphillis	104	1.2	purely medical

The total death rate of 8,445 individuals was equivalent to 71 in every 1000 inhabitants.

Source: Nieuwenhuis, 'Stadshygiëne door bodem-assaineering'.

to introduce centrally organised and, as much as possible, normalised sanitation measures for the entire country. With a view to the liveability of the urban areas (large and small cities and villages) in general and to public health in particular, and to designing sanitation facilities for urban areas – especially in order to reduce faecal-oral micro-biological infection, parasitic worm diseases and insect-vector illnesses – attention was to be paid to:

- avoiding having stagnant water or, more to the point, avoiding the negative hygienic effects of stagnant, fresh or brackish water found in permanent or semi-permanent inundated areas (*sawas*, fish ponds, swamps) within or immediately bordering on the urban areas;
- the collecting and processing of faeces and household waste water;
- the storing, collecting and processing of non-domestic waste water;
- the draining away of local precipitation in the urban areas;[24]
- the controlling of the groundwater level in urbanised regions;
- the collecting and processing of solid waste materials.

In order to find out more about the important developments that were then taking place in the field of education and research in relation to the matter of sanitation the reader is referred to the section entitled 'Public Water Supply'.

Preventing stagnant water and its related effects

Apart from 'killers' such as the plague and various intestinal diseases such as cholera, typhoid and dysentery (water-borne diseases) it was malaria which, until deep into the twentieth century, affected people in the low-lying and marshy areas. Malaria is transmitted by the anopheles mosquito (a fact not definitely proven until 1898 by the Italians: Bignami, Battista Grassi and Bastianelli) which abounds in stagnant water where the conditions are ideal for the eggs, the pupae and the hatching out of the larvae. It is thus a classic 'insect-vector disease'. In brackish water along the coast, in morasses and in fishponds the type found is the Ludlowi and in the freshwater *sawas* it is the type known as Aconita that prevails. The particularly fatal type is 'malaria tropica' where people suffer a single attack accompanied by extremely high fevers. 'Malaria tertiana' involves three-day attacks accompanied off and on by periods of high fever that can recur every three weeks. As the bigger cities started to spring up predominantly in the low-lying, swampy and coastal regions, ever larger numbers of people were subsequently affected by malaria. As late as in 1920 it was malaria that proved to be the biggest killer for the inhabitants of Semarang (see Table 8). What was rather curious was the fact that the places Padang, Pontianak and Palembang were amazingly malaria-free.

In the mid-nineteenth century the discovery of the medicine quinine was certainly a boost but it did not provide a complete cure for malaria. People therefore endeavoured to eradicate the breeding places of the anopheles mosquito and there were three main ways of doing that, either in isolation or in combination:

1. by keeping the breeding areas clean (pools, fish ponds),
2. by completely draining the pools and fish ponds,
3. by 'petrolising' the breeding places.

The ultimate aim when it came to the matter of keeping the breeding places clean was to either completely eradicate the numbers of eggs, pupae and larvae or to at least keep the numbers down to a very minimum. For that purpose water plants and all sorts of plant growth along the banks was removed, the usually brackish fish ponds were regularly flushed clean (where possible with fresh sea water) and biological control was implemented by introducing various fish sorts, like for example:

- the *bandeng* to control vegetation growth,
- the *kepala timah* and the 'millions fish' from the West Indies (lebistes reticulates), all predators of the pupae and larvae.

There were two common ways in which the pools and ponds were frequently dried out:

- by simply filling them with soil,
- by poldering (i.e. drying them out).

As far as it is possible to ascertain, rehabilitation of the morass area between the Kali Mas and Priangan to the north of Fort Hendrik in Soerabaja was mainly achieved by landfill.[25] It was certainly something that had to be integrally and extensively tackled because otherwise various breeding places would remain. For instance, in the *De Waterstaats-Ingenieur* journal of 1920, people complained about the fact that in the vicinity of the filled in Oedjoeng area in Soerabaja, in conjunction with the creation of a harbour, actual and potential Ludlowi breeding places could still be found.

The petrolisation method (involving covering water surfaces with a layer of petroleum) was an effective method but one that was only sporadically and often only temporarily implemented in conjunction with the accompanying disadvantages attached to the approach, which were these:

- it affected the entire ecosystem of the expanse of water in question (it practically prevented any absorption of oxygen from the atmosphere);
- the system was sensitive to wind;
- in conjunction with evaporation and oxidation, the petroleum had to be regularly replaced;
- partly in conjunction with cost, it was only possible to apply it to small water areas.

At the beginning of the twentieth century Batavia tackled the problem very thoroughly. Already, in the era of Daendels, by creating the southerly and higher-lying residential area of Weltevreden for part of the Batavian population – chiefly the European sector – the death rate from malaria was brought down. However, people situated in the lower city area of Batavia still suffered from the presence of all the malaria breeding places along the coastal strip and wherever there was stagnant water in *sawas*, fish ponds and morasses there was the danger of malaria. Led by the engineer H. van Breen, who was head of the irrigation and sanitation works in Batavia, who later went on to become a professor at the Bandoeng Technical College, who made his name as the 'small water management works' man and who took care of *bandjir* protection in Batavia, a start was made in the early 1900s on tackling the coastal region. Parallel to the coast and for a total length of 4 km, the low-lying area to the north of the city was dried out as much as possible: chiefly by means of improved direct draining during periods of ebb. At that time there were no large-scale polders. Partly through the successful introduction of irrigation to the *sawas* and the cleaning up of the fish ponds Batavia became virtually completely malaria-free a quarter of a century later, by the thirties.[26]

When reclaiming any land there is always the inherent danger that – just as with irrigation projects – one might inadvertently actually introduce malaria by adhering to technically inadequate water management systems and thus trigger what is termed man-made malaria. Indeed this was precisely what did happen when people set about reclaiming the low-lying zone between the coast and the island of Noesa Kembangan at Tjilatjap. The problem was ultimately resolved by flushing out the water passages in the polder with seawater. When it came to introducing irrigation to the Tjihea plains, malaria was similarly introduced and then subsequently eradicated. Regions that had once been famed for their malaria problems, such as places around Sibolga, Semarang and Probolinggo, were made malaria free by a combination of reclamation – in the case of Sibolga through extensive drainage – and the

Table 9. **The death rate in Probolinggo**

Year	Deaths per 1000 people	Year	Deaths per 1000 people
1915	52	1920	40
1916	54	1921	29
1917	67	1922	25
1918	64	1923	24
1919	50	1924	29

Source: Brandenburg, 'De centrale drinkwatervoorziening van de gemeente Probolinggo'.

cleaning of fish ponds (as in the Probolinggo area). In Sibolga the people did not even require quinine medication any longer. The relationship to the drop in the death rate is striking as is illustrated by the data relating to Probolinggo (see Table 9). The drop in the incidence of malaria in the post-1921 era, there, was mainly attributable to the drying out of dilapidated fish ponds.

Collection and treatment of human and domestic waste

In the nineteenth century it was still common to deposit human excreta in the ground (latrines, cesspits) or to allow it to flow directly into open waterways[27], both in the *kampongs* and in the city and town areas that were rapidly being created. It was also the custom to still simply dump any household waste water. It should be noted that once buried underground, pathogenic organisms are destroyed and provided that the cesspits and the clean water wells are situated far enough apart – groundwater requires around 30 days to filter through – then that way of disposing of faeces and urine in the ground need not pose particular problems or constitute health threats. Furthermore, for the time being the dumping of faeces, urine and household waste in various streams and rivers in the higher situated areas was neither particularly problematic nor a potential health hazard. The fast-flowing water quickly transported away the waste material while the natural self-purification process of the rivers took care of the further breaking down process.

It was really in the lower-lying areas that the first problems started to emerge, starting with the defecating and depositing of raw sewage and waste water in open waterways such as in: watercourses, pools, fish ponds and swampy depressions. When natural water quantities were insufficient in the dry season and when the flow velocities were too low in the rainy season such areas would be characterised by an offensive stench and would constitute a danger to public health in the form of infection resulting from direct contact with the infected water or disease transmission via organisms (insects, worms, etc.) In the tropics, the rapid and rampant growth of weeds in water expanses and ditches formed – and still forms – another problem. The first action that therefore needed to be taken amounted to quickly finding ways of flushing clean those water areas and channels and, where possible and financially feasible, to lining banks and channel beds. It was thus completely logical and justifiable to see that in the public works regulations of 1910 the national priority, directly after drinking water provision, was that of water circulation (flushing) which even came before industrial requirements, irrigation, hydropower and shipping.

At the time of the introduction of the water closet, inspired by house-to-house drinking water facilities, the cesspit no longer sufficed for the containing and proper treatment of waste water. The designers then opted, in line with the 'open East Indian city' concept[28] for the septic tank. It was a solution that might be compared with the current way of dealing with domestic waste water in the Netherlands in sparsely inhabited areas. It is preferable for the effluent from the septic tank to filter into the soil. In instances where the groundwater level was too high, people were advised to allow the effluent to pass through closed pipes before releasing it into open water where the waste could then be sufficiently dispersed. Design regulations for septic tanks and for the infiltration provisions appeared in the 1918 and 1919 editions of the *De Waterstaats-Ingenieur*. The engineer C.A.E. van Leeuwen, head of the

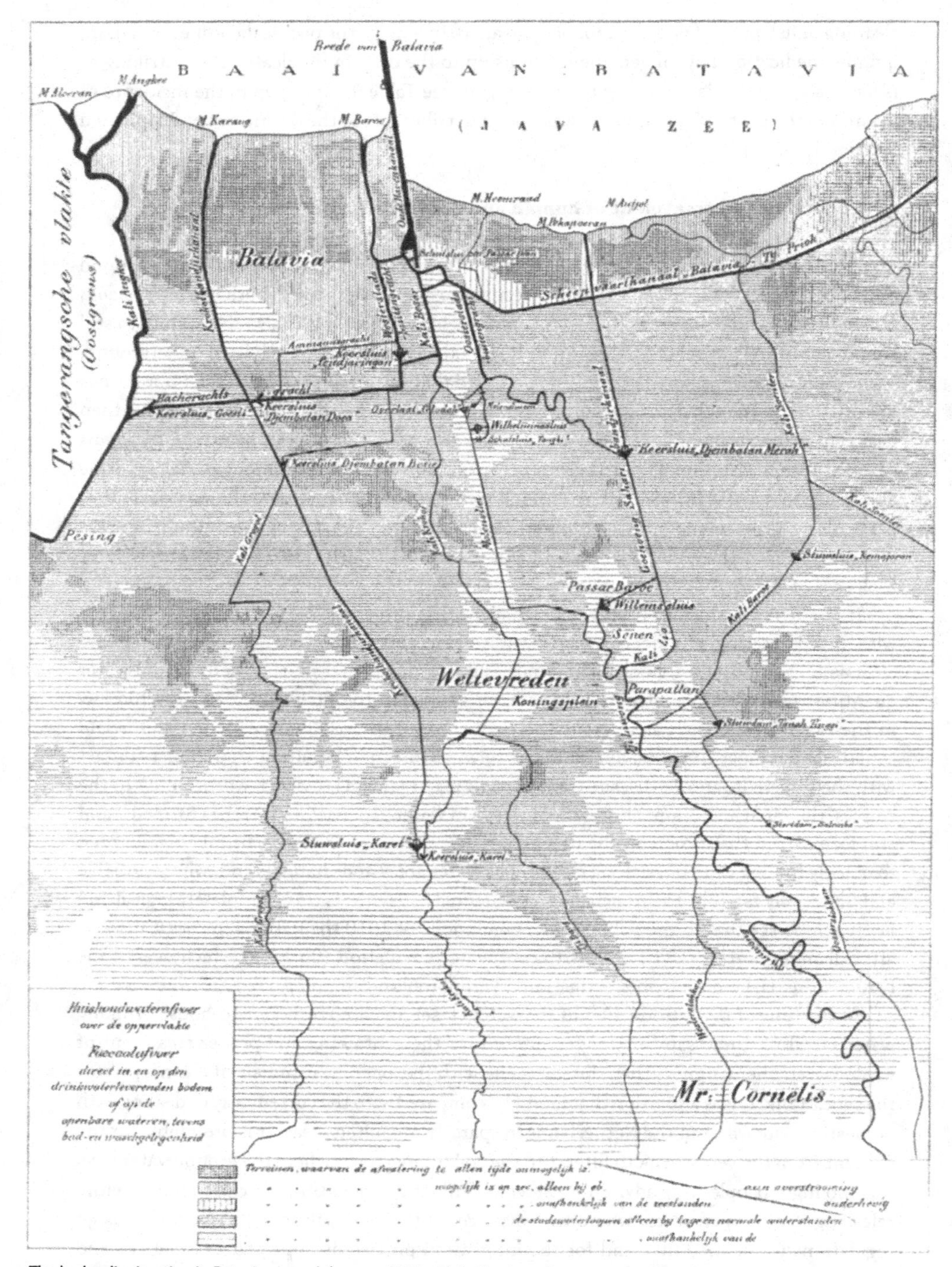

The hydraulic situation in Batavia around the year 1900 with indication of areas prone to flooding.

Sanitation Works section for BOW in Weltevreden published an article with the following title in the 1919 edition of *De Waterstaats-Ingenieur*: 'The principles to be taken into consideration when designing sewerage and drainage works in large municipalities' (see Appendix 5). In the 1920 edition of *De Waterstaats-Ingenieur* he devoted an extensive article to the sewerage problem in the Dutch East Indies (see Appendix 5). In 1934 N.D.R. Schaafsma, who was affiliated to the Experimental Station for Water Purification in Manggarai, published an article in the specialist journal *I.B.T. Locale Techniek* (Local Techniques) on 'Septic tanks and cesspits for faecal waste disposal in the tropics' and in *De Ingenieur* (The Engineer) of 1937 standard designs were still being presented for septic tanks and infiltration facilities.

As far as discharging the effluent of the septic tanks was concerned, it was still the case that such effluent flowed directly or mixed with bath, washing and cooking water into open gutters, drains and ditches. It was presumed, however, that these waterways would be periodically flushed out. Such waterways were invariably situated at the back of premises where the fire escapes were generally also situated. In the interests of hygiene those gutters, drains and ditches were often separated from the dwellings - in the European living quarters at least - by high walls, bamboo fencing or hedges. Yet calamities could never be completely ruled out as evidenced by the following quotation, which admittedly applied to cesspits but could just as well have been applied to septic tanks.

> *An investigation carried out in the Embong quarters of Soerabaja in 1906 revealed that almost all European toilets led through cesspits with overflow pipes to collecting gutters, drains or ditches and that the water was also used by the native servants to wash the crockery and clothes of the not so observant European inhabitants. Where there were still kampong areas the local population also used the same water for toilet purposes, bathing and washing. In Semarang the whole water supply and waste system was improved when the flushing method was improved.*[29]

Where sewerage systems were in place, which was something very sporadic, human excrement was channelled away with other waste water. In general, a separative sewerage system was then implemented in view of the vast amount of precipitation in the tropics and the great intensity of the downpours. It was only in the city centre of Cheribon that it was the 'tout à l'égout' (the mixed system) that was opted for with a complete pumped pipe system and emergency overflows leading to open water. There people obviously preferred the inconvenience of water in the streets during heavy showers - which constituted a mixture of rain and polluted drain water - to the danger of a separative system with the inherent risk of incorrect connections resulting from the possible unprofessional connecting of waste water sewers and rain water sewers. In Sibolga the separative system was introduced by profiting from the repair work carried out after a huge fire in 1919. The waste water was not treated but pumped into the sea.

Partly thanks to the huge influx of Europeans in the twenties and thirties, large European residential areas sprang up in Bandoeng and the need for proper sewer systems grew. In north-east Bandoeng the system of open plastered gutters and ditches was maintained, flushing the faeces, bathwater and washing water through series of passages that were fenced off by 2 m high walls. In south-east Bandoeng (the old city) they switched in 1926

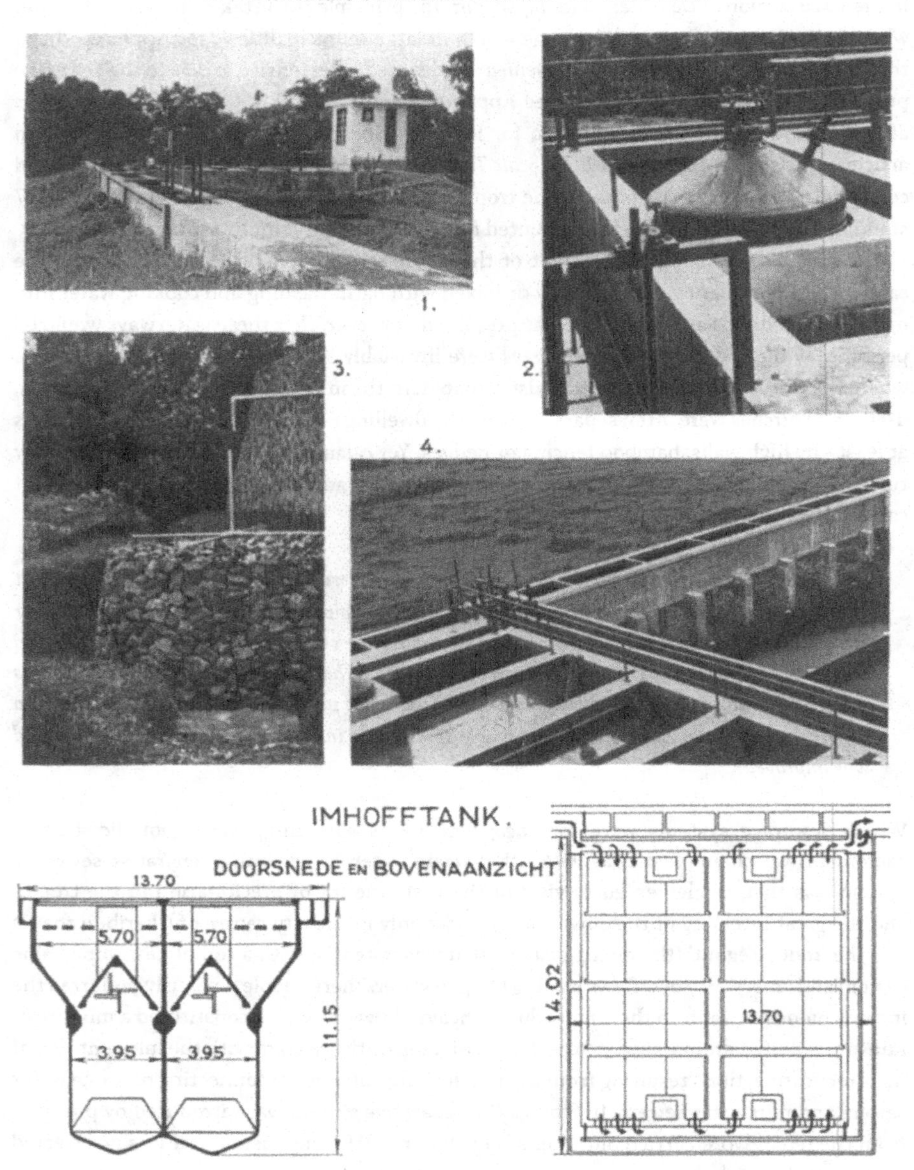

Photo 1. Overview of the plant with the influent main and centre for the dosage of chemicals

Photo 2. Gasdome

Photo 3. Trickling filter

Photo 4. Effluent line and sludge drying beds

The sewage treatment plant and Imhoff tanks in Bandoeng, 1936.

to a fifteen-year building programme for separative sewerage systems based on a design produced by H. Heetjans in 1925.[30]

By around 1938 roughly 20 percent of the municipality of Bandoeng had this style of sewage networks. The system fed into one main sewer on the south side of the city. Until October 1935 'raw'(untreated) sewage was thus deposited in the Tji Tjepoe, but after that date it passed through two parallel linked Imhoff tanks with a settling capacity of 400 m^3, an 800 m^3 decomposition space and domes that produced methane gas of circa 300 m^3 (2,550 kcal) per twenty-four hours.[31]

The report for the year 1933 produced by the Experimental Station for Water Purification in Manggarai finally reported for Djokjakarta that the (separative) sewerage system 'that had been initiated a number of years before' was to be provided with a treatment plant in the form of an Imhoff tank and that to that end a test installation would be constructed on site.

As far as can now be established, no other separative sewerage systems with supplementary treatment plants were introduced during the colonial period – possibly other sewerage systems that dealt with raw dumping (in Medan for example) – but for the rest virtually everywhere it was the septic tank with open or closed flush pipes that was in use. Much the same applied in Batavia and still applies in present-day Jakarta where, up until today, there is no sewerage system with supplementary treatment. To conclude, it may be asserted that where there were no sewerage systems in urbanised areas before 1942 all domestic wastewater was:

- either directly emptied into cesspits or into open channels that may or may not have been lined and may or may not have been regularly flushed clean;
- or discharged via septic tanks and from there into groundwater or open waterways.

Where sewerage systems were present, faeces and waste water were channelled into the waste water sewer, though discharges of polluted waste water into rainwater drainage systems occurred.

The storing, collecting and treatment of non-domestic waste water

Industrial water issuing from hotels, restaurants, garages et cetera generally passed through sand traps, grease traps, and such like before being discharged. Many such matters were dealt with and regulated in municipal statutes. Industrial waste water, derived from a number of large factories situated in the city, would often be simply dumped in sizeable waterways (rivers, canals) - which were to be regularly flushed - after first undergoing light preliminary treatment (sludge removal, grease trapping, neutralisation and dilution). The way in which industrial waste water was discharged from large state-run factories or private enterprises in inland areas, depended very much on the particular local circumstances. Raw dumping, for instance, was and still is a common practice in the big rivers in Sumatra.

The conveyance of local precipitation in urban areas

When it came to the matter of water management in urban areas the following sub-issues could be distinguished:

- the conveyance of local precipitation in the relevant urban area;

- the controlling of the groundwater level which was directly affected by local precipitation in the relevant river catchment area and by the possible influx of groundwater originating elsewhere;
- the protecting of the urban areas against the influx of surface water from elsewhere, the so-called *bandjir* protection system; see below in the section 'Bandjir protection in urban areas'.[32]

In the flat coastal regions where most of the larger cities are situated: Kota Radja, Medan, Sibolga, Padang, Palembang, Telok Betong, Pontianak, Bandjermasin, Makassar, Menado, Merauke, Den Pasar, Batavia, Cheribon, Semarang and Soerabaja these places are subjected to periodical heavy rainfall levels and at such times the rain intensity can be great (see Table 10). In mountain areas the precipitation is higher but in those places it is often possible to profit from the typically much steeper slopes enabling a speedy transportation of the water that falls. However, high plains, like the plain near Soerakarta can – by the same token – lead to just as many problems.[33]

For many years the lower city area of Batavia had its canal system and the Tji Liwoeng river for water to flow into whilst Semarang had the Semarang river and Soerabaja the Kali Soerabaja. All the same, as soon as cities started to expand and as soon as the density of certain districts increased, especially the areas populated by Chinese tradesmen and other Foreign Easterners, it became increasingly more difficult with the relatively small ground level variations in the flat coastal area, for local precipitation to be transported away fast enough. An additional problem that had to be contended with was that of the large rivers that were fed into. These rivers have large flows of water and silt in the wet season (*bandjirs*) coupled with extremely high water levels and sustained inundations. It is a problem that is resolved:
- either by leading the rivers in question around the relevant urban area;
- or by allowing them to continue to traverse the urban area while making sure at the same time that the urban area is properly embanked and that local precipitation is diverted elsewhere or is temporarily stored.

Relevant examples of this are presented in the section about urban *bandjir* protection.

In view of the hugeness of the precipitation quantities, most authorities were inclined to opt for separate conveyance systems as far as local precipitation was concerned so that the transportation of rain water and waste water was clearly separated. As has already been mentioned, the old centre of Cheribon constituted a slight exception to the rule. In cases where the water conveyance channels were open and unlined, there was also the problem of the fast-growing wild plants in the channels, hampering the flow and thus reducing the discharge capacity. The problem was further compounded in areas where the population also disposed its solid waste and faeces into such waterways. Although often classified as 'small- scale water management projects', the relevant projects were of a substantial size and, together with urban planning, were complicated as far as their design aspect went and often required major investments. Just to illustrate this point: between 1905 and 1910 554 ha of urban area in Semarang was completely cleaned up in line with the 'small scale water management' approach involving a system of channels and after 1909 a further 114 ha in West Semarang was cleared. Despite all of that it was claimed, in 1918, that: 'the implementation of proper sanitation through adequate water management in the north-west, north and north-eastern parts of Semarang still remains a hopeless task'.[34] Around

Table 10. **Maximum rainfall in Batavia in mm***

Occurring on average once per:	**1 hour**	**1 day**	**2 days**	**5 days**	**10 days**	**15 days**
2 years	57.2	112	143	210	297	357
5 years	69.9	152	199	304	425	509
10 years	76.6	179	235	367	508	610
25 years	86.4	212	280	445	613	734
50 years	93.5	237	313	500	686	821
100 years	110.8	262	346	557	765	915

* Determined by means of the 'floating limits' method for the 1879-1916 period (38 years); NEDECO, 'Master plan for the drainage and flood control of Jakarta'. To compare: the average annual rainfall recorded in De Bilt (in the Netherlands) is around 825 mm.

1940 the total area of Batavia that fell under 'small-scale water management' was some 3,400 ha and encompassed more than 90 km of canals and main waterways.

The central regulations on those kinds of projects also reflected a clear evolution (see Appendix 5). At first the design discharges were of an empirical nature. Then, especially during the thirties of the twentieth century, calculations started to be made on a stochastic basis, that is on the basis of compiled data from the continually measured and registered rainfall (available in Batavia since 1879!!). The data were compiled in such a way that series of 5 minute precipitation intervals were presented in tables. All those calculations had, of course, to be made 'by hand'. It was the name of the engineer J.P. der Weduwen that became inseparably attached to that Sisyfus labour.[35] It goes without saying that the capacity of the system had to be in line with the design discharges. In connection with hygiene it was furthermore important that after the rainwater had been transported away the gutters were either left completely dry or were periodically flushed clean to prevent water from stagnating. Such stagnant water could become anaerobic (stench!!) and attract unwanted insects and vermin. The flushing clean requirement demanded a degree of inventivity from the designers if they were to achieve that without the help of pumps. Where possible, swilling water was sought, especially in the dry season, from backed up sections of the so-called *bandjir* canals that were often led around cities as diversion canals.

The conveyance channels (gutters, drains, ditches) were frequently lined, often by means of river-cobble masonry. By lining these waterways it became possible to make do with a smaller hydraulic profile and the whole cleaning process was vastly simplified: no unwanted plant growth, no build up of dirt and easier clearing of silt and dumped waste.

One conditio sine qua non for a good functioning rainwater transportation system is the availability of a well-organised collection and processing service for solid household waste. This was something that was certainly recognised and was subsequently effected as well as possible. In Bandoeng, for instance, in both the European districts and the *kampongs*, which had been rehabilitated by means of a *kampong* improvement plan, domestic waste was collected on a daily basis.

The Tanah Abang rain water drainage channel in Batavia before and after having been lined, 1923.

The controlling of the groundwater level in urban areas

In the lower-lying coastal areas where the ground was incapable of supporting much, pile foundations had to be created, like in the Netherlands, when building structures of any considerable size were built. Before steel piles and screw piles were invented and before the time of concrete piles, it was simply wooden piles that were used. The old town hall in Batavia and the VOC warehouses in Pasar Ikan, both of which were built on wooden piles bear witness to this. Wooden piles do not rot as long as they remain under water. For that reason it is important that the groundwater level is not allowed to sink too low in those particular areas. A too low groundwater level, especially in the dry season, can be prevented by carrying out specific surface water level control, for instance by placing weirs in gutters, channels and canals and possibly also by artificially forcing water into the soil, for instance with the aid of drain tubes.

On the other hand, the groundwater level must not be too high because the surface has to be sufficiently sturdy for such things as roadways. Finally, a generally low groundwater level makes it possible for surplus rainwater to be temporarily stored in the earth, something which is desirable from the point of view of water regulation. A low groundwater level is also desirable from the point of view of public health because, provided that the soil is sufficiently permeable, the effluent from the septic tanks and cesspits will be able to penetrate before evaporating, thus giving rise to the biological breaking down and destroying of pathogenic organisms. Finally, muddy soil surfaces created by groundwater levels being too high constitute breeding grounds for illness-spreading germs and insects (eggs, larvae, etc.). One such disease-carrier which is in this respect well known in wet tropical parts is the hookworm which transports the water-related disease known as anchylostomiasis. Especially in the wet season, it is possible to prevent groundwater levels from becoming too high by creating open waterways such as trenches, ditches and canals and by having subterranean pipes (drainage tubes).

In its regulations the Public Works department provided recommendations for the various local authorities (examples of which are provided in Appendix 5). From written references and from investigations carried out on the spot, it is evident that groundwater level control is something that was widespread and standard in the main coastal cities such as: Batavia, Cheribon, Semarang, Soerabaja and Sibolga.

The collecting and processing of solid waste

Prior to 1870 the spatial set-up of the smaller and 'larger' cities was more or less similar. Each village or small town had at its centre an open area known as the maidan or aloon aloon. The small number of Europeans and the affluent native inhabitants lived in freestanding properties situated in spacious grounds. The ordinary native population dwelt in and around the urbanised area in kampongs and desas, in other words, on small premises in separate houses made of bamboo and roofed with pan tiles or Nipa palm leaves. In reality it was only the Chinese, the Arabian population and other Foreign Easterners who lived in denser more crowded and urban-like districts.

The solid waste that needed to be disposed of was largely of an organic nature. In the kampongs, the houses and yards were cleaned virtually every day and the rubbish dried on the spot and then burned. The resultant waste was usually buried somewhere on the premises. Other waste was usually dumped in ditches, streams or rivers. In hilly regions that was not a problem but in flatter urban areas it was. However, there the problems created by channels becoming blocked remained small-scale and resolvable. In the European districts of the cities and the areas inhabited by wealthier people of native origin, domestic waste was collected on a daily basis, though even in those parts the organic waste originating from the home and premises was dried and burnt every day. In the urban areas inhabited by the Chinese, Arabs and other Foreign Easterners there were considerable problems surrounding the collecting, dumping and processing of solid waste. The rubbish that was collected was usually dumped outside the city perimeters and covered with soil.

As the cities continued to grow, population-wise and geographically, during the course of the twentieth century the urbanised areas became denser which meant that more waste was generated per km^2. The composition of the waste also changed as it started to include glass, tins, et cetera: luckily plastic had not yet appeared on the scene. The daily collection rounds also had to embrace greater areas. Inevitably the problem of dumped rubbish in waterways also grew in proportion to such population expansion. It was especially problematic where the relevant waterways were located in low-lying parts and where the water subsequently hardly flowed. The thrown away bamboo cones and baskets did not rot under water and thus became obstacles that backed up water levels. The waste that was collected rather than burnt on site was, as far as possible, processed by means of the 'sanitary landfill' [36] method. Between Batavia and Tandjong Priok, for instance, 'work became dual purpose' by using solid waste to raise the low-lying land areas, swamps and former paddy fields in accordance with the sanitary landfill method.

Bandjir protection in urban areas

Once again, in view of the massive rainfall levels and the wetness of the tropics, this was immediately a sizeable project. In order to illustrate the problem and demonstrate the

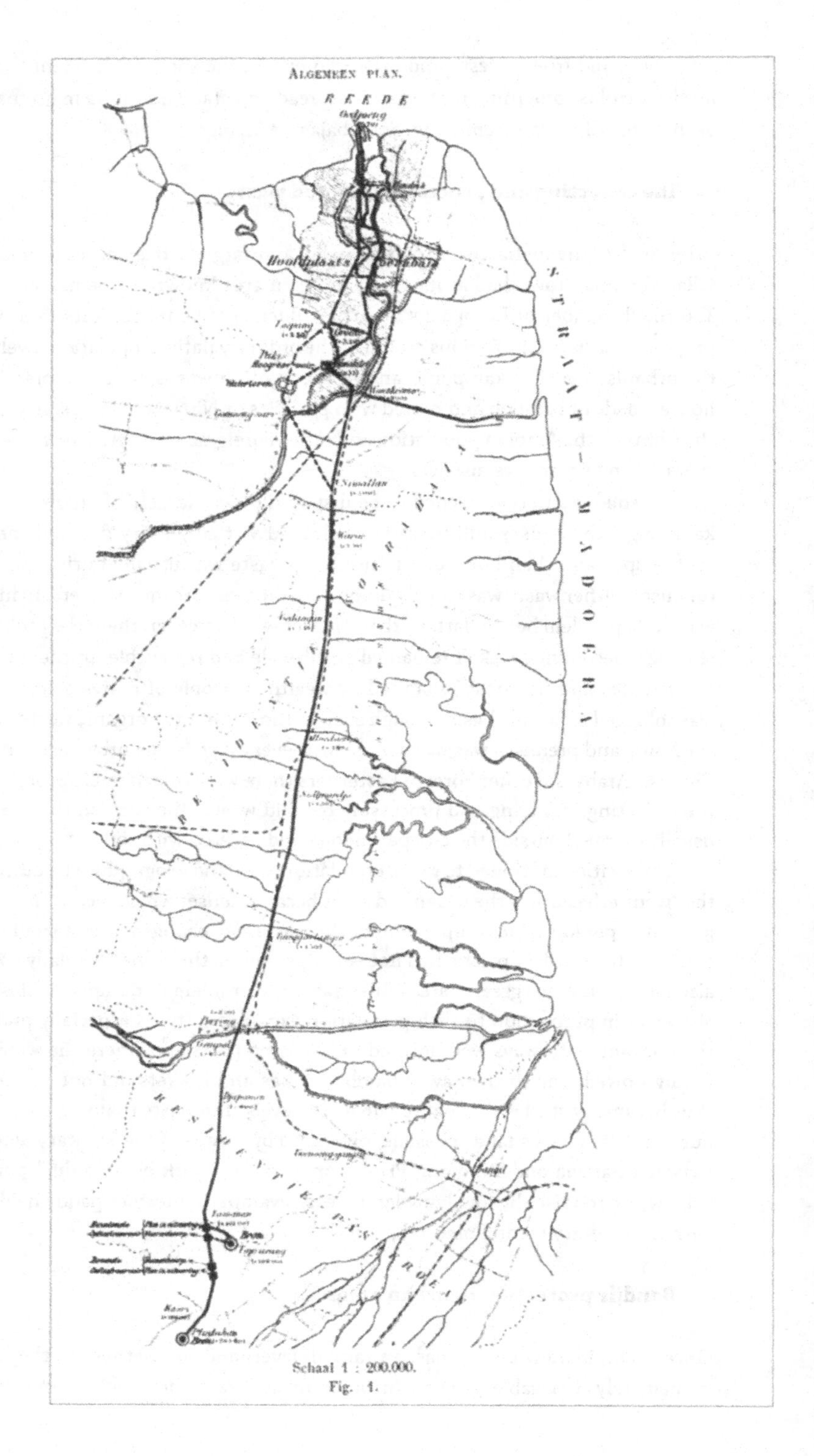

The hydraulic situation in Soerabaja around the year 1903.

approach taken, the examples given pertain to the following relevant cities: Soerabaja, Soerakarta, Semarang and Batavia. In order to find out more about the important developments that were taking place in that area in the field of education and research in relation to urban *bandjir* protection the reader is referred to the details given in the section entitled: 'Public Water Supply'.

Soerabaja

In the first place, this was a city that was situated close to the sea which made it easy to directly discharge surplus water into the sea. In the second place the city was expanding parallel to the Kali Soerabaja which, for a considerable distance, flows practically parallel to the Straits of Madura and is merely separated by a strip of land around three kilometres wide. Already at the beginning of the second half of the twentieth century (around 1856)[37] a short cut was made near Wonokromo. By means of a weir with moveable gates both the water levels and the flow rates in the diversion canal and in the downstream part of the Kali Soerabaja could be controlled which meant that:

- in the rainy season the excess water could be diverted to the Straits of Madoera,
- in the east monsoon season but also in the drier periods of the west monsoon season, sufficient clean swilling water could be directed towards the lower city region.

An important extra side effect of the short cut described above was (and is) the fact that it not only served to reduce the high river levels in the lower city areas but it also kept the transport of silt towards the mouth of the Kali Soerabaja and the anchorage spot down to a minimum.

Soerakarta

The fact that *bandjir* protection is not just something that affects the bigger cities in the lower-lying coastal areas becomes clearly evident if one examines the *bandjir* protection measures in place for the city of Soerakarta.[38] Soerakarta is situated at the confluence of the Kali Pepe and the Solo river (or Bengawan Solo). The city was affected by floods caused by the waters of the Kali Pepe, especially at times when there were simultaneous *bandjirs* in the Kali Pepe and the Bengawan Solo. The solution that was found to that problem was to divert the Kali Pepe upstream by means of a *bandjir* canal with an amazing 800 m^3/s capacity and to channel all that water through the valley of the smaller Djebres river. The moveable weir in the Kali Pepe made it possible for Soerakarta to be provided with flushing clean water, up to a maximum of 40 m^3/s. The design also included dykes or embankments designed to protect the *desa* Koemplang and the Soerakarta-Semarang railway from *bandjirs* deriving from the Kali Pepe. The other accompanying structural works were a bridge for ordinary traffic and tram traffic passing from Soerakarta to Malang-Djiwan and a bridge along the Soerakarta-Semarang railway line belonging to the Netherlands East Indian Railway Company. The costs of that latter bridge were budgeted at 64,000 guilders on top of the total project cost estimate which was 426,553 guilders. The project was completed around 1910.

Soerakarta was furthermore provided with an encircling dyke as a protection against the *bandjirs* emanating from the Bengawan Solo. Some time later, according to a plan authorised on 2nd January 1908, the crest of the dyke was raised to 1 m above the highest recorded

The large mobile weir in Wonokromo, Soerabaja; the situation around 1930.

bandjir of 1861. Because of the encircling dyke a kind of polder situation arose in the lower parts of Soerakarta. The cut off section of the Kali Pepe, the remaining part of the Kali Pepe in the poldered city area, was thus given a floodgate sluice where the encircling dyke was bisected. That meant that when the water in the Bengawan Solo was high, the lower areas of the polder could become inundated thanks to local precipitation if they were not separately pumped dry. That pumping was sometimes also neglected because 'pumping the lower areas of Soerakarta – which was virtually exclusively populated by native inhabitants – was not believed to be well-founded'.[39] The temporary inundation of the polder was accepted in conjunction with water storage and on the basis of the design criterion allowing for 119 mm of local precipitation in three whole days, presumably the maximum time required for the *bandjir* to pass in the Bengawan Solo.

Semarang

The creation of a *bandjir* canal in the city led, in 1879, to improvements in the peak discharge rate of the Kali Semarang through the city of Semarang. It was still, however, insufficient and so in 1888 a start was made on the construction of a 'déversoir' in the Kali Semarang and the construction of the Western Bandjir Canal. Those works were completed in 1890. When finally the Eastern Bandjir Canal was finished, the *bandjir* prevention project in Semarang could be said to have reached completion. In the flat, low-lying old city region the

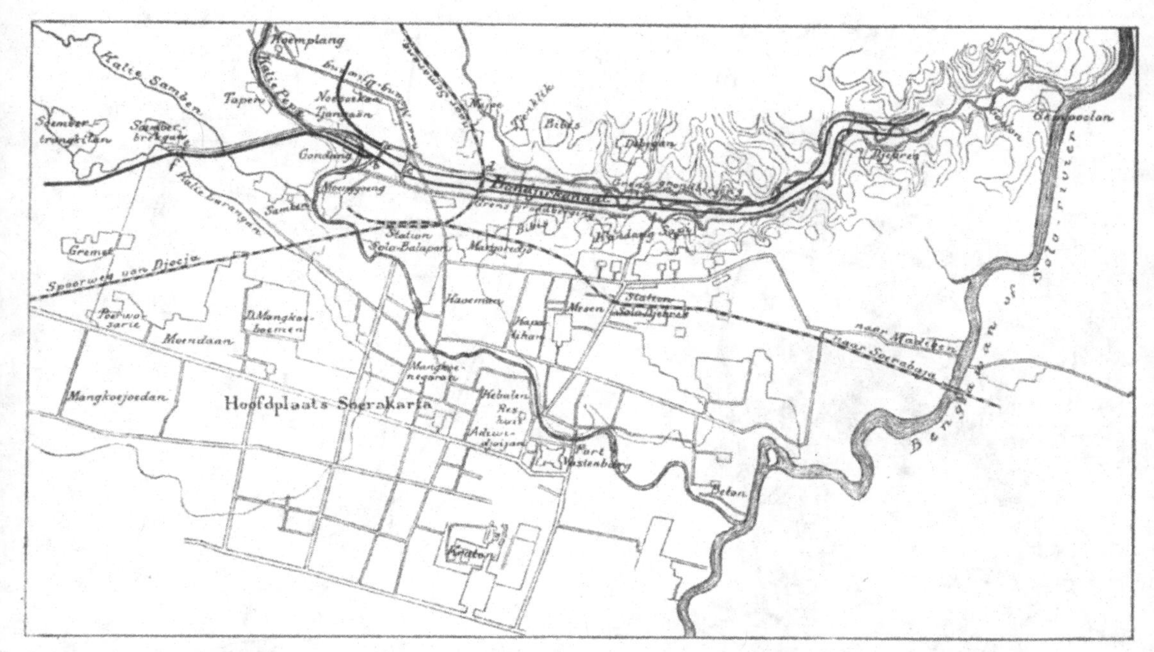

The bandjir canal from the Kali Pepe to the Solo river protecting Soerakarta (Central Java), 1904.

effective transportation of local precipitation to the sea by means of the specially developed and executed 'small-scale water management system' remained a problem, however. The affluent sector of the population – predominantly European – made sure to keep their feet dry by moving to a newly created 110 m higher situated residential area known as Nieuw Tjandi that was directly to the south of the 'lower city'.

Batavia

Batavia initially opted for a partial solution by diverting the Tji Liwoeng around the old city centre via a specially excavated canal, the Goenoeng Sari Canal, and by straightening and normalising the Kali Krokot. A certain amount of banking up with dykes alongside of the Kali Angke and the Kali Pasanggrahan on the west side of the city completed the safeguarding projects against bandjirs originating from the rivers of the vicinity. Simultaneously the creation of the Sentiong canal to the east side of the city protected future building plots against the too high water levels emanating from the hills and mountains to the south and southeast of the city. The Kemajoran airport was later to be situated on a part of that terrain. The discharge from the Sentiong canal passed through a siphon that went under the Antjol shipping canal and out into the Java Sea. The diverted Kali Soenter also fed into the Sentiong canal. Conveniently that meant that the fast expanding harbour of Tandjong Priok was thus relieved of extra undesired excess water coming from the south.

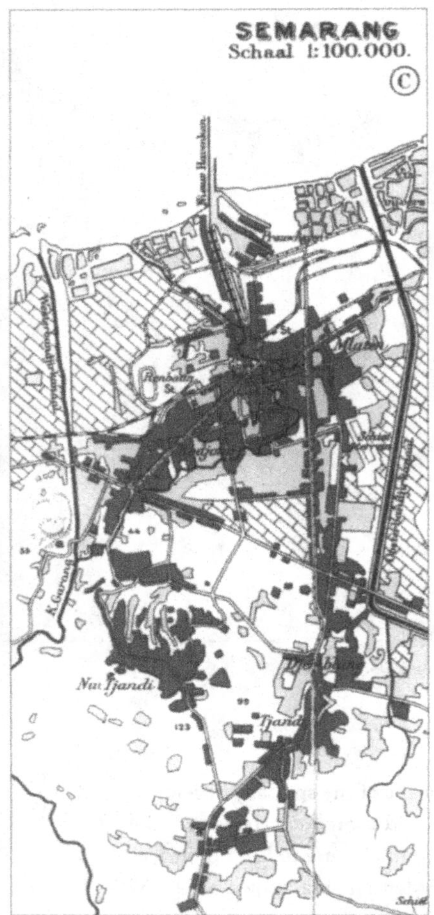

The hydraulic situation in Semarang (Central Java) around 1940.

The hydraulic situation in Batavia around 1923.

Under the leadership of the engineer H. van Breen a complete solution was later introduced which, until 1942, was to protect the city from *bandjirs* issuing from the Tji Liwoeng and a number of smaller rivers like: the Tji Deng, the Kali Krokot and the Kali Grogol. The solution, the Bandjir Canal, consisted of a diversion canal leading off from the Tji Liwoeng that encircled the city starting in a westerly direction. The Bandjir Canal was completed in 1918. The Bandjir Canal design was such that it allowed for a maximum *bandjir* discharge of 300 m^3/s to be directed to the Bandjir Canal itself via a sluice complex constructed at Manggarai whilst 80 m^3/s could be diverted to the downstream region of the Tji Liwoeng.[40] Van Breen obtained the normative flow rates of the Bandjir Canal by establishing direct discharge measurements in Tji Liwoeng. The clever thing about the whole design was that the water volumes in the wet season did not cause the Bandjir Canal to erode

The Van Breen commemorative plaque on the Manggarai weir.

its bed and banks but they were at the same time sufficient to flush away possibly silted up parts originated at low flows. The Bandjir Canal has since established itself as a 'natural' stable waterway.

Since the construction of the moveable weir, Karet, in the Bandjir Canal it has become possible, by impounding water in the dry season, to compensate the low flow of the Tji Liwoeng which can fall to 2 m^3/s so that water can still be flushed through the city. As has already been mentioned in the section about drinking water provision, after 1950 – when the Bandjir Canal was supplemented via a separate connection with water from the Tarum Barat, a canal that obtained its water from the Tji Taroem – the impounded water could also be used as a source of 'raw water' for the new surface water collection company created to provide drinking water for the capital. What was quite remarkable about the whole hydraulic design for the city of Batavia was the provisions made for transport by water, provisions which today – superseded by road transport – have almost entirely disappeared from the city's landscape. Locks were especially constructed for the bamboo rafts and the proas that came from the Mookervaart canal and passed down the Antjol canal on their journey to Tandjong Priok. With the completion of the Tangerang lock, which was finished at the same time as the large moveable weir in the Tji Sedane (1929) which had been created for irrigation purposes, it remained possible for ships to navigate between the Tji Sedane and the Mookervaart.

The citizens of Batavia showed their appreciation for everything that Van Breen had done for them by placing a commemorative plaque on the Manggarai sluice but that was not until after the first *bandjir* had successfully passed through the brand new canal.

CHAPTER 9 **FROM TECHNICAL ASSISTANCE TO GOTONG ROYONG**[1]

Hydraulic engineering, sanitary engineering and bridge building in new Indonesia

The end of the Second World War heralded the period of large-scale colonial state decline. For Great Britain the end began in 1947 when India and Pakistan became republics in their own right. For the Netherlands that was in 1949 with the establishment of Indonesia, for France in 1953 in the case of Cambodia and for Portugal it came in 1975 with the independence of the Cape Verde islands and Angola (aside from the Indian annexation of Goa in 1962). Even Russia was unable to escape this process when, after the disintegration of the Soviet Union, republics such as Kazakhstan, Uzbekistan and Turkmenistan were created after 1989, thus putting an end to Russian colonisation of those areas. Apart from being confronted with the problem of having to set up proper functioning administrative organisations, armies, police forces, and education and public health systems, all these new countries also had to take care of plans pertaining to the creation, rehabilitation and maintenance of civil public works for the purposes of agriculture (irrigation), road traffic, rail traffic, navigational traffic, drinking water provision, sanitation[2] and for the combating of erosion and flooding. For political but also for economic and practical reasons, the 'mother countries' were not always able to adequately support their former colonies in the resolving of such matters despite the fact that invariably the bilateral contacts were still intact.

The United Nations, established in 1945 with all its affiliated institutions, was thus expected to help to fill that gap by offering a helping hand. In principle the structure appeared to be sound. All the aid was professionally organised. Basically, the political independence of the newly created countries could not be affected and financial support and technical expertise drawn from all over the world could be deployed. The bodies and organisations that were of importance to civil public works at that stage were the following:

- the International Monetary Fund (soft loans);
- the Reconstruction and Development Bank/the World Bank (combination of soft loans and the availability of project expertise);
- regional banks such as the Banque Africaine de Développement and the Asian Development Bank (regional expertise);
- the United Nations Development Programme (the UNDP, *the* bureau for technological aid);
- the World Health Organisation (the WHO, for drinking water facilities and sanitation) including the separate sub-organisation known as the International Reference Centre (the IRC)[3] directed at the initiation of projects, including knowledge transfer, in the field of drinking water provision and sanitation in towns and rural areas;
- the United Nations' Food and Agriculture Organisation (the FAO, for irrigation!);
- the UN's Educational, Scientific and Cultural Organisation (UNESCO, for knowledge transfer, the position of women especially in conjunction with sanitation projects and erosion prevention);
- the Economic Commission for Asia and the Far East (ECAFE, for regional knowledge and education, notably in relation to hydrology, hydrodynamics and hydraulic engineering).

The end of 1945 saw the start of the exodus of some 250,000 Dutch citizens (European and Indo-European) from the Dutch East Indies, many of whom returned to the Netherlands. Finally, when in 1958 Soekarno expelled the last sedentary Dutch inhabitants, the exodus was complete. Some of the departing Dutch residents emigrated to countries such as

Canada, the United States of America, Brazil and Australia but the majority returned to the Netherlands.

In the post-Second World War period and up until the time of the transfer of sovereignty, in 1949, Dutch intervention in civil public works in the Dutch East Indies centred chiefly on the substantial rebuilding programme linked to all the various works that had been badly damaged or even destroyed during the Second World War, in the *bersiap*[4] and in the period before, during and after all the police actions.[5] Immediately after 1949 all that intervention – insofar as it related to the territory of the Republic of Indonesia – gradually decreased and virtually came to a standstill after 1958 because of the New Guinea question. The Dutch contribution to the recovery efforts manifested in private enterprises: plantations, petroleum winning and refining, etc. had to be halted because of matters linked to New Guinea. Subsequently, when New Guinea was then handed over to Indonesia, all Dutch contributions to the planning, execution, control and maintenance of civil public works in the area also ceased. It was not until after 1968 when Soeharto officially came into office as president of the Republic of Indonesia that Dutch involvement in civil public works in Indonesia was once again resumed.

The Netherlands: the organisation of development cooperation

The Kingdom of the Netherlands, totally cut off from Indonesia and reduced to two percent of its original land surface area had become degraded to a small country within Europe with a poverty-stricken colonial remnant in South America comprising a total population equivalent to a medium-sized metropolis and was thus forced, in the interests of survival, to a much greater degree than previously to turn to foreign countries outside Indonesia. It therefore looked mainly to Europe – notably to Germany which, through the Marshall Aid Programme, had managed to develop quite rapidly (the 'Wirtschaftwunder') – but also to the USA, Australia and Canada.

Then something remarkable happened. Thanks to the influx of Dutch people from Indonesia, the Netherlands had suddenly gained a wealth of middle and higher-level civil engineering experts and a number of agriculturalists. The knowledge possessed by those experts was of course based on their extensive work experience in rainy tropical areas. Alongside of the existing battery of experts oriented very much towards foreign conditions (one need only think of dredging companies working on land reclamation projects in South Korea, Hong Kong, Singapore, etc.) the Netherlands suddenly gained, in the early fifties, a whole range of new manpower that could aptly be deployed in new areas and especially, of course, in rainy tropical parts of the world but also in areas where there was a heritage of renowned colonial projects, such as the large irrigation tracts projects in the Indian subcontinent and in North Africa.

The United Nations offered both the opportunities and the means for these groups of Dutch individuals to be deployed in countries that were largely 'underdeveloped' (a term coined in those days). For civil engineers in particular that meant having to assist with project plans, create designs and specifications, supervise project progress, assist in repair and maintenance work, impart knowledge and train personnel. Before very long the Dutch were to be found occupying prominent United Nations posts such as at the FAO head office in Rome and in the head office of the Asian Development Bank in Manila. During the 1950s Professor W.J. van Blommestein, former divisional head of the East Indian Public Works department was the initiator and FAO team leader of the extensive Ganges-Kobadak Multi-purpose (Irrigation and Food Control) Project in East Pakistan (later Bangladesh).[6]

In 1950 the Ministry of Foreign Affairs established the International Technological Help (ITH) division, alternatively known as the Netherlands Bureau of Technical Assistance (NEBUTA) which was, at first, particularly focused on the sending out of experts (and associate experts) via United Nations bodies. Later the ITH was to become the International Technical Help Directorate (DITH) after which time Dutch foreign aid – then divided into bilateral and multilateral aid – began to display explosive growth and that was partly due to the important internationally endorsed resolution asserting that foreign aid extended to underdeveloped countries from developed areas should amount to minimally 0.7 percent of the Gross National Product.[7]

As the new countries started to realise their full potential – moving, in the process, from the position of being 'underdeveloped countries' to the status of 'developing countries' – and as the relationship between the donor countries and the aided countries started to level off[8] the technical 'help' started to be dubbed 'cooperation' and, correspondingly, the DITH was upgraded and labelled the Directorate General for International Cooperation (DGIS). Later on it came to be known as the Ministry for Development Cooperation but it fell under the Ministry of Foreign Affairs umbrella. A certain degree of unchecked growth was to be detected in international cooperation in the Netherlands when other departments started setting up their own international cooperation projects alongside those of the Ministry for Development Cooperation. Those projects were variously bilateral, initiated via the World Bank or via the West European Union ('Brussels') and either manned by own staff or by hired manpower. A government resolution of 1988 decreed that in the interests of the state, the Ministry of Economic Affairs should be given the power to somewhat curb all the autonomous development cooperation initiatives of other government departments and streamline the various efforts. Since 1977 the foreign aid expended by the different ministries has been brought together in what is known as the Homogeneous Group for International Cooperation (HGIS).[9] Furthermore, all the development people working in the field who have shifted from development aid to development cooperation have been subjected to rather erratic policy changes emanating from The Hague, policies that have been swayed by numerous political hobby-horses and whims. Over the course of time the onus in development cooperation has sometimes switched fairly abruptly and has been variously laid on: programme help, project help, the direct combating of poverty, the indirect combating of poverty, female emancipation, the environment, urban development, rural development, and so on which, obviously, has had all kinds of dramatic effects on ongoing programmes and projects, certainty regarding the way in which cooperation is established with the country at the receiving end and mutual harmony with other donor countries and organisations.

One important development was the emergence of organisations employing low-paid volunteers heralded by the establishment, in 1961, of the Peace Corps in the USA and rapidly followed by the Dutch Association of Volunteers in the Netherlands and the creation of Non-Governmental Organisations (NGOs). These tend to be organisations that work in close proximity to the population in question, bear the motto 'Small is Beautiful' in their banners and do a great deal with 'appropriate technology'.[10] In the case of the Netherlands such organisations are partly funded by BUZA, that is to say, the Ministry of Foreign Affairs. In 1978 BUZA set up PUM, the Sending Out of Managers Project,[11] aimed at achieving effective and short-term knowledge transfer of a direct and practical nature (small industries, hospitals, schools, etc.) by sending out unpaid and often pensioned experts in all types of fields and drawn from all walks of life.

Where most mainstream bilateral and multilateral aid was concerned companies, notably engineering bureaux, saw possibilities for participating in the various projects. Generally these opportunities involved tapping the know-how of the experts returning to the Netherlands from the East Indies and getting them to participate in the various projects. Special 'foreign' departments or special daughter enterprises were created for such purposes: International Land Consultants (ILACO, 1952) part of the Royal Dutch Heidemaatschappij Company (now ARCADIS), Euroconsult (1974) a collective enterprise of the Royal Dutch Heidemaatschappij and GRONTMIJ limited companies. In 1951 a 'public-private partnership' avant la lettre even emerged: the Netherlands Development Consultants bureau (NEDECO), when the engineering bureaux entered into a partnership with the Dutch Public Works agency (Rijkswaterstaat) so that all the knowledge and manpower available in the country could be combined in order to stand up to all the foreign competition existing within the framework of the commissioning bodies of the United Nations. The academic world also entered into the flow. The Dutch education system found it useful to offer courses to alumni returning from developing countries so that academic frameworks could thus be cultivated for such countries. One additional advantage was the fact that in that way a network could be created between the Netherlands and the relevant developing countries in that particular domain and at that level. It was to be a network that would bring with it reciprocal tropical knowledge input, science and experience and which was to prevent the source of direct information from the former colonies from drying up. Regarding the relationship between the Netherlands and Indonesia, such a development was extra desirable as far as civil engineering was concerned because after1958 all Dutch connections with the Bandung Technical College, by then known as the Institut Teknologi Bandung, had been broken, which automatically meant that the main route for technological knowledge exchange between the two countries had also been terminated.[12]

In 1951 the International Training Centre for Aerial Survey (ITC) was established and that was swiftly followed in 1952 by the Netherlands University Foundation for International Cooperation (NUFFIC) and the Institute for Social Studies (ISS). The latter institute constituted an indispensable social component where technological projects were concerned, especially any sizeable projects.[13] In 1953 the Institute for Land Reclamation and Improvement (ILRI) was set up. It was affiliated to the country's main Agricultural College and therefore also to the Ministry of Agriculture and Fisheries.[14] On 1st November 1956 the Civil Engineering division of the Technical College in Delft initiated an International Course in Hydraulic and Sanitary Engineering. Later it was to develop into the Institute for Infrastructure, Hydraulic and Environmental Engineering (IHE)[15] offering one and two-year courses at both bachelor's and master's level and even the opportunity to complete a doctorate at what had, meanwhile, changed from the Delft Technical College to Delft

University of Technology in 1984. The IHE even started to actually establish branches in various developing countries. With the help of the Dutch government the IHE educational institute BIPOWERED was established in Bandung to mark the renewed spirit of cooperation with Indonesia after 1967. In keeping with the renewed attention being devoted to the work of civil engineers abroad and the special need for applied technology, the Civil Engineering division within what was then still Delft Technical College created, in 1973, something that was both a civil engineering college for developing countries and a bureau for project work in developing countries whilst simultaneously establishing intensive cooperation with various local technological universities. The initiative in question was known as the Centre for International Cooperation and Appropriate Technology (CICAT).[16] So it was that in the Netherlands the foundation and framework was created for what was in the first place envisaged as a foreign aid organisation but for what was later to extend to international cooperation.

Where development cooperation between Indonesia and the Netherlands is concerned we shall restrict ourselves here to the 1968-2000 period (indeed, it would be impossible to go back further than 1968). The civil public works field discussed will pertain mainly to irrigation and water management, bridge building (road and rail bridges) and to public drinking water facilities as Dutch intervention in the concrete realisation of other works and structures within the civil engineering sphere was of lesser significance. Although, within the framework of public health, the collection, conveyance and processing of waste

Illegal shacks on the levees of the Bandjir Canal in Jakarta; the situation in 1970.

water normally constitutes an indispensable aspect of public water supply systems while the collection and treatment of solid waste remains an integral facet of urban drainage and *bandjir* prevention facilities, these matters became secondary due to financial shortages. In the field of urban water management, or urban *bandjir* protection, an exception was made in the case of one sizeable bilateral developmental project that continued for more than ten years, a project that will thus be described here and which was known as the Jakarta Drainage and Flood Control Project. Directed at the restoration of sea connections and the accompanying revamping and modernisation of the harbours, a team of Dutch advisers was employed by the Indonesian Department for Connections (Perhubungan). The team's work terrain was very wide and included, amongst other things, making recommendations in relation to scheduled services, warehouse complexes, layout and the structure of the harbours and harbour areas, planning, and the installation and maintenance of harbour equipment (e.g. cranes). The activities of the commission were recorded in annual reports.

The dams already proposed by Van Blommestein in his welfare plan for West Java of Cirata (Tji Rata) and Sangguling (Sanggoeling) in the Citarum (Tjitaroem) were built with Japanese assistance and without any Dutch intervention. The building of the hydraulic power stations in the Asahan river in Sumatra where the Wilhelmina waterfall (now known as Siguragura) and the Tangga waterfall are situated, commenced in 1939, to be finally completed in 1982 as a Japanese aid project.[17]

Indonesia: the development in independence

Indonesia as it was prior to 1968[18]

What exactly was the state of affairs at the starting point in 1968 and what were the precepts, premises and employment conditions for the 1968-2000 period? In the Soekarno era the young republic was characterised on the one hand by a new élan and, on the other hand, by disorganisation and decline. After the transfer of power in 1949 the sprawling island kingdom was converted into an independent Indonesian united state at the expense of a number of costly military actions, both in terms of money and human life, in Sumatra, Sulawesi and the Moluccas. However, things went even further. In an endeavour to also annex Sarawak, which was originally part of the Dutch East Indies, yet more brief and abortive military campaigns were undertaken. The attempts to try and make Irian Jaya (Papua) a part of Indonesia also consumed much military and diplomatic effort and money. Meanwhile Indonesia created for itself a prominent position among unattached countries. Batavia became the new capital known as Jakarta and it soon turned into a metropolis with impressive monuments, avenues, hotels of international allure and tall, imposing office blocks.[19] Much money was pumped into education in an endeavour to eradicate illiteracy and likewise, much money was expended on sizeable projects such as the building of the 80 m high Jatiluhur (Djatiloehoer) flood-control dam in the Citarum together with its accompanying hydraulic power station and reservoir with a capacity to take three billion m^3 of water. That particular project was completed in 1967.[20]

One of the negative side effects of all of this – and there were a number of such consequences – was the subsequent neglect of civil public works maintenance. Nation-wide

The Saluran Surabaya drainage canal in Jakarta-Menteng, completely filled with refuse; the 1970 situation.

the deterioration witnessed in the irrigation works, the roads (both regarding road surfacing and structural works) and the railway network was dramatic. The population proliferation in the major cities, boosted by sharp rises in local numbers (a high birth rate) and the influx of people from rural areas (those without land, those seeking their fortune) led in particular to the general illegal occupation of land and to the illegal occupying of areas set aside for *bandjirs* (dykes, inundation zones) but also to insufficient expansion, control and maintenance of public drinking water supplies and urban sanitation facilities: rubbish collection services stagnated, drainage canals were blocked up with garbage and faeces, and structural works were not kept in a proper state of repair. The result was that where public water supplies, hygiene and sanitation were concerned in the bigger cities, degrading situations arose.

Indonesia under the Orde Baru

Unlike the populist direction followed by Soekarno (mobilisation and political participation on the part of the people) the military leaders, of the post-1967 era, pinned all their hopes for the economic recovery of their country on a Western-style model; they wanted to

The Ciliwung in Jakarta-Kota around 1970.

modernise and industrialise within a free market economy. As an OPEC country Indonesia was able to profit from its oil reserves which, notably in the seventies when world oil prices rose sharply, brought in a lot of foreign currency. This was followed in the eighties by a sharp drop in oil prices which, together with the falling dollar rate, meant that drastic cutbacks had to be made. Under the New Order (that of *Baru* as opposed to *Lama*, the Old Order of Soekarno) Indonesia had come to depend on foreign investment and aid. The Western World, which supported the military coup because of the overt aversion to communism – it was after all the Cold War era – was only too glad to help Indonesia. In 1967 there were 14 donor countries that united under the chairmanship of the Netherlands to form an Inter-Governmental Group (the IGGI) on Indonesian affairs and which, together with the World Bank and the International Monetary Fund, supported the developments continuing in Indonesia. However, relations with the former coloniser became more and more strained leading, in 1992, to the termination of such formal ties with the Netherlands. The main cause of the rift was irritation on the part of Indonesia about Dutch intervention in the human rights situation (following the suppressing of riots in East Timor).[21]

As of 1969 the government drew up a series of five-year plans (or Repelitas) for the

Dr. Suyono Sasrodarsono, civil engineer and Indonesian Minister of State, while being congratulated by Professor W.J. van Blommestein during the reception of his honorary doctorate degree (TU Delft) in 1985. Visible in the background: the initiator, Professor H.J. Schoemaker.

development of the country's economy. Originally agriculture was central. In terms of indigenous agriculture the aim was to become self-sufficient where rice was concerned. This was effected by means of an agricultural modernisation and rationalisation programme. In the process, the state introduced the new technology to accompany the green revolution in question. That involved introducing better varieties of rice derived from Philippine laboratories ('wonder rice') together with the relevant production requirements such as fertilizer and insecticides. Later the emphasis was to shift more towards the establishing of modern industries and the construction of the necessary infrastructural works. The government also resolved to do something about the steady population increase by helping people with birth control. In order to take pressure off Java the government accelerated Javanese migration to South Sumatra and promoted Javanese migration to other parts of Indonesia like Kalimantan, Sulawesi and Irian Jaya. Following the economy drives of the latter half of the eighties, the expensive migration programme did become considerably constrained!

As a result of all the efforts, Indonesia was rewarded with a period of rapid economic growth. In the 1979-1989 period the Gross National Product rose from 49.2 to 95 billion dollars. The population also continued to expand by more than 2 percent per annum in the 1965-1987 period though in 1980 it decreased to 2.4 percent and then to 2.1 percent in subsequent years. The population of Indonesia in 1987 was 171.4 million people, 60 percent of whom lived in Java. Despite everything the per capita GNP increased by 4.5 percent in the 1965-1987 period (for the Netherlands that was, for instance, 2.1 percent in the same period). The agricultural policy, notably the introduction of the green revolution technology, culminated in vigorous crops so that by 1984 the coveted self-sufficiency level for rice had been achieved.

Just as often occurs in history, the successful planning and execution of the agricultural policy and, in particular, the revitalisation and expansion of irrigation works, water control works and water management could – to a large extent – be attributed to the inspiration, energy and organisational capacities of one man: the engineer Dr. Suyono Sasrodarsono. In 1965, alongside of being assistant manager to the Indonesian General Directorate for the Development of Water Supplies the civil engineer known as Suyono Sasrodarsono also held the position of Komandan Projek Pentjegah Bandjir Djakarta Raja, a semi-military project focused on the harnessing of the flood problems in the country's capital. In 1969 he was made director-general of the above-mentioned general directorate. In 1982 he was made secretary general of the Department of Public Works (Pekerjaan Umum, PU) and between 1983 and 1988 he was also the Public Works minister. Under his inspired leadership the master planning and the execution of major structural works (canals, storage reservoirs, weirs and pumping stations) for the urban drainage and flood protection of Jakarta came into being. Under his leadership the reinstatement of all the existing irrigation systems was rigorously tackled, together with the creation of new irrigation systems, so that by the 1980s Indonesia could boast of being perfectly self-sufficient in the field of rice production. He also provided the impetus for the necessary studies to be carried out throughout Indonesia per river catchment area (there were ninety such areas) in order to arrive at a system of integrated water management.[22] It was therefore most appropriate that in 1985 the Delft University of Technology saw fit to bestow upon him an honorary doctorate. That doctorate had certainly been well earned.[23]

Irrigation and water management

Restoration and modernisation

After 1967 and in conjunction with the five-year plan (Repelita) the first thing to be rigorously undertaken was the restoration of all the abandoned irrigation works. It amounted to a total of 780,000 ha, 90 percent of which was situated in Java (see Table 1). Much support was received from the World Bank and from the special and separately set up International Development Agency (IDA). The projects were thus also listed as PROSIDA (Proyek Irrigasi IDA) projects and were carried out under the auspices of the Director General for the Development of Water Supplies within the Department of Public Works (see Table 2).

The main concern when it came to all the repair work was to get the main structure of the irrigation systems restored and once again in proper working order. It was the method

Table 1. **Budget for the rehabilitation of irrigation works in millions of US $**

Five-year plan	Period	Budget
Repelita I	1969-70 / 1973-74	124
Repelita II	1974-75 / 1978-79	333
Repelita III	1979-80 / 1983-84	602
Repelita IV	1984-85 / 1988-89	145

Table 2. **The PROSIDA programme**

PROSIDA	Project area	Land area (in ha)	Region
Series A	Way Seputih	25,000	South Sumatra
	Cisedane	42,000	West Java
	Rentang	92,000	West Java
	Glapan-Sedani	42,000	Central Java
Series B	Ciujung	24,000	West Java
	Pemali-Comal	123,000	Central Java
	Sadang	55,000	Sulawesi (Celebes)
Series C	Pekalen-Sampean	229,000	East Java
Series D	Rentang-Cirebon	190,000	West Java
Series E	Madiun	140,000	East Java

Source: Ankum, 'Management of irrigation rehabilitation projects. An experience from Indonesia'.

once implemented by the Netherlands that was adhered to. It was only when the third five-year plan was introduced that attention could be given to the repair and innovation of the detailed irrigation works in the tertiary sections and attention could be given to new ways of distributing water which, more so than in the past, were based on the wishes of the farmers. At that stage the influence of foreign advisers and foreign education (in the US, Great Britain, Japan and France) and also the United Nations (FAO) became unmistakable. Systems such as the 'trial-run' Integrated Irrigation Development Approach (IIDA)[24] were introduced in conjunction with the relevant farmer's irrigation associations and were termed P3A.[25] The influence was adapted in a positive way within the framework of Indonesian-Dutch cooperation in the field of tertiary irrigation works, especially where bilateral cooperation was linked to projects in Sedeku, Cidurian and Pompengan but also by Dutch people working in the field on projects that were not financed by the Netherlands.

It is therefore remarkable that when extending the irrigation acreage for the construction and maintenance of the main systems (primary canals, secondary canals) right down to the tertiary drawing off level, the Dutch method was preserved despite the large influx of non-Dutch advisers. Evidently it was a method – characterised by upstream water supply regulation with manual controls based on hydraulic principles – that suited the demands of irrigation staff and farmers, despite the disadvantage that it was an irrigation method that was very sensitive to water level fluctuations and demanded a high level of management with a lot of personnel (very frequent manual level control and flow checks). As one might expect, computer-steered 'real time control' might well have been found too avant-gardist and too vulnerable but it is curious that the French hydraulic regulation and distribution systems, which are less sensitive to water level variations in the primary and secondary supply canals, did not catch on.[26]

In the nineteenth century the water management works in the river basin areas were chiefly devoted to and equipped for a controlled kind of water supplying and draining for the purposes of irrigation in the wet season. In the twentieth century it also became necessary to have adequate water provision for irrigation purposes in the dry season. That meant that supplies had to be stored in reservoirs. What was new was the idea of deploying the reservoirs for hydroelectric power and largely allowing the costs to be covered by selling the electricity generated to households and industry. It was an idea taken from a similar project launched in the 1930s involving the Tennessee Valley works in the USA.[27] With the building of the reservoirs came also the need to contemplate irrigation and electricity production interests as well as the reservation of reservoir capacity for the abatement of high river flood levels (*bandjirs*). Especially in Java, water requirement or even water demand parameters were a direct result of increasing industrialisation that was partly attributable to the explosive population growth. The big demand for drinking water had everything to do with population increase and heightened water-related awareness. Ultimately water pollution and erosion were important environmental factors in the weighing up of interests.

It was inevitable that the policy factors 'Integrated Water Resources' and 'Integrated Water Management' would be included in the planning and controlling of the 90 main river areas in Indonesia. The part played by the Netherlands in this integrated approach was important, both in bilateral and multilateral senses but also from practical and academic points of view. It is particularly worth mentioning the intensive cooperation in this field between the Indonesian Department of Public Works, the Bandung University of Technology (ITB) and the Hydrodynamics Laboratory (WL) in Delft. There were two people who were central in all of this: Suyono Sasrodorsono, mentioned above, and E. van Beek of the WL, the engineer who was the project leader and who later became professor of integrated water management modelling at Delft University of Technology.

The Dutch contribution to water management in the 1968-1992 period that was angled specifically towards civil engineering (canals, structural works) could be divided into three categories:

1. bilateral, the contribution made by the Netherlands was for 100 percent financial and technical,
2. multilateral, the contribution made by the Netherlands was financial and technical for less than 100 percent,
3. non-lateral in that the contribution made to projects by Dutch experts was made in conjunction with the United Nations or certain other countries.

Table 3 gives an overview of the projects that fell into category 1. Generally speaking such projects pass through the phases of assessment, defining the required steps (time planning included), measuring, analysing, studying, preliminary designing, final designing, drawing up of specifications, contracting out, execution (possibly just supervision) and follow-up work. As well as making everything operational and maintaining operation during the handing over period, the final phase almost always includes training (transfer of knowledge). It has seldom been the case that bilateral cooperation has included all project phases. It has, however, often been the case that due to financial or other considerations what may

Table 3. **Overview of the bilateral irrigation and water management projects 1970-1987**

Project	Region	Description	Commencement
Jratunseluna	Central Java	Improvement of the irrigation system including the introduction of flood control dams in the basins of the Jragung, Tuntang, Serang , Lusi and Juwana.	1970
Cimanuk	West Java	Master plan for the entire river basin.	1970
Demak	Demak + East Semarang	Feasibility study.	1970
Pasang-Surut	All of Indonesia	Preliminary research.	1971
Serayu river basin	Central Java	Hydrological and geomorphological studies.	1973
Juwana + Walahar	Central Java	Rentability calculations in conjunction with drainage.	1974
Jratunseluna (BTA- 26)	Central Java	Operation & Maintenance.	1974
Luwu (BTA-38)	Sulawesi	Technical advice for the irrigation service.	1974
Glapandam	Central Java	Feasibility study for the dam in the Tuntang in con-nection with the Jratunseluna development project.	1974
Luwu	Sulawesi	Regional development.	1975
Pasang-Surut	All of Indonesia	Measurement programme for one million ha.	1975
Rawah Sragi (B-46)	Lampung	Marshland reclamation of 1,300 ha to 6,000 ha plus a development plan for a further 2,300 ha.	1975
Pompengan	Sulawesi (Luwu)	Irrigation project; using water from the Lemasi rivier.	1979
Irrigation Java + Madura	Java + Madura	Counter expertise regarding the plan conceived by Professor W.J. van Blommestein to divert surplus water from West Java to East Java and Madura for irrigation purposes (mainly for rice cultivation).	1981
Sampean Baru	East Java	Rehabilitation and extension of the irrigation area.	1984
Citarum-Bekasi (BTA-144)	West Java	Crash Programme aimed at improving rural water management for the entire Citarum and Bekasi river basin areas.	1984
Sampean Baru (B-154)	East Java	Trial-run project.	1985
Cisadane-Cimanuk Integrated Water Resources Development (BTA-155)	West Java	Pioneering research into the integrated development and management of the river basin areas in the northern part of West Java (from Cibanjar in the west as far as Cisanggarung in the east) the main area of attention being on the Citarum river basin. The study was also of great importance for water supplies for the urban conglomeration known as JABOTABEK (Jakarta-Bogor- Tangerang-Bekasi).	1985
Water provision	West Java	Water supply measurements conducted in the river basin areas of the Citarum, Cikao, Cibeet, Cikarang and Bekasi rivers.	1986

Source: Ministerie van Buitenlandse Zaken, *Projektlijst bilaterale projekten in het kader van de Nederlandse financiële en technische samenwerking met ontwikkelingslanden.*

Pompengan (Sulawesi): fixed weir with two-way inlet works and ejection sluices, 1984.

Pompengan: tertiary irrigation with a module à masque variant, 1984.

originally have started out as a bilateral project has invariably ended as a multilateral project involving, for instance, the World Bank or the Asian Development Bank. In a number of projects the Netherlands functioned as a project booster or catalyst. It was also sometimes the case that after Dutch initiation a project would be entirely taken over by another donor country, like for example Japan.

Examples of multilateral projects (category 2) that were often part-financed by the World Bank and the Asian Development Bank, were the Pasang-Surut works in the river basin of the Musi in South Sumatra (World Bank) and the Wadaslintang project in South Kedu in Central Java (Asian Development Bank). Famous projects that fell into category 3 were: the East Java Irrigation Project (World Bank), [28] the Central and West Java Irrigation Project (World Bank), the Citagompor Irrigation Water Management Project (Asian Development Bank), the Simalungun Irrigation Project (Asian development Bank)[29] and the irrigation works in West Pasaman, West Sumatra (German Federal Republic).

Several projects[30]

The activities in the Jratunseluna area, the region to the east and south-east of Semarang, constitute a good example of the first kinds of rebuilding projects aimed at the renovating and rebuilding of irrigation systems. The engineering works for irrigation and high water protection that already existed went back at least as far as 1850 and were known as the 'Demak waterworks'. In 1967 the Indonesian government asked for assistance with the creation of an integrated development plan for the region in the field of water management. The Dutch government provided that assistance, for instance in the form of a delegation sent out to draw up plans. That delegation, composed of Dutch and Indonesian experts, arrived in 1968. There were three Dutchmen in the team and they really embodied the ties between the Netherlands, the Dutch East Indies and Indonesia where the fields of irrigation and water management were concerned. Those individuals were: Professor Schoemaker, then professor of irrigation in Delft, Professor H. Vlugter, former professor in Bandung but at that time technical advisor for GRONTMIJ Ltd. and Professor G.A.W. van der Goor, former staff member of the General Agricultural Experimental Station in Bogor and later Professor within the agricultural faculty of Universitas Indonesia, Bogor and also the Gadjah Mada University in Jogjakarta who was then working for the ILRI in Wageningen, in the Netherlands.[31]

If we look at the Wadaslintang project in South Kedu, Central Java, embarked on in the 1980s, we become aware of the attention paid to water management that was characteristic of the wave of reconstruction projects that started around 1980. This large-scale project, executed in one of the areas that competed for the origins of Dutch irrigation technology prize, was also an exponent of the great attention devoted in Indonesia to the development of reservoirs for simultaneous use for the purposes of irrigation, drinking water supplies and hydropower. The South Kedu plain is situated on Central Java's southern coast. The densely populated plain is approximately 15 km wide and extends along the coast for some 50 km. During the colonial era irrigation systems had been created, each of which drew water from one of the four rivers located in the plains. In the case of the Wadaslintang project, named after the reservoir in question, an infrastructural system was realised that had to supply the different systems with extra water via a supplementary canal. In total some 35,000 ha of land would thus be provided with irrigation. In November 1982 one of the things NEDECO became involved in was the task of, where necessary, redesigning Wadaslintang's various main canals. In the first design the Dutch consultant proposed change. In the original tender design it was presumed that there would be discharge measurements structures known as Parshall flumes to gauge water volumes, both at the Pejengkolan dam and at the outlets in the main canals. What was proposed for the smaller outlets was constant head orifices: both were actually American designs. As far as the first flume was concerned all that the NEDECO report stated was: 'change Parshall flume in broad crested weir'.[32] Though no specific allusion is made to Romijn measuring weirs (a certain type of a mobile broad crested weir), the striking preference is curious. Regarding the measuring points at the inlets of the individual systems the explanation is somewhat more elaborate:

> *At most turn-out locations the available head is large enough for the operation of a broad crested weir. Experience with this last type of flow meter in Indonesia has shown that it is easier to operate and to construct than a Parshall flume or a constant head orifice. Therefore the consultant [i.e. the Dutch one,] proposed to use a broad crested weir as a measuring device for the turn-outs.*[33]

The Pasang-Surut projects amounted to a completely different story altogether.[34] Indonesia has 7,000,000 ha of freshwater swamps that are tidal, 2,345,000 ha of which are on Sumatra, 2,268,000 ha on Kalimantan (Borneo), 84,000 ha on Sulawesi (Celebes) and 2,303,00 ha on Irian Jaya (New Guinea, Papua).[35] What that means is that when the tide comes in the freshwater flow in rivers is backed up whilst during the ebb tide it drains away. By means

A Pasang-Surut settlement in South Sumatra, around the year 1988.

of a simple system of inlet and outlet works it thus becomes possible for freshwater to be allowed in and drained out of the suitable swampy areas – mainly the riparian embankments skirting the big rivers – in a controlled fashion. Already in colonial times those areas were seen as possible development spots to accommodate immigrants from Java and Madoera. In South Sumatra a start was even made on such a project. In modern Indonesia presurised by the continued population rise in Java and the need to compensate for all the people being forced to move because their land was being inundated for water storage purposes (storage and ordinary reservoirs) in, for example, the Citarum catchment area (Jatiluhur, Cirata, Sangguling) and the Solo basin (Wonogiri), such tracts of land were simply claimed for irrigation, especially in South Sumatra and Irian Jaya. On arrival, the colonists discovered 2 ha of jungle, one of which had been reclaimed, a simple wooden dwelling, seeds (rice) and a supply of husked rice that would be sufficient to keep a family of six for six months.[36]

Bridges[37]

The repair and replacement of road bridges

The Netherlands decided to participate in the Indonesian bridge programme to which, through financial and technical help, the World Bank, Australia and Canada were already contributing by delivering prefab bridges for the replacement and new construction of road bridges on Sumatra, Java, Kalimantan, Sulawesi and East Timor. To that end a joint cooperation contract was signed between Indonesia and the Dutch government in1979. According to the contract, the Netherlands was to be responsible for supplying the steel superstructure for 161 A and B class road bridges, each with two traffic lanes. The bridge spans varied between 25 and 60 m. It was decided that the harbours of Tanjung Priok (Java) and Belawan (Sumatra) were suitable for the delivery of the steel bridge components. The Netherlands was also to be responsible for the transmission of knowledge regarding design and the final erection, new construction or repair work of the foundations (an extremely sensitive and essential aspect of bridge construction), the substructures (piles, land abutments) and the assembly of the superstructure.

On the Dutch side it was the Directorate General for International Cooperation (DGIS) that was responsible for the materialisation side of things and for Indonesia it was PU. Indonesia was particularly responsible for the design and materialisation of the foundations, the substructure and the concrete driving surface as well as for logistics regarding the transport and assembly of the steel superstructure components. The supply of the steel constructions was contracted out to a consortium of five Dutch steel companies and it was Hollandia Kloos that was the 'leading partner'. In order to provide the necessary transfer of know-how, not just in relation to the assembly process but also the other essential and sensitive areas, like the above-mentioned foundations, an agreement was signed with the limited company PT Triweger, a totally Indonesian-owned company in which the Dutch engineering bureau, F.C. de Weger, held shares. One of the reasons for choosing that particular bureau had been the availability of knowledge and experience in the field of foundation technology. Via Hollandia Kloos a Dutch advisor was also permanently stationed in the area whose job it was to supervise the assembly process.

Temporary reinforced bridge over the Cisedane river (West Java) before it was replaced, photographed around the year 1980.

The irreparable old bridge over the Cicati (West Java) and the new bridge, around 1980.

The design of the superstructure was taken care of by Hollandia Kloos in cooperation with the State Public Works which was an advisor to DGIS. The components were made relatively lightweight in anticipation of possible local transport problems. The design of the connections was based on the use of pre-stressed bolts. Particularly in the early years of the programme it was necessary to ensure that the correct degree of pre-stressing was achieved in the bolted connections. It was a method that was hardly known in Indonesia and for a number of reasons it was, in practice, impossible to pneumatically effect the pre-stressing. It was therefore necessary to develop special ways of doing that with straightforward manpower, no simple task if one bears in mind that the average Indonesian man is too lightly built for such heavy manual work. When the programme began the bolts were still being fabricated in the Netherlands but under great pressure from the Indonesian authorities the factory was eventually moved to West Java. An Indonesian company was then established especially for that purpose with which Hollandia Kloos closely cooperated. In view of the fact that all the steel components had to be thermally galvanized, a galvanizing workplace was also built: it was the first such privately-owned large galvanizing workplace.

In most cases the construction of the substructure and the assembly of the various bridges took place under difficult circumstances; the vast majority of all the bridges were constructed outside Java in remote areas. Accessibility to the workshops was often poor and invariably the local conditions were not particularly conducive either, for example because of bad foundations and/or the fast flow speed of rivers. *Bandjirs* also had to be taken into account. On a number of occasions they were responsible for the destruction of all or part of a substructure under construction. Fortunately that never happened to any of the superstructures. Work was often carried out in close cooperation with local constructors who were not only ill-equipped to take on large consignments but who were also unfamiliar with that kind of work. This meant that the input from the supervisory organisations was important if minimum quality standards were to be met.

All the bridges were assembled on site which meant that much ingenuity had to be demonstrated and a whole range of methods had to be integrated. The most usual method involved using construction aids mounted in the river. In the case of deep rivers where such constructions would be much too expensive, use was made of what was known as the free extension method (a cantilever system). This involved first building a back-up bridge on the bank side which operated as a counterweight from where the permanent bridge over the water could be extended out and so built with or without the help of other construction aids, even sometimes incorporating the existing old bridge that would afterwards be dismantled. A hybrid form that was only used on two occasions was the method which involved using one support construction in the river and allowing the bridge – which had first been assembled on land on rollers – to be drawn over the river by means of cables. One particularly memorable feat was the free extending of the spans for the bridge over the Sungai Mahakam, a bridge with a main span of 100 m, a free navigation height of 20 m and five side spans, each 60 m long. This bridge lies in East Kalimantan not very far from the city of Samarinda. Once the bridge's truss work had been mounted then the driving surface could be framed and all the reinforcement sections could be laid so that finally the concrete could be cast.

The PT Triweger bureau not only checked the state and the strength of the existing foundations that had to carry the new bridge but it also provided advice on the actual construction work and on the possible need for new foundations. In addition to that the bureau monitored the state of existing bridges, many of which proved to be in a poor state of repair due to lack of maintenance and due to the increased pressure (overload!) and wear and tear of the simultaneously increased volume and weight of traffic. It was only the original masonry and stonework bridges that proved to be an exception to this particular rule.

All these various discoveries led to a revision of the original bridge allocation programme. It was especially the 'heaviness of traffic' parameter (in both senses of the word) that gave cause for the adjustment of the necessity criteria or priority criteria. In total 120 bridges were given a different or better location.

All in all, it may be concluded that this Indonesian-Dutch programme of cooperation was a success. It gave substance to the great need for better or new river bank connections in densely populated and remote regions. The degree of flexibility both as regards design and assembly, together with the great freedom of choice regarding locations permitted by the Indonesian authorities, were important contributory factors. As a result, the programme was frequently expanded, like for instance when 48 extra traffic bridges were delivered which of course meant that the contracts with PT Triweger and the assembly advisor were prolonged. The work pressure for PT Triweger was furthermore intensified and burdened by charging the bureau with the design of the substructure for a large number of bridges. In addition to that a number of special designs were introduced in order to meet various location-specific requirements. One good example of this was the already mentioned bridge over the Sungai Mahakam. By the time of the termination of the project, in 1992, some 210 bridges with a total length of around 10,000 m had been constructed and assembled and a considerable fund of knowledge on the matter of bridge building had been imparted.

The Railway Bridge Project

From the very start, the management of the public railway network[38] in the Republic of Indonesia, the maintenance of rolling stock included, had been in the hands of one national company.[39] Over the course of time the infrastructure in question had been maintained as well as possible but, owing to financial shortages, a maintenance and new construction backlog had developed. Ultimately for quite a number of the bridges speed restrictions had been imposed in order to guarantee the safe passage of trains. As a result of train traffic increases it became necessary to modernise both the rolling stock and the material – in this case the infrastructure. For the bridges that meant having the capacity for roughly 30 percent greater mobile stress and a greater fatigue stress allowance in conjunction with the projected greater traffic frequency. It was clear that the existing bridges, which were generally made of steel, would no longer suffice and that the foundations would have to be inspected.

For a number of railway lines in Java and Sumatra an elaborate plan was therefore developed in the early 1980s by the railway company of the day, the Perusahan Jawatan Kereta Api (PJKA). The PJKA was, however, unable to independently finance the plan. Together with a number of donor countries the Railway Bridge Programme was therefore set up. DGIS adopted part of the programme and thus commenced with what was known as the Railway Bridge Project. In 1988 DGIS commissioned the Dutch engineering bureau

Grabowsky & Poort as chief consultant and the F.C. Weger bureau to conduct the following advisory activities:

a. a study into the state of repair of the substructure of 574 bridge locations located in Java and South Sumatra;
b. an evaluation of the suitability of the foundations for heavier loads and stresses;
c. in support of this, an extensive soil survey of all foundations of the relevant sites;
d. to draw up designs and specifications documents relating to all the adaptations required for those foundations or, as the case might be, for new structural activities;
e. to draw up designs and specifications documents for the 361 new railway bridges[40] with spans varying from 3.70 to 79 m;
f. to ensure that all the necessary knowledge be transferred as efficiently as possible;
g. to support and train PJKA personnel in the business of project management.

Points a to d really had to be completed within 12 months whilst the activity outlined in point e had to be rounded off within a year and a half. The referred to transfer of knowledge was something that would occur gradually and as the project progressed.

In close cooperation with PJKA all the above-mentioned activities were executed most efficiently and speedily. Great ingenuity was expected from all parties in view of the extreme shortness of the time schedule. In order to speed things up, a special software package was, for instance, developed so that the soil mechanical stability of the foundations could be assessed on the basis of the available data. Some of the soil inspections were carried out alongside the land abutments and piles using the Standard Penetration Tests (SPTs) that applied for Indonesia but in the case of a large proportion of the land abutments the measurements were taken directly from the railway line itself. A special system had been developed which made it possible to carry out these SPT measurements from a wagon and in a very short space of time between all the scheduled rail service traffic. What emerged from the foundations inspections was that more than 50 percent of all the foundations were to a large extent in need of adaptation or replacement. The required specifications were then drawn up.

In 1990 part of all the superstructure work was outsourced to the Indonesian company Kratama Belindo International (KBI), a company also participated in by Hollandia Kloos. The remaining part involving the large bridges was contracted out to companies in the Netherlands. At the request of the Indonesian government all Dutch development aid projects in the country were halted in 1992 and that included the Railway Bridge Project which meant that the further modernisation of the Indonesian railways network had to be completed without any help from the Netherlands. It was only the share of the work that had been contracted out to KBI that actually materialised.

Public drinking water

The projects completed in the area of public drinking water fell largely within the framework of bilateral development cooperation between Indonesia and the Netherlands. Almost from the word go the Dutch State Institute for Drinking Water Supplies (RID) played an important part in all of this from the state angle. Similarly, the WHO's International Reference Centre,

Table 4. **Overview of the bilateral drinking water provision projects 1975-1987**

Project	Region	Description	Commencement
Palembang water supply (GTA-22)	South-Sumatra	Rehabilitation and expansion	1975
Water supply 6 cities (GTA-18 en J-11)	Java	Master plan	1976
Rural water supply (GTA-33 en J-7)	West Java	Kecematan improvement (40 villages)	1976
Solid waste Bandung, Makassar, Pontianak	Indonesia	Preliminary design	1976
Water supply Balik Papan	Kalimantan	Rehabilitation and expansion	1977
Sewerage Bogor-Tangerang-Bekasi	West Java	Feasibility study	1978
Management training piped water companies	Indonesia	Study	1979
East Medan water supply	N-Sumatra	Design and execution	1979
Water supply 15 small cities (J-30C)	West Java	Design and execution	1979
Water supply 11 cities	N-Sumatra	Design and execution	1979
Purification plants	Indonesia	Standardisation	1979
Small community water supply	West Java	Design and execution	1980
IKK Crash programme (GTA-52 en J-37)	Indonesia	Design and execution	1981
Indramayu (OTA-33)	West Java	Small-scale drink. water provision	1982
Tasikmalaya (GTA)	West Java	Rehabilitation	1984
Bandung (GTA)	West Java	Feasibility study + rehabilitation	1984
Sukabumi (GTA)	West Java	Direct improvement + training	1984
IKK review programme (GTA-52)	Indonesia	Study	1985
Bogor (GTA-67)	West Java	Rehabilitation and expansion	1985
Rural water supply (GTA-J)	West Java	Second phase	1985
Immediate Sukabumi (GTA-J52)	West Java	Rehabilitation	1985
Bandung (GTA-J49)	West Java	Rehabilitation	1985
Monitoring drinking water projects	Indonesia	Measuring programme	1985
Classroom training drinking water projects	Indonesia	Workshops and training	1985
Ground water research (GTA)	West Java	Measuring programme	1986
IKK-Maluku	Moluccas	Sanitation improvement	1986
GTA-68	Indonesia	Testing and certifying dw-material	1986
KIWA	Indonesia	Providing of flow rate limitations	1986
Small towns water supply	West Java and N-Sumatra	Follow-up 15 cities West Java and 11 cities North Sumatra (+ Aceh)	1986

Source: Ministerie van Buitenlandse Zaken, *Projektlijst bilaterale projekten in het kader van de Nederlandse financiële en technische samenwerking met ontwikkelingslanden.*

Table 5. **The waterworks companies in Indonesia in the year 2000**

Region	Number	Region	Number	Region	Number
Aceh	11	Jakarta	3	Sulawesi Utara	7
Sumatera Utara	17	Jawa Barat	26	Sulawesi Tengah	5
Sumatera Barat	14	Yogyakarta	6	Sulawesi Selatan	23
Riau	8	Jawa Tengah	35	Sulawesi Tenggara	5
Jambi	6	Jawa Timur	33	Bali	10
Sumatera Selatan	13	Madura	4	Nusa Tenggara Barat	6
Bengkulu	4	Kalimantan Barat	7	Nusa Tenggara Timur	12*
Lampung	5	Kalimantan Tengah	6	Maluku	5
		Kalimantan Selatan	10	Irian Jaya	9
		Kalimantan Timur	7		

* Seven of which were BPAM

established by RID employees, was inextricably linked to all the different programmes. After the RID merged with the State Institute for Public Health (the RIV) the Bureau for Development Cooperation emerged as part of the State Institute for Public Health and the Environment (the RIVM). The Bureau took care of the technical advisory side of the pragmatic approach to this type of international aid and cooperation whilst also checking progress where the Dutch aid and cooperation programme in Indonesia was concerned in the field of public water supplying.[41] (See Table 4).

Thanks to the own in-house levels of expertise, Dutch consulting firms were able to react appropriately to all the various project components such as project exploration, the gathering of measurement/survey data, the developing of processes and technical facilities, the drawing up of plans and specifications, assisting in the supervising of work processes and in operational management, the institutionalisation, the requital and the training of personnel. It was furthermore possible to profit from the advantages emanating from direct and intensive cooperation between Dutch and Indonesian public water supply companies in what were termed twinning projects where invariably the bulk of the financing was provided by the relevant Dutch company. With such twinning programmes the coordination was simplified by the presence of the umbrella organisation known as the Association of Water Supply Companies in the Netherlands (VEWIN) and the Association of Water Supply Companies in Indonesia 'Persatuan Perusahaan Air Minum Seluruh Indonesia' (PERPAMSI).

The International Drinking Water Supply and Sanitation Decade Programme set up by the United Nations was responsible for the sudden improvement in the standard of drinking water supplies in developing countries in the 1980-1990 period. Those improvements were manifested both in urban and rural areas and went hand in hand with sanitation, that is to say, the collection, transport and processing of waste water (mostly domestic) and, where possible and feasible, the adequate drying out of flood-sensitive low-lying urban areas where often slums and shanty towns abounded. In this area of work the manuals written by John

Intake of raw surface water for the drinking water supply of Sukabumi (West Java), 1989.

The Bandung hand pump in action, around the year 1988 (West Java)

Kalbermatten and the practical hand pump guidelines issued by the World Bank proved to be indispensable.[42]

The Indonesian water supply companies

By the end of the year 2000 there were no less than 297 public water supply companies in Indonesia, 285 of which were government-owned (the Perusahaan Daerah Air Minum, PDAM), four of which were privately-owned companies also responsible for public water provision,[43] one was a so-called PDA Unit (Yogyakarta) and seven were semi-autonomous companies, that is to say, in the process of becoming autonomous (the Badan Pengawas Air Minum, BPAM), all united in the PERPAMSI (see Table 5). One can distinguish two main types of water supply companies:

- urban water supply companies serving the capital of Jakarta (the Ibu Kota), sizeable cities and the capitals of provinces, like for example, Banda Aceh, Sabang, Bandung and Sukabumi *(kotamadya)*, capitals of districts (*kabupaten*) and sub-districts (*kecamatan*) and possible other smaller cities;
- regional water supply companies created to serve districts and sub-districts.

The *Direktori 2000* of PERPAMSI gave, per company, a condensed overview of the state of affairs in the year 2000 in relation to such matters as staffing, raw water sources, infrastructural provisions (pumping stations, water towers and such like), the capacity to supply, the quantities supplied, the water losses and the financial details.[44] It was then still not possible to indicate what proportion of the urban population could be certain of receiving piped water with the quality 'drinking water' (*air minum*) – probably more than 50 percent – and what percentage could be certain of receiving piped water with the quality 'clean water' (*air bersih*). Much the same went for the rural population (*desa* inhabitants and those in isolated parts): however, probably less than 50 percent.

The twinning of Indonesian and Dutch water supply companies

Twinning, or intensive cooperation between Indonesian water supply companies and Dutch water supply companies, could take various forms but was chiefly manifested in the exchanging of personnel, the trips made to each other's countries, personnel training and assistance (on the part of Dutch companies) in the revamping and expanding of Indonesian companies, the detecting and reducing of leakage loss, the optimisation of water conservation and purification and, assistance in the institutionalisation and optimisation of the requital system and operational management. Within the context of this book the treatment of this subject is restricted to a simple summary of the relevant Dutch companies (see Table 6).

The deployment of Dutch consulting firms

The kind of services required of consulting firms were extremely diverse and extensive. Out of that huge package of requirements the activities were tailored to every individual drinking water company depending on the established (local) requirements or to a conglomeration of drinking water companies (e.g. in a given district or a particular province) or possibly to the steering and controlling authority (e.g. the province). An endeavour has been made to give an idea of the executed projects in one single overview in which special attention is devoted to the civil engineering provisions or alternatively to the relevant structural works (see Appendix 6).

Even though civil engineering is in actual fact nothing other than appropriate technology – that is to say adapted to the local circumstances to build, manage and maintain as economically and as sustainably as possible – a special interpretation and description is given of 'adapted technology' in the development programmes, especially where small systems and structures are concerned. Examples of what are to be found in the area of drinking water are:

- simple water wells with hand pumps,
- small tanks for the storage and purification of water (drinking water, rain water),
- small transport and distribution systems,

and in the area of waste water:

- latrines,
- sewage systems of a limited diameter (small bore),
- oxidation ponds (lagoons).

What was interesting in this connection was the application of ferro cement or, in other words, concrete reinforced with wire netting or even bamboo to create tanks and reservoirs for the purification of water derived from irrigation canals, for reservoirs, for the storage of purified water and for the storage of rain water (to bridge the dry periods). Ferro cement was even used for water tower structures.

The Jakarta Drainage and Flood Control Project

By the end of the Soekarno era the situation in many parts of Jakarta regarding public hygiene and urban water management had become virtually intolerable. The waterways

Table 6. **The Dutch waterworks companies involved in the twinning programme (1965-2000)**

Area provided with water	Dutch waterworks company (former names)	Dutch waterworks companies (as they were known in 2000)
Jakarta	Gemeente Waterleidingen (Amsterdam)	Gemeentewaterleidingen (Amsterdam)
Tangerang	Watertransportmaatschappij Rijn-Kennemerland (WRK)	Watertransportmaatschappij Rijn-Kennemerland (WRK)
Bogor	Provinciaal Waterleiding-bedrijf van Noord-Holland	NV PWN Waterleidingbedrijf Noord-Holland
Bandung (kotamadya)	Gemeentelijke Drinkwater-leiding Rotterdam (DWL)	Waterbedrijf Europoort
Bandung (kabupaten)	Watermaatschappij Zuid-West Nederland (WMZ)	DELTA Nutsbedrijven
Sukabumi	Watermaatschappij Zuid-Holland Oost (WZHO)	Hydron Zuid-Holland
Medan	Waterleiding Maatschappij Gelderland (WMG)	Waterbedrijf Gelderland
Palembang	Waterleiding Friesland (WLF)	NUON Water
Balik Papan	Waterleiding Maatschappij Noord-West-Brabant	Waterleiding Maatschappij Noord-West-Brabant
Makassar	Watermaatschappij Zuid-West Nederland (WMZ)	DELTA Nutsbedrijven

The storage of rain water in ferro-cement tanks (West Java), designed by IWACO and built in 1979.

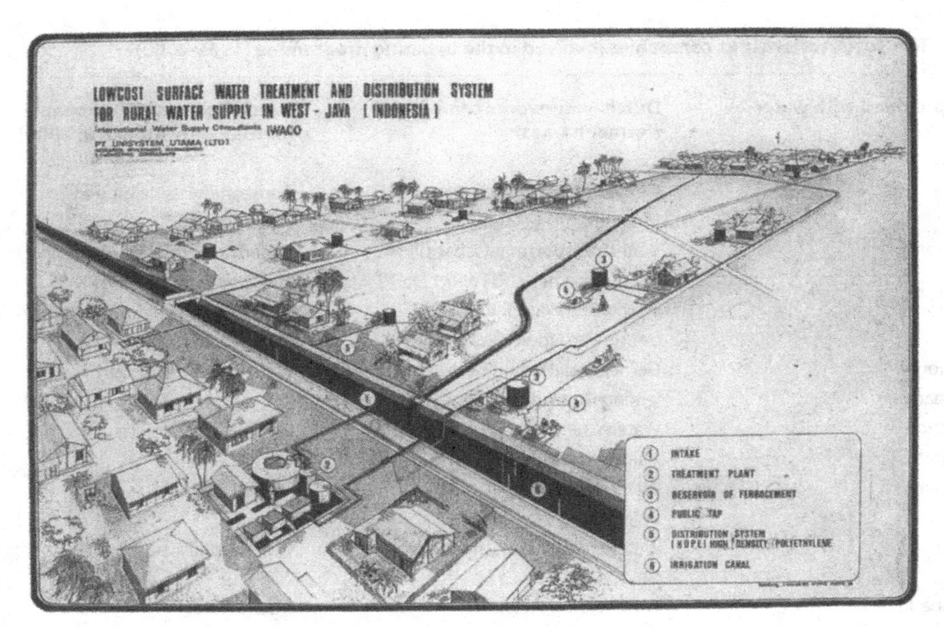

Low-cost surface water treatment and distribution system for rural water supply in West Java; with ferro-cement tanks and polyethylene pipes, 1985.

(ditches, canals and rivers) had all become blocked up with solid waste (household waste, building materials, etc.). Insofar as they still flowed in the dry season the waterways were filled with putrid anaerobic waste water. Hydraulic structural works (sluices, weirs, and so on) were either not maintained at all or only to a minimal extent. Berms and low-lying terrain intended for the temporary storage of water were propped full with illegal slums. In the rainy season there was therefore little or insufficient space for all the rainwater pouring down from the mountains and for the local precipitation. The duration, extent and intensity of the inundations therefore increased.

Pengendalian Bandjir Djakarta Raja[45]

In 1965, immediately after the political upheavals, the water problem became clearly acknowledged and a concrete start was made on a fixed plan of campaign by founding a semi-military organisation established by the central government which was known as Komando Projek Pentjegah Bandjir Djakarta Raja (the Project Command for the combating and prevention of floods in Greater Djakarta) and fell under the active and capable leadership of the engineer Suyono Sasrodorsono. In the case of a number of existing low-lying housing districts and open ground areas earmarked for new housing, businesses, offices, hotels, etc. it was the polder option that was immediately selected to deal with local precipitation and to keep out externally originating water. The poldering was effected by means of polder pumping stations and storage reservoirs (drainage ponds). The areas particularly focused on were the Tomang polder along the road between Jakarta and Tangerang, the Setia Budi polder (the polder where the present Dutch Embassy and the Erasmus House are situated) to the south of the Bandjir Canal between the Manggarai and Karet mobile weirs and the Pluit polder,[46] the area situated immediately on the coast to the west of the mouth of the Bandjir Canal that was given a drainage pond of no less than 80 ha. A start was also made on restoring the hydraulic ('wet') sections of the drainage canals. One of the main problems there was that most of that work had to be done by hand as large equipment was lacking and what was dredged up (household waste, constructional waste, bamboo baskets, plastic, silt, etc.) could not be dumped next to the waterways or in the direct neighbourhood due to the (illegal) hovels dotted everywhere and so it had to be transported elsewhere.

The Crash Programme and long-term planning

Immediately after the heavy flood of 10th February 1970 the Indonesian government sought the help of the Dutch government for the creation of what was termed a Crash Programme. It was a short-term solutions programme involving aid provided in kind in the form of large equipment and a long-term programme (Master Plan) aimed at combating and preventing flooding in conjunction with the already operational development plan for Djakarta.[47]

The mechanical dredging of waterways in Jakarta, in about 1975.

The aim of the Crash Programme was to clean up vital waterways, to immediately rebuild collapsed dyke sections and to repair any damaged or dilapidated structural works.

For the cleaning up action, equipment was deployed that was especially suited to the dredging of the virtually inaccessible waterways and attachments were designed for the transportation and dumping of what was dredged up; attachments such as drag lines and mechanical grabs mounted on pontoons in the canals that were made from locally constructed bins floating on oil drums so that the spoil could be taken away and dumped elsewhere. The dredged up material was used to raise the level of low-lying ground. One factor that complicated the repair of the dykes was the presence of illegal dwellings. Just to repair the Bandjir Canal dykes downstream of the Karet Weir, some 70,000 illegal occupants along that 4 km stretch of canal had to be evacuated to other parts of the city.

The Master Plan for the Drainage and Flood Control of Jakarta

In the interests of the Master Plan a special deliberation body was formed so that all the drainage, *bandjir* protection, groundwater level control, construction, traffic and other area planning components could be coordinated. Extensive land surveys were carried out and it emerged that in a short period of 25 years large parts of Jakarta had subsided in relation to sea level and that the land had settled. That applied especially to the old inner city and to the strip of land close to the Java Sea. The most probable reason for this was the persistent extraction of volumes of ground water for the purposes of industries and hotels.

The Master Plan was completed by 1973. It was geared to a future population in Jakarta of eight to ten million inhabitants bordered on the west by the Kali Angke and extending on the east to around 10 km east of Tandjong Priok. The southern border would remain flexible.

For the low-lying districts of Jakarta, lying roughly below the 10 m level line, the Master Plan had come up with the following series of measures:

a. to provide protection against the *bandjirs* coming in from the south of the city via the Ciliwung and other small rivers by diverting water to the catch canals (i.e. the existing Western Bandjir Canal) which would then continue from the Karet Weir to the mouth of the Kali Angke to protect the central part of Jakarta (the Ibu Kota) and the old and new western sections of the city. A completely new Eastern Bandjir Canal would be dug to serve the largely new eastern part of the city;
b. to create large canals, also sometimes known as drains, with direct connections to the sea for the areas still lying sufficiently above sea level so that local precipitation could simply accumulate and flow towards the sea;
c. to polder in and cut off from the sea (with or without pumping facilities) the remaining low-lying areas;
d. to flush the waterways in the city areas surrounded by catch canals by releasing water from the backed up sections of the catch canals;
e. to interchange knowledge so that Indonesian and Dutch know-how and experience could be exchanged and properly coordinated, plus the training up of Indonesian technical staff so that after a time project execution and management and further planning could continue without the need for Dutch intervention.

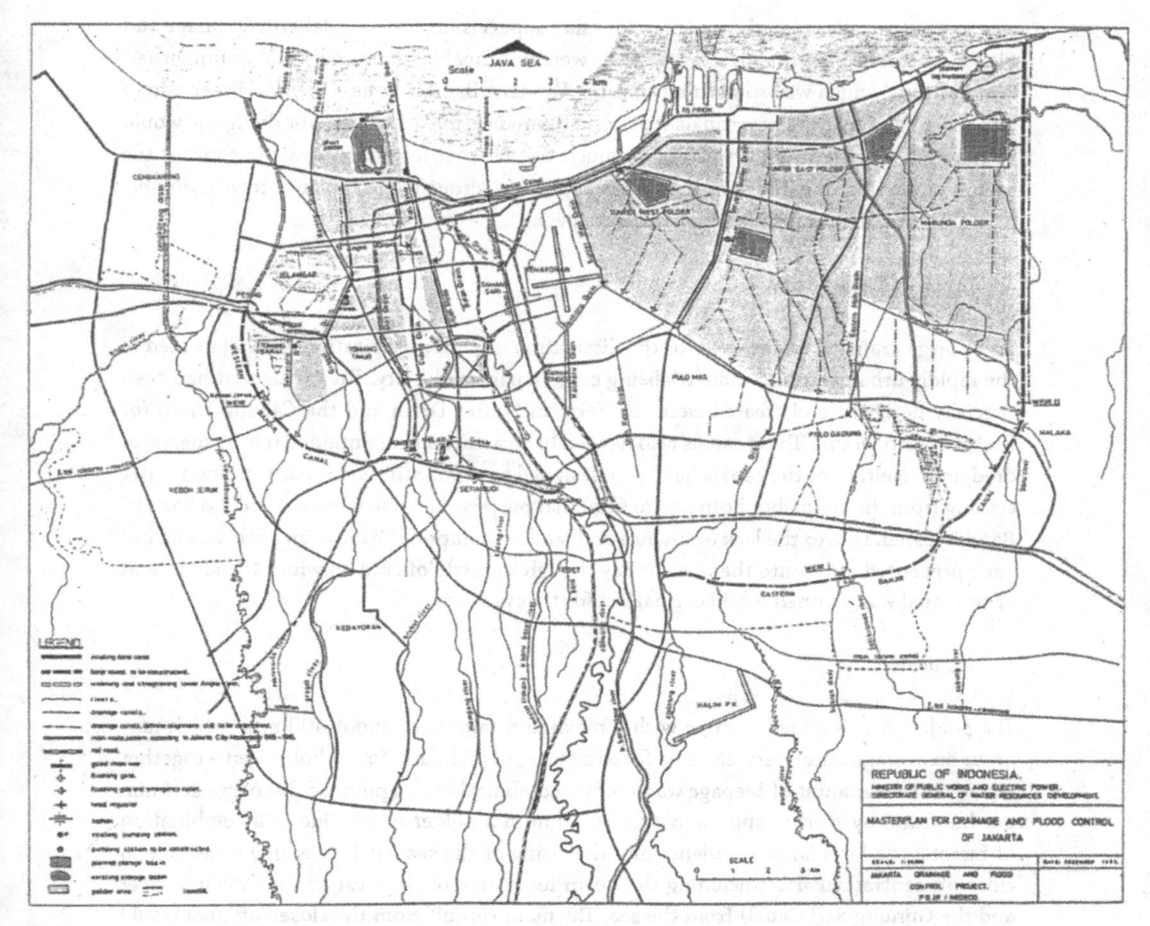

The 1973 Master Plan for the Drainage and Flood Control of Jakarta, with adaptations after the extension of the (Western) Bandjir Canal had been 'eternally postponed'.

The large bandjir canals

Speculations about land use, which arose in the 1970s, made it impossible to implement plans to extend the Western Bandjir Canal. An alternative plan was therefore chosen which, on the one hand, involved allowing water from the upper reaches of the Kali Angke (together with the Kali Pesanggrahan) and the Mookervaart canal to pass through a new drainage canal (the Cengkareng Drain) and directly out into the sea. On the other hand, it was decided that a catch canal should be excavated at a lower level to serve the area to the north of the Jakarta-Tangerang road which would also direct water coming in from the Kali Grogol and the Kali Sekretaris towards the Kali Angke. The designing of the canals and the relevant structural works and the supervising of progress was to be overseen by an Indonesian-Dutch team that was to be especially created – under Indonesian supervision – for the Jakarta Drainage and Flood Control Project. The new large canals were not capable of naturally maintaining their 'wet sections' which was something that the Western Bandjir Canal (the Van Breen canal) was able to do.[48] For the new canals special maintenance, including periodic dredging, would be required. Furthermore it was not so much the silt that formed a threat but rather the random dumping of garbage in the canals and the vigorous tropical plant growth, notably that of the water hyacinth (Eichhornia crassipes).

The large drainage canals

These large drainage canals used for the dissipation of local precipitation were all located in the rapidly urbanising and industrialising eastern part of the city. The canals that had been specially designed and created were the Terusan Sunter Drain and the Cakung Drain (or Eastern Main Drain). These canals also needed to be artificially maintained (i.e. by means of dredging). Neither of the canals had been designed to cope with the pressure of flood water coming from the south, but both had to fulfil that purpose because there was still no Eastern Bandjir Canal. Due to the lack of finance and land speculation the creation of this canal has been prevented right onto the present day though the strip of land on which the alignment of the canal was planned is still earmarked for that purpose.

The polders

The polders are designed to cope with a maximum rainfall of about 400 mm, which may occur on average once every 25 years for a duration of 96 hours (four whole days) – together with a certain amount of seepage water – by a combination of a pumped discharge at 4 mm per hour and by storing approximately 140 mm in a polder pond. Due to a combination of factors: the local land subsidence and the rising of the sea level it became necessary to close off central Jakarta (including the down town area to the west of the Ciliwung river and the Gunung Sari Canal) from the sea. The main run-off from the closed off area would henceforth be directed to the Pluit Polder area and pumped out to the sea via the existing pumping station that was due to be enlarged. This involved creating a canal that could pass through the old down town area. It was largely thanks to the vigorousness of Ali Sidikin, the governor, that this was made possible and that the old VOC buildings were preserved. A sea sluice (or floodgate) also needed to be constructed at Pasar Ikan that now dams off the navigation canal of the 'Great River' which Jan Pietersz. Coen once sailed up. Furthermore, partial pumping-stations were situated in the 'old city', all of which discharged into the Western Bandjir Canal. They consisted of a Dutch-made floating pump in the Grogol district, a pumping station with fixed pumps in the city polder of Melati to the north of the Bandjir Canal upstream of the Karet Weir and a pumping station in the Saluran Siantar with a 50 m^3/s capacity (Dutch-Indonesian design, built with Japanese financial aid),. The latter was specifically aimed at keeping dry the Merdeka Square and surrounding area, including the spot where the Presidential Palace is situated.

The Pluit polder

The Pluit polder, which has an area of 2760 ha, is home to some two million people. One of the facilities in the Pluit polder is the above-mentioned 50 m^3/s pumping station which not only serves to keep the Presidential Palace dry but also the Merdeka Square (formerly the Koningsplein, i.e. King's Square), part of the business district of Jalan Tamrin and part of the Menteng residential area by leading water away and directly into the Western Bandjir Canal. One particular aspect of concern in the Pluit polder remains the upkeep of the polder's large water storage reservoir (pond). Thanks to the large quantities of waste water – and unfortunately also the rubbish that continues to be dumped in the drainage canals – it is convenient in the dry seasons to conduct the anaerobic water around the reservoir to the discharge point (via the main pumping station). In that way the reservoir does not become unnecessarily choked with waste and sludge, it remains aerobic and it continues to simply serve its storage function when run-off and evolving drain discharges are high. In the case of the Pluit and all the other new polders this is an aspect that has, in principle, been taken into account. When the implementation of the Master Plan started the Pluit polder already had a pumping station and the pond already existed but as the pond was very polluted and silted up and since the capacity of the pumping station was too small it was obvious that works encompassing repair, adaptation and expansion had to be urgently undertaken. The Pluit pond was dredged and the two million m^3 of spoil was dumped in the sea in conjunction with land reclamation. A whole residential area has since been created on that site. The original open outlet from the polder area into the Pluit reservoir was turned into a diversion canal around the pond (capacity 16 m^3/s) leading directly to the pumping station, plus an aerated siphon (capacity 130 m^3/s [49]) which would act as an emergency outlet to the pond during discharges from the polder area higher than 16 m^3/s. An automatically operating access sluice to the diversion canal was created to make sure that when water levels are low adjacent groundwater levels do not fall too low as that would lead to the rotting away of wooden pile foundations.

The polders in East Jakarta

The creation of polders in eastern Jakarta (Sunter, Marunda), together with all the accompanying canals, sluices, storage reservoirs and pumping stations, was unable to completely keep abreast of the explosive housing and industrialisation expansion. It was often especially factories that did not wait until the poldering was complete and so just went ahead and raised their own terrains with laterite soils which sometimes hindered the draining of other low-lying terrain. Moreover, as has already been mentioned, it is unfortunately often still the case that the drainage and diversion canals in the polder regions are perpetually hampered and threatened by the vast quantities of garbage and other waste that continue to be dumped in these waterways, as elsewhere in Jakarta.

The flushing of the waterways

It is partly because waste water was (and still is) simply discharged unto open waterways that the Master Plan determined that the waterways within the city areas surrounded by

The Pluit reservoir in Jakarta: the approaching main drain, the aerated siphon and the inlet sluice leading to the diversion canal used for dry weather flow, 1982.

Illegal dumping of waste in Jakarta, around the year 1975.

the canals and (future) catch canals have to be flushed clean as often as possible. It was calculated that roughly one litre per second per hectare of drained area would be required for that purpose which is roughly equivalent to a complete refreshing (in other words replacing) each 24 hours of the entire canal water content. Coincidentally that virtually corresponds to the irrigation water volume originally required to irrigate the area. In technical terms, it is perfectly possible for the higher sections of the *bandjir* canals and for future catch canals to continually provide sufficient flushing water for the city. Sometimes, when in the east monsoon season or during dry west monsoon periods the water volumes in the supplying rivers (e.g. the Ciliwung) are insufficient, these volumes can be supplemented with water from the Tarum Barat irrigation canal. However, practically, that is to say organisationally, the implementation of such canal flushing still appears to bring with it a number of obstacles.

Indonesian-Dutch cooperation

Between 1970 and 1982 people worked together intensively in mixed teams on the planning, design and supervision of the many works being realised at that time and what was collectively

A new floodgate constructed according to Indonesian-Dutch design, which forms a part of the Jakarta Drainage and Flood Control Project, around 1985.

learned was applied. When the Dutch co-financing of the project came to an end it was the Japanese government that stepped in to perpetuate international cooperation. The skilled and meanwhile experienced Indonesian mechanical engineer M.A. Lanti was the man to take firm control of the execution and further development of the Jakarta Drainage and Flood Control Project after the period of Indonesian-Dutch cooperation had come to an end.

CHAPTER 10 **INDIGENOUS OR INTERNATIONAL**

The evolution and significance of East Indian civil engineering

MAURITS ERTSEN

Rice fields in Lampung, 2003.

> *For the Netherlands the East Indies was a strange world. If the colonial project was to be a success it would be necessary to amass much information and accumulate a great amount of knowledge. It was always primarily knowledge that was of direct importance to policy, either on the part of the government or on the part of large organisations and institutions: intelligence. The volume of such information accumulated and processed on the East Indies was extensive. As everything there was literally strange, sometimes profitable or sometimes threatening and since the indigenous community proffered little or no useful information it was necessary for the coloniser to build up an entire system of observation, analysis and knowledge dissemination […]. A whole range of tropical science was to emanate from that such as: tropical medicine and health care, East Indian or tropical economy, East Indian law, Islamics, Indonesian language studies and archaeology, botany and volcanology.*[1]

In his book *De laatste eeuw van Indië* (The last century of the East Indies), Van Doorn does not mention in his summing up of such sciences any of the scientific matters dealt with in this book, namely civil engineering. As the previous chapters have variously demonstrated, civil engineering played a very important part in the development of the East Indian colony. Various aspects of the large conglomeration of civil public works have been examined in this book. We have been taken along roads and railways that traversed difficult mountainous regions, we have viewed works designed to control and utilise the ever-present water and finally we have seen the buildings and cities created by Dutch engineers. All the different structural works were realised by engineers who made use of their civil engineering knowledge. The question dealt with in this chapter is whether or not the civil engineering works developed in the East Indies can be characterised as 'tropical science'. This chapter reveals that to a large degree that is indeed the case.[2] We shall also try to discover what traces of the civil engineering knowledge and experience established in the colony are echoed in the Netherlands, other parts of the world and in modern-day Indonesia. As the subject is rather extensive it obviously needs to be somewhat restricted. The analysis will confine itself to two representative areas in the wet domain, namely irrigation and hydropower, domains that have become the most important professional fields of action of Dutch engineering.[3]

A laboratory?

Even though there are distinct differences from subject to subject, the general impression emerging from the contributions in this book is that Dutch civil engineers were really quite successful when it came to realising public works in the East Indies despite the difficult or, to say the least, very different circumstances compared to in the mother country. The colony was different in terms of its natural aspects, geography, distances, population et cetera. The engineers' technical baggage, based on knowledge accumulated in the Netherlands thus did not, to say the very least, allow for straightforward extrapolation. This was something that applied particularly to the fields of irrigation, road and railway construction, and bridge building. Compared to the Netherlands, where obviously progress had not stood still either, a need for new knowledge to be developed emerged which then, before being applied, had to be processed and adapted to local needs and circumstances and further developed where necessary. In the East Indies the technical challenges were thus definitely different from in the Netherlands and in some ways greater. In that respect the comparison drawn with British India is quite telling:

> *Service in the challenging environment of India was almost a traumatic experience for British officers of intelligence and ability. For some of them it sharpened their appreciation of engineering techniques and undoubtedly influenced their attitude when they returned to Britain again.*[4]

Armytage even went as far as to coin the term 'Indian laboratory'. Technology plays an important part in the laboratory metaphor. Van Doorn in particular conceives of technology as a fully developed concept that is imported, applied and launched as an externally originated project. The examples he gives include military tactics, railway networks and plantation economy, all of which are tried and tested concepts introduced by experts educated elsewhere and backed with the necessary capital.[5] It is not self-evident that such a view of matters in which technology is seen as something 'diffused' from elsewhere can be

Lampung: irrigation outlet structure, 2003.

applied to all the engineering facets dealt with in this book. Conversely one might assert that Dutch engineers went to the colony with Dutch knowledge and returned to the Netherlands with East Indian knowledge. That is the impression which, for instance, strongly comes across from the chapter on irrigation and hydropower. It is probable that in certain areas of expertise engineers developed a distinct East Indian approach whilst in others they did not.

What, then, is an East Indian approach? Indeed, is it in fact possible to speak of a Dutch-East Indian type of civil engineering? To answer these questions we shall look especially at the education that the different engineers obtained and we shall see that there is good reason to answer these questions in the affirmative, at least where a number of essential technical disciplines are concerned. On the basis of the same questions the epilogue will further consider technology as such. All the knowledge and skills available in the field is laid down and ultimately re-emerges in the form of educational programmes. We shall therefore establish whether there was an East Indian influence in technical education in the Netherlands and in the East Indies and, if so, how that was manifested. Eventually Dutch engineers were able to apply East Indian technology or more precisely: could give shape to civil engineering works with knowledge and skills derived from East Indian experiences. Before that they only possessed Dutch knowledge and technology gained during the course of their Dutch technical education. Later educational programs were developed with a special East Indian focus, all of which was to culminate in the establishment of the Bandoeng Technical College, an educational institute devoted primarily to training engineers for the East Indian colony.

Educating for the colony

The first training especially for the colony, given at the beginning of the nineteenth century, was not intended for engineers at all. At that time they formed neither a large nor influential professional group. After Dutch administration had once again been imposed in the East Indian archipelago in 1815 and after that same administration had been considerably centralised and expanded, a need arose for more civil servants. Knowledge of the local language, or at least of Javanese, was considered to be extremely important. In 1818 six civil servants were seconded to the military college in Semarang in order to be educated in Javanese. It seemed that the British example was being followed. There, already in 1806, the East Indian College had been established to provide academic training for future civil servants in India. In the East Indies it took somewhat longer for all the plans to materialise partly, for instance, because of the Java war and the subsequent shortage of funds. Eventually, though, training for civil servants was set up, not at Leiden University (although that had frequently been suggested), but in the East Indies itself at the Javanese Language Institute in Soerakarta, Java. The program commenced with twelve participants who were taught all about the Javanese language, history, laws and national institutions.[6]

It was not long before people started to criticise the curriculum: the results were thought to be poor and it was claimed that the courses were too theoretical. At the same time, there was growing opinion in the Netherlands that the controlling of both the colony and the education for the various offices should be based in the Netherlands. In 1984 the Minister for the Colonies, J.C. Baud, decreed that the higher administrative positions should be reserved for individuals who had been educated in the Netherlands. He maintained that the control of the Cultivation System could only be guaranteed if it was retained in white (i.e. Dutch) hands.[7] The monopoly for the training of civil servants was to be given to the academy that was to be established in Delft. Only a certificate gained from that institute would give people access to the higher administrative echelons. The Royal Academy was established in Delft on 8th January 1842 by Royal decree. Civil engineers could also be educated at that academy. The East Indian elite tried to retain their hold on the higher administrative positions and one of the ways in which they did that was by sending male family members to the Netherlands to be educated. Of the 305 students who passed through the academy in Delft between 1842 and 1857, 195 of them were from the East Indies whilst a mere 110 had been born in Europe.[8]

The academy gave students the opportunity to complete five different four-year engineering programs and to graduate as hydraulic engineers, mining engineers (the only specific program oriented to the East Indies), architectural engineers, mechanical engineers or commercial engineers. Incidentally, originally the title 'engineer' could only be given to people who had graduated in military construction and who, since 1600, had been trained in fortress-construction at an institute established by Prince Maurits in Leiden. The influence of that institute was clearly evident within the exact sciences. For instance, in the Dutch Higher Education Act of 1814 'the application of mathematics to hydrodynamics and hydraulic engineering' was still viewed as something that belonged within the mathematics and physics faculties. It was in 1789 that the government decided to establish artillery schools. One such school was established in Amersfoort in 1805. In 1814 an Artillery and Military school was set up in Delft, which moved in 1829 to the Royal Military Academy in Breda. The curricula, both in Delft and Breda, provided training for artillery and military officers, for the pontoneers and sappers battalion and for Public Works engineers. In that one type of institute, military and technical education programs were thus united.[9]

Civil engineers

The division between military and civil technical education came in 1842 with the establishment of the Royal Academy in Delft.[10] The academy fitted into a whole series of such educational institutes that had already been established, chiefly in Germany, like for instance the polytechnic colleges in Karlsruhe (1825), Munich (1827), Dresden (1828), Stuttgart (1829), Hannover (1831) and Darmstadt (1836).[11] On January 4th 1843 the Royal Academy, situated in the buildings of the former Artillery and Military School at the address known as Oude Delft 95, was ceremoniously inaugurated.[12] The move of the military part of the college to Breda at least gave a number of lecturers the incentive to commit their knowledge to paper because military academy lecturers were obliged to support all the courses they gave with textbooks. One such lecturer, D.J. Storm Buysing, was the person to produce one of the first standard works published in the Netherlands on hydraulic engineering.[13]

During the first year there were 46 students who registered at the academy in Delft, 10 of whom were destined for administrative service in the East Indies and 36 of whom chose the General Studies department (which included all the technical studies); by 1843 there were 117 students, 142 in 1844 and 170 in 1845.[14] It was civil engineering that drew the most interest in those early years. In 1864 it emerged that of the 207 to graduate in the technical department, 183 had selected that specialisation.[15] Despite the Delft academy's relative success it was subjected to much criticism. There was a perpetual shortage of funds, the discipline was too military (not surprising in view of its history) and the standard of lecturing was not always viewed as being particularly high.[16]

By 1848 the Royal Academy's future had already been hanging in the balance for a while because the National State provided it with too little funding. Nevertheless, the academy provided educational programs for engineers and civil servants. In the first two years of the programs there was a high degree of overlap between both factions. Future civil servants also had to be schooled in technical subjects. In the Cultivation System the civil servant was not so much an administrator as a production manager. Engineering works were also executed under the supervision of civil servants who thus required a degree of technical know-how. When the various programs were divided up in the mid-sixties the administrative program became more oriented towards 'real' administrative matters; the emphasis came to lie more on legal subjects and Indonesian languages. In the wake of the new education act, effective as of 30th June 1864, the academy was closed down and replaced by the Polytechnic College. The monopoly of Delft on educating civil servants ended and administrative programs were initiated in Delft, Leiden and Batavia.[17]

The above-mentioned department of General Studies became the Polytechnical School, an educational institute with a four-year program for technicians. The introduction of that new program also marked the disappearance of the military character and a greater degree of specialisation became apparent. Some six new programs were set up and hydraulic engineering fell under the civil engineering section. It was the first time that the title 'civil engineer' had been reserved for a profession which previously had been labelled 'public works engineer'. Until 1905 the Polytechnic College fell under the secondary education sector but in that year it was turned into a Technical College and all the education provided there gained an academic status and was fitted into a five-year programme. In that way technical higher education had been placed on a par with university education so that the profession of engineer suddenly gained academic status. This gaining of academic status was also something that was occurring on an international scale. After 1900 in most engineering fields the empirical approaches were slowly making way for theoretical approaches or for applications of mathematical-physical laws to technical problems.[18] This move was, however, gradual because:

> *In the Education Report of 1934 [brought out by the Delft Technical College] (p. 314 onwards) a long explanation is still included in which it is remarked that in civil engineering and architecture it should no longer be the custom for lecturers to base their lectures only on experience but also and more specifically on scientific research.*[19]

Dutch technology in the East Indies

East Indian Public Works was clearly affiliated to Delft education. In 1874 it became compulsory for people seeking a position in the department to possess a civil engineering degree obtained from the Delft Polytechnic. After the mid-nineteenth century an average of 24 to 30 percent of all Delft graduates went on to seek work in the East Indies. As has been explained in previous chapters, it was increasingly engineers who were being deployed in the East Indies. A certain percentage of them did later return to Delft. For instance, N.H. Henket, later to become professor of hydraulic engineering at the Polytechnic was one such person. He was involved in constructing railways and tramways in the East Indies. Later (1895-1897) the same Professor Henket was also to become interim director at the Polytechnic. It was not long before the East Indian demand for Delft-educated engineers started to exceed the availability.[20]

Those engineers who went out to the East Indies had a very strong Dutch-oriented background. Van Doorn pointed out that such engineers displayed a lack of experience with and knowledge of the East Indian situation, which sometimes led people to make significant

The Bandoeng Technical College in West Java: the front view 1920.

errors of judgement (as illustrated, for example, by the Sampean works). Dirkzwager reports on the legacy of the Dutch engineers who went out to the East Indies with all their knowledge about the construction of dykes, polders, shipping canals and locks, mentioning that in Demak (near Semarang in Java) until deep into the twentieth century it was still possible to find locks with mitre gates waiting for ships that never came. Though that was a phenomenon that was especially apparent in the first half of the nineteenth century, it should also be noted that in the second half of the nineteenth century the Polytechnic paid no systematic attention to 'the land and people of the colonies, to the East Indian soil, to tropical building materials and building methods, to implements and machinery or to transport means'.[21]

In 1908 it was engineer P. Grinwis Plaat who became the first extraordinary professor to hold a chair in Delft in the field of hydraulic engineering specifically oriented towards the East Indian colony (and predominantly towards irrigation). [22] This relatively late appointment is explained by Van Doorn as being attributable to the tendency in Delft to pay ever less attention to the creation of a work circle in the East Indies. In making the program more academic the attention paid to professional preparation had become less pronounced. Apart from this phenomenon, irrigation formed a discipline in which Dutch hydraulic engineering had 'undergone a rejuvenation process', in other words, a new terrain which, prior to 1900, Dutch hydraulic engineers had little or no experience of and a field which, at the time of the starting up of the Technical College, was not officially in the curriculum.[23] As was illustrated in Chapter 6, it was not until the end of the nineteenth century that a more systematic approach to irrigation construction was to emerge. Engineers succeeding Grinwis Plaat were A.G. Lamminga and C.W. Weijs, another founder of Dutch irrigation technology in the East Indies. Weijs was to become the first full professor in Delft to be in charge of hydraulic engineering, sewerage systems and water supply. Hydropower was another field to later

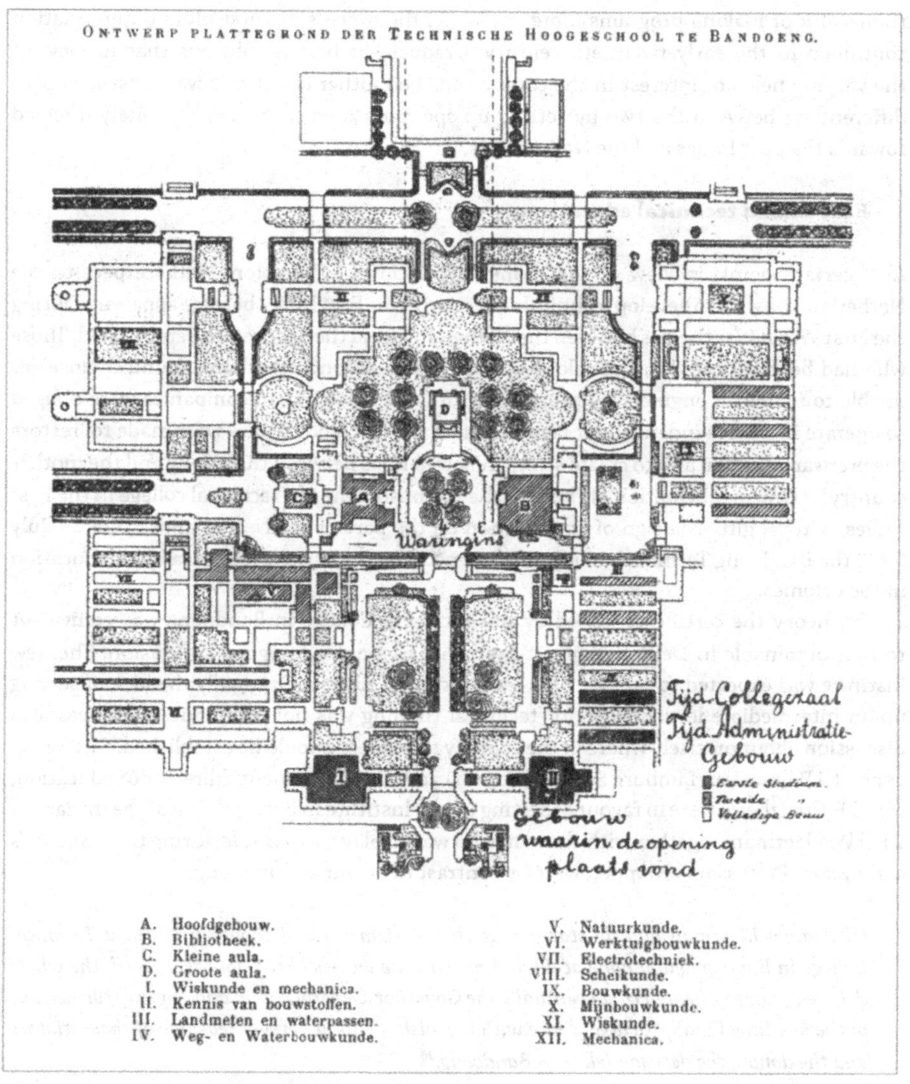

The Bandoeng Technical College in West Java: the ground plan.

permeate into programs offered at the Technical College in Delft. The first professor in that particular field was engineer G.H. van Mourik Broekman who was appointed in 1924. He had gained experience in the field in Chile.

Around the year 1900 there were not yet all that many specialist areas pertaining to the field of civil engineering. The college produced broadly educated civil engineers who were thought to be suited to a wide variety of areas. Parallel to, but more probably within the framework of making programs more academic, the process of curriculum differentiation continued in the early twentieth century. Gradually it became obvious that in view of the varying fields of interest in the colony and the mother country it was most logical to differentiate between the two by setting up specialisations that were separately directed towards the East Indies and the Netherlands.

East Indian technical education

At a certain point, in view of the changing position of the colony with respect to the Netherlands, a wish to develop technical education in the East Indies became apparent. During the First World War the ties between the Netherlands and the East Indies deteriorated. Those who had been educated in the colony at East Indian secondary schools found themselves unable to follow an engineering programme in Delft. East Indian companies were obliged to operate in a more independent fashion. After the war, endeavours were made to 'restore the pre-war situation and to create strong renewed links between the colony and the mother country'.[24] One manifestation of this was the establishment of a technical college in the East Indies, a 'royal gift' as a sign of appreciation on the part of the Netherlands.[25] On 3rd July 1920 the Bandoeng Technical College was opened, the first institute for academic education in the colonies.

In theory the certificate issued by the Technical College in Bandoeng was equivalent to that obtainable in Delft though, at first, that was not a foregone conclusion. The new institute was expected to provide high-level technical tuition. Originally the idea of setting up an intermediate level college for technical training was not ruled out. There was also discussion about precisely where in the country the college should be established. Governor-general J.P. graaf van Limburg Stirum (a count) and the government's director of education, Mr. C.F. Creuzberg, were in favour of setting up the institute in Batavia. One of the initiators, Dr. J.W. IJzerman, together with the man who was to play a major role during the institute's early years, Professor J. Klopper, were by contrast in favour of Bandoeng:

> *Ultimately IJzerman bluntly stated that the gentlemen could choose between a Technical College in Bandoeng and none at all, in line with the wishes of the donors. Indeed, the whole debate became so heated that eventually the Governor-General got up and angrily stormed out of the meeting. Finally, in accordance with the wishes – for whatever reason – of the initiators and the donors the decision fell with Bandoeng.*[26]

A curriculum was drawn up by an education committee. Klopper, who was at the time professor of applied mathematics and mechanics in Delft, played an important part in the entire process. Unlike in Delft, where there was at least a little leeway for free choice, it was compulsory to follow the entire study programme in Bandoeng. The only freedom that the committee could conceive of was that of perhaps allowing the students who were further in their studies to focus in technical drawing on the subjects that interested them most: 'Such freedom as this will more probably be advantageous rather than disadvantageous to the study'.[27]

In view of the fact that the Bandoeng programme would be a year shorter than the equivalent programme offered in Delft (four as opposed to five years), the committee maintained that concessions should especially be made where mathematics and general formation were concerned. It would be impossible to cut back elsewhere if the courses in Delft and Bandoeng were to remain convincingly equivalent. The number of hours and the number of lectures given in Bandoeng would not have to be less than in Delft but, obviously, with a programme that was a year shorter that was something that was difficult to realise. Where mathematics was concerned an exception to the rule was made on the grounds that it was:

> *generally believed that in this field at the Technical College in Delft demands are made that are higher than would be absolutely required for an engineer to be entirely fit for the practical exercising of his profession.*[28]

The mere fact that the mathematics courses given at the Technical College in Delft were so advanced was (and today still remains) one of the main arguments justifying the institute's academic status. Apparently the committee did not maintain that the programmes provided in Bandoeng could really classify as proper higher education courses, a theory that corresponded with Van Doorn's conclusion that the colony did not have a true academic climate.[29] Anyway it was, apparently, acceptable to tolerate restrictions in the mathematics programme and so the committee asked Klopper how this compacting of the programme could best be effected. Klopper indicated that there were realistic ways of reducing the amount of mathematics taught but he even went a step further. He proposed that most of the theoretical subjects, including much mathematics and mechanics, should not be given in the first years but rather in the fourth year. In that way the students with less theoretical capacity could be released into society after three years as intermediate level technicians so that the more talented could study for an extra year towards the title of engineer.[30] Klopper's conclusion was that for many students in Delft it was precisely mathematics that formed a stumbling block:

> *they struggle on, time and again resitting examinations but displaying little real progress until finally they pass. The examiners do not really know how to cope with the situation and so in the end they just give in and award such students their final certificate.*[31]

Klopper maintained that such practices should not be allowed in the East Indies in view of the fact that there individuals who failed would prove to be 'an even greater burden to society'.[32]

Right from the very start the new technical college was a resounding success. Its first faculty was the civil engineering faculty. It was not long before competition with students in Delft became apparent and so in the mid-thirties a studies committee concluded that the options for engineers originating from Delft to work in the colony had been reduced to the very minimum. Just before the Second World War the Technical Colleges of Delft and Bandoeng were placed on an even footing both regarding personnel matters and certificate recognition and the right to offer doctorate degree courses. The establishment of the Bandoeng Technical College marked the beginning of a period in which civil engineering in the East Indies began to develop an own identity. A number of laboratories and experimental stations were set up, top quality articles were written and several standard works appeared pertaining to a diversity of technical fields.[33]

The Bandoeng Technical College was devoted to educating engineers for the East Indian colony. Graduates were, of course, free to go and seek employment elsewhere in other (colonial) areas but the East Indies remained the frame of reference. The Bandoeng course descriptions were directly comparable to those drawn up for the Technical College in Delft which was, at that time, the only place in the Netherlands that provided higher education courses in civil engineering. Even though they are quite brief these course descriptions provide us with an excellent idea of what made East Indian public works 'typically East Indian'. Together with the descriptions that were adapted to East Indian circumstances the overview included in Appendix 7 provides an integral impression of the guidelines drawn up by the relevant sub-committee concerning the contents of the various courses. It is a diverse view that emerges from the summing up of the various courses. In the case of courses such as road construction and geodesy it was emphasised that extra attention should be devoted to geographical circumstances so that a closer, more East Indian slant, could be given to the subject.

With subjects such as business management and geodesy what becomes clear is that the East Indian engineer, much more so than his Dutch counterpart, had to be able to actually execute the works designed by himself. Underlying the attention paid to hygiene and building materials was not only the fact that these were important matters but also the issue that the colony consisted of relatively virgin territory. By contrast, for administrative law the latter argument provided grounds for paying less attention to precisely these disciplines and to even abolishing a specific course in the field of urban planning, especially when no suitable lecturers were to be found. It seems that East Indian hydraulic engineering had become a relatively independent field of study as a new component within the general hydraulic engineering field which, originally, had only been directed towards the Netherlands. Similarly architecture could also be seen as a relatively independent area, a point clearly borne out by the developments seen in Batavia. It was through the intervention of Daendels (at the beginning of the nineteenth century) that the further expansion of the colonial capital continued in an inland direction so that a new city district known as Weltevreden was created on higher-lying ground. This was all because of the unhealthy living conditions in the original settlement. If (old) Batavia was 'a typical Dutch city', at least from the point of view of layout, then Weltevreden was 'completely East Indian'.[34] Even Menteng, the European residential area developed in Batavia during the first half of the twentieth century, had a typically Dutch-East Indian flavour, both from the point of view of layout and architecture.[35]

East Indian technology?

In his inaugural speech as professor of hydraulic engineering given in Delft in 1919, Haringhuizen discussed to what extent East Indian hydraulic engineering could be distinguished from Dutch hydraulic engineering. Haringhuizen who had worked in the East Indies as an irrigation engineer for 20 years (for six of those years under Lamminga's supervision) and who focused exclusively on engineering in the hydraulic domain, maintained that the differences were especially to be sought in:

> *situation, soil conditions and climate but furthermore in the available materials and manpower and – to an extent – in national character as well as national political and social institutions.*
> *A number of those factors have an effect on the types of works produced, the intended goal and the purpose to be fulfilled. Other factors relate more to the construction of works and to the way in which they are prepared, executed and maintained.*[36]

One important discrepancy between the Netherlands and the East Indies was the function fulfilled by the various works if one bears in mind that in the Netherlands it was predominantly all about works being designed to reduce water hindrance whilst in the East Indies water provision works were essential too. A striking difference was the fact that in the East Indies structural works were not terribly important for inland waterway shipping in the form of proa vessels and bamboo rafts. Java also had very few shipping canals. There were, however, two very famous canals known as the Mookervaart (in Batavia) and the Prauwvaart canal (in Semarang) but they were more precisely proof of the fact that such works no longer had priority as soon as East Indian knowledge and experience increased.[37] Another big difference when compared to civil engineering works in the Netherlands was the fact that most East Indian public works projects were executed under the direct control of the government institutions concerned. That meant that East Indian engineers were not only responsible for design but also for the organising of all aspects of the job itself, from the recruiting of labourers to the organising of accommodation for all concerned. Even though generally speaking in the engineering profession such management skills are important, very emphatic demands were placed upon these skills in the East Indies.[38]

For certain elements or aspects of technology a process of diffusion may be said to apply. Various ready-to-use technologies were imported, like for instance with the railways where the locomotives tended to be shipped into the colony. In the chapter devoted to the railways it was made clear just how relatively interchangeable the nature and knowledge of individuals affiliated to the railway community really was. Components of the technological range of parts used and implemented in hydroelectric power development, like electrical equipment, also fell into the diffusion category. What has not been discussed at all in this book is the sugar refining industry. That is also an area that provides good examples of diffusion processes.[39] Generally, therefore, it was all about processes and equipment that were more or less immediately deployable and which – at the very most – required different implementation because of the natural circumstances existing in Java or elsewhere in the archipelago and it was about machinery that would be more rapidly written off or which could not operate at full power. One need only think, for example, of the problems that certain steep slopes posed to locomotives.

All in all, it may be asserted that the picture sketched by Van Doorn of colonial technological development is too one-sided. Van Doorn is not the only one to believe that diffusion constituted the main stimulus for colonial technological and scientific development. What the diffusion model implies is that science and technology developed independently of political circumstances.[40] The diffusion of blueprints must therefore be the mechanism behind dispersal and development if one bears in mind that no colonial stamp could be placed on the work in hand. By contrast, it could be asserted that there is definitely evidence of a colonial influence upon science and technology. It was precisely in colonial regions that there was a convergence (or sometimes a clash) between European and indigenous circumstances and views.

A study by Van Berkel supports this argument.[41] Though he conceives that the difference between Dutch and colonial natural sciences was especially to be sought in the supplementary character of the latter type, he bases his study of Dutch natural sciences on the notion that, also within the pure sciences, there is evidence of traditional tendencies. It is especially the way in which scientific research is organised that has a determining effect upon such traditions. Generally speaking the pure sciences derived particularly from and were developed through existing practices, invariably technical or medical, in areas where yet other scientists were active in the applied sense. The development of knowledge and technology in the colonies also evolved within a specific administrative system, namely the colonial project of Van Doorn. If one views the colonial project as a whole in that particular light one finds that what was noticeable about the Dutch colonialism variant, at least in the East Indies, was the detailed degree of control and management. That, at least, is what foreign experts found particularly striking about it.

Regarding these tendencies we see evidence of such development in East Indian hydraulic engineering, especially where irrigation was concerned. As emerged in Chapter 6, the engineers first occupied themselves with major works, the construction of which did not at first sight seem very different from similar works in the Netherlands. With such works it was not unusual to find Dutch building methods and materials that came from the Netherlands like, for example, bricks that were transported cheaply in the hulls of sailing boats as ballast. 'Before very long pure East Indian techniques were developed which involved using materials available in the East Indies and constructions that matched the materials'.[42] In his reflections Mazure reports on developments in the field of civil engineering and on how, after getting off to a hesitant start, the dimensions and perfection of irrigation works in Java soon began to rapidly increase. From the international angle it was particularly the design and exploitation of the distribution works that was impressive, which was not surprising in a colony where control was an important factor.[43] What was important in this context was the establishment of hydrodynamic laboratories in Bandoeng and Semarang in the nineteen-twenties, which focused on the designing of irrigation and other water harnessing works by creating hydraulic models for waterways and structural works. Alongside the mobile broad-crested Romijn measuring weir the designs emanating from the laboratories that became especially well-known were the hydraulic flap gates (such as the Vlugter and Begemann gates) for upstream control. Professor Vlugter who was active in Semarang also came up with design rules for stilling basins at the downstream side of weirs and sluices.[44]

A rift

The Second World War and the independence of Indonesia represented a rift in the development of public works constructed by Dutch engineers. For Dutch engineers there was no longer that much to organise and to manage in Indonesia. Already in the nineteen-thirties, a time of stringent economic measures, the colonial demand for engineering graduates had decreased. In the late forties and early fifties it was evident that Dutch engineers who, before the war, had worked in the colony in large numbers, had lost much of their work terrain.

Future-oriented discussions were also taking place in the Agricultural College in Wageningen which, like Delft, was like a 'court supplier' when it came to colonial engineers. The brand new irrigation professor in Wageningen, W.F. Eysvoogel, reported that in the 1946-1950 period there were many who believed that because the East Indies had been lost all the tropical-based courses could be given up.[45] In many different fields of study, however, it remained possible to specialise in tropical subjects. The tropical land and water management curriculum emerged from a combination of a new field known as land and water management and from irrigation-oriented subjects derived from the old colonial agriculture course, the remaining component of which continued as a subject known as tropical crop cultivation. Similar kinds of discussions were also continuing in Delft where the main concern centred on the future of Delft graduates.

> *After the war there was nobody in Delft who really considered the East Indian need for engineers in the ensuing reconstruction period or the destiny of the Bandoeng sister institute; at least nobody is recorded as having aired concerns. In objective terms there may not have been much call for concern.*[46]

After the war, the East Indies/Indonesia did have a place in the Delft curriculum but only to a limited extent. In 1930 there were two main courses at the Delft Technical College, one in Dutch hydraulic engineering and one in East Indian hydraulic engineering. The core subjects of the two courses varied slightly and both provided leeway for several optional programme components. By 1955 there were no less than seven graduation fields in the domain of civil engineering: general hydraulic engineering, polders, irrigation and hydropower, bridges and roads, commercial and industrial building, sanitary engineering and theoretical sciences. Of the seven different graduation options there was one that had pursued the East Indian studies tradition and that was the irrigation and hydropower department, the 'non-indigenous branches of practice for a civil engineer'.[47] Between 1938 and 1954 it was S.H.A. Begemann (of the Begemann gate) who occupied the irrigation and hydropower chair. From 1954 onwards the chair was divided into two separate chairs. It was F.M.C. Berkhout who became the irrigation professor in 1954 and A.R.H. Brouwer who obtained the hydropower professorship in 1955.[48]

In his inaugural speech Brouwer stated that it was remarkable that hydropower engineering had been part of the curriculum at the Technical College in Delft since 1924 in view of the fact that 'in the Netherlands the topographical circumstances do not generally lend themselves to the exploitation of hydropower works'.[49] He also sketched how, with the loss of the East Indian colony, the general situation had changed. Though he estimated that in Surinam and New Guinea the need for hydropower would arise or increase, it would not be sufficient to constitute a complete work terrain. Brouwer therefore maintained that

Dutch experts should start to branch out and play a part elsewhere in the world since

> *everywhere in the world but especially also in areas that are not yet sufficiently developed there is a noticeable incentive to [...] make use of national resources, one of which is definitely hydroelectric power.*[50]

There was, however, one obstacle: engineers who had not yet graduated also needed to be given the opportunity to gain experience. In the circumstances of the day that was something extremely difficult to realise. Within the new policies set out by the Dutch government, though, Brouwer saw some possibilities and openings:

> *In this context I would like to remind you that the Dutch Government is keen to offer services abroad and in less developed areas, also in the field of engineering, both in the form of individual services and via activities undertaken by different engineering firms.*[51]

On the grounds of this open reference to the post-war development aid programmes, Brouwer then went on to lobby for 'special measures' so that those who had recently graduated could be helped to gain the work experience they so badly needed. He therefore suggested that 'promising engineers should be given the opportunity to gain experience in the relevant countries and that, to that end, the Government should make certain means available'[52] or, to put it in present-day terms: it was proposed that an expert-assistant programme should be initiated. Indeed, the rapid rise of development aid triggered a boom in the technical activities set up by the Dutch abroad. Many of the projects in question were carried out by Dutch companies which in 1951 had united, for that specific goal, under the umbrella organisation known as NEDECO. Also organisations affiliated to the United Nations, such as the Food and Agriculture Organisation, the World Bank and the Asian Development Bank, searched for experts who could give shape to various projects. In conjunction with the whole development aid programme, the Ministry of Foreign Affairs initiated numerous projects in countries such as Bangladesh, Indonesia, India, Pakistan, Nigeria and Columbia. The geographical scope of the various projects was therefore no longer confined to Asia but extended to Africa, South and Central America and even to Southern Europe. Literally hundreds of junior and senior experts, including many who had worked in the former East Indies, were given the opportunity to gain practical experience and to apply their knowledge.[53]

Internationalisation

This internationalisation drive fitted into a longer, existing international tendency on the part of Dutch engineering. Even before 1250, notably in North Germany, but also in central East England use was made of Dutch knowledge in the field of water management and Dutch colonists were also active. Between 1250 and around 1600 the Dutch went to many areas that now lie within Polish perimeters. From the end of the sixteenth century it was no longer groups of colonists but increasingly individual hydraulic engineering experts who were active abroad. Van de Ven dubbed them the first 'firms' to provide worldwide

Begemann flaps in the Kano region (Nigeria), c.1974.

advice on matters pertaining to hydraulic engineering.[54] Here are just some examples of Dutch hydraulic engineers who took the step of sharing their acquired knowledge with people abroad: Michiel Schmidt (construction of a sluice near Berlin in 1657), Humphry Bradley (died in 1625, responsible for the construction of a canal linking the rivers Seine and Saône) and Cornelis Janszoon Meyer (active between circa 1670 and 1680) who, in 1669, wrote a book about his experiences in Italy. Even in the nineteenth century Dutch hydraulic knowledge and technology remained an export product. For instance the famous French engineer, Ferdinand de Lesseps, attributed great importance to all the Dutch engineers who cooperated in activities such as the Suez and Panama Canal projects.[55]

Apart from focusing solely on the Dutch East Indies, Dutch civil engineers were also active in other parts of Asia. In China Dutch engineers put their energy into dyke reinforcement projects along the Yangtze and the Hoang He rivers.[56] More recently, within the framework of 400 years of relations between Japan and the Netherlands, the names were revived of all the hydraulic engineers who are still revered in Japan, such as Johannes de Rijke, Cornelis Johannes van Doorn and Anthonie T.L. Rouwenhorst Mulder, for all they did in connection with the taming of rivers, the constructing of irrigation canals and the creating of polders as well as harbour construction and improvement works.[57]

Johannes de Rijke made use of knowledge gained in the East Indies.[58] The Dutch East Indies did not only manifest itself as a rich source of knowledge in the field but also as a repository for engineers who could be simply uprooted and deployed elsewhere. One such engineer was J. Homan van der Heide who worked in Siam, or present-day Thailand, at the beginning of the twentieth century and who became the first director of the Royal Thai Irrigation department before finally returning to the East Indies.[59] Another Dutch engineer to become internationally famous in the field of hydraulic engineering and who was firmly rooted in the East Indies was Van Blommestein who was referred to in Chapters 6 and 9.

Just like Van Blommestein there were many other Dutch experts who became active all over the world. In order to do their work properly they made use of all the civil engineering knowledge gained either in Delft or in Bandoeng. In the 1970s, for example, NEDECO engineers were busy in Nigeria where they were involved in the Kano project in the north of the country where a reservoir in the River Kano and a 24,000 ha irrigation system was realised. The preliminary research and design of the system became a NEDECO responsibility via the Haskoning engineering bureau. One of the most important design aspects had to do with the way in which water was distributed and the actual physical design of such distribution. The region demonstrated great variation in soil types and in the kinds of crops grown, also within the tertiary irrigation sections, which meant that over the course of time the water demand within the area varied tremendously. The irrigation system had to be geared to coping with variability (just like in the East Indies), without requiring a massive organisational structure (of the sort which had been available in the East Indies). The final choice was made on the basis of a test carried out in an experimental station where three potential distribution works were contemplated.

The three combinations that were tested out for use in the secondary canals were: (1) a so-called meter gate with a permanent weir in the secondary canal, (2) a Crump weir with an adjustable weir with beams in the secondary canal and (3) an orifice with a hydraulically controlled flap gate of the Begemann type.[60] It was the third option – a Dutch-East Indian invention – that best met the demands and was thus chosen for the Kano project. It was Vlugter, the engineer referred to above, who carried out tests on the Begemann gate system in the 1930s in the Hydrodynamics Laboratory in Semarang and it was the design nomograms emanating from that system that were afterwards used in the designing of the Kano project system.[61] A World Bank publication listing successful examples of simple but effective water management methods cited this one as an example of a top quality solution.[62] In later more refined forms the Begemann gates also found their way into irrigation districts in California.[63]

Even in the Netherlands itself there is a traceable comparable legacy of colonial artefacts to be found. The Hobrad weir is, in effect, a modified Romijn weir enabling water supplied in open canals to be measured and regulated. The history of the Hobrad weir starts in the 1970s in the Amsterdam Waterworks raw water abstraction plant 'Vogelenzang'. There an overflow weir was required which could regulate and measure discharges varying between 0.1 and 0.6 m^3/s in such a way that the required head loss would not be more than several decimetres, just like the requirements that were laid down for the hydraulic control of irrigation works in the flat plains of the Dutch East Indies. As there were a number of slight imperfections in the original Romijn weir certain adjustments were made. The sloping gradient of the Romijn weir top, for instance, was 1 : 25 and upwards but it was reduced to zero.[64]

In conjunction with development aid Dutch engineers also returned to Indonesia after 1968. At the end of the 1960s, when relations between the Netherlands and Indonesia had

The Crump-De Gruyter overflow weir in Lampung, 2003.

been restored, a major irrigation rehabilitation programme was started up. Use was made of East Indian civil engineering know-how although in certain respects, ironically enough, irrigation history began to repeat itself. At first, attention was only paid to structural reconstruction and the building of new facilities, but not to management. The fact that this latter facet also required attention was something that soon became all too evident. There was a large number of systems that had not been rehabilitated whilst the necessary improvements in water distribution could well have been made if only management measures had been taken. In many irrigation areas there was furthermore evidence of widely varying circumstances because of the growing population which thus instigated a changing use of land. In a large number of cases it was also a matter of turning non-irrigated land (*tegalan* or *pekarangan*) into irrigated land (*sawah*). There was therefore reason enough to start viewing water management in a more specific light and that is precisely what happened with the projects that began to take shape at the end of the seventies and which were, for instance, directed towards the rehabilitation of small and middle-sized systems.[65]

Conclusion

In the Indonesian-Dutch development programmes of the 1965-2000 period a great deal of energy went into the restoration of irrigation works, macro planning for the exploitation and the management of water reserves, drinking supplies, transmigration projects and – within that framework – the development of low-lying tidal regions. What has become most apparent in recent years is the fact that there is increasingly less evidence of a specific East Indian (or colonial) Dutch activity basis abroad. More and more Dutch engineers are participating in international projects and questions where they benefit collectively from the excellent reputation established worldwide in relation to successful Dutch water management, as recent experience abroad testifies. It is not unlike the situation that existed in the sixteenth and seventeenth centuries. Although at this point in time it would be a little premature to draw any sweeping conclusions, it might be possible to now cautiously assert that the direct relationship between the Netherlands' water sector, the Dutch-East Indian past and international activities has become blurred in recent years. This is a trend that is reflected in current education at Delft University of Technology. As Chapter 9 and the Epilogue demonstrate, the Dutch legacy in Indonesia itself, especially in the hydraulic engineering sector, is still very apparent, not only in terms of the works but also in terms of education.

Discussions concerning the selection of structural works for canal systems, like in the case of the Wadaslintang project, were actually able to take place because various consulting firms run by engineers of different nationalities, together with their different design preferences (irrigation traditions) were active in Indonesia at that time.[66] An attempt to stall the possible proliferation of structural works and all the solutions offered succeeded when in 1986 the *Irrigation Design Standards* were published.[67] That series, which appeared in 13 parts (see Table 1), gives a number of norms for the design of irrigation systems. The fact that this series of publications is still used in civil engineering education programmes in Indonesia surely testifies to just how strong this set of standards has become. Future Indonesian irrigation engineers are trained to operate according to those standards. Anyone who cares to consult the *Irrigation Design Standards* will immediately be struck by the fact that much of what is asserted there derives from approaches originally developed by Dutch-East Indian engineers. Designers are advised to opt for the Romijn broad-crested measuring weir or, if that does not suit, to select a Crump-De Gruyter orifice. Should they need to calculate river discharges or drainage capacities then the designer is expected to settle for either the Melchior or the Der Weduwen approach.[68] If one wishes to check whether these design standards are also used in design practice then one can take the example of Lampung in southern Sumatra to illustrate this point because there a Japanese engineering bureau has designed an irrigation system with explicit reference to the above-mentioned design standards.[69]

Table 1 **Irrigation Design Standards (1986)**

Title	Code
Design criteria	
Irrigation system design	KP-01
Headworks	KP-02
Canals	KP-03
Structures	KP-04
Tertiary units	KP-05
Structural parameters	KP-06
Drawing standards	KP-07
Technical specification	
Irrigation system design	PT-01
Topographic survey	PT-02
Geotechnical investigations	PT-03
Hydraulic model testing	PT-04
Drawing	
Typical irrigation structures	BI-01
Standardized irrigation structures	BI-02

Source: Nippon Koei, *Way Sekampung irrigation project. Design report on Bekri and West Rumbia irrigation systems.*

A railway bridge in the sawah landscape in Preanger (West Java). It was partly thanks to colonial civil engineering that the Indonesian unified state actually materialised. Dutch engineers linked up regions and islands, thus uniting the various population groups in the Indonesian archipelago. The technological and social evolutionary process witnessed in Indonesia in the past 200 years is characterised by the irreversible up-scaling of infrastructures.

CONCLUDING CONSIDERATIONS

WHERE THE HIGH MOUNTAINS RISE AND THE KALI RUMBLES FROM AFAR

The unique character of technology in the Dutch East Indies

WIM RAVESTEIJN

Probably few inhabitants of the prosperous Western parts of the world fully appreciate the fact that they do actually live in a realised technological utopia, at least when compared to the criteria of some 400 years ago. Modern Europe with all its advanced technology exudes the atmosphere of Francis Bacon's (1561-1626) Nova Atlantis, published in 1626.[1] In his epistle, the English statesman and philosopher sketches a utopian society in which making the populace happy and, in that connection, bearing responsibility for social order and harmony, is what constitutes the most important values and administrative goals. The pursuit of those dreams occurred in Solomon's House where all kinds of scientific and technological ideas were developed in an endeavour to satisfy all kinds of human requirements. As a result, the people of Atlantis enjoyed prosperity and could expect to live long and happy lives. The backdrop to Bacon's utopia was the Renaissance, the return to ancient times that was to be followed by the Scientific Revolution, the Industrial Revolution, the Glorious Revolution in Britain and the French Revolution. All those various social upheavals did, indeed, pave the way to the New Atlantis in the form of the European welfare state.[2]

The shadow side to this utopia, some would even call it the reverse side of the coin, is the inescapable reality that most of the world's inhabitants are forced to tolerate considerably worse conditions. Paradoxically it was the world of the non-Western peoples that constituted a source of inspiration for Bacon. His utopia not only embodied the traits of rebirth and renewal that echoed classical antiquity but also the wealth of experiences derived from pushing back geographical borders when the voyages of discovery began. Thanks to the fact that for most of his life Bacon lived in the England of Elizabeth I (1533-1603), the greatest world power of the day, he was free to push back the barriers. His search for a location for his utopia led him to an imaginary island in the Pacific Ocean, in what was then largely uncharted territory.

The non-Western peoples of the South Pacific and, for that matter, other areas of the world would, eventually, be transported down the road to modernity, though at first the vehicle of globalisation was a colonial and imperialistic one that would have negative consequences for them. In fact, even today we still see that the West (together with countries that have adopted Western technologies and lifestyles) is more privileged than the rest of the world. Technology played a vital part in all aspects of modernisation and globalisation and technological advancement was simultaneously a result of and a prerequisite for both those processes. This book illustrates that very point with regard to civil engineering efforts in the Dutch East Indies and in Indonesia.

Our traditional view of colonial technological development holds that everything in the field of Western science and technology is somehow superior to anything that did or could ever emerge from non-Western parts and that such knowledge therefore simply had to be exported to the rest of the world. Within the field of colonial history of technology, this diffusion model was first defined by George Basalla.[3] Since then there have been a number of authors who have modified the assertions originally made by Basalla.[4] The theories about how technological development did actually progress in colonial areas are still being formed and so this book can certainly contribute. From cultural-anthropological sources on the issue of diffusion,[5] what emerges is that in the case of cultural matters we usually see a mixture of pure diffusion and pure local ingenuity. Products that come from elsewhere invariably have to be adapted to the local circumstances and can also be dispersed via stimuli which then go

Colonial technology and indigenous culture: the goddess Shiwa in the guise of the destroyer decorates an inlet sluice near a fixed weir in the Brantas area (East Java), 1993.

on to make further development necessary. The fact that similar situations can also occur in the field of technology has been amply exemplified by the range of examples given in this publication that pertain to civil engineering in Indonesia over the past 200 years.

Chapters 2 to 8 demonstrate the combined Dutch-European and indigenous character of East Indian technological development in the colonial era (sometimes also termed 'hybrid' technology), though as far as the exact balance between such influences goes this varies tremendously depending on whether it is roads, railways, bridges, harbours, irrigation and waterpower, city design and kampong improvements, or works pertaining to health and public protection that one is considering. Chapter 9, which is devoted to development cooperation work between Indonesia and the Netherlands, as well some of the most closely involved international organisations, and the Epilogue which deals with postcolonial civil engineering activities from the Indonesian perspective, show us how work was resumed after independence. When considering all these facets the emphasis lies firstly on the external diffusion viewpoint and secondly on the own local development angle. In neither case is the mixed character of Indonesian technological development denied. If we compare technological development in the colonial and postcolonial periods then it would appear that the centre of gravity does shift towards Indonesia. Whereas it was first a matter of adapting Dutch and European technology, we see that later there is a shift to more local, independent development. Viewed from the cultural-anthropological angle, the move leans towards own development on the basis of external stimuli. Insofar as it is possible and desirable in a globalising world, this technological autonomy of Indonesia would still appear to be an ongoing process.

The themes of diffusion and endogenous development were discussed in Chapter 10 on East Indian road and hydraulic engineering where it could clearly be seen that Dutch and European skills and knowledge were adapted to meet local needs. There the emphasis was upon scientific and technological knowledge and knowledge interchange as manifested in the education of civil engineers. We have seen that it was on the basis of European and Dutch know-how that East Indian civil engineering was developed which then ultimately came to influence engineering projects in the Netherlands and elsewhere where Dutch engineers were active. As far as this final point is concerned, we touch here on an interesting phenomenon that is, in itself, difficult to reconcile with the simple 'from the West to the rest of the world' diffusion model. It is the phenomenon of transference to the Netherlands coupled with the internationalisation of knowledge developed in an East Indian context.

This point of diffusion in combination with local developments will be returned to but then the emphasis will be upon material technology, that is to say, the artefacts. By drawing especially on the information presented in the technological chapters we shall trace the unique character of East Indian civil engineering technology in this particular chapter which is devoted to various concluding considerations. In that way we shall gain more insight into patterns of colonial technological development in general. We shall trace the development of Dutch technology and ascertain to what degree locally developed technological methods were integrated into the process. The discussion presented here is not just interesting from the point of view of theory forming in conjunction with colonial technological development but also from the point of view of determining to what degree the colonial inheritance was perpetuated in independent Indonesia. We shall contemplate both the doors that were opened by colonial development and the doors which that same process closed.

Apart from providing various new details this chapter will give a thematic summary of the information presented in the technologically oriented chapters whilst at the same time rounding off the discussion on diffusion begun in Chapter 10. Parallel to the matters raised in Chapter 9 we shall especially highlight what the technological infrastructure was like when Indonesia first became independent, while simultaneously anticipating the Indonesian contribution provided in the Epilogue.

It was not just the area of technological knowledge that saw diffusion but also other areas of knowledge, like for instance the process of state formation described in the Introduction (i.e. the exporting of the national state from the Netherlands) and the affiliated developments within East Indian Public Works described in Chapter 2: the East Indian equivalent to the Dutch Public Works agency (Rijkswaterstaat) that played a mediating role in the diffusion of knowledge whilst developing initiatives that would ultimately lead to situations adjusted to East Indian circumstances.

In our character sketch there will be two main threads of argument. In the first place East Indian civil engineering will be examined within an analytical framework borrowed from appropriate technology. The angles of approach will be nature, society and culture. In the second place we shall employ the technological-dynamic perspective. In the Introduction we spoke about the 'socio-technical system' and the 'technological regime' concepts. Those are the concepts that will steer the arguments given here.[6] Finally we shall demonstrate how the material emanating from this study may be used for a broader view of theory formation on technological and social issues than what actually takes place within history of technology and technology dynamics studies.[7] First, though, we shall start with a brief synthetic characterisation of East Indian civil engineering.

East Indian technology

When Dutch engineers ventured into the Dutch East Indies they took with them considerable baggage in terms of knowledge on water management. Their knowledge was, of course, largely derived from the centuries-long battle against the water in their home country. Major problems relating to the regulating of the water of the rivers had, for instance, been quite an issue in the nineteenth century. In fact that was what had led (under French rule) to the establishment of Rijkswaterstaat in 1798. Dutch knowledge was very useful in areas such as drainage and dike construction in order to prevent flooding. However, the East Indian rivers with their perennial *bandjirs* (floods) did pose a whole new challenge to Dutch engineers. In the East Indies it was even necessary to have to fight to obtain water and that was something

Vlugter gates in the Lampung region (South Sumatra), 2003. Gates and other distribution works systems were a definite product of Dutch irrigation intervention in the East Indies.

that was new to the Dutch. The Dutch East Indian engineers were compelled to gain their knowledge by trial and error which, in the end, they did succeed in doing though that led to technological failures and disasters. Slowly a Dutch East Indian irrigation tradition began to emerge that was clearly different from similar traditions in other parts ruled by the British or French. Generally speaking, Dutch water management had traditionally been chiefly directed towards 'water level control' (i.e. polder level) whilst in the East Indies, in connection with the importance of irrigation, it was more a matter of water level control and discharge control. One special characteristic of the East Indian irrigation tradition was that of the research and development work done to support the measuring and the regulating of the supply of irrigation water to secondary crop allotments. One of the ingenious resultant inventions was what came to be known as water level and discharge control using the movable Romijn broad-crested weir. Another striking characteristic was the innovation of automatic upstream water level control in irrigation canals, effected in part by the Vlugter automatic hydraulic gates for upstream control. (See Chapters 6 and 10).

Similar kinds of developments were also witnessed in other technological areas, like for instance in the area of railway engineering, where the Chapelon modernisation was apparent. This improvement programme for steam locomotives aimed at making them operate more efficiently and more effectively. It was not implemented in the Netherlands but rather in the East Indies where the condition of the terrain gave plenty of reason for such measures. This package of measures may be viewed as a concrete component of East Indian railway technology. Other such typical East Indian innovations were the experimentation carried out with rail gauges and various management forms (i.e. private or state operated).

The real test , for all locally developed knowledge, was to see if it could stand its ground alongside international applications, like for instance what W.J. van Blommestein developed in Surinam in relation to the Wageningen Rice polder and the Brokopondo project (the Van Blommestein storage reservoir with a hydroelectric power station in Afobaka).[8] An earlier example that may be cited is that of the exportation of East Indian irrigation technology to Thailand where, at the request of the Thai king, J. Homan van der Heide designed extensive irrigation works at the beginning of the twentieth century. It was not, however, until much later (in the fifties) that those same works were actually made operational.[9] Alongside of this international diffusion there was also such a thing as 'national' knowledge transfer back to the Netherlands. One may think, for instance, of the Romijn weir which, in a modified form, was introduced to the Dutch river Linge region and, in more general terms, of the irrigation chair at Delft University of Technology. This chair arose thanks to the efforts of colonial engineers overseas and ultimately it was through knowledge in the field of irrigation that various other technical domains evolved. Delft irrigation expertise has created for itself a position in the wider terrain of land and water management and is internally linked to such areas as polder control.[10]

Innovatory know-how and the internationalisation thereof was also apparent in the harbour construction sector. One need only think, for example, of the caisson construction principle adopted by the Hollandse Beton Groep (Dutch Concrete Group). The experience gained in the East Indies, in Batavia and Surabaya, with caisson elements was later to be replicated in the world's number one entrepôt, Rotterdam. In fact caisson construction was a truly international phenomenon where one could, for instance, see experience gained in

Chile being put to good use in the East Indies and everything learnt in Rotterdam being implemented elsewhere in the world.[11]

Appropriate technology avant la lettre

The examples given above demonstrate that civil engineering technology in the East Indies was both fed by and had consequences for technology elsewhere in the world and so developed its own identity. Technological practice in the East Indies was thus a type of appropriate technology avant la lettre that supports the mission of this critical technological movement of the eighties in the twentieth century.

Paradoxically it was in the sixties, when the technological utopia had ostensibly been realised, that Western technology was to come under heavy attack. The problems in the industrialised world then centred on soil, air and water pollution, on what was termed a kind of alienation, and on the nuclear power race. Lurking in the shadows, though, was an even bigger problem. The technological-social advancement philosophy, the belief in progress, that had been strong in Europe since the seventeenth century suddenly found

When constructing the Tandjong Priok harbour near Batavia one of the innovations implemented after 1920 was rectangular caissons. In an improved version, the Priok caisson was later also used in Rotterdam.

itself in a state of crisis in the twentieth century thanks to the two world wars, a crisis that persisted despite the short-lived optimism of the reconstruction period after the Second World War. The poverty in the Third World coupled with the failure of both the Western and the communist development models to gain a hold in those parts constituted a second justification for the genesis of appropriate technology, which in this connection sought to offer an alternative to the modern science and technology that was so dominant in both models. Appropriate technology advocated a kind of technology that is friendly both to people and the environment and upheld three types of design criteria pertaining to technology, organisation and development.[12] In its extreme forms it was a kind of ideological, utopian technological movement directed at improving the world. In its technological orientation it constituted a kind of revival of Bacon's utopia.[13] In view of the fact that the problems at which appropriate technology was directed still persist, it may be said to still live on in watered-down forms. In the spirit of the design criteria of development experts, appropriate technology has been integrated into their 'technological regime'.[14]

In a certain respect, the foundations of appropriate technology amounted to a total theory of human existence including an aspired-to technological-social ideal. This theory included the insight that people are basically confronted with three fundamental dilemmas: how to physically survive, how to organise themselves as groups and what to think and find in relation to the world as a whole. There are then three corresponding basic relationships:

- the man-nature link: the fundamental ecological problem
- the man-man link: the fundamental social issue
- the man-world link: the fundamental cultural problem (or the 'world mystery' view).[15]

This theoretical model still offers scope for comparative analysis and assessment. In the ensuing three sections we shall sketch the unique character of East Indian civil engineering using the dimensions of the three above-mentioned fundamental problems. Tracing the man-nature axis we shall discuss, in the 'Tropical engineering technology' section, the special nature-related circumstances with which engineers in the East Indies were confronted. We shall show how they resolved those problems and what were the resultant technologies. In the East Indies it was especially the colonial class society that embodied the social dimension. In 'The tools of power' we shall see what consequences that had for technological development. Finally it will be the cultural aspect that will be contemplated in 'Technology and culture: the West in the East' where we shall consider how the interacting cultures of East and West influenced technological development in the East Indies.

Tropical engineering technology

The Dutch engineers cannot be accused of lacking courage: they came, saw and conquered tropical nature. In general terms, the geographical features of the East Indies were of great significance to technological development, the main characteristics of which were the following:

- The rough, mountainous terrain, especially in the inland areas. In the coastal regions the Dutch found in the plains what was, for them, a much more familiar landscape, even though it was the scene of sedimentary silt depositing of unprecedented size.

A temporary bridge in Plampangan (East Java), c.1922. The special character of East Indian bridge building techniques was something that was directly related to the unique terrain of the archipelago.

- The tropical climate with its relatively high temperatures; Java was beset by biannual monsoon winds, thus leading to a dry east monsoon period and a rainy west monsoon season. It was particularly the latter weather conditions that led to all kinds of problems in the form of *bandjirs*.
- The variable nature of the landscape as a result of volcanic action, landslides and erosion.
- The high silt content of the rivers as a result of the erosion and what were known as *lahars* (sand and silt streams emanating from volcanic eruptions and downpours).
- The social-geographical circumstances: the large population growth of Java and the thinly populated Outer Regions.
- Especially in the Outer Regions it was precisely the unspoiled character of nature in those kinds of areas that could be problematic: jungle (complete with dangerous species; nowadays predominantly endangered species).
- The incredible extensiveness of the archipelago (50 times the size of the Netherlands) accompanied at first by time-consuming communication systems resulting from the still primitive infrastructural facilities and transport means.

The local population's lifestyle was completely adapted to these geographical conditions as illustrated, for instance, by the agrarian village communities that could be found in great numbers in, for example, Java and Sumatra. In many cases, the cultivating of wet or irrigated rice crops constituted the main economic activity. Obviously the damp conditions of the *sawas* were optimal for rice cultivation, though there was not the need for such great quantities of water in all phases of the cultivation cycle. In the dry season the rice was substituted with other crops. In many respects the people of the village communities of Java and elsewhere were largely self-sufficient and also relatively independent, though they sometimes formed parts of larger political units at regional level or in the form of large or small kingdoms ruled by a monarch. The Hindu empires[16] that first developed were agriculture-based but the later Islamic empires were also trade-based. The Dutch started as traders but finished as empire-builders and so, in that respect, became heirs to the Hindu rulers and sultans.

The colonial engineers were all confronted with one or more of the above-mentioned geographical features in a whole range of technological areas. The road and rail engineers had to rack their brains when it came to surmounting the steep inclines and hairpin bends, the bridge-builders were intimidated by the problems ensuing from the deep ravines and the fast-flowing rivers, the spirits of the harbour engineers sank when they saw the vast quantities of silt in the water, and the irrigation engineers had to cope with either too little or too much water though they were able to make a virtue of necessity by harnessing the excess for hydroelectric power. Meanwhile the public health and the urban development engineers were up against rapid population growth. What amounted to entirely new technologies for the Netherlands started to pop up everywhere, like in the cases of irrigation and waterpower whilst in the case of existing technologies creative, adapted solutions were sometimes required, as with the Chapelon modernisations.

A nice example of the way in which the natural landscape influenced technology comes to us from the area of bridge building where it was notably the construction method that demanded a degree of creativity and adaptation. As the *bandjirs* and the vast heights that needed to be bridged often made upright support structures in the water impossible, the Dutch developed what came to be known as the 'free extension' technique which meant that during construction there were tensions that were not later surpassed when the bridge was in use. From the field of bridge construction, there are many more examples to be found. What was typically East Indian was the suspension bridges. In order to preserve the banks, use was made of gabions (*brondjongs*), baskets or mattresses made of woven bamboo (and later also of telephone wires or galvanised wire) which were then filled with stones from the river. This was a technique that would later be imitated in the building of the Delta works in the Netherlands.

In the area of hydraulic engineering there was great continuity. The Dutch started by excavating canals in the Batavia region for transport ends and in order to circulate the water in the city but these canals were also used for irrigation purposes. In the nineteenth century the Dutch followed in the tracks of the former Hindu rulers when they embarked on large irrigation projects, first by utilising chiefly traditional methods and the help of the local population in the form of forced or corvée labour initiated by and led by inland administrators but later this was increasingly executed by engineers. The engineers had a preference for large works enterprises designed to irrigate a large acreage of land mainly in

A train on a railway bridge above a dry river area in the Tengger region (East Java), c. 1937. The East Indian railway companies were particularly

the coastal plains which thus became more attractive residential and work areas, in the first place for the foreign plantation owners, and in the second place for the local population. The kind of technology used there was predominantly modern. They also built smaller works designed either to connect with or improve existing local works in order to support established rice cultivation, especially in mountainous areas. Modern and local technologies thus came to reinforce each other. After periods in which one approach took precedence over another, an integral approach to large and small works was adopted in the interests of improving and expanding indigenous agriculture and sugar cane cultivation. In that way the irrigation tradition that developed was absolutely unique to the East Indies.

When the irrigation engineers arrived they were confronted with a 2000-year tradition and that was something that could not be ignored, though sometimes they would have liked to believe it could. In a few other technical areas there were also traditions, like in the case of bamboo bridge building but also in the road and harbour sectors. In these cases the way in which the old influenced the new was less apparent. In yet other fields, absolutely everything was new, like in the case of the railways, hydro-engineering, urban design and urban water provision. When the Dutch first arrived in the East Indies there were no cities at all. That was something that accompanied the development of colonial society where population expansion, urbanisation and a certain degree of economic development founded largely on the basis of cash crop cultivation and the extraction of minerals from the earth were all things that went hand in hand. So it was that East Indian society slowly developed together with a colonial version of the modernity that was emerging in Europe.

The tools of power

Old Hindu and Islamic empires in the Indonesian archipelago were of an Eastern despotic character with a very rigid class structure. This society with its villages, sovereign powers and later its merchants, became ingrained in the class structure of the colonial era. The colonial engineers constituted part of the European segment of society that placed itself above the local population. This class society comprised Dutch people, other Europeans and their equals, and the local population. Alongside that there was also a kind of middle class comprising Chinese and other foreign Eastern immigrants. It was a society that evolved in the course of time with the colonial, foreign element becoming socially differentiated on the one hand, also in a downwardly mobile direction, whilst on the other hand the local element began to gain more social weight and to penetrate the higher echelons. All in all, the basic social structure was retained and even consolidated during the depression of the nineteen-thirties. The colonial class society had features that were important for civil engineering developments.

In the first place the colonial state played a key part as an agent of imperialism (see the Introduction). Initially the East Indian state leaned heavily on the local administrative structure (indirect rule) insofar as that existed. Over the course of time, a process of state 'embedding' unfolded whilst simultaneously the local element gained more of a say in matters and so 'rose' in the ranks. Here again it was definitely the case that the basic political structure remained in place: the form changed but not the content. The technological influence exercised on the part of the state fitted in with the changing interests of the state but retained an imperialist character. In the case of road structure we clearly saw in the developments that it was originally the political and military interests that dominated (the Great Post Road), that it was then economic interests that came to the fore and finally social interests that began to partly influence matters. A similar kind of progression was to be detected with the railways. Ultimately the train would take its place as an important and valued means of public transport. The first routes, especially in the Outer Regions, were very oriented towards the transport of army people (e.g. the Atjeh Tram) and export products, such as coal and agricultural produce. What was particularly striking was the extensiveness of the narrow gauge network of the enterprises (i.e. the plantations exploited for cash crops).

The colonial class society manifested itself in other ways. What was perhaps most noticeable was the fact that the local population was viewed as a category and not as an entity of individuals like the Dutch or other Europeans.[17] It was a view that influenced matters such as the way in which drinking water facilities were organised with collective water tapping points for the indigenous population. Differences could also be seen in the way in which architectural issues were tackled: state structures destined for Dutch inhabitants were individually designed whilst it was often 'standard designs' that were used for the local population. For the Europeans, new housing areas were planned and created whilst the locals had to make do with *kampong* improvements.

One thing that was very typical of the East Indian tradition when it came to the construction of public works was the deployment of mass manual labour, notably in the form of corvée labour. The state thus made use of and perpetuated an institution that was readily available in various forms in the existing society. In principle, engineers were opposed to the system and preferred to have people work for payment. It was not until after the introduction of the Ethical Policy that forced labour for the construction of public works was finally abolished. In the 1882-1916 period corvée labour was supplanted by a minimum wage system before finally being phased out altogether in 1927.[18] Normal paid labour was invariably too expensive. Obviously the extensive availability of cheap labour in many places was something that only promoted the tendency to develop and use labour-intensive structuring and production methods. A typical East Indian-style bridge that is clearly the product of such thinking is the type constructed on a foundation with steel screw piles that have to be 'fixed in place by hand'.

Technology and culture: the West in the East

Did such a thing as an East Indian culture develop? In East Indian elite circles something did develop which was later perpetuated in the Netherlands and came to be known as *tempo doeloe* or 'the good old days'. It all had to do with the common-or-garden culture of the Dutch in the East or rather, the way in which that was nostalgically recalled once back on home ground. In other words, it was connected with the East in the West. When the Dutch first made their inroads into the East Indian archipelago it was the West that originally contributed to the East. The Dutch or European influence was nothing more than yet another influence in the patchwork quilt of cultures already present in the archipelago. The development that followed could be compared to what had occurred earlier as a result of the introduction of the Hindu complex of values and later the Islamic norms and values: the culture of those in power was distributed. In the archipelago this ultimately led to political integration in the form of a nation state and economic integration with the developing world trade network. Gradually the customary dualism of village and monarch would crumble. The village, according to the reports of Western observers organised on the basis of tradition (*adat*) and a place where collectivism was stronger than individualism, would, in the long term, lose its cultural autonomy as well as its economic and relative political autonomy, though it was not until the time of the Indonesian nation state that that finally happened.

The culture brought by the Dutch and the Europeans was especially a technological culture, certainly after the time of the Industrial Revolution in the West; the first such revolution linked to steam technology. It is clear that East Indian technological development was especially influenced by Western technological know-how that was sometimes adapted to local natural phenomena and social circumstances. We have also seen that local East Indian knowledge occasionally played a part, like in the case of irrigation. On the one hand the Dutch improved things but they also integrated old methods, like in the case of the area of the lower reaches of the Solo river. The small reservoirs that had existed there for numerous years constituted an incentive for the reservoir technology employed by engineers to develop the Solo Valley after the major irrigation project there had been aborted as the budget had been exceeded and other technological problems had loomed large. Wherever possible the Dutch also made use of local institutions, like for instance in the area of water management where the traditional *oeloe oeloe* (water overseer) was called upon, though the

The complete cementing up of a sunken pit in conjunction with the building of a bridge over the Moedjoer river (East Java), c.1900. Much use was made of forced labour whenever it was necessary to construct public works.

structure was slightly modified, that was to say, he was no longer attached to the village administration. There were other local cultural elements that were similarly valued and preserved, and in many areas endeavours were made to arrive at cultural compromises. At a somewhat higher level of abstraction (e.g. in the writings of De Kat Angelino)[19] attempts were even made to reconcile the East and West in such a way that the best would be drawn from both traditions and synthesized.

Still, in the confrontation between Eastern and Western cultures in the East Indies the former had less chance of success than the latter, not so much because of the inherent inferiority of the East to the West but more because of the balance of power within the imperialistic framework. The fact that a process of westernisation was ultimately imposed – despite Dutch openness to local technology and culture – was something that happened because it was so heavily stimulated by colonial policy. This was largely based on a strong sense of civilisation and the need to civilise derived from a nineteenth century belief in evolutionary progress in which Europe wore the crown of human civilisation. Just as the French had their 'mission civilisatrice' and the British their 'white man's burden', so the Dutch unfolded its Ethical Policy which was aimed at improving the living conditions of the local population and developing East Indian society by 'elevating the East Indian brother'.

Just like all the policies of other colonial powers, this was a politics founded on a 'modernisation view' in which technology played an important part. Ultimately it harked back to the view formulated by Van Hogendorp around 1800 in which the moral duty of the Netherlands was encapsulated. The way in which irrigation technology had developed in the nineteenth century had been exemplary and was to become one of the central technological domains of the Ethical Policy movement, despite the problems manifested in the Solo Valley at the time of the policy change. Electro engineering (electrification) was a second key technology. Europe (including after the Russian Revolution the Soviet Union) was then the scene for the second Industrial Revolution, one that was characterised by the generation, distribution and use of electricity. Another characteristic of that second revolution was the deployment of science for technological advancement and planning. Irrigation plans (including the 1907-1911 General Work Plan), road plans, laboratories in a diversity of fields, technological education and technocratic administrative tendencies were all matters that featured in the accelerated growth of technological development as witnessed in the East Indies in the years after 1901.

Finally there is one point that should be noted. The dissemination of civilisation could not exactly be divorced from all kinds of relevant practical implications and that, too, had its consequences. The colonial modernisation mission did little to stimulate own development, for instance in the form of industrialising. The Netherlands shielded its own industries from the East Indies. One need only think of the cotton from Twente or the sugar and margarine processing factories dotted around the Netherlands. In many respects both world wars disrupted much of this. During those periods the East Indies was very much left to its own devices which led to independent economic ventures. When the Netherlands became occupied in 1940 many own industrial initiatives quickly sprang up in the East Indies. Car tyre factories were started up, chocolate making, jam factories et cetera, shipyards and metalworks expanded. It is true to say that imperialistic relations went hand in hand with modernisation but this was very much confined to 'dependent development'. In actual fact, with regard to technology this was less clear than in the case of industry. One need only think of the establishing of the Technical College in Bandoeng. A national revolution was needed to release the country's plentiful resources (e.g. oil) and deploy them for more or less independent industrial developmental means.[20] This was a process that was definitely made possible by the preceding East Indian technological advancements but it was also hampered by those same advancements! East Indian technology was system-based which meant that in that way any kind of burgeoning independence was somehow held back. The precise forces that were at work will be clarified in the following section.

East Indian socio-technical systems

Nature, society and culture in the East Indies may be seen as the basic ingredients of East Indian development in the field of civil engineering: they were the framework within which technological development occurred. Modern history of technology gives us the instruments to contemplate such technological development as something other than a pure derivative of the above-mentioned factors and to further characterise such development in ways that not only have scientific significance but also practical implications. As was indicated in the Introduction, civil public works are not viewed here as separate artefacts but rather as components of a socio-technical system. It is true to say that East Indian civil engineering did develop within the framework of the given natural, social and cultural specifications but alongside that it had its own special dynamics; a dynamics that was linked to the system-based character of the infrastructural networks that were created. Various railway, irrigation, and so on, systems have been considered. Once different developments had been set in motion they permeated and, on the whole, developments in the different public works areas tended to gather momentum or develop their own dynamics. That could be seen for example with the railways. Experiments were carried out with different rail widths and eventually it was difficult to agree to a standard gauge, precisely because of the different systems. However, once the East Indian standard gauge had been established it was virtually impossible to later deviate from it.

The type of momentum evident within developments, which manifests an own force, is a technological-dynamic mechanism that develops socio-technical systems from within. Two other similar kinds of development mechanisms are the 'reverse salients' phenomenon (where underdeveloped factors are turned into critical problems to which technological effort is devoted; a term from military jargon originally used to refer to the trailing component of an advancing front) and something known as the 'load factor' mechanism (i.e. the tendency within a system to seek maximal application).[21] In the case of the railways, the negotiating of steep gradients was such a reverse salient and one which undoubtedly helped to trigger the Chapelon modernisations. Another reverse salient is exemplified in the area of irrigation where the matter of management development lagging behind became one of the reasons for deciding not to pursue the Solo works project. Around 1900 there were no managing means available for the organising of agricultural activities over a wide area. A good example of the 'load factor' is to be found in the harnessing of waterpower. From the point of view of rentability it was essential to optimally utilise the power stations.

In this connection there are two interesting conclusions to be drawn where socio-technical systems are concerned. What first emerges is that public works systems compete with each other: for instance irrigation works competed with the railways (locally produced rice or supplies from elsewhere) whilst the railways was in competition with roads and shipping, two other modes of transport. In the case of irrigation, there was competition with health-focused public works (drinking water provisions) and hydroelectric power which depended on what essentially amounted to scarce water reserves. It was not free market competition but a kind of competition within the framework of a colonial state, which meant that political-administrative considerations dominated all the decision processes. In the second place there were, in effect, compounded infrastructural systems because harbours, steam shipping and railways were all interlinked. Homogeneous networks dependent on certain technologies became knotted together with such things as harbours and railway stations functioning as junctions. In the case of East Indian public works there are roughly two heterogeneous systems that can be distinguished: a transport system with roads, railways, bridges, harbours and inter-insular shipping and air connections. Then in addition to that there was a collection of systems centred around water. The most important systems in the latter group were: irrigation, public health works, urban *bandjir* protection and electricity generation. Waterpower was a binding technological factor: it provided electricity for the trains.

The hypothesis that one might attach to this is that heterogeneous socio-technical systems function in just the same way and develop in just the same way as homogeneous systems, making use of the same system mechanisms. We might assert, for example, that the harbours constituted a reverse salient within the transport system: whilst in the second half of the nineteenth century sectors of the whole system, such as the roads and railways were developing fast, the harbours remained behind.[22]

The particular constellation of works and systems which came into being is what gives East Indian civil engineering its own unique identity. Technology (in this case public works) and the social-cultural context (in the East Indies notably the state machinery) were woven together in socio-technical systems by means of the relevant interest groups (East Indian engineers, agriculturalists, inland civil servants et cetera). It is a well known fact that when it comes to the forming of socio-technical systems, social and cultural circumstances play an essential part whilst later those systems, in turn, have quite an impact on their respective environments.[23] We could see here an explanation for the position of Indonesia as a developing country. One cannot escape from the fact that to a large extent East Indian technology was a mere reflection of a political system directed towards control and exploitation. The interests of the local population were really only taken into consideration in later building projects. The transport system was predominantly developed for administrative and military purposes, and for the transport of goods. In the latter case that was chiefly angled towards taking cash crops and raw materials out of the country. Regarding water systems, it should be noted that irrigation works were initially mainly created for the production of sugar. Quite soon, though, irrigation was to gain a more diversified character so that increasing attention would be devoted to the welfare of the local population. However, this was never to become an overriding concern. Because of the system-based nature of works as a whole it became generally more and more difficult to make any major civil engineering developmental adjustments and that same system-based trait was also what formed the foundation for foreign technological aid after the time of independence, a case in point being the rehabilitation of the irrigation works. The character of East Indian technology could be further sketched by taking the technological regime angle. In that way various points can also be clarified in relation to the colonial technological legacy of present-day Indonesia.

East Indian technological regimes

'Socio-technical systems' and 'technological regimes' may, in effect, be seen as technology-dynamic twin concepts or as the two sides of the coin of technological development. What is important with both these concepts is that they not only allow technology to get embedded in the environment of nature, society and culture within which technology features but that they also allow technology to relate to all of that in an intrinsically meaningful fashion. On the one hand socio-technical systems are strongly linked to the natural environment but on the other hand they are developed in – and in interaction with – a certain social and cultural context. Especially when it comes to determining what kind of influences the socio-cultural environment has on technological development, technological regimes can be very useful. Technological regimes are collections of rules laid down to serve experts involved in the 'making of technology'. They may therefore be said to steer technological development. To a great extent these regimes are a product of the socio-cultural context, both in terms of knowledge and fundamental values. It is the knowledge status quo that determines the available range of technological possibilities whilst the values steer the choices that are made. Even here the relationship is dualistic: technological regimes do, in their turn, influence the socio-cultural environment.

The colonial state steered technological developments via the two general technological regimes distinguished in the Introduction: the exploitation and the development regimes which respectively expressed the imperialistic interests and the 'modernistic' interests of the colonial state. We have seen how these regimes operated in the various technological terrains. What is interesting here is that the two complexities of infrasystems referred to above do correlate to a degree with the two technological regimes, the transport system with the exploitation regime and the water system with the development regime. We might thus assert that transport technology served imperialistic interests whilst water technology served modernistic interests. Later on, though, transport technology also became a kind of developmental technology whilst water technology was at first purely there for pragmatic ends. There is thus absolutely no one-to-one correlation to be detected in all of this.

By taking the regime concept approach there is another way in which we can link East Indian technology to broader political-social developments. The differentiation that has been distinguished between the two general technological regimes reflects a Dutch-colonial perspective: the East Indies as a conquered land. If we view technology from the point of view of the creation of an own, potentially independent state – a viewpoint that also became popular among the colonial powers, at least in formal terms – then we could go on to distinguish three technological regimes:

1. a political-national regime directed at conquering, protecting and smelting the Indonesian archipelago into a unified entity with roads, railways, bridges and harbour works as the leading technologies
2. a rural regime directed both towards the plantations and subsistence agriculture with irrigation works and transport provisions
3. an urbanised regime focused on urban development, *kampong* improvement and public health works.

By contemplating this three-way structure it becomes easier to see not just how Dutch engineers contributed to the development of a prominent part of the Dutch empire but how they laid the foundations for an independent Indonesia. Even though the East Indian state definitely unfolded initiatives in the direction of independent development and industrialisation that certainly did not happen in a systematic fashion and it could never be claimed that a 'technological-industrial regime' had developed as had been the case in the Netherlands, largely on the basis of the initiatives of the king, Willem I, who was very active in that respect. The East Indian rail and tramway workshops worked wonders but, with a few exceptions, the locomotives came from elsewhere, including the Netherlands (Werkspoor) and much the same could be said of other technological fields. To give just another example, in the colonial era the harbour of Tandjong Priok did not really manage to establish itself as a centre for industrial businesses in the way that harbours situated in the Western part of the world had done. In educational areas matters are confirmed for us. The courses provided for engineers in Delft were based on notions that the Netherlands should develop independently and industrially so that, as a consequence, the range of fields continued to expand: civil engineers, mining engineers, mechanical engineers, railway specialisation et cetera. Bandoeng, by contrast, did not progress beyond civil engineering.

The continuity visible in the field of civil engineering developments in the East Indies and Indonesia may not only be explained in terms of the socio-technical system concept but also by taking the technological regime concept. One aspect of this was knowledge in the form of training courses and manuals, as in the case of the engineering college in Bandoeng and the 'BOW manual' (i.e. the East Indian Public Works department manual) used in everyday practice.[24] In this connection it is worth mentioning that many plans made in the period of colonial administration were subsequently carried out after independence. One example is the reservoir plans made by Van Blommestein that resulted, for example, in the construction of three storage reservoirs in the Citarum that were to be important in the areas of irrigation, energy generation and the provision of drinking water. One of Van Blommestein's objectives had been to develop industry which, at the end of the colonial era, ultimately became a serious subject of interest.[25]

A final theoretical consideration

Colonial technology does not only introduce the non-Western world to the history of technology and to its theoretical counterpart, technology dynamics, but also globalisation. The technological-social processes previously described and analysed in Dutch history of technology studies[26] turn out to have an important East Indian dimension and to be part of a worldwide framework. This revelation, for which the present study provides a basis, not only accesses new empirical terrains but also facilitates a broader interdisciplinary approach than has, up until now, been customary in the referred to fields.

Technology dynamics has profited greatly from concepts and insights that come to us from the social sciences. That is evident from terms such as 'technological regime' and 'socio-technical system' the strong social-scientific character of which is clearly recognisable. Conversely this study makes it possible to rectify or at least complement various social-scientific notions, for instance where the matter of regimes and long-term processes in the history of mankind are concerned, like those dealt with in the regime and figuration sociology of the 'Amsterdam sociological school'[27] The views of the authors concerning ecological and other regimes,[28] and historically long-term processes,[29] are highly synthesis-oriented and for that reason alone extremely attractive since they provide a welcome supplement to the prevailing scientific analyses which, though reliable, tend to present extremely fragmented views of matters. The high abstraction level does, however, give rise to problems. One problem of special relevance here is the one concerning the suggestion that there is a simple diffusion of social acquisitions, such as a new ecological regime.

If one thinks in the same terms as these sociologists then civil engineering in the East Indies might be seen as part and parcel of and derived from the transformation of the Netherlands from an agrarian to an industrial society, when the country shifted from an agricultural to an industrial regime (in the 1850-1870 period). In this optimistic presentation of matters, strongly allied to a belief in progress, justice is especially done to the modernisation process in the East Indies while the imperialistic context fades into the background. The unique characteristics of technological development in the East Indies fall outside their field of vision, not only because their thinking is too macro but also because, for that very reason, it probably tends more towards simplification. Long-term regime sociology and figuration sociology may well profit from the introduction of a principle that is generally accepted within the history of technology field and has certainly not led to any great disappointments, namely that of the combination 'thick description' (i.e. detailed) at micro level and technological-social analysis at macro level (e.g. in relation to historical processes). This study demonstrates that whilst diffusion is definitely an important process it is certainly not a one-sided process in which technology is simply imported from elsewhere and implemented. It is just as much a process which, in relation to local natural, social and cultural circumstances, brings with it adaptations and new developments thus leading to renewed diffusion, both with the country of origin of the technology in question and elsewhere in the world. In other words: globalisation of technology may be said to lead to localisation of technology and such localised technology can, again, become the subject of globalisation.[30]

A framework for a broader view of the development of technology and society that does justice to shorter and longer term developments on micro and macro levels might perhaps be found in the trusted evolution theory. A good starting point is perhaps the fascinating and refined evolution theory for technological development formulated by George Basalla, a theory in which Basalla in fact surmounts his own diffusion model.[31] The core of the theory lies in the always and eternally occurring processes of variation in technological artefacts and the subsequent selection processes of a divergent nature which lead to the disappearance of some of the alternatives while yet others are integrated into the technological-social

The machine room interior at the Ketenger (Central Java) hydroelectric power station, 1940. In the East Indies hydropower works formed the link between 'water systems' and 'transport systems'.

developmental process. One guiding factor in all of that is the ultimate ideas of a people as to what constitutes a good life. In this comprehensive and inspiring theory Basalla is once again one-sided as he confines his analyses to material artefacts and considers especially the micro level. The integration of his thinking with, on the one hand, the technological-dynamic system and regime thinking and, on the other hand, insights drawn from long-term regime and figuration sociology, could give the history of technology field an impetus, thus resulting in a new evolutionary approach to technology and society that might be socially-scientifically sound and attractive.[32]

With its detailed description of the development of civil engineering over a period of some 200 years in the Dutch East Indies and Indonesia, the placing of those developments in their broader context of colonial state formation and world trade integration, with its views on modernisation and globalisation and with its employing of technological-dynamic concepts, this book has hopefully brought that approach a step closer.

A large irrigation reservoir with an inlet tower (the Pacal waduk) in the Solo Valley (East Java), 1993, that was completed in 1935. When constructing such reservoirs, East Indian engineers were inspired by the small waduks in the Solo Valley that had already been there for many years.

The Jalan Tamrin in Jakarta. A symbiosis of old and new Indonesia.

AN INDONESIAN VIEW AND EPILOGUE **A TALE OF TWO KOTAS**

The impact of the Dutch architectural and civil engineering legacy on Surabaya and Malang

HARRY PATMADJAJA

Assisted by: Budisetyono Tedjakusuma, Handinoto, Handoko Sugiharto, Indriani H. Santoso, Jones Syaranamual, Pamuda Pudjisuryadi, Paulus H. Soehargo, Ruslan Djajadi, Aylanda Dwi Nugroho

The Gubeng Bridge leading to Simpang Street, now Pemuda Street, in Surabaya, 2002.

The Peneleh Bridge in Surabaya, 2002.

The toll motorway between Surabaya and Gempol, 2002.

In 1602 the Dutch Republic granted the Dutch United East India Company (Verenigde Oost-Indische Compagnie, VOC) the monopoly on Dutch East Indian trade. During the next 200 years trade flourished and the VOC became the largest trading company of its kind. Towards the end of the eighteenth century, though, the VOC got into great financial difficulty. Despite the support received from the Dutch Republic there were not enough funds, ships and personnel to sustain the flourishing trade with the Far East. When the VOC went into liquidation in 1798, the Dutch government started to rule in the Dutch East Indies (Indonesia) and that marked the beginning of the Dutch colonial period.

In 1870 the Dutch parliament passed two acts that were to mark a turning point in colonial history: the Sugar Act and the Land Act. Not only were they very important for the economic development of the colony but indirectly they were also to greatly influence Java's urbanisation process. More specifically, the Sugar Act heralded the abolition of the 'Cultuurstelsel' (Cultivation System), a Dutch law determining the exact agricultural production output of Javanese farmers. The Land Act opened up the Dutch East Indies to free enterprise and private capital, thus putting an end to Dutch governmental monopoly.

Most colonial products passed through the larger ports of Batavia, Semarang and Surabaya. Many mercantile houses, banks and other enterprises were therefore established in those cities, or company branches were at least opened there. Inevitably a transport network was developed to link the hinterland with the harbours. Infrastructural development (i.e. roads and railways) led, in turn, to faster city growth. In 1855 there were only 28,000 Europeans in the Dutch East Indies. That number had tripled by 1905 (84,000) and by 1920 the European population had risen to 240,000. As a result, urban infrastructural development (bridges and structures included) together with public housing, became the Dutch government's top priority.

Today, that city infrastructure has become part of the colonial heritage from the Dutch East Indian era. The continuing population growth is affecting trends in urban infrastructural development in Indonesia. The population rose from 119.2 million inhabitants in 1971, 147.5 million in 1980, 179.4 million in 1990, to 195.3 million people in 1995. By 2000 the population of Indonesia was approaching the 220 million mark. The average density is 116 people per km^2 and the population growth rate is 1.6%. Two-thirds of all Indonesians live in Java, making it one of the world's most densely populated regions. Though the island

Table 1. **Population growth in Surabaya and Malang**

Municipality	1992	1993	1994	1995	1996	1997
Surabaya	2,259,475	2,286,359	2,307,911	2,329,598	2,344,520	2,356,486
Malang	677,608	684,608	702,733	711,673	707,790	714,324

Source: The East Java Province and the Government of East Java, Statistics, *East Java in Figures in 1998*

only covers 7% of the country's total land area, its economy contributes more to GNP (gross national product) than the total economy of all the other islands put together. All the cities have also grown, including the two major cities of Surabaya and Malang, both of which are located in the eastern part of Java. The population growth in both these cities is reflected in Table 1. Because of the rapid population growth, rice fields around cities like Surabaya and Malang were soon turned into housing areas. In the process, many of the working irrigation canals of the colonial period were converted into drainage channels.

This chapter will discuss the development of the city of Surabaya between the colonial period and the present from the point of view of civil engineering infrastructures, city planning and building projects. Similar developments in the city of Malang, with regard to just some of the city's works, will also be traced. Some general technical explanations, especially where codes for concrete, steel and engineering education are concerned, will follow the various descriptions.

The growing need for transport: roads and bridges

In 1808 the governor-general was Herman Willem Daendels. He turned Surabaya into a small fortified European-style city which flourished as a trading area. In that colonial period, Daendels started on the construction of roads, bridges, buildings and harbours. In order to accelerate East Indian economic growth, he came up with the brilliant idea of creating roads to speed up the transport of trade commodities (mostly spices) which could then be exported to Europe from Java. Daendels therefore produced plans for an arterial road known as the Groote Postweg (Great Post Road) in which Surabaya was to feature as the trading centre. This famous Great Post Road connected large and small cities along Java's northern coastline. It started in Anyer (West Java) and ended in Panarukan (East Java), and was constructed on the basis of forced labour (herendienst). Apart from the economic considerations, the Great Post Road was designed to improve communications in Java and to protect the island from possible British invasion.

The Great Post Road was a two-lane road and it was 7.50 m wide. Most of the surface consisted of flexible paving of the Telford type which involved layers of large crushed stones with diameters of 20 to 30 cm (for the sub-base) being placed directly on the underlying ground surface (the sub-grade). Above the sub-base were layers of smaller crushed stones varying in diameter from 5 to7 cm (the base course). The uppermost layer comprised the surface course which covered the base course. The material used for the surface course was usually of the mixed macadam penetration type, then the standard surfacing material employed in the construction of main roads. The function of this surface layer was twofold: it served to carry, absorb and distribute the wheel loadings in a proper way and it constituted a waterproof cover so that rainwater could not seep below the surface.

The various stone layers were laid manually and rollers (8-10 tons) were used to compact the surface (there was no such thing as an Asphalt Mixing Plant at that time). A simple static theory was used to determine the road surface structure: wheel load divided by contact area was compared to allowable ground base stress and single-axle load (i.e. not dynamic load) was taken as the wheel load stress level.

After 1908, Indonesia became a member of the Paris road design organization known as the Permanent International Road Congress. The developing of the main roads in Surabaya during the period of colonial government commenced with the construction of a road leading from Tanjung Perak Harbour in the north to the Wonokromo area in the south. At that time, Wonokromo was situated on the outskirts of Surabaya near to a very fertile farming region. The main aim behind Surabaya's rapid development was to accelerate the distribution of agricultural and industrial commodities (such as sugar and rice) to be marketed outside Surabaya. Such commodities were destined to pass through Tanjung Perak Harbour.

Generally rocks brought in from Gresik and Madura were broken down and used as road material. Those two places were not very far from Surabaya which meant that the transport costs were low. The main roads in Surabaya that were built in the colonial period are still in use today. To meet the needs of increasing traffic volumes, only road widening and overlay jobs were carried out and the old main structure of the road was not removed.

In the colonial period, bridges were an important part of large city development. In 1910, the traffic density in the large cities of the East Indies increased rapidly. In fact it was not only traffic density that increased but also vehicle weight. Therefore the colonial government (through the Department of Civil Public Works, BOW) published codes for bridge structure loading specifications.

Not only did bridges serve as structures in the transportation network but they were also intended to be aesthetically pleasing as is borne out by the Gubeng Bridge structure (Goebeng-brug), located on Simpang Street (now Pemuda Street) that spans the Kalimas river or the Kali Mas. That bridge, inaugurated in 1924, was designed to have lighting in order to make it more aesthetically interesting. The Gubeng Bridge was renovated by the colonial government because the old structure was not wide enough. The main structure of the bridge consisted of a Gerber system (i.e. a simple support system with three spans and four supports). The middle part of the spans were less thick than the other sections to allow barges to pass under the bridge. The total span of the bridge is 48 m (13 m + 22 m +13 m) while the lane width is 12 m with 2 m pavements on either side. Another bridge that is similar to the Gubeng Bridge is the Wonokromo Bridge. It was renovated by the colonial government and completed in 1932. Its total length is 66 m and it consists of five spans of equal length. It is 12 m wide and has two pavements to the right and left, each of which is 2 m wide. Both the Gubeng and the Wonokromo Bridges were used for electric trams in the colonial period but today they are only used for ordinary vehicles.

The bridge on Panglima Sudirman Street (Palmenlaan), the Gubeng Bridge on Pemuda Street and the Gubeng Flyover above the railways are three concrete structure bridges that are still in use today. Besides these bridges, the Peneleh Bridge, which is a steel structure that is based on a foundation of steel screw piles (schroefpalen), is also still in use today.

Between 1945 and 1960 road construction stagnated, largely because in the early post independence years the Indonesian government was still struggling with political issues. Moreover, in the wake of the Second World War there was an economic crisis and the struggle for independence had begun. For those reasons roads were simply maintained at that time and it was not until the 1960s that proper road development was resumed. Such development initiatives were boosted by the arrival in Indonesia of foreign investors and loans from the World Bank and the Asian Development Bank.

Road development in Surabaya not only progressed from north to south but also from west to east. Local human resource development also improved as local Indonesian road engineers became more readily available. In 1975, a Postgraduate Programme on Motorway Engineering was established at the Bandung Institute of Technology in cooperation with the Department of Public Works. The programme was funded by the World Bank and the Asian Development Bank. An organisation in motorway development (the Indonesian Motorway Development Organisation) was also established. In order to anticipate the exploding traffic volume in Surabaya, a toll motorway that was approximately 39 km long was constructed between Surabaya and Gempol in 1980. At that time, the traffic volume on Ahmad Yani Street rose to 52,000 pcu (passenger car units) per hour, which clearly exceeded the street capacity. The toll motorway that was constructed is 14.4 m wide and consists of two lanes, each with a hard shoulder that has an average width of 1.5 m and a verge that is 2.5 m wide to the left and to the right.

Railways: not the first priority

The colonial government established railway companies in order to meet the economic needs of the government. It was the Industrial Revolution that motivated the Dutch to start exporting harvested commodities to Europe from Indonesia. As far as Java and Madura were concerned, there was one state railway company, the Staatsspoorwegen (State Railways), by the end of the colonial period (1942) but a further twelve private railway companies were still operational at that time (see Appendix 8).

Between 1942 and 1945 all the railway networks that were owned by the Dutch government were taken over by Japan and re-named 'Rikuyu Kyoku' (subsequently Tedsudo Kyoku). The head office was situated in Bandung. During the war, the railway network was used for political purposes and became a logistic facility for the transport of soldiers.

After the Indonesian declaration of independence, Angkatan Muda Kereta Api (AMKA) regained authority over the Indonesian railways. On 28th September 1945 at Balai Besar, Bandung, AMKA took over authority from the Allied Army. Every year that particular day is still commemorated as Railway Day. The national Railway Company changed its name a number of times. These are the various names it has had: DKA-RI (Djawatan Kereta Api Republik Indonesia, 1945-1950), DKA (Djawatan Kereta Api, 1950-1964), PNKA (Perusahaan Negara Kereta Api, 1964-1971), PJKA (Perusahaan Jawatan Kereta Api, 1971-1990), PERUMKA (Perusahaan Umum Kereta Api, 1990-1998), and PT KAI (PT Kereta Api Indonesia, Pesero, 1998 up until the present).

Gubeng Railway Station in Surabaya, 2002. At first it was intended as a mixed passenger and freight transport station but it became the main railway station for people in Surabaya.

In 1993 Indonesian railway development activities were aimed at increasing transport capacity and improving service quality so that the railways could function as a proper means of public transportation. According to a Government Ruling that came into effect on 31st October 1990, PJKA was changed to PERUMKA. That new status was to increase the chances of improving services. One of the aims of the railway company was to modernize the railway system in relation to national transportation systems. By incorporating other transportation means, passengers were encouraged to use the railways alongside of taxis, rental cars and buses. In order to increase productivity it was hoped that railway companies would optimise all the available tools and facility operations. In accordance with a 1998 Government Regulation, PERUMKA was changed to PT KAI.

Indonesian train transportation is divided into several operational areas. Surabaya, which consists of twelve stations, is included in Operational Area VIII. Gubeng and Surabaya Kota Railway Stations handle a large volume of passengers and a small amount of cargo (expedition), while Pasar Turi Railway Station deals with large volumes of both passenger and freight traffic. Wonokromo, Tandes, Sepanjang, Benowo and Kandangan Railway Stations only serve economy class passengers. In addition, Kandangan Railway Station also deals with cattle transport. Kalimas Railway Station just handles cargo while Benteng Railway Station is used for oil transportation. Sidotopo Railway Station operates as a cargo terminal and is also the place where locomotives are repaired. The Masigit Railway Station, which was once used to transport passengers, is only used as a security control point these days. The large number of railway networks present in the city demonstrates that train transportation once played a major role in Surabaya. However, recent developments show that the function of many railway networks and stations has changed.

Gubeng Railway Station, located in Gubeng Masjid Street, is a legacy from the colonial period. It was built with eleven railway lanes, six of which were used for passenger transport while the remainder were used for train shunting activities. The available buildings were: the station master's office, the office of the rail cargo manager (Pemimpin Perjalanan Kereta Api), the counter, the warehouse, the signal house, restrooms and a water tank tower for the purposes of cleaning the locomotives. The train operation system was a double track and manual signalling set up.

After the end of the colonial period, the six railway lines continued to be used for passenger transport and a platform (i.e. peron) was also provided. The routes covered by this station are: Surabaya-Jakarta, Surabaya-Bangil, Surabaya-Blitar, Surabaya-Kiara Condong, Surabaya-Bandung, Surabaya-Yogyakarta, Surabaya-Kertosono, Surabaya-Banyuwangi-Denpasar, Surabaya-Purwokerto, Surabaya-Lempuyangan, Surabaya-Jember, Surabaya-Malang, and Surabaya-Jombang. Every day there are sixty scheduled trains including executive, business and economy class services. The station currently consists of the following buildings: the Operational Area VIII office, a new station (built in June 1995), the old station (consisting of the station master's office and housing his administrative staff), the signal house (no longer in operation), a small mosque (*musholla)*, a platform, a canteen, restrooms, a counter, an administration room, a parking area and a water tank tower for the cleaning of locomotives (which is no longer operational). The train operation system already made use of electric signals.

Growing economic activities: the harbour of Surabaya

The Straits of Madura separate the island of Java from the island of Madura. Tanjung Perak, the port of Surabaya, is situated on the south side of the straits. The global coordinates are 112° 45' E - 07°12' S. The main access to the port from the Java Sea is via the north, passing through the Western Fairway and the distance from the port to the open sea is approximately 45 km. The natural depth in the western access channel is adequate for vessels with -9.0 to -9.5 m LWS (Low Water Spring) drafts and a maximum overall length of 210 m. The access channel is no more than 100 m wide. The Eastern Fairway leading to the gulf of Madura is too shallow and can only be used for small craft. The port forms the northern part of Surabaya and it was constructed on the mud flats between the mouths of two rivers: the Kali Mas to the east and the Kali Perak to the west. The naval harbour, which is approximately 106 ha big, is located to the east of the Kali Mas. To the west of the Kali Perak the extensive mud flats continue for around 15 km, extending as far as Gresik.

Malang Railway Station, 2002 (a) exterior (b) interior. The Malang Railway Station, at the end of the Daendels Boulevard and to the east of Alun-Alun Bundar, is the only station in the city. Constructed in 1930 it was based on the State Railway (Jawatan Kereta Api) design and its architects were from the Country's Building Service (Jawatan Gedung Negara).

Tanjung Perak Harbour in Surabaya, 2002

In the early colonial period (1875) the port was designed for serving *perahus* (local boats) from the Kali Mas. Large ships could not unload directly onto the quay but they could anchor in the outer harbour and be unloaded with the help of *perahus*. The Kali Mas is 30 m wide. Its berths are still generally used for sailing *perahus* with a length of 22 m and a draft of about 2.0 m. Motorized *perahus* with an approximate 50 DWT (Dead Weight Tonnage) are generally directed to the central berths. The total length of the quay for the *perahus* is 1,170 m. Local vessels, ranging up to 250 DWT, generally moor at the downstream berths nearest the mouth of the Kali Mas. The local quay is 830 m long and it has a design depth of 3.8 m.

In 1906 the engineer W.B. Van Goor, who headed the BOW department between 1908 and 1911, made a very thorough study of the harbour. In 1910 two Dutch engineers, Professor Jacob Kraus and G.J. de Jongh, recommended building a harbour with quay walls. The modern port of Surabaya was thus further developed in 1917. A choice had to be made between developing a system whereby ships could be unloaded exclusively with the aid of *perahus* and a system involving direct unloading in the harbour quay area. In view of the fact that Surabaya possessed a safe natural harbour, it was a difficult decision to make. Ultimately it was the second system that was opted for.

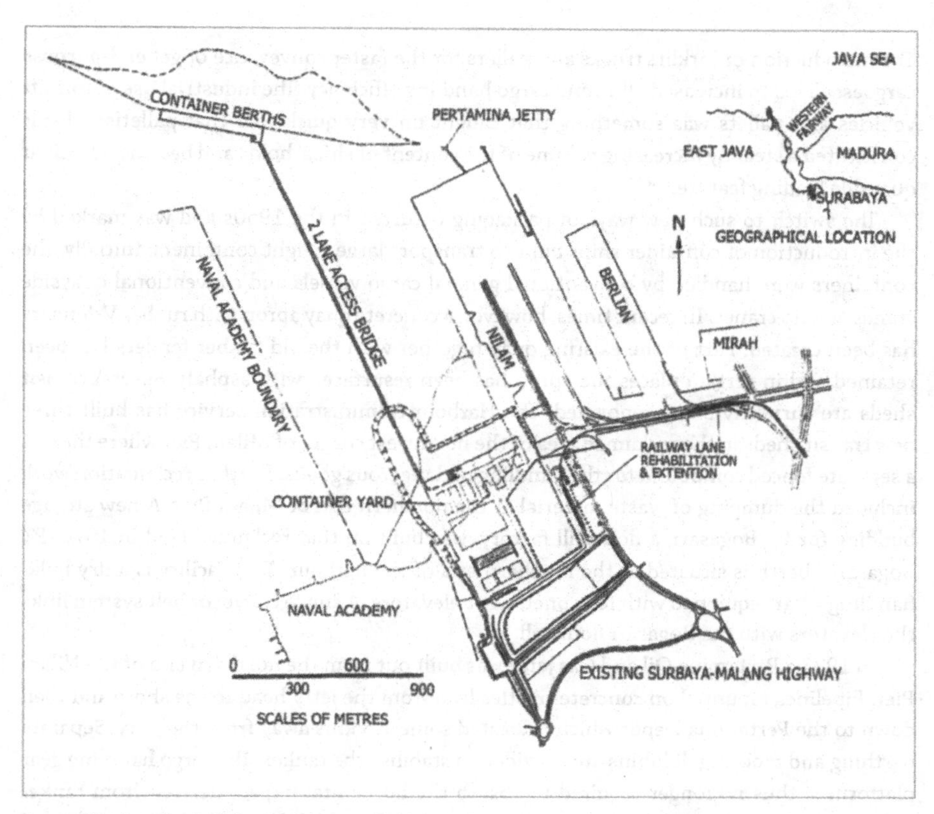

A map of Tanjung Perak Harbour in Surabaya, 2002.

Caissons were implemented. Caisson construction facilitated the erection of closed structure gravity quay walls. It was an engineering method that rapidly caught on and turned into a production line style of construction. Concrete caisson building is a type of wall structuring that requires firm, non–eroding foundations. A rock blanket was created over loose ground to prevent scouring. The Hollandsche Beton Groep Ltd. introduced 'tilting caissons', a concept that directly originated from the Valparaiso designs created by Professor Kraus and Dr. A.C.C.G. Van Hemert. The company also used another type of caisson, the type known as the Rotterdam caisson which had been developed and used in the Dutch harbour of that name. This demonstrates that Dutch civil engineering techniques were adopted in the Dutch East Indies but the reverse also applied: experience gained in the Dutch East Indies was later transferred to Rotterdam. Before the period of independence, Tanjung Perak Harbour consisted of several quays, all of which are described in Appendix 8.

After the dawn of Indonesian independence, the whole goods transportation by water sector speedily expanded in line with world trade developments and the increasing technical ability to build larger ships and cargo-handling facilities. Bigger ships transporting more goods started arriving at the port which meant that better handling facilities were required.

The introduction of forklift trucks and pallets for the faster conveyance of general-purpose cargoes, all led to increased all-round cargo handling efficiency. The industrial use of forklift vehicles and pallets was something that caught on very quickly so that palletised loads constituted a steadily increasing volume of the content of ships' holds and became a familiar quayside loading feature.

The switch to such new ways of packaging occurred in the 1950s and was marked by the introduction of container ships built to transport large freight containers. Initially, the containers were handled by conventional general cargo vessels and conventional quayside cranes or ship cranes. In recent times, however, a concrete quay apron with rubber V-fenders has been created. Part of the existing quay together with the old timber fenders has been retained and in certain places the apron has been resurfaced with asphalt. Several transit sheds are currently being renovated. The Harbour Administration service has built three new transit sheds in the reclaimed area of the northwest corner of Milam Pier where there is a separate fenced compound for the handling of dangerous goods. Further reclamation work included the dumping of waste material at the northern end of Milam Pier. A new storage building for PT Bogasari, a flourmill factory, was built on that reclaimed land in 1970. PT Bogasari's berth is situated at the northern end of the harbour. That facility is a dry bulk-handling wharf equipped with four pneumatic elevators. A single conveyor belt system links the elevators with the Bogasari flourmill.

In 1973 a Pertamina Oil and Gas jetty was built out from the northern end of the Nilam Pier. Pipelines mounted on concrete trestles lead from the jetty head to the shore and then down to the Pertamina Depot which is located some 1.5 kms away from the jetty. Separate berthing and mooring dolphins are provided to stabilise the tanker. The cargo handling gear platform is thus no longer required to absorb the horizontal impact derived from tanker berthing as the tanker only makes contact with the berthing dolphins. The dolphin is formed by a cluster of raking piles covered with a concrete pile cap that serves to hold their tops in place while a series of rubber fenders lined up on the berthing side provide the necessary resilience.

Excluding bulk petroleum import, the traffic that passed through Tanjung Perak in 1975 was 3.92 million tons and the provisional total for 1980 was 5.76 million tons. This means that in the 1975-1980 period the average traffic growth proceeded at a rate of 10.8 percent per annum. The average occupancy of the 40 berths in December 1980 was 82%. Container traffic has increased rapidly since 1976. The number of TEU's (Twenty feet Equivalent Units) handled in 1980 was approximately tenfold that handled in 1976. The container conveyance level in 1980 was 19,230 TEU's, and the cargo volume in terms of containerised goods was 139,015 tons in 1980. Virtually all container movements in Surabaya pertain to foreign imports and exports.

In 1981 eight new inter-island berths were constructed along the Perak and Mirah Wharfs and an additional 640 m of quay space was created with a berth-side water depth of 6 m. The supplementary facilities were four transit sheds with a covered storage area of 13,700 m^2 and 15,612 m^2 of open storage space together with 110 m of quay length for the Navigation Aids Division and related buildings such as workshops, storage sheds and offices. Open structures with suspended decks such as flat slabs with drop panels mounted on prestressed concrete piles function as reserve wharfs positioned adjacent to the protected natural slope.

A short-term plan was conceived in order to accommodate the rapidly rising container traffic within the existing port. The Berlian Pier was converted into a container berth without a shore crane and the Nilam Pier was turned into two combination berths. The engineering work on both the Nilam and the Berlian quays purely involved surface work, i.e. the demolition of the transit sheds, the reinforcement of the surface, new paving for container handling, electrical services, external services and alterations made to the existing roadway. All those projects were completed in 1983.

In 1986 a container terminal – accessed by a jetty – was built in the offshore area situated to the west of the existing commercial port. Two new container berths provided 500 m of quay with a quayside water depth of 10.5 m. The container yard was located 2 km away from the jetty, had an access bridge approach of some 1.5 km and was 200 m away from the causeway. The soil within this zone was found to be 'very soft' to 'soft'. In conjunction with the soil condition it was decided that the most economical plan would be to build an offshore jetty for the container berths that had a deck suspended on tubular steel pipe piles. The access bridge used an open structure for the suspended slab with a prefabricated beam and an *in situ* cast slab on tubular steel pipe piles. The causeway and the reclaimed area covered a total area of 12 ha. All the necessary soil improvement conditions were met by installing vertical drains in conjunction with preloading.

The Container Terminal provided supporting facilities such as a Container Yard, a 10,500 m^2 Container Freight Station, an Administration Building with a total area of 2,160 m^2, a water tank, a water tower, a workshop and various substations. An extended container berth that was 500 m long was built to the west of the existing berth. A similar structure to the previously constructed container berth was completed in 1997.

Developing water management: irrigation and drainage systems

Surabaya

In Surabaya water management relies on the big river that flows through Surabaya: the Surabaya river, a branch of the Brantas river. In the Brantas river system development plan, the water of the Brantas river is largely diverted to the Porong river – not to the Surabaya river – by means of the Mlirip gate and the Gedeg sluice.

At first, the full flow of the Surabaya river used to be discharged into the Straits of Surabaya via the Kali Mas. In 1856 the Wonokromo Canal and the Jagir Weir were built. These structures were designed to create a shortcut from the Surabaya river to the sea so that the Kali Mas would be prevented from flooding Surabaya. When there are high discharges, the Wonokromo sluice is closed and the Jagir Weir gates are opened. Another function of the Jagir Weir is to provide raw water for the PDAM in Surabaya – in other words, the city's local drinking water company – which is later processed to make it fit drinking water for people in the city of Surabaya.

The Surabaya draining basin is divided into three main departments composed of the Western Surabaya, the Southern Surabaya and the Eastern Surabaya basins. These drainage basins are subsequently divided into the following sub-basins:

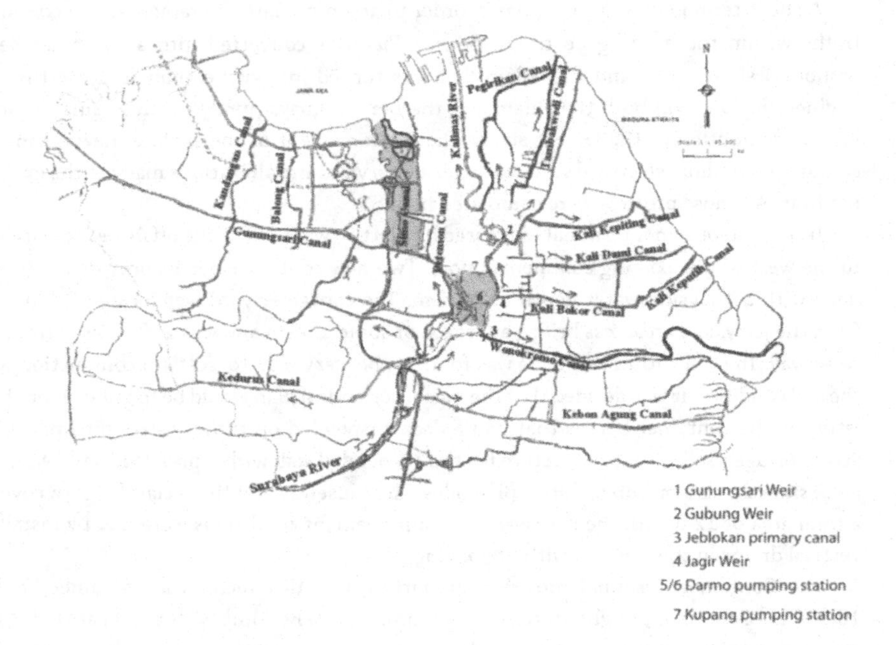

A map of the irrigation and drainage system in Surabaya, 2002.

1. The Western Surabaya basin (including Kalimas, Greges, Gunungsari, Kedurus and Karangpilang),
2. The Southern Surabaya basin (including Wonorejo, Kebon-Agung and Perbatasan),
3. The Eastern Surabaya basin (including Medokan, Bratang, Kalibokor-Keputih, Kalidami, Kalisari-Kalikepiting, Kenjeran, Kenjeran Utara-Kedung Cowek and Pegirian-Tambaksari).

The Kali Mas basin includes the Wonokromo sluice and the Gubeng Weir with its three pumping stations: Darmo (2.97 m³/s), Kupang (3.94 m³/s) and Gunungsari (0.5 m³/s). The Kali Mas itself is the main drainage canal for the housing area through which it passes. As this was a situation that created a threat to public health, pumping stations were built at Darmo, Kupang and Gunungsari to give the water higher respective elevations.

During its development, the Darmo pumping station's capacity proved to be inadequate because it received extra water from the satellite city, water that was supposed to drain into the Gunungsari Canal. Besides, the original Darmo pumping station site on Diponegoro Street was not a favourable site for the discharging of water into the Kali Mas. The Darmo pumping station was therefore relocated to Darmo Kali Street in 1982 and the pump capacity was increased to 7.5 m³/s. Similarly, the Kupang pumping station increased its capacity (to 6.5 m³/s) in that same year. Meanwhile, both the catchment area of the Gunungsari pumping station and its capacity were increased to alleviate the Darmo pumping station.

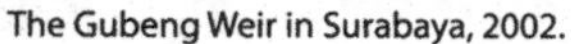
The Gubeng Weir in Surabaya, 2002.

The Gunungsari Weir in Surabaya, 2002.

The Kalibokor intake in Surabaya, 2002.

The Morokrembangan Retention Basin, which is located in the downstream reaches of the Greges Canal serves the western part of Surabaya, the Morokrembangan area, Greges Canal, Kali Dupak and the Pesapen pumping station (1.51 m^3/s). The storm runoff water originating from Western Surabaya is discharged into the sea through the Morokrembangan Retention Basin. By maintaining sufficiently low water levels in the Morokrembangan Retention Basin, which comprises a reservoir and a tidal gate, gravity flow is made possible in the Greges Canal. When the water level in the reservoir rises above the sea level, the tidal gate is pushed open and the water automatically flows into the sea. Conversely, when the sea level is higher than the water level in the reservoir, the sea automatically shuts the tidal gate. The Morokrembangan Retention Basin still functions effectively today.

Major sections of the Greges Canal, including the Petemon and Dupak Canals, have been improved to reduce inundation in the areas of Wonorejo, Embong Malang, Petemon and Pasarturi. The rehabilitation involved widening the canal beds by around 1.1 m, without allowing that to affect the Embong Malang Street catchment area. The Kenari pumping station was introduced to serve the Embong Malang Street catchment area.

In Surabaya the idea behind the irrigation plan was to make use of the water issuing from the branch rivers of the Brantas, i.e. the Surabaya river and the Kali Mas. At first, there were two main systems of intake structures that provided irrigation water for Surabaya, i.e. the Gunungsari and the Gubeng Weirs. The Gunungsari Weir, situated in the Surabaya river, was designed to raise the upstream water level in order to supply six irrigation canals. Those canals were the Gunungsari Canal, the Rowowiyung Canal, the Karah Canal, the Jambangan Canal, the Kebonagung Canal and the Simowau Canal. The Gunungsari Weir also supplied water to other intake canals such as the Kedurus intake canal, Bogangin intake canal, the Kemlaten intake canal and the Wonokuli intake canal.

The Gunungsari Canal supplied four secondary canals: Petemon Canal, Simo Canal, Balongsari Canal, and Kandangan Canal. Today, the irrigation function of these four canals has become a drainage function instead. By maintaining the original irrigation intake structures such converted irrigation canals – or present-day drains – can in principle be flushed with fresh water in the dry season, during dry spells and in the wet season, thus making an important contribution to water quality control. A renovation programme was initiated in 1980 when the Gunungsari Weir was provided with four sets of fixed single gates and one set of double gates to control water levels. Between 1982 and 1987, the Department

Table 2. **Areas and discharges of the Gunungsari Weir**

Name	Area (ha)	Discharge (l/s)
Gunungsari	678	100
Karah	25	30-80
Jambangan	50	40-130
Kebon Agung	499	100-1070
Simowau	97	40-200
Kedurus	49	-
Bogangin	214	-
Kemlaten	27	30-70
Wonokuli	24	0-90

of Public Works at the Wonokromo section recorded the discharges and irrigation area in the Gunungsari Weir system. (See Table 2).

The other weir system, the Gubeng Weir system, was also designed to elevate the upstream water level of the Kali Mas. This system comprised one main structure, which was built around 1907. The weir included a needle weir and a concrete weir controlled by stop logs. The Gubeng Weir was renovated in 1993 and provided with an inflatable rubber dam (a so-called bellows weir), the main structure being a steel gate. The Gubeng Weir was designed to provide water for two irrigation canals: the Jeblokan primary canal, with its intake just upstream of the Gubeng Weir and the Kalibokor primary canal, with its intake around 2.5 km upstream of the Gubeng Weir.

The Jeblokan primary canal was designed to provide discharges as substantial as 400-450 litre/second (l/s) in the dry season, and 250-300 l/s in the rainy season. The Jeblokan primary canal was divided into the Wonosari secondary canal, the Kaliondo secondary canal and the Rangkah secondary canal. At a later stage, the Jeblokan Canal was converted into a drainage canal to serve the industrial areas that replaced tracts of farmland. In 1988, the irrigation area supplied was only 260 ha. The other canal, the Kalibokor primary canal, was designed to provide discharges as large as 300 l/s in the dry season. It was divided into the Manyar secondary canal, the Medokan secondary canal and the Keputih secondary canal. Like the Jeblokan Canal, the Kalibokor Canal was also changed into a drainage canal. In 1988, the area it irrigated was only 70 ha.

The Gunungsari Canal discharges storm drainage water that falls in the eastern part of Surabaya and derives from the northern slopes of the Gunungsari hills. At first, that area was still an open plain (with no housing), which meant that there was no need for an urban drainage system. The only canal in the area was the Gunungsari Canal, which was specifically designed for irrigation purposes. This canal followed the natural contours of the area, which is slightly hilly. After Gunungsari had been turned into an industrial housing area, the Gunungsari Canal started being used as a drainage canal. Up until now, the Gunungsari Canal has not yet been renovated. It discharges water into the sea with the aid of four other drainage canals, all of which are former irrigation canals. They are the Petemon Canal, the Simo Canal, the Balongsari Canal, and the Kandangan Canal.

The Kedurus Canal drainage system discharges storm drainage water originating from the south of Gunungsari hill and from the flat Gunungsari hill terrain into the Kali Mas. The Kedurus Canal functions as the main drain by directing water into the Surabaya river. During its development, the Kedurus Canal catchment area increased as a consequence of the rapid population growth in the Kedurus Canal vicinity. For that reason, the local government in the province of East Java planned its rehabilitation.

The Pegirian Canal, which commences on the east bank of the Kali Mas in Ngemplak, extends 8 km to the north, joins the Tambakwedi Canal at its mouth and enters the Madura Straits in Kampung Bulak via the Tambakwedi tidal gate. Starting at the Ngemplak sluice gate on the Kali Mas, the Pegirian Canal passes through the Gembong and Jatipurwo sluice gates. The boundaries of the Pegirian Canal catchment area are the Kali Mas (to the west) and the Gubeng railway leading to the Semut Railway Station (to the east) with a total area of 902.07 ha. The Pegirian Canal functions as a drainage canal in the wet season and as a thoroughfare for wastewater disposal in the dry season. When the land use switched from farmland to residential land, the design discharge of the Pegirian Canal rose from 4.87 m^3/s to 52.2 m^3/s. A rehabilitation programme was set up in 1982 and the slope of the canal bed was changed from 0.00001 to 0.00026, while the width remained constant (15 m). The canal was also lined in order to increase its discharge capacity, prevent plant growth on the sides, stop dirt adhesion and facilitate cleaning and maintenance (especially after sedimentation).

The Tambakwedi Canal, with a catchment area of 1016.81 ha and a length of 5 km, starts from Kapas Krampung Street. It first joins the Pegirian Canal before finally entering the Madura Straits via its two tidal gates. The catchment area of the Tambakwedi Canal is bordered by the Jeblokan irrigation canal (to the east) and a railway line (to the west). Much like the Pegirian Canal, the Tambakwedi Canal is a means of waste water disposal in the dry season. Due to the population growth factor the design discharge increased from 11.4 m^3/s to 61.15 m^3/s. A rehabilitation programme was completed in 1982 and the slope of the canal bed changed from 0.00016 to 0.00026, while the width remained constant (20 m).

The Kalidami Canal catchment area, which is 1500 ha, is surrounded by the Kali Mas (to the west), the Kali Kepiting (to the north), the Kali Bokor (to the south) and the Madura Straits (to the east). The Kalidami Canal used to receive discharges from the Kalibokor irrigation canal but after the Kalibokor irrigation system ceased to exist it became a drainage canal. The catchment area dropped to 945.7 ha. The length of the Kalidami Canal is 3.93 km, the canal bed width is 22.8 m and the canal bed slope is 0.00033 with a discharge capacity of 44.8 m^3/s.

Three canals were constructed to transport waste water. This is particularly important in the dry season when water availability is limited and bad smells abound. These canals are the Jeblokan Canal, the Kalibokor Canal and the Pegirian Canal; they all receive water from the Kali Mas.

Malang

Malang lies 367-380 m above sea level and it is surrounded by the Semeru, Kawi, Anjasmoro and Arjuno volcanoes. Most of the city area (96.3%) is rather flat with a slope of 0-15%, and the remaining part (3.7%) consists of land with higher slopes (10-16%). The Brantas river

Grahadi on Simpang/Pemuda Street in Surabaya, 2002. Dirk van Hogendorp, former-governor of East Java, constructed this building between 1794 and 1798. It was later expanded by governor-general Daendels, who gave it its Empire Style motives with Doric columns. The building was renovated in 1870 and became Surabaya's Mayoral residence. In the 1900s the then mayor, Kroesen, built a park in front of the building. It was named the Kroesen park. Nowadays the building is the governor's office and it is used for receiving guests.

Stadhuis van Soerabaja/City Hall of Surabaya on Ondomohen Street, 2002. This building was planned in 1916-1917 but its construction was not completed until 1925. At the front, the building was 102 m wide and had a concrete frame as its main structure with a steel truss for the roof. It should be noted that pile foundations were used to support the columns. The newly built city hall only constitutes the rear part of a larger building plan. Due to financial problems the front part was never realised. At the same time that the city hall was built, the new mayoral residence was also constructed on the same street.

The Simpang Society building on Pemuda Street in Surabaya, 2002. This building, which was constructed in 1907, has a tower in the shape of a dome. A terrace was constructed around the whole building. The roof was made of corrugated zinc-coated steel, which was supported by circular steel pipes in the form of columns. In 1930 the annex – adjacent to the main building – was built for parties and receptions. The building is now used for cultural exhibitions.

passes through the city, as do its tributaries: the Bango, Metro, Sukun and Amprong. These rivers have a dual function because they serve as natural drainage channels and irrigation supply arteries for Malang. The drainage system of Malang was developed in the colonial period. During that period the Dutch government constructed 'drainage pools' as a main support for the natural rivers. These drainage pools were designed to act as retention basins in order to attenuate floods *(banjirs)* in the rainy season.

In the post-colonial period, the local government in Malang endeavoured to retain as much as possible of the Dutch structural heritage, including the existing drainage systems. However, in connection with all the urban development plans, the Dutch drainage pool areas had to be converted into residential, commercial and industrial areas. The rivers are therefore practically the only remaining parts of the original main drainage system.

Apart from the natural rivers, which serve as the main drainage system, there are some secondary and tertiary drainage systems that take the run off from the urban catchment areas and direct it to the rivers. These secondary drainage systems function as rainwater, domestic waste water, and industrial waste water transport systems between these areas and the primary drainage systems (i.e. the rivers). Because of the limited space available for drainage channels, the drainage system used in Malang is actually a combined system, which means that waste water and rainwater flow through the same channels. The disadvantage of this system is that the design discharge happens to be the chosen maximum discharge level for waste water and rainwater. In the dry season when rain is scarce, only waste water will

The Second Roman Catholic Church (Birth of the Virgin Mary) on Tempel Street, now Kepanjen Street, in Surabaya, 2002. This building, which is now known as the Kepanjen Catholic Church, was constructed in 1899. At the front, two towers were constructed, one to the left and one to the right. The main feature of the building is its tall stone columns.

The Amsterdam Trading Society on Komedieplein, now Merak Street, in Surabaya, 2002. This building is now used as the PTP XXIV and PTP XXV building (for the nation. forestry companies). It took five years to construct (1920-1925) and was the largest concrete structure in Surabaya in its day. Some 3000 m³ of concrete was required for t construction. The foundations and structural consultant was Oltman, chief engineer a the Semarang-Joana Steam Tram Company (the SJS). A steel construction was used fc the roof, which was built by a construction company in Surabaya. The roof and floor t were imported from the Netherlands. Belgian marble was used for the entrance hall, stairs and corridors. The ceiling panels were made of locally grown teak. The building surrounded with galleries to prevent rainwater seeping in. As a consequence, around whole perimeter of the building the roof was supported by columns.

flow through the channel and, because of the gradual slope, sedimentation easily builds up on the channel beds. The drainage system for Malang's waste water makes use of open and closed channels (most of which are a hangover from the Dutch era). That waste water flows on until it links up with the natural main drainage system (i.e. the rivers). Malang has closed channels (for sewage) with a total length of approximately 445 km, 78 km of open channels (ditches) and about 367 km of underground ditches. According to topographical classification Malang's drainage system can be divided into five catchment areas.

The rapid population growth has changed the land use of the city. What was once agricultural land has become a residential area. That change has also affected the city's water management system, which means that in certain places flow coefficients have changed. Channels that were formerly used as irrigation canals have been turned into drainage canals. For example the Kadalpang Canal, which was previously geared to irrigating farming areas as large as 1284 ha, has now become a city drainage channel.

Accommodating more people: buildings and city planning

In Indonesia building development commenced in the 19th century. On the whole, the kinds of buildings constructed were government and administration offices or forts designed to reinforce the city's defence systems. In East Java, such development was concentrated in strategic big cities like Surabaya and Malang. The development in those two cities grew rapidly but in Surabaya the developments did not involve an urban planner; so urban planning was not well controlled. It was different from developments in Malang where the well-known engineer, Thomas Karsten, was the urban planner. Many structures built in these two cities are still well maintained as evidenced from the photographs shown on these pages. Precisely how the two cities developed will be described in the following sub-sections.

City development and building structures in Surabaya

The development of Surabaya started in the downtown area located around Jembatan Merah (Roode Brug) where, in the colonial era, many buildings were constructed. At that time, buildings were constructed with the aid of tall stone columns. These buildings had gable walls at the front and were constructed in many different styles. Buildings of the gable construction style were usually built near the Kali Mas and the Roode Brug. Directly in front of the Roode Brug, the Dutch resident's office was erected (at that time it was known as the city hall). In front of that building there was a sizeable park known as Willemsplein (nowadays Jayengrono Park). That building was demolished in 1930 and the governor's office on Pahlawan Street was constructed in its place.

Generally it was colonial style buildings that were constructed where the main structure rested on stone columns and everything was in the 'Empire Style'. Early in the 20th century, because of the limited area downtown and thanks to the availability of new materials, many buildings started being constructed with cast iron as an alternative to the large dimensions of stone columns. Corrugated zinc-coated iron was used as the roofing material, so that the slope angle could be reduced. As a result, all the newer buildings tended to have cantilevers made of corrugated zinc and cast iron cantilevers with corrugated motives and the more slender cast iron columns. Apart from anything else, buildings with cantilever structures were more appropriate for Surabaya because of its wet tropical climate. Furthermore, in connection with the climate the buildings were equipped with ventilation appliances and awnings above the windows and the doors to provide extra protection against the elements (wind and rain). Another common feature of colonial architecture was all the towers of various shapes and sizes (circular or rectangular) with little roofs on top. In present-day Surabaya certain colonial buildings still have such towers.

All these buildings were constructed under the strict supervision of the BOW department. Besides designing state buildings (government employee housing, bureaux, the state court, schools, post office buildings, hospitals et cetera), this department was also responsible for the repair and maintenance of buildings as well as for all the hydraulic structures. Before 1926 Dutch engineers were often itinerant and would frequently be involved in the construction of different buildings. It was not uncommon for an engineer to construct hydraulic structures in one province and buildings in another. Because of such mobility, the local knowledge or experience of such engineers could not be fully utilized. Furthermore, the quality of the materials was poor, because there were no rulings on standards. On 1st January 1926 that situation changed when the Dutch administrative area was divided into provinces. In each province one engineer was put in charge of all constructions for that specific area. The repair and maintenance of state buildings therefore became more intensive. Those measures also triggered changes in the water management divisions within the provinces.

City development and building structures in Malang

In the 19th century, Malang was just a small, undeveloped regency (kabupaten) of Pasuruan. When, in 1914, it was declared a municipality the infrastructures began to develop. The city council remained in charge of many aspects such as city expansion, construction and road maintenance, structural supervision, housing arrangements and renovation, and other developments.

From the beginning, Malang grew up in the area centred around the city's town square (alun alun). Many public buildings, such as religious buildings and government offices, were built there. From 1926 onwards Malang expanded from north to south under the supervision of the Dutch town-planning consultant Karsten who was in charge of reviewing the regulations relating to Malang's old buildings. Building permits were based on building specifications and classifications in the planned locations. With the help of such classification methods, technical regulations and administrative work, it became possible to construct and plan the various necessary infrastructures.

Malang is an example of a municipality that was created on the basis of the city development regulations laid down in the Ordonansi Pembangunan Kota (City Development Order), which was later legalised and known as the National Building Code (Peraturan Bangunan Nasional). From the design, it can be seen that Karsten planned the expansion of the city to not only be centralized around the town square but that he also created a new city centre known as Alun-Alun Bunder to the east of the Brantas river. The buildings around the town square were effectively maintained.

One of the colonial buildings in the old city centre was the Catholic Church of Hati Kudus Yesus on Basuki Rachmat Street. This building was constructed in 1905 when Malang was still a regency. The church was not built in the form of a crucifix – which was common for Catholic churches – but rather as a rectangle with sides measuring 11.4 m and 40 m. Due to financial and constructional problems the two towers did not appear at the front of the church in the early constructional phases; in fact their construction was finally completed on 17th December 1930. The church's constructional style is that of a truss construction with a ceiling made of iron.

The new city centre, designed by Karsten, was located to the east of the Brantas river, because he wanted the river to be an integral part of the Malang landscape. Later the area became known as Alun-Alun Bunder and it also became Malang's civic centre. It was for that reason that, in 1927, the construction of the city hall was started to the south of Alun-Alun Bunder. The municipality took responsibility for the construction work. The unique thing about the city hall was the way in which the building's corridor was located to the rear so that only windows could be seen at the front. Additional awnings were placed above the windows to keep out rain and sunlight. The interior of the building included the meeting hall, the mayor's office and the town clerk's office, which were designed by C. Citroen. The building was completed and officially inaugurated in November 1929.

Another building in the Alun-Alun Bunder district is the HBS/AMS (secondary school) which is located to the northwest of the town square. Just so that it would not be out of keeping with the city hall the school was deliberately designed to resemble a villa. It was decided that the building should be one to two storeys high. Hard wood and tin plate

The 'Hati Kudus Yesus' Catholic Church on Basuki Rachmat Street in Malang, 2002.

obtained from the American Sheet and Tin Plate Company were used for the roofing. It took two years to construct the building (1930-1931).

All in all, many colonial buildings were built between the turn of the century and the 1930s, especially in the 1920s and 1930s. In that period, many of the buildings that were constructed used concrete for the main structure and steel for the roof. Cantilever constructions made of concrete with a span of 2 m or more started to come into fashion. Flat concrete deck roofing also began being used as an alternative to the classical sloped roofs, which frequently made use of roof-tiles, hard wood or corrugated zinc.

After the 1930s, though – in conjunction with the global economic crisis – the pace of construction became slower than it had been in the previous three decades. There was practically no actual construction work carried out after 1940 as the Dutch government was totally preoccupied with the Second World War. Unfortunately, during Indonesia's War of Independence (1945-1950) many colonial buildings were destroyed by bombs. In the first years of the post-war period building maintenance was very limited, due to the scarcity of funding. Unfortunately, new building construction also stagnated at that point.

Serving the population: drinking water systems

Surabaya

As the groundwater in the area is mainly brackish, Surabaya needs a drinking water supply system that continuously provides sufficient healthy drinking water. There are still some people, however, who draw groundwater from shallow wells for their clean water supply. The depth of these wells does not usually exceed 5 m, and the water is uncontaminated. In some areas where the groundwater is salty, ponds are used to conserve rainwater for domestic purposes.

In 1890 the drinking water for Surabaya derived from a spring located in Purut, Pasuruan and was transported by privately owned shipping and rail companies. Between 1901 and 1903, Birnie Ltd. installed a pipeline so that water could be conveyed from two different springs in Pandaan. In 1903, the regional drinking water company was established and its first director was the engineer Van Bouven. By 1906, the company had 1588 customers which testified to the growing awareness among citizens of the importance of healthy drinking water.

The Pandaan springs comprise sixteen small springs located in the Pandaan area in Pasuruan, including the Sumber Rejo, Duren Sewu, Plintahan and Sumber Soko districts, with a total discharge of 220 l/s. The water from the springs flows down to a reservoir in Tamanan which is located 38 km to the south of Surabaya and has an elevation of 103 m above sea level. The springs are protected against floods and may not be tapped for public use. The discharge is not affected by seasonal change. The quality of the water is so good that it does not need to undergo a treatment process. It only needs to be purified with hypochlorite. The water from the reservoir is transported by pipeline (a pipeline with a diameter of 450 mm) to the Wonokiri Reservoir in Surabaya.

In line with the increased demand for drinking water, a drinking water purification plant named Ngagel I was built in 1928 which drew raw water supplies from the Surabaya river. The capacity was 80 l/s. During the Japanese period (1942-1945), the capacity of that plant was increased to 180 l/s.

In 1932, the Umbulan Spring (which is located approximately 30 km to the east of Pandaan and 75 km to the southeast of Surabaya) was used to supplement Surabaya's drinking water supply. The Umbulan Spring elevation is 25 m above sea level and it is located in the Winongan district, Pasuruan. The water supply rate from this spring to Surabaya is 110 l/s. This water is pumped to Gempol through pipelines with a 400 mm diameter and then on to the Wonokiri Reservoir in Surabaya through a 450-mm-diameter pipeline. Here again the quality of the water supplied is good, so a simple purifying process is the only form of treatment needed. In the past some of the Umbulan Spring discharge was used to irrigate 800 ha of rice fields. The overall discharge of this spring is about 5000 l/s.

The drinking water company in Surabaya was handed over to the government of the Republic of Indonesia in 1950 and the overall capacity was increased to 350 l/s in 1954. In 1959, the national government funded the construction of an additional drinking water purification plant (Ngagel II) which had a capacity of 1000 l/s. This structure was designed and constructed by the French Dégrémont firm. The capacity of the Ngagel I drinking water purification plant was raised from 350 l/s to 500 l/s in 1977, and further increased to 1000 l/s in 1980. In 1982, with the rise in Surabaya's population and subsequent increased demand for drinking water, the Ngagel III drinking water purification plant was built on licence by Neptune Microfloc, America. The construction was finished on 1st June 1982 and the capacity was 1000 l/s. In the interests of further improvement, another drinking water purification plant was built, in 1990, again with funding from the World Bank. It was named Karang Pilang I and it had a capacity of 1000 l/s. In the following decade, the capacity of all these plants was increased and the Karang Pilang II drinking water purification plant was created.

At present, more than 90% of the fresh water production for Surabaya comes from the Surabaya river and its tributaries. It is processed in five water purifying plants in Ngagel and Karang Pilang, the total production of which is 7,500 l/s and the water supplies 70% of the population (which is approximately a total of three million people). There are plans to build the Karang Pilang III and IV drinking water purification plants in the future.

As well as the above-mentioned plants, there is also a temporary plant in Kayun with a capacity of 100 l/s. It obtains its raw water from the Kali Mas. To meet the standard quality demands for drinking water, a complete process needs to be adhered to in order to treat the raw water from this river.

The pipes installed for the PDAM Surabaya distribution system vary in diameter as well as in type and material. In general, pipes installed before the 1970s have diameters of 400 mm to 800 mm and are made of steel or ductile iron. Where the pipes or diameter is less than 100 mm such pipes are usually made of galvanized iron. Since 1990 the material for pipes used for house connections has been replaced by Polyethylene (PE).

The whole distribution system within the service radius of the PDAM Surabaya now has a total length of approximately 4500 km and consists of various piping materials with diameters ranging from 80 to 1600 mm. Most of the secondary lines use steel piping for the large diameter pipes whilst for most of the tertiary lines PVC pipes are preferred.

In 1998, the population of Surabaya was 2,806,132. The drinking water supply was 7,500 l/s (236,520,000 m^3/year) and losses – due to leakage – 85,463,000 m^3/year (36%). The fee for drinking water in Surabaya varies; Table 3 shows just how it is determined.

Malang

Malang's clean water comes from springs in the city's vicinity. Since 1915 (during the time of Dutch rule), Malang has drawn all its clean water supplies from the Karangan Spring. Later, due to increased demands for clean water, other springs such as the Sumbersari and Binangun Springs also started being used.

Until 1980, various mains supply lines were created, as well as reservoirs, e.g. the Betek and Dinoyo Reservoirs. The drinking water distribution pipeline network was extended too. Several years later the PDAM, Malang's local drinking water company, increased its capacity when it started drawing water from the Wendit Spring (with an effective discharge that rose to 500 l/s). Later, in order to meet the demand for clean drinking water,[2] for around 60% of the city's population during the next ten years, the PDAM started developing and making use of the Ngesong (II) and Banyuning Springs, which together have a capacity of 140 l/s.

Table 3. **Retribution**

Customer classification	Description
Class I	Public facilities such as fire hydrants, public conveniences and places of worship
Class II	Public facilities that generate funds (such as hospitals and orphanages)
Class III	Entrepreneurs, small industries, state institutions and embassies
Class IV	Trade and industry
Class V	Harbours
Class VI	Housing

Table 4. **Water capacity of each spring**

Spring	Elevation (meters)	Production	
		m3/year	L/s
Binangun	+ 840.12	6,773,932	215
Karangan	+ 721.35	1,009,152	32
Sumbersari	+ 756.84	409,968	13
Wendit	+ 756.84	15,768,000	500
Ngesong and Banyuning	+ 428.00	4,415,040	140

Source: PDAM of Malang city 1993

Those two springs are located in the administrative district of Batu, near Malang.

On the basis of its topographical situation and the system devised for supplying clean city water, Malang is divided into two areas: the Upper and Lower areas. The Lower area draws its water from the Binangun and Karangan Springs, while the Upper area gets its water from the Wendit Spring. The water capacities of each spring for the city of Malang are given in Table 4.

Most of the water from these springs is transported to the distribution reservoirs. Some of it is the distributed directly to customers. The distribution reservoirs are located on Dinoyo Baru, Dinoyo Lama, Betek, Mojolangu and Buringbawah. The lines of distribution are elaborated upon in Appendix 8. In 2001 the population of Malang was 894,563 only 356,865 of whom (40%) benefited from piped water supplies. The annual raw water supply level was 24,371,020 m^3.

The development of concrete and steel codes in Indonesia

In 1955, Indonesia published its own concrete building construction regulations which came to be known as the 1955 Indonesian Concrete Code (Peraturan Beton Indonesia 1955). This code adopted the Dutch 1950 GBV (Gewapend Beton Voorschriften 1950) stipulations for reinforced concrete. At universities, the structures that civil engineering students were

Surabaya Post building, 2002.

The Sheraton Hotel in Surabaya, 2002.

taught to design were based on this code. It is evident that the colonial influence is still very strong in Indonesian civil engineering. Even though Indonesia had its own code in the 1960-1970 period the actual construction side was very limited. However, since then the maintenance of existing buildings had improved.

In the early 1970s Indonesia embarked on the first stage of a five-year development plan. Many new buildings were erected, especially in the larger cities like Jakarta and Surabaya. The building codes continued to develop, a situation that was marked by the establishment of the 1971 Indonesian Reinforced Concrete Code (PBI, 1971). Most sections of the 1971 PBI still adopted the 1962 Dutch GBV and, to a small extent, other regulations were adopted such as those of the CEB (Comité Européen du Beton), the FIP (Fédération Internationale de la Précontraint) and the AIJ (Architectural Institute of Japan). With the establishment of this code, the designs of new structures were forced to follow stricter guidelines. In civil engineering courses at Indonesian universities it was that new code that was adhered to when it came to the matter of structure design.

One example of the buildings constructed using this PBI 1971 in Surabaya is the administration office of the Surabaya Post, a local newspaper in Surabaya. The building was built in 1979 with concrete covered steel (composite) as the main structure. The foundation was bored pile, due to the low bearing capacity of the soil. The structure span of this five-story building is 6.5 m. Concrete covered 'bondex' (i.e. a composite floor deck) was used for the floor. Structural bolt connections were designed according to British Standards.

From 1980 to 1990, the construction of new buildings in Indonesia grew even faster than before. This was partly due to the arrival of Indonesian experts who had studied abroad. Many Indonesians studied abroad, especially in the United States, to get their doctoral degree in civil engineering in the late 1970s. When they returned to Indonesia in the 1980s, they played an important role in the development of the Indonesian concrete code.

In 1991 the Concrete Code was renewed with the establishment of Tata Cara Perhitungan Struktur Beton untuk Bangunan Gedung (Concrete Structure Design for Buildings) SKSNI T-15-1991-03, which was mostly adopted from the ACI (American Concrete Institute). The major differences with PBI 1971 are the load and resistance factor and shear analysis in which SKSNI T-15-1991-03 takes account of the matter of equilibrium and compatibility torsion.

For three years, starting from 1994, the government published the application of this code. For the construction license proposal, the local government required the structural design using SKSNI 1991. This condition ensured that lecturers in universities taught their students to design structures based on the latest code. In 1990-1996 the high-rise building construction in Indonesia, particularly in Jakarta, reached its peak. This phenomenon was supported by the availability of advanced structural analysis programs such as ETABS and SAP90.

In Surabaya, many high-rise building constructions appeared in the 1990s. One of the high-rise buildings located in the centre of the city is the Royal Regency Condominium, which was built in eighteen months, starting in December 1994. The main structure is made of shear wall, core wall, prestressed and reinforced concrete, fixed by raft foundation on piles to the hard soil layer. The building consists of 25 stories and 2 basement stories. Another high-rise building is the Surabaya Sheraton Hotel, which was built in 1996. This 28-story building was designed with shear walls as the main structure.

The pace of development of steel multi-story buildings in Indonesia was not as fast as for concrete multi-story buildings. Before 1983, steel buildings were designed according to the Dutch code VOSB 1963 (Voorschriften voor het Ontwerpen van Stalen Bruggen 1963). This code was generally adhered to in Indonesia for bridge and other steel structures. The steel structure code was renewed with the establishment of PPBBI 1983 (Peraturan Perencanaan Bangunan Baja Indonesia, i.e. Indonesian Steel Construction Code 1983). This code adopted the AISC (American Institute of Steel Construction) Code and the Australian Standard AS 1250.

For road design and construction, the Department of Public Works established some standards or codes, such as the Standard Specifications for Flexible Pavement Design (1983 and 1987), Standard Specifications for Rigid Pavement Design (1985), Standard Specifications for Rural Roads (1990) and the Indonesian Highway Capacity Manual (1997).

In the early stages of Indonesia's post-independence history (between 1945 and 1970), practically no amendments were made to bridge codes. The colonial codes (including the 1932 AVBP and the 1963 VOSB) were still in force during that period. In the 1970s, highway design codes started being published by the government (the Directorate General of Highways and the Department of Public Works), such as various codes and specifications on highway design, building design and bridge design. Indonesia also began to adopt codes from other countries like the United States of America, Britain and Australia. In 1971, Indonesia established a Concrete Design Code (Peraturan Beton Indonesia 1971), which also partly applied to bridge designing.

In the following two decades (1971-1989), there were practically no significant changes in bridge codes except in the area of load stress. Stipulations on tee (T) sections and composite materials were introduced to the codes. In the 1989-1992 period Indonesia started working on a new bridge design code in collaboration with Australia. Those efforts culminated in a code that was established in 1992 and came to be known as the Bridge Management System (BMS'92). The BMS'92 was based on limited state analysis. It only took into consideration bridges that were less than 100 m long and the load derived from it was excluded. Another code, the Standard Steel Bridging code for Indonesia, which was established in 1990, was also influenced by the related Australian code. In November 2000, a code for concrete bridge design, the Tata Cara Perencanaan Struktur Beton untuk Jembatan (an Early Concept), was established. This code was very much influenced by the related American code (certain parts of which were adapted from the AASHTO, i.e. the American Association of State Highway and Transportation Officials).

Engineering education

In Indonesia civil engineering education has seen great innovations. In the Soekarno era (1945-1967) it became, over the course of time, a recognised undergraduate programme. Soekarno, the first president of Indonesia, was an engineer who had graduated from the Department of Civil Engineering within the Bandoeng Technical College, a well-known Indonesian institute that was established in 1920 and gained university status in 1959 when it became the Bandung Institute of Technology (ITB).

During the following Soeharto era (commencing in 1965), the Postgraduate Highway Engineering Programme was established at ITB in 1975 in cooperation with the Department of Public Works. The programme was established with the support of the World Bank, the Asian Development Bank and several foreign investors. The underlying aim was to turn out competent locally educated engineers at considerably less expense than what would have been incurred had they been sent abroad for the same degree. About 50% of the lecturers came from foreign countries such as the United States of America, Great Britain, the Netherlands, Australia and Japan. Later on, other universities in Jakarta, Surabaya and Malang started to follow suit. Like ITB, they all established postgraduate programmes in civil engineering. In 1990 it also became possible to gain a doctorate in civil engineering in Indonesia.

From time to time the civil engineering curriculum itself has also undergone change. From 1945 to 1970 there were only the two specializations of 'wet' (e.g. irrigation) and 'dry' (e.g. structures) engineering. When further adjustments were made (1970-1985), civil engineering was broken down into the three fields of: construction engineering, transportation engineering, and hydraulic and irrigation engineering. Since 1985, the range of specialisations has extended to embrace five main areas: construction engineering, transportation engineering, hydraulic and irrigation engineering, construction management engineering and geo-technical engineering.

Conclusion

In this chapter Indonesia's Dutch legacy – in terms of architecture and civil engineering – has been discussed, particularly in Surabaya and Malang. Most of the colonial constructions described are still in use today, even though some of them serve different purposes than those originally intended. Indeed, that is something that especially applies to irrigation and drainage systems in the two cities. It is perhaps in certain of the design codes and in civil engineering education that the Dutch influence is most apparent. Other countries such as the United States of America, Great Britain and Australia have also, however, influenced developments in architecture and civil engineering.

The fixed weir in the Tji Liwoeng river (West Java) with its ejection sluice and suspension bridge from where irrigation water can be tapped for the Oosterslokan, 1970.

AFTERWORD

Engineers and the colonial legacy of the Dutch East Indies

In the introductory chapter to this publication, the two editors remind us that a collection of articles were published in 1941 reflecting on, at what turned out to be its eleventh hour, the colonial project known as the Dutch East Indies. In the book with the somewhat triumphant title *Daar wèrd was groots verricht* (Great things were achieved there) the former chairman of the colonial People's Council, W.H. van Helsdingen, considered what precisely the Netherlands had achieved overseas.[1] The title, a reference to Jan Pieterszoon Coen - the man who had founded Batavia and who had proclaimed that 'great things [could] well be achieved in the East Indies' - was powerful in its simplicity.

Many copies of the book must have been printed because for several decades second-hand bookshops had stocks of it and were able to supply it at bargain prices. One should also not overlook the fact that the book was published at a time when optimistic literature was welcome in Dutch society. The rapid German invasion had done much to damage the country's national pride, and simply being able to look back on a time when, thanks to its extensive colonial realm, the Netherlands had – at least in its own eyes – played an important part in world politics as a medium-sized power was something that undoubtedly did much to alleviate the pain inflicted by defeat and occupation.

Such reflections (and these were clearly the desire of many of the authors who had contributed to the publication) could also underline how progressive and selfless the modern Dutch colonial regime had actually been, and how much better than the regimes of many other countries. It should not be forgotten that, in the 1920s and 1930s, various English and French colonial administrators had been sent out to 'our East Indies' to see just how the Dutch managed to do things by, on the one hand, 'elevating' the indigenous population while, on the other, hugely increasing colonial production levels while maintaining political stability in the region. It was thus with a certain degree of smugness that Van Helsdingen concluded that he could look back with satisfaction on what had been achieved whilst, at the same time, encouraging people to indefatigably pursue the same goals, not realising how brief the continued Dutch presence in the East Indies was to be.

Science and technology as a colonial project

What then had actually been achieved in the East Indies over all those years and, especially, since the dawn of the nineteenth century? Perhaps, rather than the 40 articles collected in the book, it is the numerous monochrome and coloured illustrations within and between the chapters that provide us with a visual answer to the question of what the people of

1941 saw as the harvest reaped during those centuries of Dutch effort. Of the precisely 100 photos and drawings included in the book, more than one-third give a very traditional view of Indonesian society: landscapes predominantly at sunset or sunrise, *sawah* terraces where hard-working farmers are busy, the staff of the princely palaces and court dancers, visitors to mosques and, naturally, temple complexes and classical sculptures. The rest of the illustrations underline the Dutch input, a few show United East Indian Company (VOC) ships and the odd colonial administrator, but the vast majority of the photographs reflect on what would seem to epitomise the essence of the Dutch presence in the East Indies: modern urban architecture, stations, harbour works and roads; the headquarters of East Indian enterprises; the churches of missionaries; and weirs, dams and other irrigation works. Clearly all of these provided the clearest representation of what Coen had once termed 'what good courage permits' and of how the Dutch had converted the archipelago into a modern society designed along European lines.

What the observant reader will undoubtedly have noticed is that there is something rather interesting about the choice of photos: whilst the illustrations portraying the Indonesian world are populated with people (on the *sawahs*, in the mosques or in front of the temples), the human component is virtually absent in the photos of the modern colony. The photos of the colonial world show technological feats of engineering: dredging machines, oil refineries, rubber estates, weirs, sluices and aircraft. Even the train carriages, shopping streets and residential areas appear devoid of people. It seems as if the modern colony only consisted of concrete and machinery, as if everything functioned autonomously and as though there were no Europeans living and working in the East Indies.

It is this 'dehumanised' view of the world of the Dutch East Indies that most clearly demonstrates what was central to twentieth century colonial thinking: it was all about the glorification of the powerful hand that had brought modernity and progress which provided the objectified evidence of that supremacy. It was a supremacy that manifested itself through the rapid development of science and technology. It would appear that the elevation of the native inhabitant and the creation of a modern colonial society were things that could only be achieved through modern science and all its applications. Reinforcing this view would seem to have been the implicit mission of the contributors to *Daar wèrd wat groots verricht*. On the one hand, there was the Indonesian East Indies with its tranquil nature, sober farmers, timeless culture and idyllic village life, on the other hand, there was the Dutch East Indies: a dynamic world of technological structures and innovations which had shortened distances and enabled production levels to soar. Seemingly overnight, this had led to the construction of comfortable homes and districts, the taming of nature and the channelling of water: all forces for good.

In his rightly acclaimed study of science, technology, ideology and Western dominance entitled *Machines as the Measure of Men*, Michael Adas emphasises that such an attitude towards technology, as an 'imposer of civilisation', was not peculiar to the Dutch but rather something that could be said to epitomise the wider colonial stance. In such relationships, feelings of racial superiority clearly play a major role with the Europeans of the time seeing themselves as maintaining a monopoly on wisdom and thus knowing what is best for those they have colonised. In Adas' view, however, colonial ideology was not primarily a form of racism that sought to anchor Western superiority in an unshakeable biological hierarchy, rather it was much more a conviction that Western achievement, largely encapsulated in science and technology, could bring enlightenment and put an end to 'lives of darkness'. Such an ideology could justify colonial domination. It was the technology that gave European powers the means and knowledge to free the 'underdeveloped' world from poverty, illness, natural disasters, war and violence. It was from such a perspective that technology was seen as the basis of, and the vehicle for, the European drive to 'educate' and 'elevate' the rest of the world (its proudly called 'mission civilisatrice').

Since the middle of the nineteenth century, this 'mission to civilise' had provided the ideological legitimacy for colonial expansion and exploitation. Even though, in retrospect, one might argue that this was nothing more than a euphemism for exploitation and a way of justifying the oppression of other peoples, there was, in reality, more to it than that. Michael Adas worded it thus:

> *Undoubtedly, claims that colonial conquests had been undertaken in order to uplift African and Asian peoples could be little more than cynical camouflage for brutal exploitation, as the Belgian king Leopold II and his rapacious agents demonstrated in the Congo in the late nineteenth century. But many of those who justified imperial expansion or colonial policies in the name of higher purposes linked to the civilizing mission were firmly convinced that they were acting in the long-term interests of the peoples brought under European rule. The civilizing mission gave a moral dimension to arguments for imperialist expansion that were otherwise limited to economic self-interest, strategic considerations, and national pride. Like most ideologies, it enabled its adherents to defend violence and suffering as necessary but temporary evils that would prepare the way for lasting improvements in the condition of the subject peoples. It lent a 'humanitarian mystique' to the nasty business of conquest and domination. It gave credence to the belief that the interests of all peoples could be equated with those of Britain and France. Because of it, nineteenth-century European colonizers could speak of conquests as a 'liberation' or 'deliverance' and of repression as 'pacification'. Politicians and writers representing all positions on the political spectrum, including at times those on the left, routinely used or accepted terminology that strikes us today as Orwellian doublethink.*[2]

Sure enough, *Daar wèrd wat groots verricht* is packed with euphemisms such as elevate, improve, educate and 'rescue from the darkness and bring into the light' when referring to a population that was deemed to be insufficiently modern. Over the course of the mission to civilise, many colonial political opinions emerged. It was to be a long process, in line with its conservative style, and one that was perhaps most clearly articulated by the governor-general B.C. de Jonge who said in 1936: 'I believe that now that we have worked in the East Indies for three hundred years it may be another three hundred years before the East Indies is perhaps ripe for a form of independence'.[3] In other words, the East Indies, or rather its native people, were viewed as being far from ready to stand on their own two feet in a modern world and were therefore perceived as requiring the 'protection' of the Netherlands for a long time to come.

It was not only in colonial-conservative quarters that such technology-based paternalism reigned supreme. Also, and perhaps even predominantly, in the more progressive social-democratic circles, where people were generally critical regarding the interests and justification of colonial trading capital, there was an unshakeable belief in the mission to be completed in the East Indies before Indonesia could progress without the help of the Netherlands. For such people, even more so than for the conservative contingent, the true vehicle for enlightenment and modernisation resided in the power of machines. The engineer Henri van Kol, who was the colonial expert of the social-democratic party in the Lower House during the early decades of the twentieth century, and who had been trained as an engineer in Delft, glorified technology as a means of 'getting the peoples of backward countries onto the road of civilisation' and of eradicating 'all kinds of primitive social wrongdoings such as cannibalism and polygamy'.[4]

In short, both the left and the right within Dutch East Indian politics pulled in the same direction, both upholding a shared belief in the civilising mission and the future prowess of technology. In a recent study, Rudolf Mrázek alluded to this world in the title of his book *Engineers of Happy Land*. If the message of the civilising mission was truly encapsulated in technology, then the colonial engineers were the natural prophets and implementers of such a message. From the mid-nineteenth century onwards, Delft engineering graduates, a relatively large percentage of whom found employment in the colony, increasingly began to take over certain tasks from the old guard of administrators. The young engineers were particularly involved in the planning, designing and construction side of things and they carried out their tasks with great verve. The various chapters of Mrázek's book reveal how these engineers not only gradually managed to establish their own position in relation to the once-supreme power of the colonial administrators but how they also managed to annex increasing authority. They saw themselves as pioneers of colonial policy, and could be seen as the precursors of the later development aid experts, who not only wanted to leave their mark on how projects were realised but also on the decision-making surrounding the direction in which developments should go. This sentiment was neatly expressed by C.G. Cramer, the first chairman of the East Indian Social-Democratic Party and a Delft engineering graduate, who described the mission of the engaged engineers in the following high-flown and pretentious fashion:

> *We engineers shall – undoubtedly more than others – through our education, competence and social involvement, be equipped to exercise a great positive influence upon the way in which society develops [...] More and more it will become evident that a significant portion of the leadership of modern society will have to be entrusted to engineers.*[5]

The mission of the engineers became the basis for what Van Doorn was in his study on *De laatste eeuw van Indië* (The last century of the East Indies) to term 'the colonial project': a centralistic and top-down regulated process of interventions directed at the gradual transformation of East Indian society in which the native population was at the receiving end of all kinds of Dutch plans and good intentions. Although the colony was supposedly under the jurisdiction of Dutch civil servants within the Interior Administration, from the mid-nineteenth century onwards it was increasingly the technologists who determined matters and who steered the direction of colonial change. It was a task they largely undertook with the best of intentions and with a great deal of idealism, and one which was based on a technocratic vision which Van Doorn characterised as:

A firm belief in the value of specialised expertise and applied science, (together with) the appropriation of large-scale technological projects in the interests of promoting general public welfare, pursuing the battle for the emergence of rational organisation and ordering and preventing the wasting of native manpower.[6]

Technologists and technocrats

The present volume, with its various contributions on East Indian technological history, takes a fresh retrospective look at the role of Dutch engineers in the East Indies. The story that comes across is different to the one told in Van Helsdingen's *Daar wèrd wat groots verricht* and that is not solely down to the increased objectivity resulting from the passage of time. Now, with the passing of more than half a century of Indonesian independence, present-day authors view the world quite differently from those who, without realising it at the time, were publishing in what turned out to be the final year of the Dutch East Indies. One advantage of the passage of time is hindsight, in this case one can look back at what was achieved in the field of civil public works and more easily evaluate the negative facets of the colonial project. It is within such a context that several of this volume's contributors have asked themselves what the technological structures imposed and realised by the Dutch engineers have come to mean to present-day Indonesia. However, it is not that easy to give an unequivocal answer to this question. Towards the end of the nineteenth century, the likes of J. Homan van der Heide, an energetic and very influential engineer[7], were convinced that their legacy was extremely valuable.

The hydraulic works scattered throughout the entire country which bear witness to the purposeful upholding of the economic interests of country and people, will certainly be the most valuable monuments which one can imagine of Dutch control in Java which will live on in history, not so much like Turkish but rather more like Roman reminders of what will ultimately be remembered as an auspicious domination.[8]

Two decades later, in 1922, the engineer P.J. Ott de Vries also chose to make some sweeping claims concerning the way in which future Indonesians would view the various structural works of himself and his colleagues.

If, as we certainly hope will not be the case, fate dictates that our nation be banished from the East Indies for ever, then the Dutch people will be able to thank Dutch engineers for the fact that even the East Indians will say: Yes, the Dutch might have gone but they have left their monuments behind.[9]

Indeed, the major hydraulic works, the crowning glory of the Dutch engineers, are still intact after all these years and many have since been renovated and extended. Many of them still fulfil a crucial role in the network supplying Indonesia with its irrigation and drinking water. However, despite this, it would be wrong to conclude that Indonesia sees its colonial inheritance in a solely positive light.

At the time of independence, Indonesia in effect not only inherited a house but also all the furnishings and household effects which had been selected and installed by the previous owners and, moreover, the house had a lived in feel to it. An extensive network of roads, irrigation works and residential areas would be in place once all the necessary post-war repair work had been done, and the effects of neglect been eradicated, which could form the basis for a national economy. However, it became clear that technology is not politically neutral. Much of what had been realised in the Dutch East Indian era had been geared towards the needs and requirements of a colonial society and a colonial economy. In practice, that had particularly involved adjusting to meet the needs of the large agricultural estates, the mining industry, trade and the colony's Dutch population.

While the local Javanese agricultural sector had received plenty of attention from irrigation engineers and from the agricultural advisory service, it had ultimately remained subordinate to the needs of the sugar and tobacco concerns. In 1963, the American anthropologist Clifford Geertz shocked people with his study on Agricultural Involution in which he showed that, while the local Javanese farmers had profited from the irrigation works, the political-economic context and the partly symbiotic and partly competitive relationship with the sugar companies had served to increasingly marginalize the small farmers resulting in a situation from which it was hard to escape. The plantation-based economy was also strongly oriented towards the production of monocultures and was over-dependent on colonial political and economic protection. Therefore, after 1950 it proved to be unable to meet the changing demands of the world market while simultaneously satisfying the wage claims of the plantation workers.[10]

On top of this, right up to the end of the colonial period, the Dutch regime had held back a substantial amount of indigenous industrialisation for the simple reason that it could have provided significant competition to those companies in the motherland that exported a large proportion of what they produced to the colonies. This explains why, in the 1950s, there was a lack of infrastructure which was to seriously hamper the industrial development of independent Indonesia for many decades to come. Education had not generated sufficient industrial and technological knowledge, supply and marketing channels hardly existed and the scale of supply companies was far too small to structurally steer the national economy in the direction of industrialisation.

Similar criticisms could be levelled at the urban development and road construction inheritance. The model adopted, of the Dutch colonial city where housing took precedence over work, appeared to be increasingly inappropriate in Indonesia's post-colonial context. The country's first president, Soekarno, himself a Bandung Technical College-educated engineer and architect, succeeded in transforming the face of Jakarta in a dramatic fashion. It rapidly went from being a colonial urban town to being a sprawling metropolis, able to compete on the world stage with other such major cities. In the 1960s, the skyline that was being developed was intended to impress and to convert the city into the capital of a proud country. This required a different approach to the residential question: Dutch colonial models were rejected and were replaced by American models and, as one of the contributors to the present volume has noted, 'by 1970 little more of the Dutch origins remained than a few physical and written traces'.

Finally, whereas in the early part of the twentieth century the Dutch had concentrated on the construction of high-quality railway lines for the transport of rice, export products and people, modern Indonesia had a much greater need for small-scale road transport systems. The colonial road network was becoming increasingly inadequate as more and more trucks and minibuses started to take business from the rail sector, and so roads had to be widened on a large scale, bridges built and new motorways constructed.

Although Dutch traces are still very apparent, and this volume provides many such examples, many of the works have become so blurred or so markedly changed in the last half century that they are, in the words of the Dutch novelist Hella Haasse, 'nothing more than scratches on a rock'. Nevertheless, there remains a discernable legacy from the colonial engineers although it should not primarily be sought in what they did or what they left behind in terms of physical constructions in the Dutch East Indies. In a much less visible area, the Dutch engineering tradition has strongly permeated post-colonial Indonesia in the technocratic views that it holds on how a society should be developed and modernised. Consecutive Indonesian governments have included adherents dedicated to this way of viewing matters.

If one were to characterise the development policy of the Indonesian New Order administration then it would be as a predominantly top-down approach ('development to order') with attention paid virtually exclusively to the 'hardware' of development projects. The projects enacted were reputedly the brainchild of what was known as the 'Berkeley Mafia', a group of economists who had studied at the University of California in the 1950s and 1960s, who later became important advisers to the Soeharto government.[11] However, one could equally argue that this developmental strategy constituted a direct continuation of the colonial policy since this had displayed a similar lack of affinity with input from the local population and decentralised planning. Even the Indonesian term 'pembangunan' (an interesting translation of the word 'development' as it literally means 'building' or 'constructing') which was coined during the New Order era[12] carries with it an implication that construction experts and designers, rather than the wider population, should be determining the direction and nature of development. Development amounted to an endless series of physical projects, emanating from the technical ministries, all of which in principle were created to meet the needs of the population but in a situation where the population has little say. Despite this approach, one should recognise that 'many of Indonesia's technocratic planners honestly endeavoured to raise living standards, to improve social welfare and to modernise the economy'.[13] Much the same of course can be said about those who drew up and implemented the plans in the Dutch East Indies. In this respect, it may be argued that independent Indonesia is not such a far cry from the colonial East Indies.

FRANS HÜSKEN

Frans Hüsken (1945) is professor of cultural and social anthropology at the Radboud University in Nijmegen. Since the end of the 1960s, he has been conducting research into Indonesia focussing especially on the rural areas of Central Java and West Sumatra. At the request of the Dutch Directorate-General for International Cooperation (DGIS) he headed a number of evaluation missions concerning irrigation projects in Java, Sumatra and Sulawesi. He gained his doctorate from the University of Amsterdam with a thesis on the social history of labour relations in Java's agricultural sector. His current research interests especially concern the development of local political relationships in Indonesia since the 'Reformasi'. His publications include: *A Village in Java. Social Differentiation in a Javanese Peasant Community* (1987), *Cognation and Social Organisation in Southeast Asia* (1993), *Development and Social Welfare. Indonesia's Experiences Under the New Order* (1994), *Beneath the Smoke of the Sugar-Mill. Javanese Peasant Communities During the 20th Century* (2000), *Violence and Vengeance. Discontent and Conflict in New Order Indonesia* (2002), and *Ropewalking and Safety Nets. Local Ways of Managing Insecurities in Indonesia* (2006).

Appendices

Appendix 1. **Annual government expenditure on irrigation works in Java and Madoera (in guilders)***

Year	Surveys/ construction**		Repair and maintenance**	Private contribution	Total government expenditure
1895	129.897	*300.392*	*128...*		558.289
1896	135.540	*823.469*	*164...*		1.123.009
1897	154.226	*451.302*	*177...*		782.528
1898	125.319	*676.879*	*195...*		1.001.170
1899	129.291	*695.934*	*213...*		1.034.253
1900	116.565	*398.669*	*226...*		741.234
1901	111.928	*1.337.495*	*256...*		1.705.423
1902	131.177	*1.792.342*	*295...*		2.218.519
1903	156.778	*1.577.190*	*331...*		2.064.968
1904	164.591	*1.669.565*	*370...*		2.204.156
1905	171.971	*1.404.247*	*404...*		1.980.218
1906	195.834	*1.345.840*	*440...*		1.981.674
1907	228.918	*1.668.074*	*480...*		2.376.992
1908	239.971	1.848.124	378.728		2.466.823
1909	234.509	1.860.624	444.769		2.538.902
1910	244.829	2.885.349	471.136		3.601.314
1911	264.840	3.878.741	451.147		4.594.728
1912	333.527	3.917.606	583.059		4.834.192
1913	336.270	5.355.885	845.182		6.537.337
1914	387.896	5.099.913	784.999		6.272.808
1915	352.826	4.014.009	1.193.239		5.560.074
1916	400.793	4.818.852	1.644.809		6.864.454
1917	304.607	5.769.253	1.488.277		7.562.137
1918	325.351	6.722.567	1.613.021		8.660.939
1919	283.537	6.990.994	1.973.307	166.056	9.081.782
1920	356.793	7.545.443	2.390.956	77.043	10.216.149
1921	344.169	7.663.923	2.249.615	170.269	10.087.438
1922	301.758	5.080.059	2.505.407	210.749	7.676.475
1923	245.220	4.892.284	3.045.180	269.096	7.913.588
1924	257.622	4.880.666	2.447.954	323.687	7.262.555
1925	266.885	6.574.534	2.566.655	2.185.742	7.222.332
1926	236.242	6.796.989	1.977.752	2.561.616	6.449.367
1927	198.641	7.192.349	1.964.702	656.382	8.699.310
1928		7.321 ...	2.929 ...	1.181 ...	9.069 ...
1929		7.417 ...	3.835 ...	381 ...	10.871 ...
1930		7.025 ...	2.959 ...	682 ...	9.302 ...
1931		5.313 ...	2.787 ...	454 ...	7.646 ...
1932		2.716 ...	2.345 ...	353 ...	4.708 ...
1933		2.138 ...	2.389 ...	311 ...	4.216 ...
1934		1.646 ...	2.282 ...	50 ...	3.278 ...
1935		1.264 ...	2.179 ..	35 ...	3.408 ...
1936				183 ...	3.505 ...

* The table is based on data obtained from the Civil Public Works (BOW) report.
** For the 1895-1907 period it is only the total sum spent on construction, repair and maintenance that is known. The separate amounts have been estimated on the basis of data gained for the 1908-1927 period. The money spent on the suspended Solo Valley Works is not included in these sums.

Source: P.L.E. Happé, 'Eenige beschouwingen over bevloeiingswerken op Java en Madoera' [Considerations concerning irrigation works in Java and Madoera], in: *De Ingenieur in Nederlandsch-Indië* [The Engineer in the Dutch East Indies] 6 (1939) 1: II. 25.

Appendix 2. **The rentability of the irrigation works situated on government land areas in Java en Madoera***

Year	Government spendings on irrigation (in guilders)**	Land rent increase (in f 1000)*** / %		Indirect land income increases / %		Total land incomes: sum total of the percentages	Average interest obtained from the fixed East Indian debt
1901	14.643.188	*140*					
1902	16.861.707	*280*					
1903	18.926.675	*420*					
1904	21.130.831	*560*					
1905	23.111.049	*700*					
1906	25.092.723	*840*					
1907	27.469.715	*980*					
1908	29.936.538	*1120*					
1909	32.475.440	*1260*					
1910	36.076.754	*1400*	*3,88*	1700	4,72	8,60	8,60
1911	40.671.482	*1540*	*3,80*	1972	4,86	8,66	8,66
1912	45.505.674	*1680*	*3,69*	2244	4,93	8,62	8,62
1913	52.043.011	*1820*	*3,50*	2516	4,83	8,33	8,33
1914	58.315.819	*1960*	*3,36*	2788	4,78	8,14	8,14
1915	63.875.893	*2100*	*3,29*	3060	4,78	8,07	8,07
1916	70.740.347	*2240*	*3,17*	3323	4,71	7,88	7,88
1917	78.302.484	*2380*	*3,04*	3595	4,60	7,64	7,64
1918	86.963.423	*2520*	*2,90*	3867	4,45	7,35	7,35
1919	96.045.205	*2680*	*2,79*	4139	4,31	7,10	7,10
1920	106.261.354	*2800*	*2,64*	4411	4,15	6,79	6,79
1921	116.348.792	*3248*	*2,79*	4683	4,03	6,82	6,82
1922	124.025.267	*3696*	*2,98*	4955	4,00	6,98	6,98
1923	131.938.855	*4144*	*3,14*	5219	3,96	7,10	7,10
1924	139.201.410	*4592*	*3,39*	5419	3,89	7,18	7,18
1925	146.423.742	*5040*	*3,44*	5763	3,93	7,37	7,37
1926	152.873.109	*5474*	*3,58*	5984	3,92	7,50	7,50
1927	161.572.419	*5922*	*3,66*	6205	3,84	7,50	7,50
1928	170.641 ...	*6370*	*3,74*	6440	3,78	7,52	7,52
1929	181.512 ...	*6818*	*3,75*	6924	3,81	7,56	7,56
1930	190.814 ...	*7266*	*3,80*	7509	3,93	7,73	7,73
1931	198.460 ...	*7714*	*3,88*	7942	4,00	7,88	7,88
1932	203.168 ...	*8162*	*4,02*	8027	3,95	7,97	7,97
1933	207.384 ...	*8596*	*4,16*	8395	4,05	8,21	8,21
1934	210.662 ...	*9044*	*4,30*	8399	4,00	8,30	8,30
1935	214.070 ...	*9492*	*4,43*	8961	4,18	8,61	8,61
1936	217.595 ...	*9856*	*4,53*	9815	4,51	9,04	9,04

* The table is based on data obtained from the Civil Public Works (BOW) report.
** The money spent on the suspended Solo Valley works is not included.
*** The italicised sums are estimates.

Source: P.L.E. Happé, 'Eenige beschouwingen over bevloeiingswerken op Java en Madoera' [Considerations concerning irrigation works in Java and Madoera], in: *De Ingenieur in Nederlandsch-Indië* [The Engineer in the Dutch East Indies] 6 (1939) 1: II.26.

Appendix 3. **The research carried out at the Manggarai Experimental Station for Water Purification in 1928***

I Organisation

* Laboratory + servants laboratory (for the preparing of Petri dishes).
* Weighing room.
* Separate installation for the treatment of waste water with active sludge.
* Rapid filtering of the Reisert system (Starkstromrückspülung).

II Personnel employed

Chemical bacteriology

Dr. C.P. Mom, N.D. Schaafsma (successor to K. Holwerda) and O.H. van den Hout (all engineers).

Analysts

Ang Goan Hoat, C. de Roode, Mohammed Arif, Raden Soekandar.

Other staff

1 clerk, 1 orderly laboratory technician, 5 lab assistants, 1 orderly for the test installation and 5 coolies.

III Work trips

Banjoemas, Tangerang, Bandoeng, Bengkoelen, Batoeradja, Manggar (Billiton), Soekaboemi, Soerabaja, Klaten, Djocjakarta, Cheribon, Tasikmalaja, Serang, Pekalongan, Salatiga, Tjimahi, Tandjong Pinang and Djambi (+ the sugar refineries Garoem in Soerabaja and Langsee in Pati)

IV Publications (most of which were Public Health Service reports)

C.P. Mom, *Coagulation processes in the purification of river water II.*
C.P. Mom, *Coagulation processes in the purification of river water III.*
C.P. Mom, *Oxidation processes in the purification of river water.*
C.P. Mom, 'The hygienic approach to water purification during the treatment of a number of large European drinking water installations', in: *De Waterstaats-Ingenieur* [The Public Works Engineer] 3 (1928).
C.P. Mom, 'Simple equipment for the sterilisation of water with the help of calcium hypo chlorite and chlorine', in: *De Waterstaats-Ingenieur* 5 (1928).
C.P. Mom and O.H. van den Hout, *On the effect of potassium permanganate on the removal of iron from drinking water.*
C.P. Mom and K. Holwerda, *On the purification of humus-retaining water by means of slow sand filtration.*
K. Holwerda, *The disinfecting of tropical swimming pools with the use of chlorine and chloramines.*
K. Holwerda, *Details on speeded up slow sand filtration in the tropics.*
K. Holwerda, *On the monitoring and the reliability of the chlorination process in relation to drinking water, also in conjunction with the chloramines process and the chlorination of water with an ammoniac content.*
K. Holwerda, *Various tests relating to the studying of changes in bacteria flora when storing polluted surface water in the tropics, especially with a view to the Clemesha method used to establish the recentness of pollution.*

V Laboratory research

Bacteriological

356 samples in connection with drinking water provision,
764 samples in connection with the monitoring of swimming pools and ice or mineral water production processes.

Chemical

211 analyses on:
5 products of mineral water production lines, 16 limestone samples, 5 black manganese samples, 5 $KMnO_4$ samples and 14 sand samples.

Other research

1. Copper water pipes for the purposes of the supply of drinking water in Batavia from artesian water from Tandjong Priok and purified water from the Tji Liwoeng river. Conclusion: pipes not suitable for calcium-aggressive water but they were suitable for non-calcium-aggressive water.

2. Faecal transportation. Doubts about decomposition in the soil according to tests carried out by Grijns and Eyken.
3. Organic substances in water pipes and the continuing growth of bacteria. Solution to the question: method 1 – chemical oxidation of nutrients that can be assimilated by bacteria, method 2 – conservation of the water in the distribution network with the aid of chloramines.
4. Corrosion of water pipe material.
5 Influence of the settling of river water in the subsequent purification processes. Coagulation principally influenced by the varying composition of silt and organic material. For a regular circulation of the coagulation, oxidation and sterilisation processes and also for the respective aluminium sulphate, potassium permanganate and chlorine the water must settle before chemical treatment can take place (content in mg/litre):

		Silt	NH_3	NH_3 proteid	Organic material ($KMnO_4$)
Raw river water	max.	900,8	0,27	0,23	65,7
	min.	13,6	trace	trace	3,5
Effluent settling tank	max.	90,0	0,12	0,23	11,4
	min.	10,4	trace	trace	3,5
Average reduction influent-effluent		36,4%	36,5%	14,6%	9,2%

VI Research in connection with drinking water provision

Bandoeng. Manganate deposits in the water pipe network, evidence of corrosion in the supply pipes from the water springs on the hills of the Tangkoeban Prahoe.
Banjoemas. Group water provisions for Banjoemas, Poerbolinggo, Poerwojerto and Soekaradja. Source: Kawoen Tjarang. Exploitation begun in 1928.
Batoeradja (South Sumatra Railways). The initiative for drinking water provision was taken by the 'Head of the State Railway and Tramway Company in the Outlying Regions'.
Source: Surface water from the Soengei Ogan.
Bengkoelen. Slow sand filtration of the water from the Doessoen Besar Waduk (i.e. reservoir). Conclusion: As raw water the Bengkoelen river water is preferable to reservoir water. It is processed in the following way: settling, coagulation, filtration, chlorination.
Buitenzorg. New sources alongside the old (Kota Batoe).
Cheribon. Replacing of the Soember groundwater drainage field by the abstraction of one or more springs in the Tjipaniis source terrain. Spring water that flows from the old lava flow layers of the Tjerimai volcano.
Djambi. Source: Soengei Batang Hari. Design (by the Technical Division of this particular Service): 2 pre-settlement basins, coagulation basins, rapid filters and chlorine sterilisation.
Pekalongan. Research into the exploiting of the Rogoselo source.
Pontianak. Source: peat water. Process:

1) coagulation by means of a chlorinated ferro-sulphate solution (200 gr. $FeSO_4 7aq$ + 12,5 gr. liquid chlorine) per m^3;
2) first sedimentation at least 6 hours, excluding pumping time;
3) second sedimentation and acid neutralisation with the help of calcium;
4) third sedimentation at least 6 hours, excluding pumping time;
5) rapid filtration;
6) slow sand filtration (biological filter), filtering speed less that 0.10 metres per hour.

Serang. Research into the water quality of the Tjikoeloer sources and the artesian wells.
Soekaboemi (completed in 1927).[1]
Zuid (South)-Soerabaja (collective supplies for Modjokerto, Djambing, Modjosari, Modjoagoeng, Krian, Sepandjang + several smaller places and estates) + the possibility to supplement Soerabaja by providing 1000 m^3 (per day). Prise d'eau: source Djoebel. Supply lines: length 108 km, diameter 150 to 200 mm, pressure 14 ato.[2]
Seroe (Salatiga). Source: Sinongko.
Tandjong Pinang (Riouw).Source: 2 drainage fields (connected together) at the foot of the Boekit

Pantjoean. Advice: water purification by means of slow sand filtration.
Tangerang. Source (not mentioned): apparently ground water, the need to extract iron was in fact discovered. Further advice on the flushing of the distribution network.
Tasikmalaja. Corrosion research carried out on water pipes. The water contained high levels of sulphate and aggressive carbonic acid. Advice: choose between intensive aeration and less intensive aeration followed by marble-filtration, calcium treatmeant (like in Djocjakarta).

VII Other advice

1) Recommendations made concerning smaller drinking water facilities for Government organisations, enterprises and private companies.
2) Drinking water facilities of the Garoem (Modjokerto) enterprise; source: Kali Abab.
3) Drinking water facilities for the Langsee (Pati) sugar estate; source: irrigation pipes.
4) Drinking water provision for Manggar (Billiton).

* This appendix is an excerpt from the 1928 annual report that was compiled by the engineer, Dr. C.P. Mom, director of the Manggarai Water Purification Experimental Station in Batavia. In this brief representation of the annual report as few as possible changes have been made to the original structure, layout, authentic style and original spellings of Indonesian place names. The footnotes introduced to clarify certain points are those of the chapter's author, Professor J.H. Kop.

Appendix 4. **The research carried out at the Manggarai Experimental Station for Water Purification in 1934***

I Publications

C.P. Mom, 'Water purification technology in the Dutch East Indies', in: *Openbare Werken* [Public Works] (1934).
N.D.R. Schaafsma, 'The assessment of a specific drinking water facility on the basis of bacteriological research', in: *I.B.T. Locale Techniek* [Local Techniques] III (1934).
N.D.R. Schaafsma, 'Septic tanks and cesspits for the disposal of faecal waste in the tropics', in: *I.B.T. Locale Techniek* III (1934).
J.K. Baars, 'The disinfection of drinking water', in: *Mededelingen D.G.V.* [D.G.V. Proceedings] (1934).
J.K. Baars, 'The use of salicylic acid and benzoic acid in the preparation of fizzy drinks and squashes in the Dutch East Indies', in: *Mededelingen D.G.V.* (1934).
J.K. Baars, 'A modern East Indian swimming pool', in: *I.B.T. Locale techniek* III (1934).
J.K. Baars, 'Biological sand filters and fish', in: *Antonie van Leeuwenhoek* (1934).

II Laboratory research

Bacteriological

523 samples for drinking water pipes and swimming pools,
136 samples for ocean-going ships,
1028 samples for ice, fizzy drinks, and aerated water and squash manufacturers,
1908 samples for special research purposes.

Chemical

p.m.

Overview of the chemical and bacteriological research

Year	Chemical	Bacteriological
1930	531	1989
1931	736	2141
1932	702	3486
1933	799	3520
1934	1150	3595

Further research

* Coagulation of organic materials by means of calcium and electrolysis,
* Binding of chlorine with phenol and the allied smell and taste consequences,
* Influence of active carbon on the dissimilation of organic materials in water.

III The purification of peat bog water with potassium

Bagan Si Api Api, Bengkalis, Selat Pandjang and Pontianak.

IV Periodical inspection of water pipes

Batavia, Tandjong Priok, Buitenzorg, Soekaboemi, Tjimahi, Tjiandjoer, Tasikmalaja, Garoet, Tangerang, Rangkasbitoeng, Tandjong Karang and Telok Betong.

V Small drinking water reserves

Places

Endek, Emmahaven, Bentjoeloeh (prisons for convicted offenders), Noesa Kembangan (prison), Tjipinang, Mr. Cornelis (prisons), Quarantainestation Onrust, Ambon,Pekalongan, Liwa, Landschapswerken Boeton, Tandjong Poera, Tondano, Soengei Liat, Kisaran, Weleri (hospitals), Semarang, Tjisaroea (sanatorium) Solo, Karang Poetjong, Tijolo, Loewoe, Palembang (abattoir), Waterkrachtcentrale (Hydropower station) Oebroeg, Algemeen Delisch Emigrantenkantoor (General Delian Emigration Office) in Central Batavia.

Water supply pipes

Municipality of Batavia, the Schieper company, the Lindeteves Stokvis company, the Pont à Mousson company.

Various

Enterprises (7), swimming pools (6), companies (15), including the Water Purification Bureau (Haarlem).

VI Research for the purposes of central water works

Index	Company	Specifications
1	Bagan Si Api[2]	Peat water
2	Balik Papan	River water (Klandaran)
3	Bandoeng	Especially the deacidification installation Tjisalada
4	Bengkalis	Well research
5	Den Pasar-Tabano	Network corrosion Recommendation: aerate
6	Djambi	River water (Batang Hari) Network corrosion Recommendation: replace in connection with calcium saturation
7	Fort van der Capellen	Source (Kiambing). Water wastage. Recommendation: place in section metres
8	Granjar	Source. Recommendation: improve the abstraction
9	Indramajoe	Recommendation: add calcium
10	Karangasem	Source research
11	Kendal	Artesian water research
12	Kloengkoen	Research source
13	Laboean Bilik	Research artesian water Recommendation: supplement with surface water
14	Langsa	Recommendation to enlarge the pre-settling tanks was successfully followed through
15	Lembang	Routine research into sources at Tjikolegede and Tjipangkoeloean
16	Leprozerie Malalajang (Menado)	Source Waroekoen Recommendation: aerate and chlorinate
17	Muntok	Water losses still too great despite improvements in the distribution system
18	Padang	Plans for the water towers (1932) continued
19	Pajakoemboeh	Research 4 sources: Aer Tabit and Aer Kapo (preference, Q = 5 l/s)
20	Pangkalpinang	Replacement of surface water by artesian water, executed by the public works department of Sumatra's East Coast
21	Pangkalpinang	Recommendations carried out in 1933
22	Pangkalan Brandan	Replacement of surface water by artesian water, executed by the public works department of Sumatra's East Coast
23	Poerwakarta	Source complex Naratjang, planning stage
24	Quarantine station Poeloe Roebiah (near Sabang)	Source water or rain water harvesting: study
25	Samarinda	Source: Mahakam river
26	Sanatorium Noöngan Menado)	Well in the river bed
27	Sigli	Routine research
28	Singaradja-Boeleleng	Source polluted with river water Recommendation: introduce slow sand filter at the high reservoir
29	Solok	Routine inspection of sources Pantjoeran Koerapen and Pantjoeran Gadang
30	Takengon	Natural source capacity insufficient
31	Tandjong Pinang Telok Betong	Existing chlorination system replaced by 0.7 mg Cl2 or 1 gram of calcium hypo chlorite per liter
32	Tandjong Pinang	Large iron difficulties Recommendation: add calcium
33	Tasikmalaja	Deacidification installation and aeration completed in 1934
34	Tjepoe	Proposal to use the Solo river as a source of raw water
35	South Soerabaja	Source: Djoebel. Network badly affected by aggression

Further details:

Palembang
Completion of the second clean water cellar (including) 2 carbon filters.
Improving of existing procedure (pre-settling of river water (Moesi), coagulation (20-30 mg/l Al_2SO_4), settling of flocculated colloids and rapid filtration) under *normal* circumstances by:
adding ammonium chloride, chlorine (± 2 mg/l), sterilising for 2 hours in the 2 clean water cellars, adding calcium, pumping out the water that still contains 1 mg/l of chloramine (the city contained 0.4-0.7 mg/l) and in the event of *abnormally high organic material content* by:
adding approx. 5 mg/l of chlorine, sterilising for a minimum of 2 hours, carbon filter (chlorine oxidises organic materials and also binds itself), adding of ammonium chloride, adding chlorine in the form of calcium hypochlorite or triseptol, adding of calcium, pumping away into the city.
Pangkalpinang
Surface water in the Aer Baik. Problems due to the changing intensity of the bandjirs in the Aer Baik river. Water too murky in the slow sand filters and the marble filters also have to be flushed out too frequently.
3 possible solutions (choice still to be made in 1934):
- Creation of a reserve wadoek; disadvantage: large reservoir (2400 m³) and expensive.
- Creation of a reserve clean water cellar; disadvantage: good but not sufficient.
- Creation of an extra (reserve) prise d'eau in a branch of the Aer Baik (too little capacity in the dry season, namely less than 10 l/s), but even then mixed intake was possible.

Furthermore one pre-settling tank to be situated in the wadoek upstream of the weir.
Samarinda
Q = 60 m³/hour. Pre-settlement: 3 tanks, each 350 m³, retention period, aro.18 hours. Mixing channel length 50 m,
Q = 40 m³/hour per channel. Coagulation in 2 Dortmund tanks, each 36 m². three filters, each 4 m².
Sterilisation and then the clean water cellar, also acting as a high reservoir.
Takengon
Supplement the insufficient source capacity with water drawn from the Takengon lake (Q = 5 m³/hour).
Process: rapid filtration, breaking point chlorination (i.e. overdosing), filtration over coal.
Tjepoe
Recommended process: pre-settlement in Dortmund tanks, coagulation by means of ferri-sulphate + calcium in the mixing channel, flocculation en settling in Dortmund tanks, rapid filtration, sterilization by means of calcium hypochlorite.

* This appendix is an excerpt from the 1934 annual report that was compiled by Professor C.P. Mom, director of the Manggarai Water Purification Experimental Station in Batavia. In 1931 Mom was inaugurated as professor at the Bandoeng Technical College. In this abridged version of the annual report, created by Professor J.H. Kop, as few as possible adjustments have been made to the original structure, layout, authentic style and spellings of Indonesian place names.

Principles to be observed when designing sewerage and drainage works for large municipalities

by the engineer C.A.E. van Leeuwen, Chief Engineer at Public Works, head of the Sanitation Works division at the Department of Civil Public Works in Weltevreden (Batavia)

1) The aim of every general drainage system must be to speedily remove – or to at least make harmless – out of the vicinity of the community, all waste products that pose a threat to public health.
2) The waste has to be divided into *dry* and *wet* waste. In other words, certain solid matter that rots less quickly, such as roadside mud, leaves and dry domestic waste can be transported on the surface in carts. Conversely, all waste that rapidly decomposes and is a threat to public health such as bath, washing and kitchen water, and faeces must be transported underground and there must be as little as possible direct contact with humans, animals (and insects).
3) Every system must be adapted as well as possible to the habits of the population and to the circumstances of the terrain on the condition that bad practices are gradually eradicated and the system must be accomplished in line with the developmental pace and resources of the population. This is especially relevant to kampong sewerage systems, the improvement of which can never be accomplished without simultaneously ensuring that the habits and housing of the inhabitants in question are improved. Particularly in the main conurbations, where housing is more dense and simple but also more solid, the cleanliness of the premises is easier to guarantee and check as the inhabitants are less burdened by the tasks and costs of maintenance.
4) Each system will have to include the entire built-up area, insofar as that can be established beforehand, which means that urban expansion must be taken into consideration.
5) The designs for decent drainage systems must go hand in hand with the building construction plans and with the establishing of the necessary *regulations* laid down for inhabitants in connection with habitation and waste disposal services.
6) From the hygienic and economic point of view it is advisable to keep the transportation of rain water *separate* from the transportation of all other waste.
7) It is therefore not a problem to use open channels for the conveyance of rain water. Such channels should have the capacity for a maximum discharge of 150 litres per second per ha.
8) It must be possible for all rain gutters to completely dry up, for instance within the space of a day after there has been a shower, so that no anopheles have the chance to develop. Similarly, the beds and sides of the gutters should not remain humid as that would encourage the development of unwanted algae vegetation. The cleaning services engaged in the collection, transport and treatment of such solid waste should be responsible for the upkeep of such gutters..
9) In order to ensure that no rain gutters or ditches remain damp it is desirable to, where necessary, introduce drainage pipes - which can freely debauch, by gravity, to open water, either by installing them below and parallel to the rain gutters or ditches or, as a last resort, directly beneath the channel-bed of these water ways.
10) Depending on the permeability of the soil, such drainage pipes may be used exclusively for the conveyance of ground water and geared to a capacity of 2½ litres per second per ha in the case of clay soil or up to as much as 6 litres per second per ha in the case of sandy soil.
11) In cases where the rain gutters are lined then drainage may be led straight into the ground by means of gratings introduced at 20-40 metre intervals which give access to the sand or gravel trench in which the drainage pipe is situated.
12) One of the most dangerous waste products is human faeces. The safest way to transport that is to take it directly from the toilets alongside of *septic-tanks* or cesspits where no other waste may be dumped, apart from the amounts of swilling water required for watering down the waste. No disinfectants may be used (apart from small quantities of quicklime).
13) The drainage of liquid from septic tanks may take place:
    1 via the soil through infiltration trenches provided that the soil has been declared suitable for the biological purification process,
    2 via drain pipes or sewers provided that they lead into very big open water areas where 1000-fold dilution is possible or into specially designated waste water treatment works.

14) In cases where the ground water level may be high it becomes desirable to install drain pipes that can empty out into all open water bodies - or water ways - provided that these bodies do not regularly run dry. These pipes may not therefore connect with rain conveyance gutters - or ditches - but they may connect with other drain lines. The drain pipes should be installed alongside of the infiltration trenches as referred to under 13, point 1, at a distance of at least 3 metres in a horizontal direction and about 1 metre below ground level, or else 50 centimetres below the discharge level of the septic tank.
15) All *domestic waste water* which includes bath, kitchen, dishwashing and scrubbing water but also stable water insofar as the preference for the latter does not lie with septic tanks may, provided it is free of solid waste and provided that the soil is sufficiently permeable, be filtered in much the same way as is possible with septic-tank effluent. The *solid waste* must go into *drainage pits*, as must fatty solids, where it must be held back to be further collected, transported and treated by the 'dry cleaning' services.
16) Domestic waste water may also be directly channelled into public waterways provided that there is enough of a flow (sufficient for a1000-fold dilution).
17) If septic tank fluid and domestic water devoid of solids is to be conveyed via sewers or drainage pipes then water will need to be regularly flushed through the pipes in question.
18) In cases where all waste such as faeces, household waste water, stable water and industrial waste water goes into the same sewers then permanent periodical swilling out of the sewers will be required; the amount of water required for such purposes is 2½ litres per second per hectare. Sewers may only ever be discharged out into the open sea or into purification works designed for that particular purpose. In view of the fact that these studies have not yet been completed, it is quite possible that in future these principles will be changed and supplemented.

From: C.A.E. van Leeuwen, 'Beginselen in acht te nemen bij het ontwerpen van rioleerings- en afwateringswerken in groote Gemeenten' [Principles to be observed when designing sewage and drainage works for large municipalities], in: *De Waterstaats-Ingenieur* [The Public Works Engineer] 7 (1919) 12, 571 etc.

Rules for the designing of a sewerage system

by engineer C.A.E. van Leeuwen , Chief Engineer at Public Works, head engineer of the Sanitation Works division within the Department of Civil Public Works in Weltevreden (Batavia)

1) Limit the use of swilling water as much as possible.
2) Separate rain water by means of open drains or gutters which are able to entirely dry out after it has finished raining and through which no other waste water or other liquids are allowed to flow.
3) Eradicate high water table levels by introducing drainage works.
4) For the collection of faeces use cesspits or septic tanks, which may or may not be accompanied by subsequent post-treatment by means of infiltration trenches.
5) Remaining waste material (bath, kitchen, washing, rinsing and factory water) is to be transported away in sewers or drainage pipes, after the solids have been separated from the liquid in settling drainage pits or cesspools.
6) All solid waste must be removed as soon as possible from roads and properties and taken out of houses by the services that collect, transport and treat (dispose of) solid waste.
7) For the transport of solid and fluid waste matter and rain, make use of the cleaning thoroughfares behind the properties.
8) Make sure that the roads have a watertight hardened surface.

From: C.A.E. van Leeuwen, 'Het riooleringsvraagstukin Nederlandsch-Indië' [The sewage question in the Dutch East Indies], in: *De Waterstaats-Ingenieur* 8 (1920) 5, 196-212

Appendix 6. **The contribution made by Dutch engineering bureaux to public drinking water facilities in Indonesi**

Period	The job commissioning bodies	Bureau	Project name and field of work
1979 1984	D.G. Cipta Karya + DGIS*	DHV	Six Cities.
1979 1986	D.G. Cipta Karya + DGIS	DHV	East Medan.
1979 1986	D.G. Cipta Karya + DGIS	DHV	11 Kabupaten Cities in Aceh and North Sumatra.
1979 1990	Min. of Public Works + DGIS	IWACO	15 Secondary Cities in West Java: Majalengka, Padalarang, Pamanukan, Pandeglang, Soreang, Jatibarang, Majalaya, Rangkasbitung, Lembang, Subang, Cikampek,Banjar, Tangerang, Bekasi, Cimahi.
1980 1991	Min. of Public Works + DGIS	IWACO	Cirebon Small Communites.
1981 1989	Min. of Public Works + DGIS	IWACO	45 Small Towns (IKK) West Java.
1982 1988	D.G. Cipta Karya + DGIS	DHV	25 Small Towns (IKK) Aceh and North Sumatra.
1984 1992	Min. of Public Works + DGIS	IWACO + DHV	Bandung (kotamadya).
1985 1991	D.G. Cipta Karya + DGIS	DHV	Sukabumi.

Tasks	Notable features (mainly technical)
Obtaining water, purifying and distributing it.	Tapping of water source, wells, lined earthen reservoirs, pumping stations.
Transport and distribution.	Transporting mains (1200 mm), lined earthen reservoir + pumping station, pipe lines (distribution).
Obtaining water, purifying and distributing it.	River tapping (inlet works) bank infiltration, water source tapping, deep wells, transportation pipe lines, pumping stations, purification installations, lined earthen reservoirs, water towers, distribution systems and networks, hydrants.
Setting up of water supply companies (and all which that involves) + training in relation to general and financial management, information systems, budgeting, requital, technical control and maintenance in relation to socio-economic and water requirement studies, raw water exploration, topographical studies, soil research, planning, technical design, specifications, assistance in the acquiring and purchasing of materials and equipment, building supervision, monitoring, manuals for management and maintenance, and for water loss reduction.	River tapping (inlet works), tapping of sources, groundwater abstraction, transport pipe lines, pumping stations, purification installations, storage reservoirs, distribution systems and networks. The design in relation to these cities varying in inhabitant numbers between 20,000 and 100,000 was based on what was termed the Basic Needs Approach (BNA), a higher facilities level than was generally applied to the capitals of sub-districts Ibu Kota Kecamatan (IKK) where the inhabitant numbers varied between 3,000 and 20,000 in West Java.
Introduction of a centralised system for 50,000 inhabitants of widely scattered villages in the Cirebon district, all tapping water from one natural source in conjunction with: community participation programmes, social-economic studies, hydrological studies, planning, technical designs, specifications, assistance in the acquisition and purchasing of materials and equipment, building supervision, assistance with starting up, manuals for management, maintenance and and retribution.	Tapping of sources, transporting mains, purification installations, storage reservoirs, distribution systems and networks, hydrants.
Manuals including information on the training of personnel for standardised designs, building supervision, management and maintenance, reimbursement and community information programmes for 45 IKK (300,000 inhabitants) with the aid of socio-economic studies, water requirement studies, etc.	Standard designs for surface water inlets, deep wells, pumping stations, distribution systems. Transporting mains (1200 mm), lined earthen reservoirs + pumping stations, pipe lines (distribution).
Economic (low cost) systems based on perpetual provision by means of a water flow rate limiting device + reservoir (mandibak) per property or house connection.	River tapping (inlet works), bank infiltration, tapping of sources, deep wells, transporting mains, distribution systems and networks, hydrants with ferrocement reservoirs.
Rehabilitation and expansion of the drinking water system for the city (including leakage reduction), pertaining to groundwater abstraction and river water collection + purification, transport, storage and distribution, on the basis of: See list of 15 Secondary Cities in West Java	Practically all relevant technical aspects. Introduction of Environmental Effect Reporting with special attention being paid to water requirement development, future water supplies with regards to quantities and quality, deforestation, erosion, sustainable agriculture, pesticides and risks to public health, fertilisation and eutrophication, social effects of dam and reservoir construction.
Obtaining water, purifying and distributing it.	Fixed weir in connection with the tapping of river water.

Period	The job commissioning bodies	Bureau	Project name and field of work
1986 1990	Min. of Public Works + DGIS	IWACO	Bogor (kotamadya).
1987 1992	Min. of Public Works + DGIS	IWACO	17 Small Towns (IKK) West Java: Pengalengan, Ciranjang, Kandang Haur, Cibadak, Cicurug, Cibeureum.
1988 1990	Min. of Public Works + DGIS	IWACO	125 Small Towns (IKK): Central Java, Yogyakarta, South Sumatra, Lampung.
1990 1992	Min. of Public Works + DGIS	IWACO + DHV	District water supply companies: East Java, West Java and Aceh.
1990 1992	Min. of Public Works + DGIS	IWACO + Haskon	Kabupaten/kotamadya Dev. Progr. ACEH Kabupaten: Pidie, Aceh Utara, Aceh Timur; Kotamadya: Banda Aceh, Sabang; WEST JAVA Kabupaten: Sukabumi, Purwakarta, Cirebon; Kotamadya Sukabumi
1992 1993	Min. of Public Works + DGIS	IWACO	150 Small Towns (IKK) Several institutional places in East Indonesia. 35 IKK in West Java, including the design and execution.
1992 1993	Min. of Public Works + DGIS	IWACO	23 Local Governments (kabupaten) in Sumatra : ACEH (Sabang, Aceh Besar, Barat, Tengah, Tenggara, Selatan) SUMATERA UTARA (Langkat, Dairi, Nias, Simalungun, Asahan. Labuhan Ratu); SUMATERA BARAT (Agam, Pasaman, Lima Pulu, Pariaman, Tanah Datar, Sawahlunto, Pasir Selatan, Solok); RIAU (Bengkalis, Indragiri Hulu, Indragiri Hilir).
1993 1994	Min. of Public Health + DGIS	IWACO	7 District water supply companies in Aceh: 2 kotamadya (Banda Aceh, Sabang), 5 kabupaten (Pidie, Aceh Timur, Barat, Besar, Utara

* DGIS = (Dutch) Directorate General for International Cooperation

Tasks	Notable features (mainly technical)
Crash Programme, Master Plan and Final Design for the rehabilitation and expansion (up to a million inhabitants) of the city's drinking water system, including the collecting, purification, storing and distribution of river water, on the basis of:: See list of 15 Secondary Cities in West Java + consumer and management information systems.	See Bandung (above). In the Environmental Effect Report (the Cisedane river basin) attention was also paid to the effects (up river and down river) of tapping and discharging at Bogor. Effective combating of water waste through better collection of bills and tariff diversification.
Setting up of water supply companies for, in total, 500,000 inhabitants, amongst other on the basis of: See list of 15 Secondary Cities in West Java.	Tapping of sources, obtaining of groundwater (deep wells, pits), river inlets. All 17 designed according to BNA, but only 6 were actually constructed due to shortage of funds.
Setting up of water supply companies, amongst other on the basis of: See list of 15 Secondary Cities in West Java.	Virtually all relevant technical aspects of water obtaining (natural sources, ground water, surface water), purification, transport, storage and distribution. Transfer to district water companies after building and entering into service.
Reinforcing the know-how and skills of provincial authorities in relation to the control and development of district water supply companies by means of: the implementing of procedures, technical training (design, management, maintenance), management training and sector studies.	Execution of the commissioned job which was particularly directed towards planning, development and management at district and provincial level by means of management techniques (conscious use of English): Institutional Development, Human Resources Development, IN-Service Support Programmes, Water Enterprise Info Systems (WEMIS), Water Enterprise Performance Assessment (WEPA).
Assistentence offered to local authorities in relation to infrastructural five-year plans including drinking water provision plans on the basis of, for instance: field studies, financial feasibility calculations, Environmental Impact Studies, community participation, female emancipation, detailed designs, building supervision, institutional development and training.	A new facet in this connection was the field of study relating to public & private partnership.
Assistance in the institutional reinforcement of the water supply companies so that they could achieve PDAM-status by receiving technical and managerial advice in the field of organisational and institutional development, information systems and personnel training.	Attention to customer-related activities and public information.
Investment programmes for urban infrastructure and final design for drinking water provision.	Planning for repayment (retribution) and for local institutional development.
Master planning (2015) + environmental and business analysis including the monitoring and protecting of water supplies.	Estimating of water storage supplies and water requirements (urban, irrigation, industry).

Appendix 7. **Overview of the study programme as proposed by the Bandoeng Technical College**

1. Theoretical and applied mathematics. When determining how the time should be divided in the mathematics course the sub-committee was entirely guided by the programme compiled by Professor J. Klopper. According to that programme a certain number of hours will be reserved for practical issues. Those hours must not be entirely taken up with the resolving of questions relating to descriptive geometry but also put aside for questions linked to other mathematical areas and to learning how to work with slide rulers and other calculation equipment.

2. Theoretical and applied physics. Where this subject is concerned, the existing Delft Technical College programme will be basically adhered to but it is still worth considering whether it might not be worthwhile for the Bandoeng Technical College to incorporate various matters that are of particular importance as far as the atmospheric phenomena in our colony is concerned in relation to trade winds, land and sea winds, rain precipitation and evaporation, the effect of light and heat on physical and chemical processes, and so on.

3. Chemistry for the engineer. This subject, given faculty-wide in Delft but not always sufficiently promoted for students, is of special importance to the civil engineer in the East Indies who is very involved in irrigation. The composition of water and soils, fertilisation and the leaching out of soils through irrigation, are all matters that give rise to interesting chemical questions for the irrigation engineer, questions that he does not have to resolve independently. He does, however, need to know the significance of it all and, via his knowledge of chemistry, sense and assess the value of the outcomes determined by others. It is a subject that should therefore certainly be more prevalent for the East Indian civil engineer than for his Delft student counterpart.. Also in the case of sanitation work - water works, sewerage systems and soil purification - chemistry is to be given greater emphasis than is the case in Delft.

4. Architecture. In the East Indies it will be necessary to approach this subject slightly differently from the way it is tackled in Delft. Sometimes it will not be necessary to explore building aspects in such depth but rather to focus on how, for instance, heat can be excluded instead of just considering how protection can be offered against the cold. Facets such as ground plans, foundations and roofing styles may be simpler; the handling of uncovered floors and walls may be more extensive. Building styles can perhaps be more guided by Eastern art forms than by Western styles. It should at least not be so that the former (Eastern) aspect is totally ignored. Particular attention should be paid to structures erected on shifting soil or in earthquake-sensitive regions. In the case of utilitarian buildings, the focus must be upon particular countrywide requirements and the demands of the extensive Government companies and, in broad outline, of the agricultural industry. It will be necessary to adapt to the customary East Indian designs adhered to for public service office buildings such as bureaux, schools, prisons, hospitals et cetera.
Attention will also have to be given to the use and the processing of indigenous materials for temporary structures, scaffolding and other aids implemented in the construction of buildings.

5. Hydraulic engineering. In the area of sluice building, shipping canal construction and seawall building as well as in the areas of river training, polder creation and land drainage, much less will need to be considered than in Delft. By contrast, though, more attention will have to be devoted to irrigation, both directly from flowing rivers and from water reservoirs and indirectly using water extracted from the ground and obtained by pumping or other means. More study aspects will have to concentrate on volcanic, erosive and alluvial soil types, on deep excavation through various soil strata and on the large-scale level fills of a diversity of soil types than would normally be the case in parallel degree courses in the Netherlands. Rainfall, evaporation, the permeability of the soil and groundwater flow rates, river regime correlation - ranging from minimal to normal and maximum (bandjir) discharges - with the above-mentioned phenomena together with the nature, shape and size of the river basins will all have to become special subjects of study. Similarly, the regulating and harnessing of mountain rivers and the way in which boulders, rolling stones, sediment, etc. move in these areas and progress through flatter regions will be important matters to consider , as well - in particular - the channelling or constraining of sand floods, the notorious *lahars* and *bezoeks (mud flows of volcanic origin)*. The hydropower question in itself, and in

connection with irrigation, will have to be dealt with in a detailed fashion. Attention will have to be paid to the required closing off of valleys and to canals along ravine sides and through tunnels. In short, hydraulic engineering in mountainous areas will have to come more to the fore, whilst such lowland or polder land engineering will fade more into the background.
There should not be less instruction on harbours and harbour works than in Delft. However, at the same time - as far as tidal factors go - large shipping channels, coastal formation but also the collecting, transporting and processing of revetment stone, and dredging work, it will be necessary to pay more attention to the colonial rather than the motherland situation without at the same time losing site of the demands imposed by the world traffic situation. Particular attention will therefore also need to be given to the world waterways situation and to large shipping canals. In all of this it will not only be necessary to think of the interests of our colonies but also to bear in mind the situations and circumstances existing in other colonial and overseas areas parts where the East Indian engineer will undoubtedly be able to find work. From that point of view it would not be unreasonable to propose that more rather than less time should be devoted to that field than is devoted to it in Delft.
In the Sub-Committee the matter of whether or not the East Indian Technical College would need to have a hydro technical laboratory for the purposes of research and for the elucidation of the many hydro-technical questions posed by hydraulic engineering, also in the field of irrigation, was a point of discussion. Ultimately what all those exchanges led to was the conclusion that all those demands could not be placed upon a newly created educational institute and that academically it should , as it were, endeavour to lead the way so that when compared to other similar institutes such types of laboratories will be found at very few technical colleges. For the time being it will not therefore be necessary to think of establishing such a laboratory at the East Indian Technical College.

6. The creation and exploitation of roads and bridges. In this field of study it will also be necessary to deviate somewhat from the approach adopted in Delft, without losing sight of the fact that composite bridge constructions, though there to serve rather different requirements in the East Indies than in Europe, will in essence remain the same in both parts of the world. In the case of ordinary roads the requirements made by heavy and dense freight traffic will have to come to the fore, as will the requirements for fast traffic with automobiles. Special attention will have to be devoted to the alignments followed by these roads through difficult and undulating terrain. Where the railways are concerned, just as much attention will need to be paid to railway construction through mountainous areas as to railway building in lower-lying areas. Tunnel construction, transport cables and narrow-gauge tracks for agricultural enterprises are all subjects that will have to be reviewed in more depth by the Technical College.
Both in the area of hydraulic engineering and in the field of roads and bridges it will be necessary to draw attention to the actual execution of works in connection with the possible extensive involvement that the East Indian engineer may experience when executing his own particular designs in projects for which the execution takes place under the direct control of the administration. Also in this connection it is fortunate that in the case of both these main civil engineering fields more time can be made available than that allowed at the Technical College in Delft.
The concept considered in Delft to the effect that in the fields of architecture, hydraulic engineering and bridge building general lectures should be given on foundation building is also something that should be considered within the context of the East Indian Technical College. In that connection the building foundation methods adopted in other colonial countries should not be ignored.

7. Mechanical technology. In this field of study the own manufacturing of building materials involving the felling and transporting of timber, alongside the handling of tools and appliances for the preparation of building materials, is something that should be prominent. What should furthermore be considered, be it merely in an overview fashion, is the kinds of product processing undergone in relation to the major East Indian plantations (estates) and the kinds of machines that need to be deployed in such processing. The kinds of agricultural crops in question are: rice, sugar, coffee, tea, rubber and quina cultivation.

8. Knowledge of and research into building materials. Obviously this subject will, in the first place, have to be directed towards the most commonly processed building materials in the East Indies, including in particular all those known as indigenous materials. Processing in conjunction with the promotion of sustainability, resistance to East Indian wood type insect attack, East Indian stone, East Indian mortar and all their qualities, origins and useful purposes are all topics that require much more research, research that will have to be pursued by the East Indian Technical College's top management. More time will thus need to be set aside for this field than is done at the Technical College in Delft..

9. Technical drawing. The idea of drawing, in the sense of by hand and in an artistic fashion, should immediately be abandoned. Technical drawing, as truly based on technical demands, either freehand drawing or with the aid of instruments, needs to be more frequently and better practised than would normally be the case for engineers. Undoubtedly it will not just be a matter of needing to practise but lectures given to students will have to make clear precisely what tools the technical draughtsman has at his disposal so that ideas relating to the construction and execution of works can quickly be put across in a form that is clear both to the eye and to the mind. This subject area must also consider the matter of just how drawings can be reproduced. Any schoolish kind of activity related to artistic drawing which is certainly not part of the tradition and is virtually useless to all potential civil engineers (in relation to the time and energy spent devoted to such activity) must be rigorously avoided from the very start. It is a subject that must be entrusted to someone who fully realises what engineering practice demands in terms of draughtsmanship and not to a lecturer who is in the first place trained as an artist.

10. Mechanics and 11. Hydraulics. At the East Indian Technical College these subjects will be approached in the same way as at the Technical College in Delft. Certain differences could, however, emerge when it comes to the matter of elucidating on the basis of exemplification.

12. Mechanical engineering. Also in the case of this subject, the way matters are dealt with at the East Indian Technical College could be broadly based on the way that that is done at the Technical College in Delft.. What must be borne in mind is the fact that the civil engineer is not a constructor but merely a user of such tools and that he only needs to know about their properties and construction in order to appreciate the demands can be placed upon them and to realise for what purposes and power applications they can be used.

13. Electro engineering. Here, too, the way in which things are done in Delft can be decisive for the path that is to be followed. Electrical energy potential obtained from hydropower will often be able to serve the civil engineer and so that is something that will have to be considered when dealing with electrical power.

14. Technological hygiene. The Sub-Committee contends that for the civil engineer this subject is almost of more importance than it is in the mother country. In fact, the consequences of sinning against hygiene when engaged in architectural and hydraulic engineering activities may be dreaded more there than here whilst, as everyone knows, in our colonies the field of sanitation works is an area that has remained more or less unexplored. In order to recognise the great significance of such works and to become familiar with the measures that are required to protect the health of labourers in the execution of major public works and in order to give instructions on the hygienic norms required in the constructing of houses, offices and workplaces in the tropics - and further to obtain some knowledge about tropical diseases, their origins, causes, propagation and control - it will be necessary to determine precisely how this subject is dealt with at the East Indian Technical College.

15. Land surveying, levelling and geodesy. This subject can probably be approached at the East Indian Technical College in virtually the same way as it is approached at the Technical College in Delft but possibly with more emphasis on the facet of working in extremely differently composed and hilly terrain. Generally, it is the case that the East Indian civil engineer, more so than his Dutch counterpart, must be a competent surveyor. It is thus important that at the East Indian Technical College more time is devoted to the assignments and exercises surrounding that subject than at the Technical College in Delft.

16. Administrative law. The East Indian Technical College's design programme puts aside slightly less time for this subject than does the Technical College in Delft. This is because the law referred to here is rather less developed in the East Indies than in our country. One need only take, for instance, the fields of labour and factory legislation. In the East Indies there is little legislation in such fields and thus little that the civil engineer has to know. Also where railway and shipping agreements are concerned, there is really little that can be said in those fields. It will therefore be necessary to approach this subject more from the angle of studying political and constitutional law and perhaps more emphasis will have to be given to the official legal relations existing in our colonies. From that point of view there is certainly enough scope for the engineer's services to be of use.

17. Metallography. This subject will be temporarily dropped [...].

18. Town planning. In the East Indies this is still a very new and developing field. It does not really need to be separately dealt with and it can be incorporated into architectural studies unless a particular lecturer with special knowledge of the field can be appointed.

19. Commercial law. This is a field that reflects a situation identical to that in the mother country and thus requires the same degree of attention.

20. Business economics and accounting. At the East Indian Technical College more time may be devoted to this subject than is the case at the Technical College in Delft, also for civil engineers themselves because they are invariably the executors and thus also the managers of projects, they are responsible for the sizeable related administration and for the great financial responsibility which that brings. This subject can be thoroughly learned by also making sure that time is made for its practical facets.

21. Political economy, ultimately an East Indian civil engineer will not be less useful than one in the Netherlands, hence the reason that this subject may definitely not be omitted from the East Indian Technical College programme.

From: S. Hoogewerff, C.W. Weys and R.A. Sandick, *Het leerplan der op te richten Nederlandsch-Indische Technische Hoogeschool. Verslag der Sub-Commissie, ingesteld door de Commissie van advies voor onderwijs van het Koninklijk Instituut voor Hooger Technisch Onderwijs in Nederlandsch-Indië* [The curriculum for the new Dutch-East Indian Technical College. The report of the Sub-Committee, established by the Advisory Committee for education for the Royal Institute of Technical Education in the Dutch-East Indies] (The Hague 1918).

Railway companies in Java and Madoera (1942)

Apart from the State Railways (Staatsspoorwegen) there were twelve private railway and tramway companies operating in Java and Madoera at the end of the colonial period:

1. NV Nederlandsch-Indische Spoorweg Mij (NIS)
2. NV Semarang-Cheribon Stoomtram Mij (SCS)
3. NV Semarang-Joana Stoomtram Mij (SJS)
4. NV Serajoedal Stoomtram Mij (SDS)
5. NV Oost-Java Stoomtram Mij (OJS)
6. NV Kediri Stoomtram Mij (KMS)
7. NV Modjokerto Stoomtram Mij (Md.SM)
8. NV Malang Stoomtram Mij (MSM)
9. NV Pasoeroean Stoomtram Mij (Ps.SM)
10. NV Probolinggo Stoomtram Mij (Pb.SM)
11. NV Madoera Stoomtram Mij (Mad.SM)
12. NV Bataviasche Verkeers Mij

Tanjung Perak Harbour in Surabaya (1942)

1. Jamrud Utara Quay (Rotterdam Kade)
   The total quay length was 1200 m with, alongside the quay, a 9.10 m design depth and a 15 m wide apron. Jamrud Utara handled six ocean-going vessels and two inter island vessels. The typical ocean-going vessel was about 150 m long, requiring an ocean-going berth of 165 m while the average inter island-vessel size was about 64 m, requiring an inter-island berth that was 70 m long. The available floor area was 22,800 m² of transit shed and about 6,000 m² of open storage area.

2. Jamrud Barat Quay (IJmuiden Kade)
   The total length was 210 m, but only 160 m was suitable for berthing vessels, due to the considerable kink in the cope line. Jamrud Barat generally handled two inter-island vessels. The design depth was 9.10 m. The transit shed and covered storage area was 2,450 m². The pilot station was located at the intersection of the Jamrud Barat and Jamrud Utara Quays with all tug activities centred at that point.

3. Jamrud Selatan Quay (Amsterdam Kade).
   The total quay length was 780 m with, alongside the quay, a 9.10 m design depth. The berths were generally used for ten inter-island vessels. Occasionally, ocean-going vessels were directed to where demand dictated. There were eight transit sheds with a covered storage area of 16,037 m².

4. Perak Wharf (Tanjung Perak Board)
   The quay length was 140 m and it was entirely used for two inter-island vessels. The design depth was 9.10 m. Due to the adjacent dock facilities, there were navigation problems when it came to berthing large vessels along this frontage.

5. Nilam Pier (Genua Kade)
   The quay length was 920 m with a 9.80 m design depth and a 15.5m wide apron. The berths were generally used for three ocean-going passenger vessels and two inter-island berths for the discharging of bulk cargo.

6. Berlian Pier (Holland Pier)
   The Berlian Pier extended out into the inner harbour basin and was connected to the south end of the pier. The pier was 142 m wide and about 780 m long. The alongside design depth was 11.30 m. The eastern Berlian berthing point was 780 m long and the berths had behind them four transit sheds with a 13,750 m² storage area. The western Berlian quay was around 750 m long and handled ocean-going vessels. There were four transit sheds with a covered storage area of 17,150 m².

Lines of drinking water distribution in the Malang area (2002)

Water is made to flow from the storage location into the distribution reservoirs through the following pipe networks:

1. From the Ngesong and Banyuning Springs a PVC pipe with a diameter of 300 mm and a length of 21,000 m is used to transport the water to the Dinoyo Reservoir by gravity.
2. From the Binangun Spring a DCIP pipe with 250, 300 and 350 mm diameters and a length of 21,000 m is used to transport the water to the Dinoyo Lama Reservoir, and then on to the Betek Reservoir through a pipe with a diameter of 200 mm which is 2,500 m long.
3. From the Sumbersari Spring a steel pipe with a diameter of 200 mm and a length of 1,500 m is used to transport the water to a military area and to the Betek Reservoir, together with water from the Karangan Spring.
4. From the Karangan Spring a DCIP pipe with a diameter of 200 mm and a length of 1,500 m is used to transport the water to the Betek Reservoir, together with water from the Sumbersari Spring.
5. From Wendit I, water is transported to the Betek Reservoir by means of a pump and three kinds of pipes (with diameters of 700 mm for 6,000 m, 600 mm for 2,520 m and 500 mm for 450 m). The pipe types are ACP, and DCIP.
6. From the Wendit II Spring water is transported – with the help of a pump – to the Mojolangu Reservoir by means of two pipes (with a diameter of 600 mm for 3,000 m and a diameter of 500 mm for 2,700 m). The pipe types are ACP, and DCIP. Water is also transported to the Buringbawah Reservoir by means of a pump and three kinds of pipes (with diameters of 600 mm for 5,500 m, 500 mm for 1,500 m, and 400 mm for 896 m). The pipe types are ACP, and DCIP.

Appendix 9. **A selection from the supplementary remarks to the Dutch edition made by Professor H.J. Schoemaker** [1]

Hydrology

After 1860 the Magnetic and Meteorological Observatory in Batavia started setting up hydrologic measuring stations all over the archipelago but predominantly in Java. Thousands of the rain gauges installed at government posts and in schools were subsequently monitored by officials from the Bureau of Public Works, later the Department of Civil Public Works, and by local teachers. After 1890 self-recording rain gauges were installed in Batavia and surrounding parts. On the basis of these recordings, Boerma developed sets of duration curves which, in 1931, were adopted by the engineer S.H.A. Begemann and reproduced in his article 'Toepassing van de waarschijnlijkheidsleer op hydrologische waarnemingen' (The application of the theory of probability to hydrological observations).[2]

In the second half of the nineteenth century the Swiss engineer Lauterberg cleverly combined the relationship between precipitation and river flow with the characteristics of river basins. Melchior applied Lauterberg's findings to the tropical conditions in the Dutch East Indies. His results, for which he was subsequently awarded the Conrad medal, were published in 1895. Melchior's method of calculating flood flows from observed rainfall was widely implemented in judicial and successful ways.

With our present knowledge of flow dynamics we know that the weakness of the method lies in the estimating of the 'concentration time' or, the flow time between the highest point in the river basin and the river section under consideration. In actual fact, even today the 'concentration time' remains a weak link in our chain of calculations.

In the well-known Unit Hydrograph method the 'concentration time' is assumed to be constant but a doctoral student of Professor D.A. Krayenhoff van de Leur's at the Wageningen University of Agriculture in the Netherlands observed that it changes depending on the precipitation intensity.[3]

Melchior had an enormous amount of data at his disposal, gathered from a large number of rain gauges, most of which have unfortunately since disappeared.

Erosion, sedimentation and sediment transport

Erosion and sedimentation have always caused and still do cause fundamental problems in the designing and managing of irrigation off-takes and outlets, weirs (mobile and otherwise), sand traps, etc. and canals. In the wet tropics the combination of heavy rainfall and soils that erode easily – e.g. volcanic ashes – often leads to the generation of large quantities of silt (which then has to be transported), heavy sedimentation and blocked channels. At the end of the nineteenth century, various ideas, notions and theories on flow dynamics, turbulence, dimension analysis, the mechanics of deformable solid matter, the dependency of sedimentation and sediment transport on the mineralogical origin of matter, water temperature and dissolved salts in water were either not yet formed or were insufficiently developed. It was therefore often the case that in those days the engineers were unpleasantly surprised when trying to control mud flows in the upper reaches of the Brantas river originating, for example, from Kloet volcano lahars or flows and sediments along the whole reach of the extensive Solo river. The river Serajoe irrigation works were plagued by the same kinds of problems. Systematic reconnaissance, investigations and experimentation were initiated to overcome these problems, e.g. with regard to the design of silt trapping works at the pumping stations of Pesanggrahan and Gambarsari. A sedimentology laboratory was even set up at Gambarsari. Two of the people heavily involved in this work were Clason and Trense.

British engineers, one of whom was Robert Gregg Kennedy (1851-1920), developed what came to be known as the 'regime theory' for their irrigation canals in British India. It was a theory which assumed that the effects of erosion and sedimentation could be compensated by changing canal discharge so leading – if properly designed – to the production of canals with stable sections.

However, the basic assumption was incorrect because erosion and sedimentation do not occur at the same locations in the wet section (the profile flowed through) and there is hysteresis between these two extremes. The 'regime theory' proved to be inapplicable to Java as that 'theory', based merely on empiricism, related to the typical soils of the Punjab, which differ in texture and structure from the volcanic soils of Java.

Slowly, special formulas for the design and management of Javanese irrigation canals – and later on for the other islands of the archipelago – were developed on the basis of new theoretical insight and empiricism. The lecture notes of Professor H.C.P. de Vos of the Technical University in Bandoeng provide us with clear evidence of these developments.[4]

Irrigation

On the whole, irrigation necessitates appropriate water management. In this respect there are two factors that are essential: plot size and the efficiency of irrigation water distribution.

As far as plot sizes are concerned, the study 'On irrigation efficiency'[5] by Professor J. Nugteren, formerly professor of irrigation at the Agricultural University of Wageningen in the Netherlands, stands out. Commissioned by the International Commission on Irrigation and Drainage he investigated and endeavoured to formulate the optimal size of tertiary irrigation units. His findings led to the establishment of an optimal size of 100 to 200 hectares with a flat minimum in between. In the Dutch East Indies the plot dimension striven for was approximately150 hectares.

The irrigation canal system with its branches (i.e. canals) and outlets requires a distribution curve that determines the distribution capacity of the various branches and outlets. Finally the Tegal distribution curve, as developed by the engineer A.G. Lamminga, emerged as the most useful type on Java. Later on it was 'exported' to other irrigation projects in the archipelago. The curve corresponds well with the French distribution curve, developed by R. Clément, which is based on the theory of probabilities.[6] Schoemaker improved Clément's curve by replacing the Gauss distribution with the more realistic Poisson distribution.[7]

One important aspect of water distribution is legislation. In the beginning attempts were made to 'export' to Java the Balinese *soebak* water distribution system, which is based on the Balinese tradition but also bears a certain resemblance to the Water Board System in the Netherlands. The project ended in failure. After many well-documented discussions in the pre-1934 era in the engineering periodical entitled *De Waterstaats-Ingenieur* (The Public Works Engineer) and, between 1934 and 1942, in *De Ingenieur in Nederlandsch-Indië* (The Engineer in the Dutch East Indies}, the General Water Code (Algemeen Waterreglement) was established in 1936.

The Inland Administration department (Binnenlands Bestuur) officials regularly consulted the General Water Code as did engineers seeking what was termed 'notice and advice'[8] , when preparing and granting water management permits. Inevitably De Kat Angelino's overview of the laws and jurisprudence pertaining to public works – available at every design and water management office – became a 'must' for the engineers.

When Schoemaker was preparing for Mr. Suyono Sasrodarsono's honorary doctorate in 1984 he stumbled upon the prevailing Indonesian Water Code of the day. It turned out to be a translation of the 1936 General Water Code including an introduction on the 'Panca Sila' and the following Soekarno quotation: 'you cannot create a revolution with lawyers'.

The irrigation systems used in the archipelago are all of the upstream control type. With that type of flow control system the downstream discharge gates to the canals and the outlet gates to the fields are tuned to the discharge set at the main inlet at the head of the system. In order to prevent unnecessary water spillage it is important that all the downstream outlet gates are properly adjusted in time and that requires a communication system that can guarantee rapid information exchange. The oldest irrigation area of Serajoe near Purworejo (Central Java) that was established in 1889 already possessed, for that specific purpose, an efficient telephone grid.

Design work for new canals and structures, as well as all the repair work, demanded the fabrication of copies from 'white' prints (i.e. brown lines drawn on a white background) to 'blue' prints (i.e. white lines drawn on a blue background), made possible by the use of direct sunlight. Each office therefore had a special workshop with specially trained personnel who used a curved glass plate which enabled exposure to the sun.

Incidentally, before 1910 all the writing was done by hand by native calligraphers (the official term in those days) who wrote the necessary letters and memorandums. After 1910 the typewriter took over. In the field offices all the calligraphers and secretaries were male. It was only in the Department of Civil Public Works that female secretaries were also appointed. High quality maps based on detailed field exploration carried out by *mantri* levellers[9] were already available at that time.

The *Grondslagen voor de samenstelling van begrootingen* (Fundamentals for drafting estimates), necessary for the tendering procedures, were mainly based on central European examples, probably because of the comparable character and size of the different works. The familiar German *Bauvorschriften* (Building Instructions) were of particular importance.

Bridges

Bridges appeared on the scene in the era before steel and concrete. Such bridges were made of rope and were wooden (i.e. from bamboo). Those cheap bridges were not suited to carrying heavy loads and they required substantial maintenance.

For military and commercial purposes Daendels' Great Post Road (Groote Postweg) required, where technically possible and economically feasible, strong trustworthy all-weather bridges that demanded little maintenance. Simple arch bridges made of natural stone were therefore constructed. In many places – recognisable by the black tarred parapets – they have stood the wear and tear of two centuries.

One outstanding example of the 19th century natural stone arch bridge tradition is the bridge over the river Lok Ulo near Kebumen (Central Java) which still stands today; see Figure 1.[10] With its one, large, 34-metre-wide, central arch and its two smaller side arches, each 19.50 meters wide and decorated with pilasters and Corinthian capitals, it was the brainchild of an engineer by the name of G.A. Pet, an excellent all round engineer who was practically a legend in his own lifetime. The Javanese called him 'Gapet' and believed that he possessed the gift of being able to make water flow upstream.

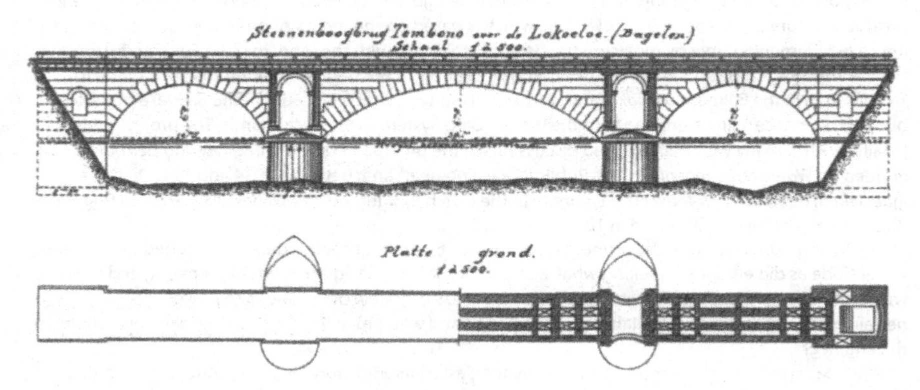

Figure 1. Natural stone arch bridge over the river Lok Ulo near Kebumen (Central Java)

In the nineteenth century the discovery of Portland cement not only led to large cement factories in the archipelago, e.g. known under the trademark of Karbouw (Buffalo), but also to the construction of even larger natural stone bridges (better mortar!) and – of course – to the introduction of reinforced concrete road bridges, soon to be followed by reinforced aqueducts, inverted siphons, etc. in irrigation canals.

The distances that needed to be covered in the archipelago were great and, to a large degree, proper infrastructure was still lacking (bridges, metalled roads, railways, etc.) The readily available manual labour resources thus meant that having prefabricated steel screw piles – many examples of which are still to be found throughout the archipelago – was very useful and most effective. Piles were available in all sizes. For the first railway bridge from Medan to Belawan (North Sumatra) even piles with a diameter of 1.80 meter (6 feet!) were used. The piles were made of 'packed' steel. Storehouses (or 'go-downs') were available which contained screw pile materials, including spare parts like coupling sleeves, tension bars with swivels, screw blades and wheel rims for the driving of piles. Complete and readily available packages of all the steel parts were delivered to the building site, with a full list of all the parts and the design-drawings of the bridge in question. Screw piles have stood the test of time very well. In wartime it emerged that often the only possible way to sabotage a screw pile bridge was by destroying the paving.

Every generation of engineers produces its progress and resistance heroes. Much the same can be said of bridge builders. Bijlaard, Van der Eb and Kist were typical bridge designers. Van der Eb was an excellent arithmetician, a real calculation expert. Kist's formulas were met with scepticism in the former colony.

Notes

INTRODUCTION **COVETOUSNESS AND VOCATION**

1 Landes, *Arm en rijk. Waarom sommige landen erg rijk zijn en andere erg arm*, 166. [Landes, *The wealth and poverty of nations: Why some are so rich and some so poor*]. See also Burger's *Sociologisch-economische geschiedenis van Indonesië* and De Jong, *De waaier van het fortuin. Van handelscompagnie tot koloniaal imperium*, 618-619.

2 Adjectives: Dutch East Indian or East Indian. We shall refer here to 'Indonesia' in cases where it is modern Indonesia that is being referred to and the 'Dutch East Indies' or simply 'the East Indies' at times when the reference is to the colonial era. When alluding to the archipelago in general the term Indonesia will also be adhered to. Similarly with place names the spelling adhered to will – where it varies – correspond to the usage current in the historical period being referred to. One will therefore come across all the following mutations: Batavia; Djakarta, Jakarta; Bandoeng, Bandung; Soerabaja, Surabaya; Cheribon, Cirebon et cetera. In 1972 official changes were made in the Bahasa Indonesian language so that: the 'tj' became 'c', the 'dj' became 'j' and the 'j' was turned into 'y'. Before that time the 'oe' had been converted to 'u'. Much the same applies to derived terms such as 'Indonesian' and 'East Indian' though here one must consider also the Indonesian or Dutch perspective. Parallel to 'Indonesians' we retain the terms 'natives' or 'inlanders' rather than the incorrect usage of 'Indians' though the term 'inlanders' is avoided as much as possible. In the colony this population group was distinguished from Europeans and their equals, such as Indo-Europeans (i.e. those with a European [Dutch] father) and the Japanese, on the one hand and on the other hand people of Chinese or other foreign Eastern extraction (Arabian et cetera). The same system is applied to other naming methods.

3 Van Doorn, *The engineers and the colonial system: Technocratic tendencies in the Dutch East Indies* and *De laatste eeuw van Indië. Ontwikkeling en ondergang van een koloniaal project*.

4 See, for instance, Van Helsdingen and Hoogenberk, *Daar wèrd wat groots verricht .... Nederlandsch-Indië in de XXste eeuw*.

5 Here it should be mentioned that civil servants (controllers, engineers and so on) did usually speak the language of the region to which they had been sent to work. Similarly, Dutch began to be spoken by the rising literates of the indigenous middle and top social classes: there were Dutch East Indian schools and secondary education was in the Dutch language. Finally, all official documents were also written in the Dutch language.

6 The pre-colonial political-feudal pattern was also a strong forming factor. See for points of continuity, for instance, De Jong, *De waaier van het fortuin*, 618-621.

7 This assertion comes from T.P. Hughes. See Van der Vleuten's: 'Twee decennia van onderzoek naar grote technische systemen: Thema's, afbakening, kritiek'.

8 These are the 2000 figures. Dutch Central Statistical Office; the *Encyclopaedia Britannica*.

9 Which also dominated areas in the West as well as in the East: Surinam and the Dutch Antilles. The Indonesian archipelago, though, was far and away the most important colony. See for the term 'colonial empire': Wesseling's *Europa's koloniale eeuw*.

10 Wesseling, *Indië verloren, rampspoed geboren*.

11 In the debate on the causes of special or deviant European developments, the ultimate cultural and thus idealistic explanation as given by Landes opposes the (ultimately) geographical, demographic and thus materialistic explanation recently provided by J. Diamond. The title of his book *Guns, germs and steel. The fates of human societies* is indeed very telling. Though Diamond does draw comparisons between Europe and Asia he is of less interest to us in the context of this research because it is very much Eurasian developments that he discusses in relation to those continuing in the rest of the world. He discusses, for instance, the great variety of cultivable plants and the variety of animals in Eurasia and talks about the climatological advantages of the east-

west axis of that continent (agricultural and horticultural activities could be 'climatologically neutrally' and thus easily expanded). He compares this to the relatively unfavourable north-south axis of other parts of the world, like for example America. In the end both explanations can be well reconciled, in the first place by analysing matters according to the time scale: in the long-term material factors predominated whilst in the short-term cultural issues came to the fore. In the second place the explanations can be reconciled by adopting a multi-causal approach: in the case of large and complex social changes there are always many factors at work. For further discussion see, for instance, the review of Diamond's book by W.H. McNeill in *The New York Review of Books*, June 26, 1997 and Diamond's answer.

12 There were also structural problems and it was very logical to separate trade and administration in relation to the aspect of colonial possessions, also in the case of other European companies. See De Jong, *De waaier van het fortuin*, 141-164 and 168 et cetera. See also Gaastra, *De geschiedenis van de VOC*. There was also continuity: the VOC resurged, to a certain extent, through the Dutch Trade Company, which was founded in 1824 by the Dutch king and which, for instance, transported and sold the products derived from the Cultivation System.

13 'State' should be taken here to mean the entire East Indian governmental machinery, the administration included.

14 This was a process that was also perceptible in other colonies. See for the Dutch East Indies and other areas: Van den Doel, *De stille macht. Het Europese binnenlands bestuur op Java en Madoera, 1808-1942*.

15 The policy and administration processes that were relevant to the development of East Indian Public Works are dealt with in Chapter 1.

16 See De Graaff and Stibbe, 'Junghuhn'.

17 The Colonial report of 1896 already mentions endeavours to set up a Government Savings Bank in order to encourage the local population to save their earnings. In the East Indian budget of 1899, 28,540 guilders had been reserved for the above-mentioned banking system.

18 Where possible in the 1945-1949 period efforts were made to repair the damage caused during the time of Japanese occupation (to plantations, factories, oil extraction locations, refineries, bridges, harbours, and so on) and as a consequence of the combat actions of the *bersiap* (campaign mounted by Indonesian fighters) and policing activities (of the Dutch army).

19 See also *De koloniale roeping van Nederland*.

20 Van Helsdingen and Hoogenberk, *Daar wèrd wat groots verricht .... Nederlandsch-Indië in de XXste eeuw*, VI.

21 De Vos, 'De strijd om en tegen het water', especially 284 and 277.

22 Boeke, 'Van vier tot vierenveertig millioen zielen op Java'.

23 Vink, 'De inheemsche landbouw', 358.

24 Van Helsdingen and Hoogenberk, *Daar wèrd wat groots verricht .... Nederlandsch-Indië in de XXste eeuw*, 232-272 (De overwinning op den afstand).

25 See for instance, in alphabetical order, Baudet and Brugmans, *Balans van beleid. Terugblik op de laatste halve eeuw van Nederlandsch-Indië*; Fasseur, *Kultuurstelsel en koloniale baten. De Nederlandse exploitatie van Java 1840-1860* and *De weg naar het paradijs en andere Indische geschiedenissen*; Kamerling, *Indonesië toen en nu*; Clemens and Lindblad, *Het belang van de Buitengewesten. Economische expansie en koloniale staatsvorming in de Buitengewesten van Nederlands-Indië*; Locher-Scholten, *Sumatraans sultanaat en koloniale staat. De relatie Djambi-Batavia (1830-1907) en het Nederlandse imperialisme* and *Sumatran sultanate and colonial state: Jambi and the rise of Dutch imperialism, 1830-1907*; Van Schaik, *Colonial control and peasant resources in Java. Agricultural involution reconsidered*; Van 't Veer, *De Atjeh-oorlog*. See for an overview of the discussion on Dutch imperialism: Kuitenbrouwer, 'Het imperialisme-debat in de Nederlandse geschiedschrijving'.

26 Especially by Van Doorn, *The engineers and the colonial system*.

27 Van Doorn, 'De eerste spoorweg op Java'.

28 Especially by Van Doorn, *De laatste eeuw van Indië*.

29 À Campo, *Koninklijke Paketvaart Maatschappij. Stoomvaart en staatsvorming in de Indonesische archipel 1888-1914* and *Engines of empire. Steamshipping and state formation in colonial Indonesia*.

30 Leidelmeijer, *Van suikermolen tot grootbedrijf. Technische vernieuwing in de Java-suikerindustrie in de negentiende eeuw*. See also H. Maat, 'Techniek en het koloniale verleden'.

31 Ravesteijn, *De zegenrijke heeren der wateren. Irrigatie en staat op Java, 1832-1942*; Ertsen, *Prescribing perfection: Emergence of an engineering irrigation design approach in the Netherlands East Indies and its legacy 1830-1990.*

32 Mrázek, *Engineers of happy land: Technology and nationalism in a colony.*

33 Compare Van den Doel, *De stille macht. Het Europese binnenlands bestuur op Java en Madoera, 1808-1942.*

34 This therefore concerns Dutch engineers employed overseas, known as Dutch East Indian or simply East Indian engineers. These engineers were educated in Delft and later in Bandoeng. Local or Indonesian engineers only became available towards the end of the colonial period and played no significant part in the civil engineering developments that took place in the East Indies.

35 Developments in the 1950-1965 period that are important for the subsequent period will also be briefly dealt with in Chapter 9.

36 East Indian Public Works did not play an important part in the area of urban development. The department did have a construction division, but that was devoted more to public buildings though former Public Works engineers were often involved in the urban planning side of things (see Chapter 7).

37 Many private works are described and recorded in annual reports and the commemorative books of companies such as the Deli Company, the Amsterdam Trade Association (HVA) and the Amsterdam Rubber Culture Company (RCMA).

38 Modern irrigation methods that were implemented on a big scale were in the past in the East Indies and still today in the Indonesian era, viewed as the way of making autarkical rice provision possible.

39 On the basis of historical considerations a conscious decision to use the word 'sanitation' for urban drainage, for the transport and processing of faeces, waste water and solid waste is maintained.

40 No attention is given to the building of aerodromes.

41 See, for instance, Fischer, *Geschiedenis van de techniek.*

42 See, for example, Paulus, 'Irrigatie'.

43 Post-colonial literature also frequently breathes the same atmosphere as that emanated by these models. See Ravesteijn, *De zegenrijke heeren der wateren. Irrigatie en staat op Java, 1832-1942*, 23-28 and for a criticism of both models, 30-32.

44 What has been especially vital to the image of modern history of technology in the Netherlands has been the work of H.W. Lintsen. His doctoral dissertation was on the rise of the engineering profession in the twentieth century. Together with other authors he has published two series of overviews of the history of technology in the Netherlands in the nineteenth and twentieth centuries: Lintsen, *Techniek in Nederland. De wording van een moderne samenleving 1800-1890*; Schot, Lintsen and Rip, *Techniek in Nederland in de twintigste eeuw*. A technology dynamics classic is: Bijker, Hughes and Pinch, *The social construction of technological systems: New directions in the sociology and history of technology.*

45 In accordance with what is customary in the field of modern technology history we do not differentiate very strictly between 'technique' and 'technology'. In both cases it is all about a combination of artefacts and knowledge.

46 Hughes, 'The evolution of large technological systems'. See also Hughes, *Networks of power: Electrification in modern society, 1880-1930.*

47 See Van de Poel and Franssen, 'Understanding technical development: The concept of "technological regime"'.

48 In the case of modern irrigation in Java, the system-based character of the network has been made plausible. The colonial irrigation efforts in Java resulted in an island-wide socio-technical irrigation system comprising all kinds of technical provisions (artefacts), a whole range of actors including for instance building companies and other elements such as control provisions, rules and legislation, and an education system for engineers and other technologists. See Ravesteijn, 'Dutch engineering overseas: The creation of a modern irrigation system in colonial Java'. See for the systematic approach in relation to water management and 'water resources development' in general: Ravesteijn, Hermans and Van de Vleuten, 'Water systems: Participation and globalisation in water system building'.

49 See, for instance, Winner, 'Upon opening the black box and finding it empty: Social constructivism and the philosophy of technology'.

19 Boreel, 'Mededeelingen betreffende de gewone wegen in Ned.-Indië en meer in het bijzonder omtrent den aanleg daarvan in de Buitenbezittingen', 813; Ott de Vries, 'Het wegenvraagstuk in Nederlandsch-Indië', 866.
20 Ott de Vries, 'Het wegenvraagstuk in Nederlandsch-Indië', 947; Stibbe and Stroomberg, 'Materiaalonderzoek', 252-253.
21 Van Sandick, 'Algemeen Ingenieurscongres te Batavia 8-15 Mei 1920', 831.
22 Ortt, 'Verslag van het Ned.-Ind. Wegencongres 23-25 Juni 1924 te Bandoeng', 866.
23 *Nederlandsch-Indisch Wegen Congres Bandoeng*; Ortt, 'Verslag van het Ned.-Ind. Wegencongres 23-25 Juni 1924 te Bandoeng', 866-867.
24 H. van Breen, 'De Nederlandsch-Indische Wegenvereeniging, 649.
25 Stibbe and Sandbergen, 'Wegverkeer. Wegverkeerswetgeving ', 486-487.
26 D. de Longh Wzn., 'Regeling van het verkeer te land in Ned.-Indië', V.115.
27 Stibbe and Sandbergen, 'Wegverkeer. Wegverkeerswetgeving', 486-487.
28 *De Ingenieur in Nederlandsch-Indië*, I.32.
29 The price of petrol, for instance in 1939, in the Netherlands was 14.5 cents per litre but in the East Indies, where petroleum was extracted and refined, it cost 23 cents per litre in conjunction with the high taxes that were levied (personal details obtained from Jan Kop).
30 Stibbe and Sandbergen, 'Wegverkeer. Wegverkeerswetgeving', 487-489.
31 Van de Linde, 'Wegenaanleg op Java', 18.
32 Ortt, 'Verslag van het Ned.-Ind. Wegencongres 23-25 Juni 1924 te Bandoeng', 894; Afdeeling Bruggen en Wegen van het Departement B.O.W, *Handleiding voor het traceeren van wegen*, 68-69.
33 Ortt, 'De ontwikkeling van de Ned.-Indische wegenbouw-techniek', A.524.
34 Ortt, 'De ontwikkeling van de Ned.-Indische wegenbouw-techniek', A.524.
35 Ortt, 'De ontwikkeling van de Ned.-Indische wegenbouw-techniek', A.524.
36 Ortt, 'De ontwikkeling van de Ned.-Indische wegenbouw-techniek', A.524-A.525.
37 Van Leeuwen, *Honderd jaar Nederland 1848-1948*, 157; Bakker, *Bali in kleuren*, 20.
38 Ortt, 'De wegen in Deli', in *Wegen* 1939, 354-355.
39 *Oosthoeks encyclopedie*, 513; *Grote Winkler Prins encyclopedie*, 135.

CHAPTER 3 **THE LOCOMOTIVE OF MODERNITY**

1 Fasseur, 'De Nederlandse Koloniën 1795-1914'. More recently, De Jong's *De waaier van het fortuin. Van Handelscompagnie tot koloniaal imperium. De Nederlanders in Azië en de Indonesische archipel 1595-1950.*
2 Van der Wijck, *De Nederlandsche Oost-Indische bezittingen onder het bestuur van den kommissaris-generaal Du Bus de Gisegnies*, 113.
3 See for data on the Cultivation System: Fasseur, *Kultuurstelsel en koloniale baten. De Nederlandse exploitatie van Java 1840-1870.* Also in what is termed the *Koloniaal Verslag* (Colonial Report), published annually after 1850 in which much information is to be found on the economic situation in the Dutch East Indies; after the eighteen-seventies that *Koloniaal Verslag* constituted a permanent supplement to the *Handelingen van de Staten-Generaal* (Proceedings of the States General). Particularly on the economy: Lindblad, *New challenges in the modern economic history of Indonesia.*
4 Figures on the East Indian credit balance in: De Waal, *Onze Indische financiën*, II, 44.
5 Mansvelt, *Handelsstatistiek van Java 1823-1873*, Tables 6 and 12.
6 De Jong, *De waaier van het fortuin*, 184; a map of the Great Post Road on p. 209.
7 See for details on travelling in the East Indies, for instance: Gevers Deynoot, *Herinneringen eener reis naar Nederlandsch-Indië in 1862.*
8 A history of the Government's Navy in: Wijn, *Tot in de verste uithoeken...De cruciale rol van de Gouvernements Marine bij het vestigen van de Pax Neerlandica in de Indische Archipel 1815-1962.*
9 On the early interinsular transport see: A Campo, *Koninklijke Paketvaart Maatschappij. Stoomvaart en staatsvorming in de Indonesische archipel 1888-1914*, 39-73.
10 Campo *Koninklijke Paketvaart Maatschappij*, 48.
11 Campo *Koninklijke Paketvaart Maatschappij*, 73.
12 Campo *Koninklijke Paketvaart Maatschappij*, 412, 444.

CHAPTER 1 **TECHNOLOGY AND ADMINISTRATION**

1 De Graaff and Stibbe, 'Krakatau'.
2 Van Bosse, 'J.A. de Gelder in het tijdperk van de bouw der haven van Tandjonk Priok', 789.
3 Van Bosse, 'J.A. de Gelder in het tijdperk van de bouw der haven van Tandjong Priok', 789.
4 Van Bosse, 'J.A. de Gelder in het tijdperk van de bouw der haven van Tandjong Priok', 789.
5 Van Sandick 'Ter herinnering aan J.A. de Gelder', 786.
6 See for further details on the construction of Tandjonk Priok harbour Chapter 5. See for more information on the history of East Indian Public Works: De Meyier, 'Waterstaat'; Stibbe and Sandbergen, 'Waterstaat'; Ravesteijn, *De zegenrijke heeren der wateren. Irrigatie en staat op Java, 1832-1942* and 'Irrigatie en koloniale staat op Java: De gevolgen van de hongersnoden in Demak'. See for the various resolutions and decrees concerning East Indian Public Works the *Staatsblad van Nederlandsch-Indië*.
7 Landes, *The wealth and poverty of nations: Why some are so rich and some are so poor*.
8 Kuitenbrouwer, 'Het imperialism-debat in de Nederlandse geschiedschrijving'.
9 Cited in Van Doorn, *Indische lessen*, 29.
10 Van den Doel, *De stille macht. Het Europese binnenlands bestuur op Java en Madoera, 1808-1942*.
11 The penetration of the Van Hogendorp doctrine into colonial policy may be seen as the heart of the modernisation process within the colonial state (see the introduction).
12 See Chapter 2.
13 Post, *Over den waterstaat in Nederlandsch-Indië*.
14 De Meyier, 'Waterstaat', 727.
15 See Chapter 6.
16 As well as the public railways network there was also an extensive network of railways – consisting of narrow-gauge track – within the agricultural enterprises for the purposes of processing the crops, harvesting them and transporting them. See for more information on railways and hydraulic power: Chapters 3 and 6.
17 Ravesteijn, 'Irrigatie en koloniale staat op Java'.

CHAPTER 2 **THE ROAD TO A NEW EMPIRE**

1 Most of the information given in this section derives from: De Meyier, 'Wegen en bruggen in Nederlansch-Indië', 301; Stibbe, 'Wegen', Stapel, *Geschiedenis van Nederlandsch Indië*, 38.
2 Stapel, *Geschiedenis van Nederlandsch Indië*, 38; Stibbe, 'Wegen', 744.
3 Government Resolution, 11th March 1835 No. 1, in: *De Waterstaats-ambtenaar in Nederlandsch Oost-Indië*, 149.
4 Stapel, *Geschiedenis van Nederlandsch Indië*, 296.
5 De Meyier, 'Wegen en bruggen in Nederlandsch-Indië' 302.
6 Stibbe 'Wegen', 747-748.
7 Stibbe 'Wegen', 743-744; Stapel, *Geschiedenis van Nederlandsch Indië*, 38.
8 Boreel, 'Mededeelingen betreffende de gewone wegen in Ned.-Indië en meer in het bijzonder omtrent den aanleg daarvan in de Buitenbezittingen', 809.
9 Dumas and Blok, 'Aanleg en onderhoud van wegen in de buitengewesten. Prae-advies BOW', 5-6.
10 Reitsma, *De wegenkwestie op Java en Madoera*, 4.
11 Stibbe and Sandbergen, 'Wegverkeer. Wegverkeerswetgeving', 477-478.
12 Cramer, 'Het wegennet', 469, 470.
13 Cramer, 'Het wegennet', 471.
14 De Meyier, 'Wegen en bruggen in Nederlandsch-Indië', 302.
15 Stibbe, 'Wegen', 746, 749.
16 Circular from the Director BB dated 21st Dec. 1880 No. 12770, in: *De Waterstaats-ambtenaar in Nederlandsch Oost-Indië*, 153-154.
17 Boreel, "Mededeelingen betreffende de gewone wegen in Ned.-Indië en meer in het bijzonder omtrent den aanleg daarvan in de Buitenbezittingen', 816.
18 Circular document issued by Public Works, 19th April 1872. No. 3705, in: *De Waterstaats-ambtenaar in Nederlands Oost-Indië*, 467-468.

13 Reitsma, *Korte geschiedenis der Nederlandsch-Indische spoor- en tramwegen*, 5.
14 Reitsma, *Korte geschiedenis,* 7. In 1839 John Dixon established an engineering works and iron foundry which, after 1841, operated as 'De Atlas' and became one of the leading engineering works in the Netherlands (Van Hooff, *De Nederlandse machinefabrieken 1825-1914*, 23).
15 For the situation in the Netherlands see: Veenendaal, *De ijzeren weg in een land vol water. Beknopte geschiedenis van de spoorwegen in Nederland 1834-1958.*
16 *Nederlandsche Staatscourant* of 6th June 1842.
17 G.H. Uhlenbeck was later to become Director of Civil Public Works in the East Indies and was, from 1862 until 1863, Minister for the Colonies in the second Thorbecke cabinet (Van Ette, *Onze ministers sinds 1798*, 36).
18 Reitsma, *Korte geschiedenis,* 11.
19 The debate in the Netherlands in: Veenendaal's *De ijzeren weg,* 26-27.
20 Thomas Johannes Stieltjes (1819-1878), artillery officer and later engineer with the Overijssel Canal Company, in the East Indies between 1860 and 1863. From 1866 until his death he was a member of the Second Chamber, first for Zwolle and later for Amsterdam. He was renowned for his caustic criticism, especially in the field of public works and military affairs. He published numerous pamphlets under the pseudonym 'an old soldier' (*Nieuw Nederlandsch biografisch woordenboek*, II, 1370; hereafter referred to as the *NNBW*).
21 Nicolaas Hubert Henket (1829-1904), land surveyor and engineer, involved in numerous railway and canal plans in the Netherlands, for instance in Overijssel under T.J. Stieltjes, was in the East Indies from 1860 until 1865. In 1866 he was made professor at the Delft Polytechnic where he became one of the biggest authorities in the field of civil engineering (*NNBW*, VI, 760).
22 Criticism on the report published by Stieltjes c.s. by 'an officer of the engineering corps of the Dutch East Indian Army', *Het rapport van den heer Stieltjes over verbeterde vervoermiddelen op Java*. Stieltjes was especially blamed for his carelessness, incompleteness and poor knowledge of the field; he was also criticised for not being sufficiently familiar with documentary sources, i.e. existing reports compiled by previous committees and engineers.
23 For the Sloet-Reuchlin franchise request see: Jonkers Nieboer, *Geschiedenis der Nederlandsche Spoorwegen 1832-1938*, 79-90.
24 Stieltjes, *Gegevens omtrent de zaak der spoorwegen op Java.* Stieltjes repeated it all in a watered down form in a pamphlet entitled, *Overzicht van hetgeen met de spoorwegen op Midden-Java is voorgevallen*. Henket in fact agreed that the line as proposed by Stieltjes would be by far and away the best solution. Henket, *De aanleg van spoorwegen op Midden-Java*.
25 Reitsma, *Korte geschiedenis*, 18.
26 *Het rapport van den heer Stieltjes over verbeterde vervoermiddelen op Java.*
27 Van Herwerden, *De spoorweg-kwestie op Java.*
28 Johan Philip de Bordes (1817-1899), originally military officer, September 1860 secretary of the Railway Construction Committee, September 1861 member of the Committee (after Sloet's departure to the East Indies) and chairman of the Board of Supervisors for Railway Affairs. De Bordes worked in the East Indies from the end of 1863 until the end of 1869 but remained active in the East Indian and Dutch railway worlds (*NNBW*, I, 413).
29 On the General Company see: Jonker, *Merchants, bankers, middlemen. The Amsterdam money market during the first half of the 19th century*, 258-59 and Hirschfeld, *Het ontstaan van het moderne bankwezen in Nederland*, 55-64, 83-88.
30 Hirschfeld, *Ontstaan van het moderne bankwezen*, 85.
31 Details in: Reitsma, *Korte geschiedenis*, 22-24.
32 See Van Doorn, 'De eerste spoorwegen op Java', 80-88.
33 A description of the NISM works based on information derived from J.P. de Bordes himself in: *Tijdschrift van het Koninklijk Instituut van Ingenieurs* and in his pamphlet *De spoorweg Samarang-Vorstenlanden*.
34 Van Doorn, 'De eerste spoorweg op Java', 88.
35 Oegema, *De stoomtractie op Java en Sumatra*, 39-40.
36 Oegema, *De stoomtractie,* 41.
37 Stieltjes, *Eenige beschouwingen over spoorwegen op Java*, 27.
38 See for details of track gauges in the Netherlands: Veenendaal, *De IJzeren Weg,* 15, 39.

13 Reitsma, *Korte geschiedenis der Nederlandsch-Indische spoor- en tramwegen*, 5.
14 Reitsma, *Korte geschiedenis,* 7. In 1839 John Dixon established an engineering works and iron foundry which, after 1841, operated as 'De Atlas' and became one of the leading engineering works in the Netherlands (Van Hooff, *De Nederlandse machinefabrieken 1825-1914*, 23).
15 For the situation in the Netherlands see: Veenendaal, *De ijzeren weg in een land vol water. Beknopte geschiedenis van de spoorwegen in Nederland 1834-1958.*
16 *Nederlandsche Staatscourant* of 6th June 1842.
17 G.H. Uhlenbeck was later to become Director of Civil Public Works in the East Indies and was, from 1862 until 1863, Minister for the Colonies in the second Thorbecke cabinet (Van Ette, *Onze ministers sinds 1798*, 36).
18 Reitsma, *Korte geschiedenis,* 11.
19 The debate in the Netherlands in: Veenendaal's *De ijzeren weg,* 26-27.
20 Thomas Johannes Stieltjes (1819-1878), artillery officer and later engineer with the Overijssel Canal Company, in the East Indies between 1860 and 1863. From 1866 until his death he was a member of the Second Chamber, first for Zwolle and later for Amsterdam. He was renowned for his caustic criticism, especially in the field of public works and military affairs. He published numerous pamphlets under the pseudonym 'an old soldier' (*Nieuw Nederlandsch biografisch woordenboek*, II, 1370; hereafter referred to as the *NNBW*).
21 Nicolaas Hubert Henket (1829-1904), land surveyor and engineer, involved in numerous railway and canal plans in the Netherlands, for instance in Overijssel under T.J. Stieltjes, was in the East Indies from 1860 until 1865. In 1866 he was made professor at the Delft Polytechnic where he became one of the biggest authorities in the field of civil engineering (*NNBW*, VI, 760).
22 Criticism on the report published by Stieltjes c.s. by 'an officer of the engineering corps of the Dutch East Indian Army', *Het rapport van den heer Stieltjes over verbeterde vervoermiddelen op Java.* Stieltjes was especially blamed for his carelessness, incompleteness and poor knowledge of the field; he was also criticised for not being sufficiently familiar with documentary sources, i.e. existing reports compiled by previous committees and engineers.
23 For the Sloet-Reuchlin franchise request see: Jonkers Nieboer, *Geschiedenis der Nederlandsche Spoorwegen 1832-1938*, 79-90.
24 Stieltjes, *Gegevens omtrent de zaak der spoorwegen op Java.* Stieltjes repeated it all in a watered down form in a pamphlet entitled, *Overzicht van hetgeen met de spoorwegen op Midden-Java is voorgevallen.* Henket in fact agreed that the line as proposed by Stieltjes would be by far and away the best solution. Henket, *De aanleg van spoorwegen op Midden-Java.*
25 Reitsma, *Korte geschiedenis,* 18.
26 *Het rapport van den heer Stieltjes over verbeterde vervoermiddelen op Java.*
27 Van Herwerden, *De spoorweg-kwestie op Java.*
28 Johan Philip de Bordes (1817-1899), originally military officer, September 1860 secretary of the Railway Construction Committee, September 1861 member of the Committee (after Sloet's departure to the East Indies) and chairman of the Board of Supervisors for Railway Affairs. De Bordes worked in the East Indies from the end of 1863 until the end of 1869 but remained active in the East Indian and Dutch railway worlds (*NNBW*, I, 413).
29 On the General Company see: Jonker, *Merchants, bankers, middlemen. The Amsterdam money market during the first half of the 19th century*, 258-59 and Hirschfeld, *Het ontstaan van het moderne bankwezen in Nederland*, 55-64, 83-88.
30 Hirschfeld, *Ontstaan van het moderne bankwezen*, 85.
31 Details in: Reitsma, *Korte geschiedenis,* 22-24.
32 See Van Doorn, 'De eerste spoorwegen op Java', 80-88.
33 A description of the NISM works based on information derived from J.P. de Bordes himself in: *Tijdschrift van het Koninklijk Instituut van Ingenieurs* and in his pamphlet *De spoorweg Samarang-Vorstenlanden.*
34 Van Doorn, 'De eerste spoorweg op Java', 88.
35 Oegema, *De stoomtractie op Java en Sumatra*, 39-40.
36 Oegema, *De stoomtractie,* 41.
37 Stieltjes, *Eenige beschouwingen over spoorwegen op Java*, 27.
38 See for details of track gauges in the Netherlands: Veenendaal, *De IJzeren Weg,* 15, 39.

39 For Queensland see: Durrant, *Australian steam*, 80.
40 Hughes, *Indian locomotives. Part 1-broad gauge 1851-1940*, 7-8.
41 At that moment a bill was being debated in the British East Indies which contemplated switching to the metric system. It was not passed and it would be a further 86 years before the East Indies would go metric.
42 Hughes, *Indian locomotives. Part 2-metre gauge 1872-1940*, 7-9.
43 Burman, *Early railways at the Cape*, 50-51.
44 The Amsterdam company Wertheim & Gompertz placed the first Rio Grande RR bond loan in the Netherlands in 1871. The Rio Grande's initiator, general W.J. Palmer, had visited the Netherlands previously in a bid to gain support for his enterprise (Veenendaal, *Slow train to paradise. How Dutch investment helped build American railroads*, 63-65).
45 *NNBW*, II, 707; Veenendaal, 'De kennisoverdracht op het gebied van de spoorwegtechniek in Nederland 1830-1870', especially 80.
46 Reitsma, *Korte geschiedenis*, 24-25. Kool and Henket's report was published (in Dutch) as *Research as to how far the narrow gauge of 1.00 to 1.10 meter would be applicable to the needs of Javanese transport and recommendable from an economic point of view*.
47 Stieltjes, *Eenige beschouwingen*, 29.
48 The *Tijdschrift van het Koninklijk Instituut van Ingenieurs. Afdeeling Nederlandsch-Indië* (1876-1914). A mine of information on engineering works in the Netherlands.
49 This committee was established by Royal decree on 22nd February 1871.
50 The Bank für Handel und Industrie in Darmstadt had not long before helped the new Netherlands Company for the Exploitation of State Railways with its financial problems and was therefore seen as a suitable partner for the East Indian Railways.
51 Van Herwerden, *Een spoorwegnet over Java*.
52 One of those opposed to a state-backed railway system had dubbed it this (Reitsma, *Korte geschiedenis*, 32).
53 David Maarschalk (1829-1886), officer in the East Indian Army was, already in 1868, chairman of the NISM. After De Bordes' departure he finished building the Semarang-Principalities and the Batavia-Buitenzorg lines. In 1875 he was appointed head of the State Railway Service in Java. He strictly supervised the construction of the first lines and was promoted to the position of inspector-general in 1878, a position held until the time of his honourable discharge, at his own request, in 1880 (*NNBW*, V, 327).
54 De Jong, *De waaier van het fortuin*, 307-08.
55 Oegema, *Stoomtractie*, 66-68.
56 Reitsma, *Korte geschiedenis*, 34-35.
57 Reitsma, *Gedenkboek der Staatspoor- en Tramwegen in Nederlandsch-Indië*, 1875-1925, 34.
58 Reitsma, *Korte geschiedenis*, 43.
59 Overview of the construction of the various lines for all the rail and tramways in Reitsma's, *Korte geschiedenis*, 106-24.
60 Gonggrijp, *Schets eener economische geschiedenis van Nederlandsch-Indië*, 147-50.
61 The discussion on the organisational form in: Reitsma, *Gedenkboek*, 92-110.
62 Oegema, *Stoomtractie*, 178-79.
63 Reitsma, *Gedenkboek*, 54.
64 The whole affair discussed at length in: Reitsma, *Gedenkboek*, 57.
65 Reitsma, *Korte geschiedenis*, 52-53, gives an overview of the lines included in the 1893 plan.
66 Oegema, *Stoomtractie*, 137-38; Reitsma, *Korte geschiedenis*, 37-39.
67 The Batavia Electric Tramway Company opened its first electric line in the East Indian capital in 1899.
68 For the history of the SJS and the related 'sister tram companies' see: *De stoomtramwegen op Java. Gedenkboek samengesteld ter gelegenheid van het vijf en twintigjarig bestaan der Semarang-Joana Stoomtram-Maatschappij*. The company spelt Samarang with an 'a' even though it was more customary to have an 'e' (Semarang).
69 The 'sister tram companies' which fell under the same Dutch directorate and worked in close cooperation in the East Indies were: the Samarang-Joana (SJS), the East-Java (OJS), the Semarang-Cheribon (SCS) and the Serajoedal (SDS).

70 It is virtually impossible to name here all the railway and tram companies active in Java. For a full overview of all the franchises and openings of railway lines see: Reitsma, *Korte geschiedenis*, 106-24.
71 The Java Railway Company established in 1885, which was backed by a great deal of British capital, had big plans for railway lines in Central Java but, apart from the 24.5 km long Tegal-Balapoelang line, nothing was to come of those wonderful plans (Reitsma, *Korte geschiedenis*, 41).
72 The largest portion of the NISM railway line between Semarang and Soerabaja was constructed between 1900 and 1903 but it was still necessary to transfer at Goendih until 1924 in connection with the wide gauge line. After that time the standard gauge Semarang-Gambringan line was opened.
73 Oegema, *Stoomtractie*, 48.
74 Oegema, *Stoomtractie*, 142.
75 De Bruin, *Du Croo & Brauns locomotieven*.
76 The takeover was rounded off in 1898 for a sum of almost four million guilders.
77 Slim, 'De aanleg der nieuwe spoorverbindingen Solo-Djocja'. The line was constructed by the NISM and rented out to the State Railways.
78 Reitsma, *Gedenkboek*, 80-82.
79 Caspersz, 'De concurrentie van het motorvervoer aan de spoor- en tramwegen in Indië'.
80 Van Helsdingen and Hogenberk (eds.), *Daar wèrd wat groots verricht...Nederlandsch-Indië in de xxste eeuw*.
81 See for the Dutch plans for electrification: Veenendaal, 'Techniek, economie of politiek? Het Nederlandse spoorwegelektrificatieplan van 1922'.
82 See Chapter 6.
83 Reitsma, *Gedenkboek*, 158-69.
84 Before 1914 engineer G. de Gelder had worked for the Chilean Electric Tramway & Light Company as an electro-technologist. It was a company that operated a number of tramways and light factories in Chile.
85 See De Gelder, 'De electrificatie der spoorwegen in Nederlandsch-Indië' and his 'De electrische Staatspoorwegen in Ned.-Indië'.
86 See Keus and Brandes, 'Electrische locomotief voor de staatsspoorwegen in Ned.-Indië, gebouwd door Heemaf in samenwerking met Werkspoor'.
87 An American interurban was an electric tramline that linked up localities and was sometimes laid on the sides of roads but which more often than not had its own lines and was used by relatively heavy and fast trams. In the early years of the twentieth century this form of transport soon became very popular but the rage did not last for long.
88 Oegema, *Stoomtractie*, 180-81.
89 On 1st November 1928 a scheduled flight service between Batavia-Tjililitan and Semarang was started up by the KNILM. The aerodrome Morokrembangan in Soerabaja was declared unfit as a landing area at the last minute which meant that passengers had to travel on from Semarang by train. On 1st November 1929, however, another suitable landing area was found in Soerabaja which meant that from that time onwards aeroplanes were able to fly on (*De koloniale roeping van Nederland*, 113-14).
90 Reitsma, *Van Stockum's travellers' handbook for the Dutch East Indies*, 158, 321.
91 Oegema, *Stoomtractie*, 193-201. During test journeys speeds of more than 110 km/hr were achieved with these engines.
92 This section is based on: Oegema, *Stoomtractie*, 66-78.
93 See for more information on these locomotives in general: Durrant, *The Mallet Locomotive* and Vilain, *Les locomotives articulées du système mallet dans le monde*.
94 Oegema, *Stoomtractie*, 78-81; Durrant, *The Mallet Locomotive*, 74-82.
95 Between 1900 and 1910 the total number of passengers transported on the SS Java network rose from 1.2 million to 2.7 million per year after and, until the top year of 1920, the number rose to 7.3 million passengers. Figures from: Reitsma, *Gedenkboek*, 178.
96 It was only towards the end of the thirties that the SS Java succeeded in finding a suitable solution to this problem.
97 Most types of locomotives have a sort name, e.g. 'Pacific' for the 2-C-1, 'Atlantic' for the 2-B-1 or 'Hudson' for the 2-C-2. As the Javanese State Railways was the first company to use a 1-F-1 engine this category was named the 'Javanic'.

98 On main lines, rails weighing 33.4 kg/m were used as of 1909 but after 1921 that changed to 41.5 kg/m (Reitsma, *Gedenkboek*, 117).
99 One of the Werkspoor engines was brought back in 1981 and can now be viewed in the Dutch national railway museum in Utrecht (i.e. *Het Spoorwegmuseum)*.
100 Oegema, *Stoomtractie*, 110.
101 Reitsma, *Gedenkboek*, 148-58.
102 Special details on the improvements made to the various series of steam locomotives are given in: Oegema, *Stoomtractie*.
103 See for these developments: Clemens and Lindblad, *Het belang van de buitengewesten. Economische expansie en koloniale staatsvorming in de buitengewesten van Nederlands-Indië 1870-1942*.
104 A brief description of the Atjeh war is given in: Stapel's *Geschiedenis van Nederlandsch-Indië*, 306-15. A present-day history is to be read in: Van 't Veer's, *De Atjeh-oorlog*.
105 Reitsma, *Korte geschiedenis*, 88-93; Ermeling, 'De genie-uitrusting van de tweede expeditie tegen Atjeh, deel III, Het expeditionair tramway materieel'; Caspersz, 'De Atjehtram; haar geschiedenis, haar aandeel in de onderwerping en pacificatie van Atjeh; technische bijzonderheden'.
106 Oegema, *Stoomtractie*, 164, claims that the first own locomotives for the 1067 mm tracks of the Atjeh Tram were not supplied by Hohenzollern (Düsseldorf) until 1880. Mr. J. de Bruin, however, in Rotterdam informed me that in 1877-78 Hohenzollern delivered four B tank-engines to the Dutch East Indies, three for the Batavian Harbour Works services and one for the Atjeh Tram line. From the time when the Atjeh line opened, two C tank locomotives fabricated by Fox Walker were in use which were later added to the SS Java inventory of stock.
107 Wouter Cool (1848-1928), was a military engineering officer who was stationed with the East Indian Army from 1878 until 1882. Later he had a career as an officer in the Netherlands and from 1909 until 1911 he was the War Minister (*Biografisch Woordenboek van Nederland* 3, 110).
108 Oegema, *Stoomtractie*, 164.
109 With thanks to Mr. J. de Bruin.
110 Stapel, *Geschiedenis van Nederlandsch-Indië*, 304; Boersma, *Oostkust van Sumatra*, deel I, *De Ontluiking van Deli*.
111 De Balbian Verster, *Gedenkschrift der Deli Maatschappij 1869-1919*. De Jong, *Waaier van het fortuin*, 302-05; a somewhat popular and patriotic description of the first years of Nienhuys and the Deli Company in: *Wat een goede couragie vermach*, 92-114.
112 See Breman, *Koelies, planters en koloniale politiek. Het arbeidsregime op de grootlandbouwonderne mingen aan Sumatra's Oostkust in het begin van de twintigste eeuw*.
113 Gerretson, *Geschiedenis der 'Koninklijke'*; Clemens and Lindblad, *Het belang van de buitengewesten*.
114 The history of the Deli Railway Company, apart from being dealt with in Reitsma, *Korte geschiedenis*, 94-97, is especially described by: Weisfelt, *De Deli Spoorweg Maatschappij als factor in de economische ontwikkeling van de Oostkust van Sumatra* and Meijer, *De Deli Spoorweg Maatschappij*. The brothers De Guigné, from France, had been working in Deli since 1872 as tobacco planters in the enterprise Soengei Sikambing, later known as the Franco-Deli Tobacco Company (Broersma, *Oostkust van Sumatra*, part 1, 49).
115 Harbour was perhaps a rather ambitious description for the mooring spot for the steamboats in the Straits of Malacca (Weisfelt, *Deli Spoorweg Maatschappij*, 37). After working for the Deli Company Jacob Theodoor Cremer (1843-1923) became a member of the Dutch parliament and Minister for the Colonies from 1897 until 1901 in the Goeman Borgesius cabinet.
116 Weisfelt. *Deli Spoorweg Maatschappij*, 39.
117 Meijer, *Deli Spoorweg Maatschappij*, 20-23.
118 The question of what constituted the most suitable harbour in: Meijer, *Deli Spoorweg Maatschappij*, 28-31 and Weisfelt, *Deli Spoorweg Maatschappij*, 54-59.
119 Broersma, *Oostkust van Sumatra*, part 2, 278-81.
120 Overview of the locomotives in: Oegema's *Stoomtractie*, 168-74.
121 The tanks were owned by the various companies but the frames and wheels belonged to the DSM (Oegema, *Stoomtractie*, 174).
122 Bonds rose from 4 percent in the first years to 7 percent in 1922 and no less than 13 percent in

1928 (Reitsma, 'De spoor- en tramwegen in Nederlandsch-Indië', 64).

123 Figures obtained from: Weisfelt, *Deli Spoorweg Maatschappij*, 82, 108 and tables.

124 This section is largely based on Reitsma, *De Staatsspoorweg ter Sumatra's Westkust (S.S.S.)*.

125 Jacobus Leonardus Cluijsenaer (1843-1932), civil engineer, first helped with the building of the State Railways in the Netherlands and was, for instance, involved in the construction of the Moerdijk bridge. After his time in Sumatra he was a teacher for a while at the Royal Military Academy in Breda, secretary of the Company for the Exploitation of the State Railways in Utrecht (1882-1887), director of the Netherlands South African Railway Company (1887-1890) and finally director-general of the Netherlands Railway Company (1890-1900) (*Biografisch Woordenboek van Nederland*, II, 94).

126 *Rapport van den ingenieur J.L. Cluysenaer over den aanleg van spoorwegen in de Padangsche Bovenlanden ter verbinding van de Ombiliën-kolenvelden met de Indische Zee.*

127 *Rapport van den ingenieur J.L. Cluysenaer over de aanleg van spoorwegen in de Padangsche Bovenlanden.*

128 *Het hellend vlak van Agudio en de stangenbanen. Rapport van den ingenieur J.L. Cluysenaer over de waarde en bruikbaarheid dezer stelsels bij den aanleg van een spoorweg ter verbinding van de Ombiliën kolenvelden op Sumatra met de Indische Zee.*

129 Reitsma, *De Staatsspoorweg ter Sumatra's Westkust*, 20.

130 Willem IJzerman (1851-1932), military engineering officer who served in 1874 under Cluijsenaer at the time of the exploration of the Pandang Highlands. After that (from 1876 until 1886) he was chief engineer involved in the building of the State Railways in Java and, whilst on leave in the Netherlands, he was put in charge of the coal line construction project in West-Sumatra which he worked on until the time of his pensioning in 1896. Later he was involved in petroleum extraction in Sumatra and he developed a great interest in Hindu-Javanese antiquity. In 1899 he became chairman of the Royal Netherlands Geographical Society which, under his supervision, embarked on numerous expeditions linked to filling in the last blank spaces on tropical maps in Dutch colonial areas. He even found time to edit a number of old Dutch travel journals for the Linschoten Association and in 1919 he was the great inspirer behind the establishment of the Technical College in Bandoeng (*Biografisch Woordenboek van Nederland*, 2, 638).

131 Reitsma, *De Staatsspoorweg ter Sumatra's Westkust*, 28.

132 In 1891 IJzerman personally undertook the investigation of possible ways of transporting coal to the Sumatran east coast and, to that end, was the first to cross the island. See his *Dwars door Sumatra. Tocht van Padang naar Siak* (Haarlem/Batavia 1895). The engineer W.J.M. Nivel mapped out a route in 1920, later described in his *Verslag eener spoorwegverkenning in Midden-Sumatra* in two parts (Weltevreden 1927). During the war, the Japanese occupying force embarked on the construction of such a railway line using prisoners of war and forced labour but the line was never operated. See for further details: Hovinga, *Dodenspoorweg door het oerwoud. Het vergeten drama van de Pakan-Baroespoorweg op Sumatra, aangelegd door krijgsgevangenen onder de Japanse bezetting.*

133 During the time that Cluijsenaer was director of the Netherlands South African Railway Company, the same Riggenbach system was chosen for a short but steep route from Waterval-Onder to Waterval-Boven.

134 Reitsma, *De Staatsspoorweg ter Sumatra's Westkust*, 35; Van Sandick, 'De Ombiliënkolenvelden'.

135 The State Railways on Sumatra's western coast had no fewer than 180 self-discharging coal wagons, each with a 20-ton load capacity and 60 such 10-ton wagons (Oegema, *Stoomtractie*, 155).

136 Oegema, *Stoomtractie*, 152-64.

137 Reitsma, *De Staatsspoorweg ter Sumatra's Westkust*, 50-56.

138 Oegema, *Stoomtractie*, 160.

139 Reitsma, *De Staatsspoorweg ter Sumatra's Westkust*, 109-11, gives an extensive overview of the arguments for and against the closing of the line.

140 A lobby for state construction had already been launched by R.A. Eekhout in 1891, old planter and railway franchise official in his pamphlet entitled *Aanleg van Staatsspoorwegen in Nederlandsch Borneo en Zuid-Sumatra*.

141 De Jong, *De waaier van fortuin*, 429.

142 Reitsma, *Korte geschiedenis,* 101.
143 Oegema, *Stoomtractie,* 161-164.
144 See, for example, the report by Middelberg and Van Hennekeler, *Tramwegen op Billiton en Banka.*
145 Reitsma, *Korte geschiedenis,* 104-05. Maps of the other lines planned for Celebes and Borneo in: Reitsma, *Gedenkboek der Staatsspoor- en Tramwegen in Nederlandsch-Indië 1875-1925,* 76.
146 During the Second World War a final railway line was constructed in the archipelago and that was the line alluded to in footnote 132 between Sawahloento on the west coast of Sumatra and Pakanbaroe, situated in the Soengei Siak in East Sumatra. The 220 km long line built by 4967 prisoners of war and thousands of Indonesian forced labourers (*romushas*) took the lives of 698 prisoners of war and more than 9000 *romushas.* The line, which was completed in 1945, was never operated and it has now fallen to rack and ruin and has been largely overgrown by jungle undergrowth (Hovinga, *Dodenspoorweg door het oerwoud*). Yet, another small railway was built to transport coal from the Baya fields in South East Bantam (Java) at that time. The construction of the line demanded the lives of thousands of Javanese *romushas* ('De Japanse bezetting', 1110; Gonggryp, *Geïllustreerde encyclopaedie van Nederlandsch-Indië,* 679).

CHAPTER 4 **FREE EXTENDING OVER THE BANDJIR**

1 Introduction [HvM, KvM, BC, JA, EY, MB], The designing, manufacturing, transporting and assembling of bridges [AK], Permanent iron and steel railway bridges, 1862-1945 [HvM], The iron bridge over the Kediri river [BC], The bridging of the Tjisaät ravine [JA], The bridge over the Ajer-Silau river on the main connecting road between Tandjong-Balei and Deli [BC], The replacing of the railway bridge over the Moedjoer [JA], The standard draw-line arched bridge and the designer Professor J.H.A. Haarman of the State Railways [HvM], The Blang Me Bridge over the Kroëng river in Atjeh [HK], The Tjisomang viaduct and the designer W.J. van der Eb [HvM], The bridge over the Kali Progo on the Bandoeng-Djokja railway line and the designer Professor P.P. Bijlaard [HvM[, Reinforced concrete in bridge building in the East Indies [JvL], The own weight of East Indian and Dutch bridges [JvL], The bamboo bridge over the Tjitandoei in Indihiang [JK], Wooden bridges in Southwest Celebes [JvL], The strengthening of the reinforced concrete bridge over the Tjimanoek in Leuwidaoen [JvL], Suspension bridges [HB], The technical activities of the Military Corps during the conflict between the Netherlands and Indonesia in the 1946-1949 period [KvM]. Acknowledgements: T.A. Adisoebagjo, engineer; M. van Ballegoijen de Jong, lawyer; Erik Bouwmeester, Major of the Grenadier Guards; P.E. Manuhutu and H. Sonnemans, engineers; Marijke Meyer, the late Professor C.J. Vos; J.A. de Vries, former Lieutenant-Colonel and Professor L.A.G. Wagemans.
2 'Bandjir: a flood of water that is the consequence of extremely heavy rainfall and a sudden rise in river level. A bandjir appears suddenly and disappears just as fast. All rivers in the East Indies display this phenomenon. Rivers are generally not navigable during a bandjir. When constructing bridges the possibility of a bandjir arising is something that must be borne in mind, as most construction scaffolding will not be able to stand up to the forces. When constructing a bridge during the time of year when bandjirs arise it is advisable to do as much construction work as possible on the banks and then quickly draw the bridge over the river in a matter of hours' (Zwiers, *Bouwkundig woordenboek*).
3 Triebart, 'Ontwerp van ene normaal-constructie voor stalen bruggen, van 12-24 meter spanning in Indië'.
4 The dimensions of the material then currently available in the retail business was something Triebart obtained from the yearbook of the Koninklijk Instituut van Ingenieurs [the Royal Institute of Engineers] (1878), dept K, permanent section, p.6.
5 Gratama, 'Mededeelingen omtrent de in aanleg zijnde lijn Goendih-Soerabaja der Nederlandsch-Indische Spoorweg-Maatschappij in het bijzonder met betrekking tot den metalen bovenbouw der bruggen'.
6 Haarman, 'De montage van de brug over de Serajoe-rivier in de lijn Cheribon-Kroja der Staatsspoorwegen op Java'.
7 The abbreviations 'Technical Bureau' and 'Central Bureau' actually referred to the same bureau

which was, in full: The Technical Bureau of the Ministry for the Colonies in The Hague.

8 See Chapter 3.

9 Reitsma, *Gedenkboek staatsspoor- en tramwegen in Nederlandsch-Indië 1875-1925*, 118.

10 Bijlaard, *Vrije uitbouw, uitgevoerde werken*, 6.

11 Calculation principles. When calculating the horizontal sections a weight of 400 kg / m^2 was taken as the standard (4 kN / m^2) based upon 'soldiers marching in closed units and in step (!) with their backpacks and arms' and the own weight of the surface construction under a permitted stress of 700 kg / cm^2 (7kN / mm^2). It was furthermore presumed that under the maximum load the girders of the five middle spans and the two side spans could sag 9 mm and 8 mm respectively. During the tests carried out on 11th March 1869 the bridge was evenly loaded with a weight of 323,65 ton (sand from the river in sacks) which corresponded to 330 kg / m^2; the measured sagging that then occurred only measured 4 mm.

12 Van Velzen, 'De ijzeren brug over de Kediri-rivier, ter hoofdplaats van de Residentie Kediri'.

13 De Jongh, 'De overbrugging van het Tjisaät-ravijn in den Staatsspoorweg Tjitjalengka-Garoet'.

14 De Muralt, 'Iets over de afvoerravijnen van den Smeroe en den bouw van de nieuwe brug over de rivier de Moedjoer'.

15 Haarman, 'De berekening van ijzeren bruggen en de richting waarin die zich ontwikkelt'.

16 Reitsma, *Gedenkboek staatsspoor- en tramwegen*, 122.

17 Haarman, 'De groote bruggen in de spoorlijn Cheribon-Kroja der Staatsspoorwegen op Java'.

18 Van der Eb, 'Het Tjisomang-viaduct'.

19 Van der Eb, 'Onderzoek van staalconstructies en vermoeiingsverschijnselen'.

20 Bijlaard, *Vrije uitbouw*, 18.

21 Bijlaard, *Vrije uitbouw*, 28.

22 Bijlaard, *Vrije uitbouw*, 27.

23 Bijlaard, *Vrije uitbouw*, 28.

24 Bijlaard, 'De brug over de Kali Progo, ontworpen volgens een nieuw systeem'.

25 Roosseno Soerjohdikoesoemo, 'Het gewapend beton in den bruggenbouw in Indië'.

26 In independent Indonesia R. Roosseno Soejohadikoesoemo became professor of concrete structures and worked, for instance, at the Bandoeng Technical College (ITB) and for the Ministry of Transport and Public Works.

27 In the article published by the engineer W.J.G. Paardekoper 'Houten bruggen in Zuidwest-Celebes' on the matter of alternative protection better wood sorts would lobbied for, good impregnation and rot prevention treatment, and extreme attention would be paid to detail in combination with regular maintenance. The awnings or roofing were also expensive and perishable. Even the piles in the form of wooden pole trestles were of limited duration, partly due to the adverse effects of water and wind, woodworm etc.

28 From the mid-thirties on there were six classes of roads (I, II, III, IIIA, IV and V) all based on vehicle weight; see Stibbe, *Encyclopaedie van Nederlandsch-Indië*, part 8, 480.

29 Since the time of the Maas Tunnel the limit in the Netherlands has usually been 1: 28.

30 TNO's structural division (TNO: Dutch Organisation for Applied Scientific Research) has done similar research at the request of the Ministry of Public Works and the CUR (the Centre for the Implementation of Research and Legislation). The results were presented in CUR reports 16A and B entitled 'Betonplaten onder geconcentreerde belastingen'.

31 Kist, 'Het eigen gewicht van Indische en Nederlandse bruggen'; Kist, Het eigen gewicht van Nederlandse en Indische bruggen.

32 The derivation of this formula may be found in Kist, 'Een algemeene methode tot het schatten van het eigengewicht van hoofdliggers van vakwerkbruggen'.

33 In the *Tijdschrift van het Koninklijk Instituut van Ingenieurs. Afdeeling Nederlandsch-Indië* of 1893-1894 appendix B is a contribution from J.E. de Meyier about a 'bamboo bridge'.

34 The description is derived from the contribution by De Meyier referred to in footnote 33.

35 In *Het bouwen in overzeesche gewesten* by J.A. van der Kloes and J.N. van Ruijven there is some information about *idjoek*: 'The names Doek, Idjoek and Gemoeti are all used to denote the black material that resembles horse hair and is wrapped between the leaf stalks and the trunk of the aren palm. It is twisted by hand to form a double stranded string known as tali-doek (rope-cloth) that is rough and very useful for bamboo connections and for lashing objects together. It is

durable, even in damp conditions, but less strong than cane. Also frequently used for roofing. The cloth can withstand white ants. The string or the raw material is also sometimes used to wrap around bamboo posts in the ground'.

36 Beversen, 'Brug over de rivier Ajer-Silau in den grooten verbindingsweg van Tandjong-Balei naar Deli (oostkust van Sumatra)'.

37 Paardekooper, 'Houten bruggen in Zuidwest-Celebes'.

38 Roosseno Soerjohadikoesoemo, *De Ingenieur in Nederlandsch-Indië.*

39 Oosterhoff, *Bruggen in Nederland 1800-1940.* On p. 89 another example that is given of a beam bridge is the tram viaduct in Berg en Dal dating from 1912 with one span of 20.7 m.

40 Oosterhoff, *Bruggen in Nederland,* 107.

41 Described in *Beton und Eisen* 1938, no. 21.

CHAPTER 5 **NODES IN THE MARITIME NETWORK**

1 I would like to thank Professor H.A. Sutherland and Dr. G.J. Knaap for their valuable contributions to the research underlying this chapter. With his comments and editorial support Dr. W. Ravesteijn contributed much to this chapter. I am most grateful to Gerry de Graaf-Veering for her help in the processing of the source material. I am sad that she was unable to see the final publication.

2 The project The Java Sea Region in an Age of Transition 1870-1970 is a collaborative effort between the Free University of Amsterdam, Leiden University, the Royal Dutch Institute for Language, Culture and Ethnology in Leiden and the Universitas Diponegoro in Semarang. The harbour histories of Tandjong Priok, Semarang and Soerabaja were doctoral research subjects for A. Veering (author of this chapter), A. Supriyono and Mr. Indriyanto. In addition, in conjunction with this project, A. Claver completed in 2006 a dissertation on European and Chinese trading relations in colonial Java: *Commerce and capital in colonial Java: Trade finance and commercial relations between the Europeans and Chinese, 1820s-1942.* In 2003 Dr. S.T. Sulistiyono completed a dissertation on the developing of trade activities and shipping in the Java Sea region (both nationally and internationally): *The Java Sea network: Patterns in the development of international shipping and trade in the process of national economic integration in Indonesia, 1870-1970s.* Dr. J. Touwen has published a book on various statistical details relating to trade, shipping and harbours in the Java Sea region: *Shipping and trade in the Java Sea region, 1870-1940. A collection of statistics of the major ports.*

3 Bird, *Seaports and seaport terminals.*

4 For a precise analysis of the A. Vigarie model the reader is referred to Kidwai's, 'Conceptual and methodological issues: Ports, port cities and port hinterlands'.

5 Meijer examines especially the urban architectural aspects of harbour cities. From the different evolutionary models he created a functional steps model for harbour development which is also perfectly applicable outside the field of urban architecture. Meijer, *De stad en de haven. Stedebouw als culturele opgave in London, Barcelona, New York en Rotterdam: Veranderende relaties tussen stedelijke openbare ruimten en grootschalige infrastructuur,* 20-28, 395-396.

6 Kidwai, 'Conceptual and methodological issues', 30-34.

7 One of the classical studies on the Cultivation System is that of Fasseur, *Kultuurstelsel en koloniale baten. De Nederlandse exploitatie van Java 1840-1860.*

8 Kok, *De scheepvaartbescherming in Nederland en Nederlandsch-Indië,* 21-25.

9 Huijts and Tils, 'De veranderende scheepvaart tussen 1870 en 1930 en de gevolgen voor de handel op Nederlands-Indië', 63.

10 De Goey, 'Water- en luchttransport in Nederland, 1814-1990', 20.

11 A Campo, *De Koninklijke Paketvaart Maatschappij. Stoomvaart en stadsvorming in de Indonesische archipel 1888-1914* (Hilversum 1992).

12 Dick, 'Indonesian economic history inside out'; Dick, 'The emergence of a national economy, 1808-1990s'; Sulistiyono, *The Java Sea network.*

13 A Campo, *Koninklijke Paketvaart Maatschappij,* 625-6.

14 Van Doorn, *De laatste eeuw van Indië. Ontwikkeling en ondergang van een koloniaal project,* chapters 3 and 4.

15 The erosion material from the volcano slopes which was transported down to the sea by the

various rivers caused the northern coast of Java to steadily grow. Right alongside of the deltas this amounted to some 50 m per year.

16 De Meijier, *Aanleg van eene zeehaven te Tandjong Priok*, 1-4.

17 Knaap, *Shallow waters, rising tide. Shipping and trade in Java around 1775*, 21-23.

18 In 1818 Batavia was the only harbour that was open to international shipping but the government soon diminished the protectionism. By 1825 international shipping was accepted in Batavia, Semarang, Soerabaja, Palembang (Sumatra), Bandjermasin (Borneo) and Makassar (Celebes). Twenty years later the number of international ports had grown again. Kok, *De scheepvaartbescherming in Nederland en Nederlandsch-Indië*, 62-72.

19 'De havenwerken in Nederlandsch-Indië'.

20 Huff, *The economic growth of Singapore: Trade and development in the twentieth century*; Dobbs, *An ecological history of the Singapore river with particular reference to the lighterage industry*.

21 Cool, *Nederlandsch-Indische havens*, 55.

22 Kloppenburg, *De haven van Batavia*, 11.

23 Van Raders, 'Historische schets bij de aanbieding aan het Koninklijk Instituut van Ingenieurs van eenige bescheiden betreffende eene voorgestelde haven voor Batavia', 120.

24 Van den Berg *Munt-, crediet- en bankwezen, handel en scheepvaart in Nederlandsch-Indië: Historisch-statistische bijdragen*, 374-377; A Campo, *Koninklijke Paketvaart Maatschappij*, 406-409.

25 *Bataviaasch Handelsblad*, 17-3-1858 and 11-3-1863.

26 *Bataviaasch Handelsblad*, mail edition July 1865. The anonymous 'letter to the editor' was possibly an editorial trick. The *Bataviaasch Handelsblad* had criticised the poor state of affairs at the harbour on a number of occasions.

27 National Archive (NA), MiKo, mail report 762 (1872).

28 *Java Bode* 25-1-1867.

29 Bruining, 'De reede van Batavia'; Bruining, 'De reede van Batavia, II'.

30 Annual Report 1873 of the Dutch Steam Ship Company, NA, SMN, inv. no. 21. The first letter is dated 8th June 1872 and the second letter 12th August 1872.

31 Royal Archive (KHA), Hendrik collection, inv. no. 219, letter from HRH Prince Hendrik of the Netherlands to governor-general Mijer, Amsterdam 8th June 1871.

32 KHA, Hendrik collection, inv. no. 129, letter from Hendrik to Mijer, 8th June 1871.

33 KHA, Hendrik collection, inv. no. 129, letter from HRH Prince Hendrik of the Netherlands to Loudon, the Minister for the Colonies, Amsterdam 12th August 1871.

34 NA, MiKo, mail report 805 (1871).

35 NA, Nederlandsch-Indische Spoorweg Maatschappij, inv. no. 457.

36 Van Raders, 'Historische schets'.

37 In 1872 the Chamber of Commerce and Industry in Batavia published a collection of documents concerning the plan for a new harbour. The documents (reports, plans and correspondence) give a good impression of the great pressure placed upon the government by the Chamber. *Eene zeehaven voor Batavia, verzameling van de officiële bescheiden uitgegeven door de Kamer van Koophandel te Batavia.*

38 *Rapport der kommissie ingesteld bij gouvernementbesluit van den 9den januari 1873, no. 28 met het doel om de kwestie omtrent de geschiktste plaats waar eene zeehaven voor Batavia kan worden daargesteld, te beoordelen*. Minutes of the meeting attached to the report as Appendices.

39 Van Doorn, *De laatste eeuw van Indië*, 105-164.

40 The battle for the franchise of the Ombilin railway line is described in Colombijn, 'Uiteenlopende spoorrails. De verschillende ideeën over spoorwegaanleg en ontginning van het Umbilin-kolenveld in West Sumatra 1868-1891'.

41 Originally the committee consisted of three people but the member J. Blommendal died before any recommendations had been produced. His remarks and notes were, however, sent to the minister by the other two committee members, together with their official recommendations. NA, MiKo, inv. no. 2775.

42 NA, MiKo, inv. no. 2775.

43 Van Rader 'Historische schets'.

44 Dussaud put in a bid for 23,000,000 guilders and added risk limitation conditions. Lee & Van Hattum submitted a quotation for 20,000,000 guilders for execution under certain conditions.

For unconditional execution Lee & Van Hattum asked 28,500,000 guilders. NA, MiKo, inv. no. 2917, public account (openbaar verbaal: OV) 13/9/1876,8.
45 NA, MiKo, inv. no. 2917, ov13/9/1876,8.
46 NA, MiKo, inv. no. 2922, ov28/9/1876,40.
47 From that time onwards the ministry repressed Waldorp's input for fear that De Gelder would remain too dependent on his mentor during the final preparation phase in Britain. 'The time has come for De Gelder to stand on his own two feet', NA, MiKo, inv. no. 2930, ov25/10/1876, 57 and 58.
48 Van Berckel, 'De zeehaven voor Batavia te Tandjong Priok'.
49 Four mechanical experts, two smiths, two bench workers, a pile-driving gang foreman, a cartwright, a carpenter, eight (chief) supervisors, a machine designer, two designers, three dredger chiefs, two engine drivers, four bookkeepers, a clerk and a mine worker chief. De Meijier, *Aanleg van eene zeehaven te Tandjong Priok,* 14.
50 Van Bosse, 'J.A. de Gelder in het tijdperk van den bouw der haven van Tandjong Priok. Een herinneringswoord'. In 1885 Van Bosse was made the first director of the harbour at Tandjong Priok. He later became director of BOW.
51 The monthly reports were published in the *Javasche Courant*. Later they were collected under the title: *Maandverslagen omtrent de werkzaamheden aan de bouw van de havenwerken van Batavia.*
52 Compare Van Gasteren, Moeshart, Toussaint and De Wilde, *In een Japanse stroomversnelling: Berichten van Nederlandse watermannen – rijswerkers, ingenieurs, werkbazen – 1872-1903.*
53 De Gelder, 'Henri Emanuel van Berckel'.
54 This section is based on the monthly reports referred to above and on De Meijier's *Aanleg van eene zeehaven te Tandjong Priok,* 15-40.
55 Van Bosse, 'J.A. de Gelder', 788.
56 In the space of just over four years the quarried stone had generated 506,225 m^3 of trachyte.
57 Van den Berg was sharply criticised for his franchise request. His critics accused him of being an opportunist. What emerges from his personal notes is that Van den Berg had bitter memories of the Tandjong Priok question. NA, N.P. van den Berg collection, inv. no. 1. For the calculations made by Van de Berg and De Gelder, see De Meijier, *Aanleg van eene zeehaven te Tandjong Priok,* 68-70.
58 De Meijier, *Aanleg van eene zeehaven te Tandjong Priok,* 70.
59 The Tandjong Priok Dry Dock Company Ltd., was established in 1891, its headquarters being in Rotterdam and its official place of operation Batavia. The company's goal was: to operate shipbuilding and repair wharves in the Dutch East Indies and to provide construction workshops and places where ships could go into dry dock. At the end of the fifties, the East India branch was nationalised. It was not until 1972, when all the open legal cases had been closed, that the company officially went into liquidation. NA, Droogdok Maatschappij Tandjong Priok collection.
60 Kraus, 'Mededeelingen over eenige Nederlandsch-Indisch havens. Voordracht gehouden in de Vergadering van het Koninklijk Instituut van Ingenieurs van 20 december 1910', 34.
61 Cool, *Nederlandsch-Indische havens,* 43; Touwen, *Shipping and trade in the Java Sea region,* 110-111, tables 8a and 8b. After Soerabaja Semarang was the second biggest sugar exporting harbour in Java. In 1899 some 153,384 tons of sugar were exported via Semarang with a value of 15 million guilders.
62 Cool, *Nederlandsch-Indische havens,* 33.
63 For a description of the naval base in Soerabaja the reader is referred to 'Werven en dokken'. See further, for example, Teitler, *Anatomie van de Indische defensie. Scenario's beleid, plannen 1892-1920.*
64 A Campo, *Koninklijke Paketvaart Maatschappij,* 625-626.
65 Colombijn, 'Uiteenlopende spoorrails'.
66 Cool, *Nederlandsch-Indische havens,* 63-64.
67 'Havenwerken', 73; Cool, *Nederlandsch-Indische havens,* 65-66.
68 'Havenwerken', 71; Cool, *Nederlandsch-Indische havens,* 82-86. More about the early stages of Sabang harbour: M.G. de Boer, *Zeehaven en kolenstation Sabang 1899-1924.*
69 Cool, *Nederlandsch-Indische havens,* 56-57.
70 A Campo, *Koninklijke Paketvaart Maatschappij,* 162-165.
71 A Campo, *Koninklijke Paketvaart Maatschappij,* 440-442
72 Kraus, 'Mededeelingen over eenige Nederlandsch-Indische havens', 2.

73 Ten kilometres up river, at Goebeng, a mobile flood-control weir with shipping locks was introduced so that the silt-bearing water from the Brantas could be diverted and used, for instance, for irrigation purposes. Via a network of gullies and flushing canals the dam helped to improve Soerabaja's sewerage system. The dam resulted in a considerable narrowing of the Kali Mas which, in turn, hampered connections between the coast and Soerabaja's trade centre. Cool, *Nederlandsch-Indisch havens*, 31-32; *Encyclopaedia van Nederlandsch-Indië*, 'Brantas-Deltawerken', part 1 (The Hague/Leiden 1917, 2nd edition) 407-408.

74 Kraus and De Jongh, *Verslag over verbetering van de haventoestanden van Soerabaia*, 4-6.

75 Kraus and De Jongh, *Verslag over verbetering van de haventoestanden van Soerabaia*, 6-9; A. Campo, *Koninklijke Paketvaart Maatschappij*, 411-412.

76 Homan van der Heide, *Economische studiën en critieken met betrekking tot Java*, 197-205.

77 The Dutch East Indian Advisory Council to the Gouverneur-General, 29th November 1907, cited in Kraus and De Jongh, *Verslag over verbetering van de haventoestanden van Soerabaia*, 16-17.

78 Kraus, 'Mededeelingen over eenige Nederlandsch-Indische havens', 9.

79 Kraus and De Jongh, *Verslag over verbetering van de haventoestanden van Soerabaia*, 2.

80 Kraus and De Jongh, *Verslag over verbetering van de haventoestanden van Soerabaia*, 3.

81 Biography of J. Kraus in: Van den Aardweg, Brugmans and Japikse, *Persoonlijkheden in het Koninkrijk der Nederlanden in woord en beeld: Nederlanders en hun werk*.

82 Lichtenauer, 'Jongh, Gerrit Johannes de (1845-1917)'.

83 Kraus and De Jongh, *Verslag over verbetering van de haventoestand van Makassar*; Kraus and De Jongh, *Verslag over verbetering van de haventoestanden van Soerabaia*; J. Kraus and G.J. De Jongh, *Verslag over verbetering van de haventoestanden van Tandjong-Priok, Semarang en andere havens*.

84 Kraus and De Jongh, *Verslag over verbetering van de haventoestanden van Soerabaia*, 10.

85 Kraus and De Jongh, *Verslag over verbetering van de haventoestanden van Soerabaia*, 11-16.

86 Kraus and De Jongh, *Verslag over verbetering van de haventoestanden van Soerabaia*, 19 and further; 'Havenwerken', 72; Cool, *Nederlandsh-Indische havens* 14-17.

87 Touwen, *Shipping and trade in the Java Sea region*, 110-111, Tables 8a and 8b. Tandjong Priok was not of any significance to sugar exportation.

88 Kraus, 'Mededeelingen over eenige Nederlandsch-Indische havens', 33-34.

89 Kraus and De Jongh, *Verslag over verbetering van de haventoestanden van Tandjong-Priok, Semarang en andere havens*, 24-25; Cool, *Nederlandsch-Indische havens*, 44. From a comparison between the Kraus and De Jongh plan and later descriptions by Cool it may be concluded that in practice the operations were even larger-scale. The proa harbours were given a water surface area of 13 ha instead of 8 and 1393 m of quay was built instead of the planned 850 m.

90 The plan for a deep sea port for Semarang did remain. In the long term, the harbour division of BOW certainly seemed keen to create such a deep basin. On the map of Semarang that accompanies the 1920 BOW publication *Nederlandsch-Indische havens* a sea harbour is envisaged to the east of the proa harbour. Cool, *Nederlandsch-Indische havens*, part 2.

91 Kraus, 'Mededeelingen over eenige Nederlandsch-Indische havens', 19. In 1909 24 ships were kept waiting in the outer harbour for moorings on the quayside.

92 Kraus, 'Mededeelingen over eenige Nederlandsch-Indische havens', 18-19.

93 Kraus, 'Mededeelingen over eenige Nederlandsch-Indische havens', 17.

94 Kraus, 'Mededeelingen over eenige Nederlandsch-Indische havens', 20-23.

95 Kraus and De Jongh, *Verslag over verbetering van de haventoestanden van Makassar*; Kraus, 'Mededeelingen over eenige Nederlandsch-Indische havens', 22-23; Cool, *Nederlandsch-Indische havens*, 58.

96 Cool, *Nederlandsch-Indische havens*, 71.

97 The findings of Kraus and De Jongh concerning Belawan are not included in the three-part series but were immediately handed over to the authorities.

98 Cool, *Nederlandsch-Indische havens*, 80-81. See for an extensive description of these works Cool, 'Belawan oceaanhaven'.

99 Kraus, 'Mededeelingen over eenige Nederlandsch-Indische havens', 25.

100 Officially no particular size ranking was introduced within the large harbour category but a comparison between the salaries paid in Tandjong Priok and Soerabaja and the BOW promotion policy would indicate that Tandjong Priok was widely viewed as the main harbour.

101 Lecture on 'East Indian harbour politics', 8-4-1923, NA, Wouter Cool collection, inv. no. 105.
102 Mesman and Vos, *The history of caisson construction within the Hollandsche Beton Groep from 1902 to 1977.*
103 Krul, 'Wouter Cool'.
104 NA, Wouter Cool collection.
105 Cool, *Technische lessen en vraagstukken op het gebied van den Indische havenbeheer.*
106 *Handleiding havenbeheer.*
107 The idea was that *Nederlandsch-Indische havens* would be the first introductory part to a series of publications where all the harbours would be discussed in detail. Monographs would then have to be written for the bigger harbours. The series never materialised, probably due to budget cuts at the port department in the early twenties. The incentive to publish would probably also have been lost after Cool's departure from BOW in 1920.
108 'Havenwerken', 70-71; Cool, *Nederlandsch-Indische havens,* 52-53.
109 Cool, *Nederlandsch-Indische havens,* 53-54.
110 Cool, *Nederlandsch-Indische havens,* 54.
111 Lecture on East Indian harbour policy. NA, Wouter Cool collection.
112 *1930 Handbook of the Netherlands East-Indies,* 415-421.
113 Touwen, *Shipping and trade in the Java Sea region,* 15, 155-156, Figures 1 and 2.
114 The report: 'Justification of dredgers for the republic of Indonesia', The J.G. White Engineering Corporation, New York 3rd November 1951, Archive of the Department of Public Works: library.
115 In the 1951-1959 period the newspaper cutting service produced a weekly *'Warna Warta Dagang Sepekan / Commercial Weekly'* that gave an overview of economic reports or articles that had been printed in the Dutch and Indonesian language media. Port problems was one of the recurrent issues.
116 *The ports of Tandjong Priok, Soerabaja and Belawan,* 5.
117 In conjunction with developmental aid the Dutch Ministry of Foreign Affairs sent out a team of experts to Indonesia in the 1969-1983 period who were known as the Kerjasamaan Teknik Shipping Team (the name later changed to: Integrated Sea Transport Planning). At first the emphasis was chiefly on inter-insular links but later ocean-going shipping was taken into account, notably the shipping company Jakarta Lloyd that was running at a loss. The committee's fields of interests were very wide and included recommendations relating to the creation and maintenance of scheduled services, the warehouse sector, the laying out and organising of harbours and harbour terrains, planning, and the installation and the maintenance of harbour equipment (e.g. cranes). All the committee's activities are recorded in annual reports. Personal report provided by R. Barendsen, former member of the Team Shipping Kerjasamaan Teknik organisation.
118 A Campo, *Koninklijke Paketvaart Maatschappij,* 619.
119 Compare with De Meyier who, after the construction of Tandjong Priok, gave clear directives on what a modern harbour should provide: 'When nowadays a harbour is created for major shipping activities it is not sufficient to simply provide the ships with safe moorings where, because of the system of beacons, the harbour is easy to reach. There are, however, numerous other things that are needed if trading in the harbour is to be successful. The kinds of elements required are: quays where loading and unloading can take place, warehouses where the freight can immediately be stored and where it can be safely kept until more is known about the forwarding destinations. The storage places must lie on rail or waterway junctions so that all the freight can be transported inland or, indeed, loaded onto ships for export. The whole terrain must be well lit at night and there must always be workmen available to help with the loading and unloading work. The ships should furthermore be able to dock, be repaired or be loaded with coal. Pilot boats must be on hand to show the vessels the way around the harbour and, if necessary, there should also be tugs available. In short, there are numerous services that need to be provided, in the interests of making the ship turnaround time as short as possible, preparing vessels for their onward journeys and handling the goods as cheaply and as quickly as possible.' De Meyier, *Aanleg van eene zeehaven te Tandjong Priok,* 68.
120 Compare Dick, 'Indonesian economic history inside out'; Sulistiyono, *The Java Sea network.*

CHAPTER 6 **LIVING WATER**

1 As far as the details on irrigation go, this chapter is largely based on the doctoral dissertations of Ravesteijn, *De zegenrijke heeren der wateren. Irrigatie en staat op Java 1832-1942*, and Ertsen, *Prescribing perfection: Emergence of an engineering irrigation design approach in the Netherlands East Indies and its legacy 1830-1990.* See also Ertsen 'Irrigation traditions, roots of modern irrigation knowledge' and 'Experimental learning in the Netherlands East Indies' and Ravesteijn, 'Controlling water, controlling people: Irrigation engineering and state formation in the Dutch East Indies'. See for an interesting overview of Dutch irrigation intervention in the East Indies information written by an engineer who was actually involved in such matters: Vlugter, 'Honderd jaar irrigatie'.
2 Vreedenburgh, 'Opmerkingen over het ontwerp van een groot waterkrachtwerk in het bijzonder in Nederlandsch-Indië'.
3 See the Introduction.
4 Repair work was either done immediately and/or in the dry season which was the east monsoon season (lasting from April until November).
5 Rietveld, 'De Sampeanstuw honderd jaar', 280.
6 Melchior, 'De toepassing van de formules van Lauterburg voor de bepaling van den grootsten afvoer van de rivieren op Java'; Rietveld, 'De Sampeanstuw honderd jaar'.
7 Lauterburg, 'Anleitung zur Berechnung der Quellen und Stromabflussmessungen aus der Regenmenge, Grösze und Beschaffenheit der Quellen-und Flussgebiete'; Van Maanen, *Irrigatie in Nederlandsch-Indië. Een handleiding bij het ontwerpen van irrigatiewerken ter dienste van studeerenden en practici.*
8 Van Kooten, *Eenige empirische methoden tot het berekenen van den maximum afvoer eener rivier uit de grootte van den regenval.*
9 Van Maanen, *Irrigatie in Nederlandsch-Indië.*
10 Van Kooten, *Eenige empirische methoden tot het berekenen van maximum afvoer eener rivier uit de grootte van den regenval.*
11 Lamminga, *Beschouwingen over den tegenwoordigen stand van het irrigatiewezen in Nederlandsch-Indië*, 5.
12 Ertsen, 'Water Distribution on Java: A revision of a traditional perspective'.
13 Lamminga, 'Pekalen-irrigatiewerken in de afdeling Kraksaän der residentie Pasoeroean'.
14 At the place where the monument once stood there is now a commemorative plaque that was put there in 1969 and is dedicated to one of the heroes of the republic, Jos. Soedarso, a marine officer.
15 See Appendix 2. The rentability of irrigation works on government land in Java and Madoera.
16 De Meyier, 'De legende der Solovallei-werken', 823.
17 Ravesteijn, *De zegenrijke heeren der wateren*, 75.
18 Melchior 1895/1896, 'De toepassing van de formules van Lauterburg', 58.
19 Van Kooten, *Eenige empirische methoden tot het berekenen van den maximum afvoer eener rivier uit de grootte van den regenval.*
20 Van Kooten, *Eenige empirische methoden tot het berekenen van den maximum afvoer eener rivier uit de grootte van den regenval.*
21 Two important irrigation areas in the Outer Regions were Sadang in Celebes (63,900 ha) and the colonisation area known as Way Sekampong in Sumatra (42,600 ha).
22 See Chapter 1, Table 2.
23 Lamminga, 'Pekalen-irrigatiewerken in de afdeling Kraksaän'.
24 Ertsen, 'Water distribution on Java'.
25 De Gruyter, 'Beschouwingen over aftapsluizen en meetinrichtingen voor bevloeiingswerken'; Van Maanen, *Irrigatie in Nederlandsch-Indië.*
26 Van Maanen, *Irrigatie in Nederlandsch-Indië*; 'Meetinrichtingen in de irrigatieafdeling Pemali-Tjomal'; Romijn, 'Een regelbare meetoverlaat als tertiaire aftapsluis'.
27 What is so clever about the Romijn measuring weir is that it comprises two movable flaps or valves: an upper one with a measuring strip and a lower valve. The periodical raising of the lover valve prevents silt from building up in the weir.
28 What should also be mentioned is the fact that in order to develop an automatic hydraulically operated upstream control system that was as efficient as possible, hydraulically operated gates

provided with floats were first created in the Semarang laboratory. A number of those Vlugter gates were then installed along the Proa Transport Canal in the Demak area.

29 Compare Graadt van Roggen, 'Plant- en waterregelingen in de provinciale waterstaatsafdeling "Pemali-Tjomal"'.

30 Nijman, 'Bepaling van de maximale afvoer van rivieren volgens Melchior'.

31 Ertsen, 'From trial and error to science. The development and application of hydrological knowledge on Java'; Roessel, 'Maximum afvoer en hydrologische empirie'; Begemann, 'Toegepaste waarschijnlijkheidsleer op hydrologische waarnemingen'.

32 Van Kooten, *Eenige empirische methoden tot het berekenen van den maximum afvoer eener rivier uit de grootte van den regenval*; Kras, 'Het berekenen van den maximum afvoer van stroomgebieden met een oppervlakte van 0-100 km^2'; Der Weduwen, 'Het berekenen van den maximum afvoer van stroomgebieden met een oppervlak van 0-100 km^2'.

33 Begemann, 'Toegepaste waarschijnlijkheidsleer op hydraulische waarnemingen'; Verweij, 'Het berekenen van den maximum-afvoer van stroomgebieden met een oppervlak van 0-100 km^2'.

34 Groothoff, 'Eenige mededeelingen over de waterkrachtindustrie in Scandinavië en over het waterkrachtvraagstuk in Nederlandsch-Indië'.

35 Van Sandick, 'De Dienst voor Waterkracht en Electriciteit'.

36 Groothoff, 'Eenige mededeelingen over de waterkrachtindustrie in Scandinavië'.

37 Weijs, afterword to Groothoff's 'Eenige mededeelingen over de waterkrachtindustrie in Scandinavië', 32.

38 See Chapter 1.

39 Wenkebach, epilogue to Groothoff's 'Eenige mededeelingen over de waterkrachtindustrie in Scandinavië', 32.

40 Groothoff, 'Een en ander over de ontwikkeling van waterkracht- en electriciteitsbedrijven voor de energievoorziening van belangrijke gebieden in Nederlandsch-Indië'.

41 Groothoff, 'Verrichtingen en verwachtingen op waterkrachtgebied in Nederlandsche-Indië'.

42 Janssen van Raay, 'Eenige mededeelingen over den bouw van grootere waterkrachtwerken door den Dienst voor Waterkracht en Electriciteit in Nederlandsch-Indië'.

43 Groothoff, 'Een en ander over de ontwikkeling van waterkracht- en electriciteitsbedrijven voor de energievoorziening van belangrijke gebieden in Nederlandsch-Indië', 777.

44 Janssen van Raay, 'Jaarverslag van den Dienst van Waterkracht en Electriciteit in Nederlandsch-Indië over 1923'.

45 Van der Ley, 'Het nieuwe onderstation Badra van den Dienst voor Waterkracht en Electriciteit en de electriciteitsvoorziening van de Bandoengsche hoogvlakte'.

46 Vreedenburgh, 'Opmerkingen over het ontwerpen van een groot waterkrachtwerk'.

47 Van Haeften, 'Drukbuizen van gewapend beton'; Goemans, 'Een merkwaardig knikgeval'; Goemans, 'Waterkrachtwerk Sengguruh'. The fact that experience gained with hydroelectric works could take on less predictable forms is illustrated by the problem of capacity reduction in the case of the plant at Kratjak. It was biological growth that caused the tunnel pipeline capacity to be reduced from 8 to only 6 m^3/s. It turned out to be growth caused by 'an as yet not identified member of the caddis fly family', and the insect was promptly taken away to be studied by entomologists at the Royal Gardens in Buitenzorg (Betz, 'Capaciteitsvermindering van de toevoerleiding van het waterkrachtwerk Kratjak, ten gevolge van biologische aangroeiing', I.28). It was a bomb attack on an aqueduct of a hydropower plant in the Toentang that gave rise to theoretical considerations concerning the phenomenon 'buckling' (Goemans, 'Een merkwaardig knikgeval').

48 Van Iterson, *De Asahan-waterkracht en de Billiton maatschappij*; Goemans, 'Waterkrachtwerk Sengguruh'.

49 Van Blommestein, 'Een federaal welvaartsplan voor het westelijk gedeelte van Java'. See also Ravesteijn, 'Willem Johan van Blommestein'.

50 An extensive description of the history and execution of the Asahan works is to be found in: Soehoed, *Asahan, Impian yang menjadi kenyataan*.

51 Van Blommestein, 'A development project for the islands of Java and Madura (Indonesia)'.

CHAPTER 7 **FOR KOTA AND KAMPONG**

1 Karsten 'Indiese stedebouw', 154.

2 Karsten 'Indiese stedebouw', 154.
3 The information on inhabitants numbers in the late nineteenth century demonstrates great variation. The figures given come from De Jong, *De Waaier van het fortuin. Van handelscompagnie tot koloniaal imperium. De Nederlanders in Azië en de Indonesische archipel 1595-1950.*
4 In addition architects prided themselves on occupying themselves exclusively with the design and not with the execution side as well.
5 Respectively Tillema, *Van wonen en bewonen. Van bouwen, huis en erf*, 19 and S., 'Architectuur van openbare gebouwen'.
6 Architects in the Netherlands in 1842 already had established the Society for the Advancement of Architecture [Maatschappij tot Bevordering der Bouwkunst]. In 1924 the Dutch-East Indian Circle of Architects (Nederlandsch-Indische Architecten Kring, NIAK) was established.
7 Letter dated 10th August 1898 from H.P. Berlage to the directors of the Life and Life Insurance Company 'De Algemeene'. Found in the GAA, ref. 580 inv. no. 5342.
8 Snuyf, 'Over bouwkunst en bouwkunde in Indië', 131.
9 Snuyf, 'Over bouwkunst en bouwkunde in Indië', 131.
10 Snuyf, 'Over bouwkunst en bouwkunde in Indië', 131.
11 S., 'Beschouwingen over bouwkunst I', 119; S., 'Beschouwingen over bouwkunst II', 151.
12 The first three municipalities to be created were Batavia, Meester Cornelis and Buitenzorg (1905). In 1906 Bandoeng, Blitar, Cheribon, Kediri, Magelang, Pekalongan, Semarang, Soerabaja and Tegal in Java, Makassar in Celebes and Padang and Palembang in Sumatra were established. The municipality of Medan and the local council for the Cultivated Lands on the Eastern coast of Sumatra (including the departments of Deli, Serdang, Langkat and Asahan) were established in 1909. Malang and Soekaboemi became municipalities in 1914. Salatiga, Tebing Tinggi, Bindjai, Tandjoengbalai and Pematang Siantar in 1917, Madioen, Modjokerto, Pasoeroean and Probolingo in 1918, and Menado in 1919 and Amboina in 1921. The local council in Lematang Ilir was established in 1918 and that of Minahasa in 1919. In 1922 16 native municipal councils (or *marga* councils'; the Sanskrit word *marga* means clan or district): they fell under the regional council of Lematang Ilir. The local councils for Barabai (Borneo), Ambon (Moluccas) and Karangasem (Bali, Lombok) were created in 1921. The first regional councils were established in 1907 and 1908 in Java and Madoera. Kerchman, 25 *Jaren decentralisatie in Nederlands-Indië 1905-1930*; Schrieke, *Ontstaan en groei der stads- en landsgemeenten in Nederlands-Indië*, 12-13.
13 Like the new name of governmental commissioner, the assistant governmental commissioner was named the assistant advisor from 1912 onwards.
14 Soesilo, 'De grondrechten en de gemeenschap', 120; Van Vollenhoven, *De Indonesiër en zijn grond*.
15 H. van Breen, *Overzicht der voorgestelde werken tot verbetering van den wateraf- en aanvoer ter hoofdplaats Batavia* (1913) 14. Found in the NA, Archive of the Ministry for the Colonies 2.10.036.04, inv. no. 1124.
16 The decentralisation ruling stipulated that the central government was and was to remain the owner in matters where management was entrusted to the municipalities. As the municipalities were not owners the government remained the only legal entity with the right to determine matters. *Bijblad* 6802 reports in this connection that the owner, in this case the country, had not transferred any rights to the manager other than those required for the control c.q. for ensuring that the manager sees to it that the matter meets its requirements and that the owner is able to gain access.
17 *Locale Belangen* was started up in 1912 at the same time as the VLB and was the association's periodical. *Locale Techniek* was started up in 1932. It was the association's periodical and dealt with technical issues. After two years the journal merged with *Indisch Bouwkundig Tijdschrift* (East Indian Architectural Journal) and was renamed *I.T.B. Locale Techniek. I.T.B. Locale Techniek* was the journal for both the VLB and the Association of Dutch East Indian Architects (Vereeniging van Bouwkundigen in Nederlandsch-Indië). In 1938 the name of the journal was changed again. It was to continue until the mid-1940's under the name *Locale Techniek Indisch Bouwkundig Tijdschrift*. What was evident from the subtitle was that the journal was the technical periodical of the VLB which was distributed free of charge to members of the VLB and the Dutch East Indian Circle of Architects that was established in 1924.

18 Tramways and railways, harbours and airports as well as drainage and sewerage were only dealt with briefly because they had to form one entity with the town but otherwise they fell outside of the urban development plan. On the matter of drainage and sewerage Karsten wrote that such things belonged 'more to the utilitarian than to the actual design-determining side of town planning'. Karsten described various situations in order to meet the views or lack of understanding that laymen and technicians had. Karsten, 'Indiese stedebouw', 204.
19 Karsten, 'Indiese stedebouw', 243.
20 Karsten observed that 'excellent smaller old native maps' were the product of a perfect combination of popular spirit and 'adat' (tradition). In view of the fact that they could include sound motivation for modern town planning he argued in favour of their preservation and protection. Karsten, 'Indiese stedebouw', 150, 154, 250.
21 Karsten, 'Indiese stedebouw', 150.
22 Granpré Molière, 'Indiese Stedebouw door Ir. T. Karsten', 226.
23 Granpré Molière, 'Indiese Stedebouw door Ir. T. Karsten', 227.
24 Later the Department of Transport and Public Works was also situated in Bandoeng.
25 In line with recommendations dated 18th June 1907 by D. Tollenaar, advisor for decentralisation, local authorities were allowed to execute public works (roads, bridges, gutters et cetera) on the land for which it was responsible.
26 From the sources consulted it is hard to determine whether and, if so, to what degree the motivation to deal with the housing shortage in an as early as possible phase was inspired by the fact that many Europeans were also confronted with limited housing reserves. It is not unthinkable that it was a consideration that did play a part.
27 Although they were in the minority, some of the houses built in the *kampongs* were for 'natives' as well as for Europeans. In *kampong* Sekip the ratio was 4 European to 36 indigenous houses, in *kampong* Djati Oeloe 30 European to 158 indigenous houses. Found in the ANRI Arsip Depo Bogor (1891-1942) BGS 6-3-1924, inv. no. 562.
28 'Housing for Indonesians' was the theme of the next decentralisation conference in 1923.
29 Up until 1922 the work division between the various departments regarding *kampongs* and housing was as follows: the Inland Administration was responsible for land affairs and the *kampong* issue, BOW for the technical side and the BGD for the hygienic side of the public housing issue, Agriculture and Industry for the materials, and Justice for the regulations. The government possessed 50 percent of the shares of the Public Housing Public Limited Companies.
30 Soesilo was very critical about *Bijblad* 11272 and the way municipalities used it. He wrote: 'Merely on the grounds of urban interests alone one was not prepared to tackle the existing institution of land rights so that the municipalities had to do things in a haphazard manner. Thus appeared *Bijblad* 11272 [...]. In practice this *Bijblad* leans too heavily on the variable drive and the variable expertise of the municipalities which are free to use or not use the facilities it provides.' Soesilo, 'De grondrechten en de gemeenschap', 134.
31 Rückert emphasised in his argument that in his role as chief civil servant of the advisor for decentralisation he visited many town *kampongs* and therefore was one of the very few European professionals who could speak from experience about the local situations. In that respect he followed in the footsteps of Tillema who similarly had been one of the very few to have taken on the physical confrontation with the nature and extent of the problems in the *kampongs* and had then informed Europeans on such matters. Minutes of the People's Council meeting of 31-7-1928, discussion of the motion brought by Rückert c.s. regarding kampong improvement, 26. Found in the ANRI, Archive of the Department of Inland Administration, inv. no. 1395.
32 The official name of the Building Restriction Committee was 'Committee for the initiation of investigation into building restrictions emanating both from general and local building regulations, and for advising the government on whether it is desirable, in view of those restrictions, to initiate legal rulings and, insofar as such a question can be answered in the affirmative, for drafting the necessary regulations'. At the time of its establishment in 1932 the committee comprised the following members: S. Bastiaans (chairman and inspector for Agricultural Affairs), the lawyer C.C.J. Maassen (inspector for Agricultural Affairs and secretary until the arrival of Verhoef), R.T.A. Abdoerachman (regent in Meester Cornelis), A. Bagchus (mayor of Semarang), the lawyer P.A. Blaauw (member of the States General for the province of West

Java), C. Citroen (architect in Soerabaja), L.G.C.A. van der Hoek (resident of Batavia), Karsten, P.C.A. van Lith (head of the Administrative Affairs division for the Outlying Regions), A. van Roosendaal (director of Building and Housing Inspection in Bandoeng), Moehamad Hoesni Thamrin (member of the People's Council), W. Westmaas (architect in Batavia), A. Poldervaart (director of Town Planning in Bandoeng) and engineer A.H. Stam (head of the Division for Sanitation and Public Housing of the Public Health Service). Maasen and Van Roosendaal left the committee in the same year of its establishment. The lawyer, H.G. Verhoef (inspector for Agricultural Affairs), took over the position of secretary until the end of 1932. In the year of its establishment the Building Alignments Committee consisted of Bastiaans (chairman), Abdoerachman, Bagchus, Van der Hoek, Karsten, Van Lith, Thamrin, Rückert and Westmaas.

33 Apart from Logemann (chairman), Bagchus, Blaauw, Citroen, Karsten, Van Roosendaal, Thamrin and Westmaas the Town Planning Committee consisted of the lawyer A.P.G. Hens (acting assistant inspector for Agricultural Affairs within the Department of Inland Administration, secretary), R.T.A. Achmad Probonegoro (regent in Batavia), Dr. A.J.R. Heinsius (acting 1st class official within the decentralisation office) the lawyer H. Fievez de Malines van Ginkel (resident in Batavia), the engineer W. Lemei (architect in Soerabaja) and the engineer J.C. de Willigen (operational head of the Division for Sanitation and Public Housing within the Public Health Service).

34 Ch. 11 § 1, Art. 8. The intent of the town planning regulations. *Stadsvormingsordonnantie Stadsgemeenten Java*, 12. The content of the ordinance that was enforced in 1948 deviated very little from the draft ordinance. The main difference was to be found in an extra chapter on activity stipulations and the addition of the transition stipulations.

35 The collecting of data on the matter of housing shortage, overpopulation, and so on commenced in 1922.

36 Despite efforts to achieve unity in the area of town planning, the draft ordinance solely addressed the municipalities of Java. The committee gave two main reasons for this constraint. Firstly, the committee contended that the deviant hierarchical relation between the government in Batavia and the lower-level authorities in the Outer Regions made the establishing of a decision-making procedure for the entire archipelago impossible. Secondly, the committee maintained that it was insufficiently aware of the actual situation in the various Outer Regions which made the drafting of an ordinance for those regions a precarious business. The committee was, however, of the opinion that by drawing up a simple supplement or a new, similar version of the Town Planning Ordinance, it should be possible to soon establish a ruling in the field of town planning for the Outer Regions as well. Provided that they were equipped with the right kind of expertise, responsibility for the execution of this assignment could be handed over to the decentralisation bureau. *Toelichting op de Stadsvormingsordonnantie Stadsgemeenten Java*, 112. The ordinance provided the opportunity to include harbours, railways, etc. in the urban development plan if this was deemed necessary. *Toelichting op de Stadsvormingsordonnantie*, 112. If the advisor believed such a thing was necessary he was empowered, by means of Chapter 2, § 1, Art. 3 of the Town Planning Ordinance and with the permission of the Provincial Executive, to allow a local council to adopt a town plan in phases. *Stadsvormingsordonnantie Stadsgemeenten Java*, 10.

37 This contrasted with the consequences of *Bijblad* 11272. In an effort to streamline urban development and other issues the government proclaimed a general ruling in 1926 with regard to the drafting and submitting of urban development plans. The differentiation between a preliminary, general plan (the master plan) and the final plan led to confusion. Another problem was that the plans were not assessed accurately which meant that various municipalities had to wait for a number of years before the plan in question could be approved. *Toelichting op de Stadsvormingsordonnantie Stadsgemeenten Java*, 117.

38 The Planological Study Group was unofficially installed in 1938 as a successor to the former Social-Technical and Social-Economic Committees of the VLB. Apart from Logemann and Karsten the following people were members of the Planological Study Group: the engineer W. van de Broek d'Obrenan, R.T.A. Abas Soeri Nata Atmadja, the engineer A.H. van Assen, Professor G.M. van der Kolff, the engineer W. Lemei, Dr. W.M.F. Mansvelt, the engineer R.C.A.F.J. Nessel, the engineer F.M. Razoux Schultz, M. Soesilo, the engineer Werner Sörensen, the engineer J.P. Thijsse, Dr. A.C. Tobi and the engineer M. Valkenburg. In 1939 Professor C.P. Wolff Schoemaker and the engineer J.L. Moens were not members of the Planological Study Group though they had asked to join the

committee. Whether or not they joined the PS after 1939 is not clear.

39 'Verslag van den Planologische Dag georganiseerd vanwege de Vereeniging voor Locale Belangen door de Planologische Studiegroep dier Vereeniging te Bandoeng op 1 en 2 juli 1939', 16.

40 'Verslag van den Planologische Dag georganiseerd vanwege de Vereeniging voor Locale Belangen door de Planologische Studiegroep dier Vereeniging te Bandoeng op 1 en 2 juli 1939', 16.

41 'Verslag van den Planologische Dag georganiseerd vanwege de Vereeniging voor Locale Belangen door de Planologische Studiegroep dier Vereeniging te Bandoeng op 1 en 2 juli 1939', 16.

42 *Staatsblad van Indonesië* 331; *Staatsblad van Indonesië* 241.

43 J.P. Thijsse, *Aantekeningen over de Stadsvormingsordonnantie* (s.a.). Found in the NAi, Thijsse Archive.

44 The Government Committee for Spatial Planning in Non-Urban Areas fell under the Department of Public Works and Reconstruction and consisted of the following persons: Professor J.P. Thijsse (CPB head, chairman), engineer J.H. Schijfsma (CPB, secretary), Professor L.G.M. Baas Becking (director of the Country's Botanical Gardens, after April 1949 succeeded by Dr. D.F. van Sloten, director of the Royal Botanical Gardens and the engineer J.H. de Haan, head of the Land Use Bureau at the Department of Agriculture and Fisheries), the lawyer P. Creutzberg (head of the Economic Secretariat of the Department of Economic Affairs), the engineer J. Fokkinga (head of the Forestry Service), Dr. J.W. de Klein (assistant resident on hand for the Secretary of State for Inland Administration, after January 1949 succeeded by K. Mantel, second assistant advisor for Agricultural Affairs at the Department of Inland Administration), Dr. D.R. Koolhaas (head of the Industry division of the Department of Economic Affairs, succeeded after January 1949 by W. van Warmelo, MSc, also head of the Industry division of the Department of Economic Affairs), the lawyer B.J. Lambers (head of the Central Housing Division of the Department of Social Affairs), R.T. Praaning (assistant resident to the resident of Batavia and chief civil servant for municipal affairs), Professor W.F. Prins (University of Indonesia), Professor G.C. Suermondt (acting chief civil servant for the Secretary of State for Justice), M. Soesilo (engineering CPB), the engineer C.A.P. Takes (Social Planning Bureau for the Department of Social Affairs, succeeded after July 1949 by the lawyer J. Gerritsen of the Department of Social Affairs), the engineer W. Vitringa (head of the Electricity Affairs Division at the Department of Transport, Energy and Mining, as of July 1949 succeeded by the engineer A.D.J. de Bergh from the Electricity Affairs Division of the Department of Transport, Energy and Mining), the engineer H. Vonk (head of the Department for Agricultural), the lawyer J.H. Weber (replaced by Seurmondt, official within the Ministry of Justice). Found in the ANRI: Inventory of the Archive of the General Secretary and the Cabinet of the Governor-General 1944-1950, inv. no. 924.

45 Anticipating the fact that he would soon be forced to step down as chairman for the Government Committee for Spatial Planning in Non-Urban Regions after the transfer of power, Thijsse handed in his resignation to the Secretary of State for Public Works and Reconstruction on 8th December 1949. As of 1st April 1950 Soesilo resigned as member of the Government Committee for Spatial Planning when it became apparent that the secretary-general at the Department of Public Works and Reconstruction had not taken up Thijsse's suggestion to make Soesilo new chairman of that committee but that the preference lay with Kartanagara Purwadiningrat. Purwadiningrat was rapidly followed by the engineer K. Hadinoto.

46 J.P. Thijsse, *Bijlage en overzicht van bijlagen voor ECAFE Report 1954*. Found in the NAi, Thijsse Archive.

47 Up until October 1950 that faculty was known as the 'Faculty for Technical Science'. It was headed by Professor engineer F. Dicke. Before the Second World War Dicke had been an art teacher at the Batavia grammar school. After the war he was appointed head of the Ministry of Housing and Construction. The University of Indonesia's faculties for medicine, law and economy were located in Djakarta, agriculture and veterinary medicine were in Bogor. Architecture could also be studied at the university in Djokjakarta (Universitas Gadjah Mada). Interview with Professor Sidharta, MSc Semarang (28-11-2000).

48 The lectures on Hindu-Javanese architecture were given by engineer V.R. van Romondt. Interview with the architect Kwee Hin Goan (16-6-2003).

49 The courses on town and regional planning were temporarily interrupted between July 1962 and July 1963. Because of worsening political relations between Indonesia and the United States the courses were once again terminated in 1965. The courses were resumed again in 1969.

CHAPTER 8 **WATER IN THE CITY**

1 According to details derived from the *Aardrijkskundig en statistisch woordenboek van Nederlandsch Indië* and Tollens, *Warnasarie: Indisch jaarboekje voor 1858*, the population of Batavia city in 1855 was approximately 55,000 whilst by 1857 there were some 1,961 Europeans ('including also women and children and the elite attached to the garrisons').

2 Between 1860 and 1940, 900,000 ha of land (mainly jungle) had been reclaimed/developed in Deli for the purposes of tobacco, tea, coffee, palm oil, rubber and agave cultivation. See Goedhart, *Het wonder van Deli*.

3 In the interests of the establishment of the Djatiroto sugar estate on East Java with then the biggest sugar refinery in the world - large expanses of swamp areas were reclaimed and developed. See in this connection Directie Handelsvereeniging 'Amsterdam'.

4 Groen, *Een cent per emmer. Het Amsterdamse drinkwater door de eeuwen heen*.

5 Source: VEWIN (The Association for Water Supply Companies in the Netherlands).

6 'Raw water': not yet purified water derived from surface water, groundwater et cetera, serving as the basic commodity for the (public) supply of drinking water.

7 Van Sandick, 'De drinkwatervoorziening van Soerabaja'.

8 Unless separately reported from the point of view of relevance, the information relating to public drinking water is derived from the following annual issues of: *De Ingenieur* (1886 until 1940), *De Waterstaats-Ingenieur* (1913 until 1926), *De Waterstaats-Ingenieur in Nederlandsch-Indië* (1927 until 1933) *De Ingenieur in Nederlands-Indië* (1934 until 1941) and *De Ingenieur in Indonesia* (1948 until 1955).

9 The groundwater in Grissee, for instance, was reported to have a temperature of 40°C at a depth of 416 m and 50°C and 58°C at the respective depths of 540 m and 730 m.

10 *De Ingenieur* 1901, *De Ingenieur in Nederlandsch-Indië* 1939.

11 This was often restricted to the supplying of clean water (*air bersih*). Even then the water still had to be boiled before it could be consumed as drinking water (*air minum*): a common practice throughout the Dutch East Indies/Indonesia at that time.

12 By making use of the energy harnessed from running water and by converting that into potential energy a hydraulic ram (bélier hydraulique, Stossheber) can succeed in pumping up small quantities of water to great heights.

13 By then the plans for Kebayoran Baru, a satellite city to the southwest of Batavia, were already on the drawing board.

14 Due to the explosive growth of Batavia/Jakarta after the Second World War, the works for the abstraction of raw water from the Bandjir Canal upstream of the Karet Weir were quickly completed. Partly thanks to the ideas presented in 'Een federaal welvaartsplan voor het westelijk gedeelte van Java' [A federal welfare plan for the western part of Java] compiled by Professor W.J. Blommestein in 1948 the too low flow rates of the Ciliwung (or Tji Liwoeng, the river that feeds the Bandjir Canal) during dry periods could be supplemented with fresh water from the Jatiluhur storage reservoir in the Citarum (Tji Taroem) after the completion of the Tarum Barat Canal, the large western irrigation canal plus tunnel that derives its water from the Citarum and the Jatiluhur reservoir. See also Chapter 6.

15 An article that throws considerable light on this subject is the one written by D.A. Koster entitled: 'Drinkwatervoorzieningen in tropische landen, in het bijzonder in Nederlandsch-Indië'.

16 In actual fact 1934 was the last year when the laboratory still existed in Manggarai. The work being carried out in the Laboratory for Technological Hygiene and Regulation in Bandoeng was really quite similar to the work done in the laboratory in Manggarai, especially in the early years.

17 Van Wiechen, 'De drinkwatervoorziening van Malang'.

18 Gompert, 'Waterverbruik door minvermogende inlanders en vreemde oosterlingen bij Indische drinkwatervoorzieningen'.

19 According to the records, more than 50 percent of the troops of T.S. Raffles stationed on Java died from typhoid.

20 Just for the sake of comparison: the mortality rate in the Netherlands in the year 2000 was 0.875 percent.

21 Just for the sake of comparison: the mortality rate in the Netherlands in 1930 was 0.96 percent (of which the cases of typhoid and dysentery amounted to 1 in the 10,000 and the related death rate was 0.1 percent).

22 A roof covered with leaves from the Nipa palm.
23 'De gezondmaking van Cheribon'.
24 According to the division of labour between the departments and various bodies involved in the protecting of urbanised areas against *bandjirs* (resulting from precipitation outside the urban area in question) and possible inundation caused by too high sea levels, these tasks were not considered to be strictly 'sanitational', though clearly there is a relation. In present-day Jakarta possible compacting of the soil and subsidence of the area resulting from uncontrolled abstraction of groundwater l is, unfortunately, a familiar phenomenon. It increases the chances of inundation caused by external factors (*bandjirs*, high sea-levels) and internal factors (local rainfall).
25 Source: *Koloniaal Verslag* 1914.
26 Van Breen, 'Kleine werken ter verbetering van den gezondheidstoestand der hoofdplaats Batavia'.
27 Even today, people still defecate in fish ponds and the waste serves as feed for the fish.
28 This is remarkable because although not very extensive, there was already evidence of compact urbanization, especially in the quarters inhabited by Chinese tradesmen.
29 Van Leeuwen, 'Het rioleeringsvraagstuk in Nederlandsch-Indië'.
30 Massink, 'De drinkwatervoorziening van Bandoeng'.
31 C. Baumgarten, 'Uit het Laboratorium voor Technische Hygiëne te Bandoeng. Over de gebruiksmogelijkheden van de bij rotting van afvalwaterslib vrijkomende gassen'.
32 Coastal cities such as: Batavia, Cheribon, Semarang and Soerabaja ought, in principle, to be defended against tides and sea waves. However, the tidal differences are slight and the wave heights small. In the nineteenth century and at the beginning of the twentieth century buildings were situated sufficiently high enough above sea level. There were a few sea quays but no real sea dykes. It was not until the second half of the twentieth century that problems related to the compacting of the soil, land subsidence and rising sea level started to become apparent.
33 See in this connection the section entitled: 'Bandjir protection in urban areas'.
34 Rückert, 'Klein Assaineeringswerk'.
35 See Chapter 6.
36 'Sanitary landfill': the controlled dumping of solid waste by means of modern soil covering methods and drainage systems.
37 See the Epilogue.
38 The following information derives from Slinkers, 'Het verslag over de burgerlijke openbare werken in Ned.-Indië over het jaar 1902' and Slinkers, 'Uit het verslag over de burgerlijke openbare werken in Ned.-Indië over het jaar 1908'.
39 Slinkers, 'Uit het verslag over de burgerlijke openbare werken in Ned.-Indië over het jaar 1908', 749.
40 Van Breen, 'Verbetering van den waterstaat van de hoofdplaats Batavia'.

CHAPTER 9 **FROM TECHNICAL ASSISTANCE TO GOTONG ROYONG**

1 *Gotong royong* means mutual willingness to aid (*gotong* means carrying something heavy together); a generally accepted Indonesian convention first introduced during the time of the Golkar regime; in this connection freely translated as development cooperation.
2 'Sanitation' means the collection and hygienic processing of faeces, household and industrial waste water and solid waste.
3 Now IRC, the International Water and Sanitation Centre in Delft.
4 'Bersiap' (being ready, springing into action); synonymous with the bloody murders of 1945 and 1946 carried out by generally disorganised Indonesian combat groups.
5 The relevant descriptions of this are, for instance, laid down in volume 2 (1950) of the *Voordrachten voor het Koninklijk Instituut van Ingenieurs* (Presentations for the Royal Institute of Engineers) by the engineer F.M.C. Berkhout (professor of irrigation at the Technical College in Delft from 1954 until 1966), 'Het waterstaatswerk in Indonesië na de oorlog', by the engineer J.W.J. Beek, 'Indische spoorwegen na de Japanse bezetting', and by the lawyer B.H.A. van Kreel, 'De Deli Spoorweg Maatschappij'.
6 See W. Ravesteijn, 'Een ingenieur met visie. Prof. dr. ir. Willem Johan van Blommestein'.
7 Resolution 2626 (24th October 1970) of the General Assembly of the United Nations. The 0.7

percentage is, in practice, only achieved or exceeded by a very small number of countries.

8 One facet of this is that the receiving countries tend, increasingly, to themselves indicate which projects and services initiated by them are worthy of aid. Such 'aid' thus slowly becomes more and more a matter of cooperation and knowledge transfer so that once a project or aid campaign has ended the benefiting country is able to continue operating independently with its own staff and experts.

9 Source: *Internationale Samenwerking*, a monthly magazine on international cooperation published by the Dutch Ministry of Foreign Affairs, The Hague, November 2003.

10 'Appropriate technology' should, in this connection, be understood to mean the implementation of equipment, machinery and instruments which can be used by ordinary people (farmers, villagers, etc.), deployed (e.g. for construction purposes) and perpetually maintained.

11 Now PUM development aid-volunteer work.

12 In 1947, after the first police actions, education activities at the Bandung Technical College were resumed. In 1949 the curriculum for the 1949-1950 courses referred to the departments of General Science, Road and Hydraulic Engineering, Chemical Technology, Mechanical Engineering, Electro Engineering and Mining Engineering of the then University of Indonesia known as the Faculty of Technological Science in Bandung's Technical College. The curriculum for the 1950-1951 course given at the Balai Perguruan Tinggi Republik Indonesia Serikat, Fakultet Tehnik, di Bandoeng hardly deviated from the previous course and alluded to the 20 Dutch full professors directly affiliated to the faculty, one Indonesian associate professor and three Dutch associate professors. After 1958 the places of the Dutch professors, lecturers and assistants were taken over by Indonesians and non-Dutch foreigners. At the request of ITB, the Bandoeng Technical College Fund with its secretariat in the Netherlands remained intact. Even today, the Fund still serves to stimulate the ITB's academic efforts and enterprises.

13 Projects of stature may be unique projects of considerable magnitude (like a large-scale irrigation scheme) but also projects that are really the sum total of a large number of similar types of projects (e.g. the organising of drinking water supplies for a large number of small Indonesian cities).

14 Now part of the 'Alterra' Agricultural Institute. The Dutch flood disaster of 1953 accelerated the foundation of the ILRI.

15 As of 1st May 2003 the IHE has become a part of UNESCO and it is currently known as the UNESCO-IHE Institute for Water Education.

16 There was also the Appropriate Technology and Development Cooperation department within the Philosophy and (Technological) Social Sciences Faculty at Delft University of Technology/the former Technical College that provided education and carried out theoretical research in the field of applied technology and development cooperation. Some of the group's publications were these: Riedijk, *Technology for liberation*; Riedijk, *Appropriate technology for developing countries*; Boes, Ravesteijn and Riedijk, *Appropriate technology in industrialised countries*.

17 See note 50 of Chapter 6.

18 1968, the year in which Soeharto officially succeeded Soekarno as president; already in 1967 he was appointed ad interim president.

19 A consequence of this being that in the space of a mere 15 years the population of Jakarta rose by some 600,000 to above the five million mark.

20 Already presented in 'Een federaal welvaartsplan voor het westelijk gedeelte van Java' by Professor W.J. van Blommestein. See Chapter 6.

21 Except for the final paragraph on the engineer Dr. Suyono Sasrodarsono, the following section consists of citations from the epilogue of the dissertation *De zegenrijke heeren der wateren. Irrigatie en staat op Java, 1832-1942* by W. Ravesteijn in which many, though not all, of the repaired irrigation works in Java of the 1950-2000 period are described. The footnotes and references have been omitted and several details have been added.

22 A number of the larger 90 river basin areas in Indonesia already have their own management organisation; they are the Brantas river and the Citarum river system. It was no coincidence that precisely in these areas management state companies (i.e. public corporations) were set up (Jatiluhur 1992, Jasa Tirta 1993).

23 Source: civil engineers Jan Yap and Jan Kop.

24 Alternatively known as the Public Sector Approach, the Participatory Approach, etc.

25 P3A: Perkumpulan Petani Pemakan Air, introduced by presidential decree in 1984.

26 The French system, as implemented for instance in North Africa and Madagascar, comprises tertiary outlets (modules à masque) with adjustable constant tapping systems that are 95 to 90 percent accurate in the case of upper reach level variations that range from several centimetres to several decimetres (depending on the module type). Moreover, the French system also distinguishes between upper reach water regulation (upstream control) in which the canal water level at the tertiary tapping or drawing off points is kept virtually constant by means of automatic hydraulic floating valves or long fixed overflows and lower reach water regulation (downstream control, on demand, not unlike domestic tapping systems) where the water level in the downstream canal sections is controlled by automatic hydraulically regulated floating valves, or even a hybrid water regulation system (i.e. 'système mixte').
27 Source: W.J. van Blommestein (verbally conveyed information).
28 Vivekananthan, 'Rehabilitation of irrigation systems in East Java, Indonesia'.
29 A perpetuation of the irrigation works in the Pematang Siantar (Deli) area initiated by the Netherlands.
30 The first two examples were provided by Maurits Ertsen.
31 'Reconnaissance survey, Djratunseluna area'.
32 De Jager, 'Wadaslintang irrigation and drainage scheme, Central Java, Indonesia', 8.
33 De Jager, 'Wadaslintang irrigation and drainage scheme, Central Java, Indonesia', 8.
34 'Pasang-Surut' literally means high tide-low tide
35 Kop, Van den Toorn, Van Veen and T. van der Zee, 'Irrigation sector mission 1989, main report and annexes'.
36 The Pasang-Surut projects included alongside housing other supplementary facilities for the population in relation to drinking water, roads and sanitation. Further information on these projects may be found in: *Proceedings symposium lowland development in Indonesia-Jakarta 1986.*
37 With thanks for the contributions made by the two civil engineers J. Brinkhorst and A.D. Maier.
38 Thus not, for example, the network of narrow gauge railways used by the various enterprises (on the plantations).
39 See the epilogue.
40 The remaining new bridges that had to be constructed were paid for by other donors.
41 Also, in principle, for sanitation but as has been reported that was marginal.
42 *Water supply and sanitation decade*; Kalbermatten, *A planner's guide*; Arlosoroff, *Community water supply. The hand pump option*.
43 Towards the end of the 1968-2000 period private companies were given rights, by means of a franchise, to supply drinking water. Thus, in Jakarta in 2000, alongside the PDAM of DKI Jakarta there was also the PT'PAM Lyonnaise Jaya and the PT Thames PAM Jaya.
44 The information given in the PERPAMSI *Direktori* must be cautiously interpreted; that is the conclusion after having compared the *Direktori* and the results of a questionnaire (prepared by the author) sent to the relevant companies.
45 Pengendalian Bandjir Djakarta Raja (is the spelling adhered to pre-1970): the harnessing of the flood water of Greater Djakarta.
46 'Pluit' derives from the word Fluit, the ship which at the time of Jan Pietersz. Coen was pulled onto the shore and subsequently served as a look-out point.
47 This became the 'Master Plan for the Drainage and Flood Control of Jakarta'.
48 See Chapter 8.
49 The design of the aerated syphon which was to gradually switch on and off in time was examined by a hydraulic model developed in the Hydrodynamics laboratory in Bandung under the Directorate for the Research of Water Problems (Direktorat Penyelidikan Masalah Air) of the Director General for State Public Works Affairs (Direktorat Jenderal Pengairan) of the Department of Civil Public Works and Electricity Supplies (Departemen Pekerjaan Umum dan Tenaga Listrik). The construction of the model and the carrying out of the accompanying tests took approximately a year. Tirtotjondro, 'Siphon Bertekanan Rendah Sunter-Jakarta'.

CHAPTER 10 **INDIGENOUS OR INTERNATIONAL**

1 Van Doorn, *De laatste eeuw van Indië. Ontwikkeling en ondergang van een koloniaal project*, 89.

2 In this chapter I do not differentiate between scientific and technical knowledge.
3 As can be read in Chapter 9. Jan Kop has augmented a number of points in the present chapter.
4 Armytage, *A social history of engineering*, 167.
5 Van Doorn, *De laatste eeuw van Indië*.
6 Van Leur and Ammerlaan, *De Indische instelling te Delft. Méér dan een opleiding tot bestuursambtenaar*.
7 De Jong, *De waaier van fortuin. Van handelscompagnie tot koloniaal imperium*.
8 Van Leur and Ammerlaan, *De Indische instelling te Delft*; De Jong *De waaier van fortuin*.
9 Kamp, *De Technische Hogeschool te Delft 1905-1955*; Ringers, *Een eeuw Nederlandse waterbouw*.
10 Van Leur and Ammerlaan, *De Indische instelling te Delft*; Van Doorn, *De laatste eeuw van Indië*; Ten Horn-van Nispen, Lintsen and Veenendaal, *Wonderen der techniek. Nederlandse ingenieurs en hun kunstwerken: 200 Jaar civiele techniek*.
11 Kamp, *De Technische Hogeschool te Delft*; Groen, *Het wetenschappelijk onderwijs in Nederland van 1815 tot 1980. Een onderwijskundig overzicht*.
12 Van Leur and Ammerlaan, *De Indische Instelling te Delft*.
13 Kamp, *De Technische Hogeschool te Delft*.
14 Van Leur and Ammerlaan, *De Indische instelling te Delft*.
15 Schippers, *Van tusschenlieden tot ingenieurs. De geschiedenis van het hoger technisch onderwijs in Nederland*.
16 Van Leur and Ammerlaan, *De Indische instelling te Delft*; Van Doorn, *De laatste eeuw van Indië*.
17 De Jong, *De waaier van fortuin*; Van Leur and Ammerlaan, *De Indische instelling te Delft*; Van Doorn, *De laatste eeuw van Indië*.
18 Stevens, 'De Nederlandse ingenieur'; Groen, *Het wetenschappelijk onderwijs in Nederland van 1815 tot 1980*; Ten Horn, *Wonderen der techniek*; Van Doorn, *De laatste eeuw van Indië*.
19 Groen, *Het wetenschappelijk onderwijs in Nederland van 1815 tot 1920*, 208.
20 Van Doorn, *De laatste eeuw van Indië*, Baudet, *De lange weg naar de TU Delft. I. De Delftse ingenieursschool en haar voorgeschiedenis*.
21 Van Doorn, *De laatste eeuw van Indië*, 115; Dirkzwager, *Water – van natuurgebeuren tot dienstbaarheid*.
22 Baudet, *De lange weg naar de TU Delft*; Van Doorn, *De laatste eeuw van Indië*.
23 Mazure, 'De verjonging van een oude techniek', 207.
24 Baudet, *De lange weg naar de TU Delft*.
25 Baudet, *De lange weg naar de TU Delft*, 389.
26 Baudet, *De lange weg naar de TU Delft*, 390. One of the initiators and donors was K.A.R. Bosscha, also known from the Bosscha Observatory in Lembang.
27 Hoogewerff, Weijs and Van Sandick, *Het leerplan der op te richten Nederlandsch-Indische Technische Hoogeschool. Verslag der Sub-Commissie, ingesteld door de Commissie van advies voor onderwijs van het Koninklijk Instituut voor Hooger Technisch Onderwijs in Nederlandsch-Indië*, 21.
28 Hoogewerff, Weijs and Van Sandick, *Het leerplan der op te richten Nederlandsch-Indische Technische Hoogeschool. Verslag der Sub-Commissie*, 17.
29 Van Doorn, *De laatste eeuw van Indië*.
30 Baudet, *De lange weg naar de TU Delft*; Klopper, *Het leerplan der op te richten Nederlandsch-Indische Technische Hoogeschool. Opmerkingen naar aanleiding van het ontwerp der Sub-Commissie, ingesteld door de Commissie van advies voor onderwijs van het Koninklijk Instituut voor Hooger Technisch Onderwijs in Nederlandsch-Indië*.
31 Klopper, *Het leerplan der op te richten Nederlandsch-Indische Technische Hoogeschool. Opmerkingen*, 3.
32 Klopper, *Het leerplan der op te richten Nederlandsch-Indische Technische Hoogeschool. Opmerkingen*, 3.
33 See Chapter 8.
34 De Jong, *De waaier van fortuin*, 260.
35 'Menteng' derives from H.W. Muntinghe (1773-1827), advisor to H.W. Daendels, T.S. Raffles and G.A.G.P. Baron van der Capellen, later member of the Council of India. In the second half of the twentieth century Kebayoran Baru, the satellite city to the southwest of Jakarta, followed suit. The European districts in Bandoeng, Semarang, Soerabaja, Malang and Medan similarly reflect in their architecture and urban planning interpretation the relative independence of East Indian architecture. Dutch-East Indian architecture is furthermore still clearly visible in the styles of

existing public buildings such as offices, banks, hospitals, universities, schools and stations. As far as the layout of the towns and town quarters is concerned, names that spring to mind are those of engineer H.T. Karsten (Semarang, Batavia, Malang) and engineer J.P. Thijsse (Bandoeng) and regarding architecture as such, among the many, stand out the names of engineer H. MacLaine Pont and Professor C.P. Wolff Schoemaker of the Technical College at Bandoeng. Before he became professor at the Technical College in Delft, lecturing in reinforced concrete, A.M. Haas designed the characteristic and now well known mushroom type floors and roofs, virtually avant la lettre. See further Chapter 7 and the Epilogue.

36 Haringhuizen, *De Indische waterbouwkunde*, 6.

37 When motorised traffic became popular all the shipping canals were made redundant. At that stage the transport of freight by means of non-motorised *proa* and bamboo raft conveyance largely switched to the relative ease of transport by trucks.

38 Haringhuizen, *De Indische waterbouwkunde*; compare with Disco, *Made in Delft: Professional engineering in the Netherlands, 1880-1940*: 'Typically they have to be technical designers at one moment, accountants the next, and personnel managers at yet another time', 117. Disco labels this heterogeneity the 'Protean habituses' of engineers, 118.

39 Leidelmeijer, *Van suikermolen tot grootbedrijf*.

40 Storey, *Science and power in colonial Mauritius*.

41 Van Berkel, *In het voetspoor van Stevin. Geschiedenis van de natuurwetenschap in Nederland 1580-1940*.

42 De Vos, 'De strijd om en tegen het water', 278.

43 Mazure, 'De verjonging van een oude techniek'; Ertsen, 'Irrigation traditions, roots of modern irrigation knowledge'.

44 See for more details on the Romijn measuring weir and the Vlugter gate, Chapter 6. The Begemann gate will be further discussed in this chapter. See the Epilogue for a more extensive description of the specific character of East Indian technology within the framework of diffusion as opposed to endogenous developments.

45 Groen, *Het wetenschappelijk onderwijs in Nederland van 1815 tot 1980*.

46 Baudet, *De lange weg naar de TU Delft*, 391. The Bandoeng Technical College was nevertheless reopened in 1947.

47 Brouwer, *Waterkracht perspectieven*; Groen, *Het wetenschappelijk onderwijs in Nederland van 1815 tot 1980*.

48 Berkhout was succeeded by H.J. Schoemaker (1966-1984). As of 1985 it was R. Brouwer who held the chair in irrigation in Delft.

49 Brouwer, *Waterkracht perspectieven*, 3.

50 Brouwer, *Waterkracht perspectieven*, 4.

51 Brouwer, *Waterkracht perspectieven*, 4. Even Mazure optimistically asserted 'that after the loss of that work terrain the task set aside for the Technical College in the field of hydropower would not be completely over – just as it would not be for irrigation – but that Delft engineers would continue to make their contribution to the useful consumption of the earth's permanent energy resources in more divergent parts of the world' ('De verjonging van een oude techniek', 207-208).

52 Brouwer, *Waterkracht perspectieven*, 4.

53 These developments are extensively discussed in Chapter 9.

54 Van de Ven, *Leefbaar laagland. Geschiedenis van de waterbeheersing en landaanwinning in Nederland*.

55 Stevens, 'De Nederlandse ingenieur'; Dirkzwager, *Water – van natuurgebeuren tot dienstbaarheid*.

56 See, for instance, Duyvendak, *De Hangende Drievoet, indrukken bij een weerzien met China*.

57 Van Gasteren, *In een Japanse stroomversnelling*.

58 See Chapter 5.

59 Homan van der Heide, 'Irrigation and drainage in the lower Menam Valley'. See also Ten Brummelhuis, *De Waterkoning. J. Homan van der Heide, staatsvorming en de oorsprong van moderne irrigatie in Siam, 1902-1990*.

60 Brouwer, 'Design and application of automatic check gates for tertiary turnouts'.

61 This story contains a curious personal detail. The idea of using the Begemann gate in the Kano region was put forward by an engineer involved in the project by the name of C.L. Begemann, the son of S.H.A. Begemann. Begemann senior, after whom the gate was named, had worked with

Vlugter in Semarang on the gate construction. The same concept was later also implemented by R. Brouwer in Lampung.

62 Plusquellec, Burt and Wolter, *Modern water control in irrigation: Concepts, issues and applications*.
63 Burt, Angold, Lehmkuhl and Styles, 'Flap gate design for automatic upstream canal water level control'.
64 Boiten, 'De Hobrad stuw: regelen en meten van debieten'; Waterloopkundig Laboratorium, 'Meetstuwen Linge-project: bepaling van de afvoerbetrekkingen voor twee types meetoverlaten in de Linge'.
65 See Chapters 6 and 9.
66 See for the Wadaslintang project Chapter 9. See further Ertsen, 'Irrigation, traditions, roots of modern irrigation knowledge' and *Prescribing perfection: Emergence of an engineering irrigation design approach in the Netherlands East Indies and its legacy 1830-1990*. See also Horst, *Irrigation water division technology in Indonesia. A case of ambivalent development*.
67 *Irrigation design standards*. Theories and points of view taken from Delft irrigation readers, notably those of Professor F.M.C. Berkhout and his colleagues are integrated into this text. This work is alternatively known as the 'BOW manual'.
68 See Chapter 6 for further elucidation on the referred to artefacts and approaches.
69 Nippon Koei, 'Way Sekampung irrigation project. Design report on Bekri and West Rumbia irrigation systems'. The Dutch company Haskoning was also involved in this system, the creation of which already commenced in the colonial period (1935).

CONCLUDING CONSIDERATIONS **WHERE THE HIGH MOUNTAINS RISE...**

1 Bacon, *The New Atlantis*.
2 Lintsen, 'Wees Utopisch!'. See also Kroesen and Ravesteijn, 'De toekomst als opdracht. Utopie, revolutie en techniek in Europa'.
3 Basalla, 'The spread of Western science'.
4 Headrick, *The tools of empire. Technology and European imperialism in the nineteenth century* and *The tentacles of progress. Technology transfer in the age of imperialism*; Adas, *Machines as the measure of men. Science, technology, and ideologies of western dominance*; Storey, 'Colonial technology: Science and the transfer of innovation to Australia'.
5 See, for instance, Grottanelli, *Het leven der volken*.
6 If we envisage 'character' as that which is formed or nurtured and 'nature' as something innate then it might, by extension, be possible to conceive of the character of East Indian technology in terms of natural environment, society and culture, and the unique nature of the relevant technology in terms of a 'socio-technical system' and a 'technological regime'. Natural environment, society and culture are the forming (nurturing) external factors that affect technology whilst the listed concepts make that same technology, as it were, transparent from inside-out. In order not to unnecessarily further complicate matters this possible shade of nuance is omitted here.
7 See also Ravesteijn, 'Between globalization and localization. The case of Dutch civil engineering in Indonesia, 1800-1950'.
8 Van Blommestein was also active in other places such as Bangladesh. As Jan Kop knows from own experience, Van Blommestein implemented there the French irrigation technique (i.e. of downstream control as opposed to the customary East Indian upstream control method) and non-East Indian hydraulic formulas. Van Blommestein used his East Indian experience as the basis and then overlaid that with French and German knowledge.
9 See Ten Brummelhuis, *De Waterkoning. J. Homan van der Heide, staatsvorming en de oorsprong van moderne irrigatie in Siam, 1902-1990*. There were earlier contacts with Thailand if one bears in mind that the Bangkok canal system was developed by the Dutch in the seventeenth century.
10 The internationalisation of Dutch hydraulic engineering was at its height in the twentieth century. It is fair to assert that in all the cases listed below Dutch engineering intervention had a distinctively 'East Indian flavour': polder development in the Rio Plata region (Argentina), irrigation in Peru, polder development in Japan, dyke construction in China, water management (both quantitatively and qualitatively) in China (still ongoing), water management in the Philippines (Manila), water management in Vietnam, urban drainage and flood control in Thailand (Bangkok), urban water management (both quantitative and qualitative) in India (Kampur, Mirzapur) and

Sri Lanka (Colombo), new irrigation systems in Thailand, irrigation in India, river training in India (Ganges, Hooghly), hydraulic engineering (irrigation, flood control, river training, urban water management) in Bangladesh, groundwater extraction and management in Jemen, groundwater extraction and management in Egypt, groundwater extraction and management in Mozambique, urban water management in Egypt (Ameria Gedida), irrigation in Turkey, Egypt, Tunisia, Morocco, Senegal, Mali, Kenya, Zimbabwe and Mozambique, irrigation and drainage works in Iraq. Twentieth-century applications of Dutch hydraulic engineering that were devoid of East Indian influences were the following: water management activities in Eastern Europe (both quantitative and qualitative), in Poland, Romania et cetera, water management operations in Russia (Newa, Wolga et cetera), polder works in Britain (Portsmouth), urban drainage works in the Villes Nouvelles (New Towns) near Paris.

11 Mesman and Vos, The history of caisson construction within the Hollandsche Beton Groep nv from 1902 to 1977.

12 See especially Riedijk, *Technology for liberation* and *Appropriate technology for developing countries*; Boes, Ravesteijn and Riedijk, *Appropriate technology in industrialized countries*. In the work of W. Riedijk, the driving force behind appropriate technology at TU Delft, these three categories correspond to: self-sufficiency, self-organisation and self-development. These objectives served as an answer to the global relations of dependency as formulated in the 'dependence theories' of André Gunder Frank, Johan Galtung and others. Self-sufficiency involves demanding technological-economic autonomy. Self-organisation relates to political autonomy and democracy, the self-ruling socialism of former Yugoslavia being an excellent example of that. Self-development related to cultural autonomy and emancipation, both in social and personal respects.

13 See, for example, the criticism of Hollick, 'The appropriate technology movement and its literature: A retrospective'.

14 See, for instance, the irrigation reader compiled by Professor R. Brouwer of Delft University of Technology.

15 These three basic problems may be said to correspond to the three questions which, according to the philosopher Immanuel Kant, were central to his work: what can we know, what do we have to do and what can we believe? (Störig, *Geschiedenis van de filosofie*). See for a cultural-anthropological view: Keesing, *Inleiding tot de culturele antropologie*. Similar basic questions and categories may be seen in the works of sociologists such as Talcott Parsons, Jürgen Habermas and Norbert Elias.

16 Hindu is used to denote '(British) Indian', that is to say, of or influenced by old Indian customs.

17 See Van Doorn, *De laatste eeuw van Indië. Ontwikkeling en ondergang van een koloniaal project*.

18 See Gonggrijp, *Schets ener economische geschiedenis van Indonesië*.

19 De Kat Angelino, *Staatkundig beleid en bestuurszorg in Nederlandsch-Indië*.

20 See, for instance, Witjes, Hüsken and Banning, *Indonesië* and Lindblad, *Historical foundations of a national economy in Indonesia, 1890s-1990s*.

21 Hughes, 'The evolution of large technological systems'.

22 Combined or heterogeneous networks can result in 'second order large technological systems', that is, systems built on systems. In that connection one may, for example, think of the whole postal system which integrates different modes of transport. See Van der Vleuten, 'Twee decennia van onderzoek naar grote technische systemen: Thema's, afbakening en kritiek'.

23 See, for example, Ravesteijn, 'Dutch engineering overseas: The creation of a modern irrigation system in colonial Java'.

24 *Irrigation design standards*; see Chapter 10.

25 In these Concluding Considerations we focus more on the social construction of civil public works and less on the consequences of the modern techniques employed in the East Indies though they did, of course, exist. Alongside of the described differentiation process there was a tendency towards segregation within colonial class society. There was a time when the Dutch and the local population lived in close proximity to each other and when – regardless of the obvious social class barriers – the Dutch and the local administrators and their families would visit each other. All of that changed when it became easier to traverse great distances and the Dutch moved, en masse, to separate residential areas ('white islands in a brown Indonesian environment'), all of which was

made possible by improved communication means, notably the arrival of the car and good road networks, and the rise of urban development. See for a colonial publication with a feel for the interplay of technology and social factors: Jongejans, *Land en volk van Atjeh vroeger en nu*.

26 Lintsen, *Techniek in Nederland. De wording van een moderne samenleving 1800-1890*; Schot, Lintsen, Rip and Albert de la Bruhèze, *Techniek in Nederland in de twintigste eeuw*.

27 Especially J. Goudsblom and F. Spier. The big example was Norbert Elias, see for instance *Het civilisatieproces. Sociogenetische en psychogenetische onderzoekingen*. Goudsblom and Spier are also 'big history' supporters and practitioners.

28 Spier differentiates between ecological, social and personal regimes and sees, in the case of ecological regimes as exemplified in human history, a consecutive series of regimes passing from the hunter and gatherer regime to the agricultural regime and further to the industrial regime which is being dispersed in the present world (Spier, *The structure of big history. From the big bang until today*).

29 Goudsblom pinpoints and typifies several long-term historical processes (in particular demographic growth, growing concentrations in permanent settlements, ever more specialisation according to social functions, the process of being organised into increasingly large units and increased stratification) and he used for that purpose 'small models', all of which served to signify diffusion in terms of social gains. In the first place, he is thinking of the ecological transformations resulting from the domestifying of fire, the introduction of agriculture and industrialisation (all of which correspond to the three ecological regimes created by Spier; Goudsblom, *Het regime van de tijd*).

30 See Ravesteijn, 'Between globalization and localization'.

31 Basalla, *The evolution of technology*.

32 Compare Goudsblom, *Stof waar honger uit ontstond. Over evolutie en sociale processen*.

AN INDONESIAN VIEW AND EPILOGUE **A TALE OF TWO KOTAS**

1 Harry Patmadjaja, MT, was the project coordinator of the Petra Christian University of Surabaya research team that consisted of the following engineers: Budisetyono Tedjakusuma (harbour design), Handinoto (building and urban expansion), Handoko Sugiharto, MT. (structural design), Harry Patmadjaja, MT (transportation), Indriani H. Santoso, MT (railways), Jones Syaranamual, M.Eng (water management), Pamuda Pudjisuryadi, S.T., M.Eng (structural design), Paulus H. Soehargo, M.Arch (architectural design), Ruslan Djajadi, M.Eng (water management), Aylanda Dwi Nugroho, M.A. (editor). Louise van Gemert edited their report and turned it into this epilogue.

2 Note that clean water (*air bersih*), although it is clear and clean, is not sufficiently treated to be suitable for direct consumption, while drinking water (*air minum*) is potable.

AFTERWORD

1 Van Helsdingen and H. Hoogenberk (eds.), *Daar wèrd wat groots verricht ... Nederlandsch-Indië in de XXste eeuw*. The title was taken from a message by Coen to the directors of the East India Company in Amsterdam: 'Do not despair, neither be troubled for your enemies. Great things can be done in India'. Four years after the original publication of the volume, an abridged English translation came out under somewhat more modest and nostalgic title *Mission interrupted. The Dutch in the East Indies and their work in the XXth century* (Amsterdam 1945).

2 Adas, *Machines as the measure of men. The science, technologies and ideologies of western dominance*, 200-201.

3 Quoted by Van den Doel, *Het rijk van Insulinde. Opkomst en ondergang van een Nederlandse kolonie*, 244.

4 Van Kol, 'Ontwerp-program voor de Nederlandsche koloniale politiek'.

5 Cramer, *De ingenieur in Nederlands-Indië op technisch en sociaal gebied*.

6 Van Doorn, *De laatste eeuw van Indië. Ontwikkelingen en ondergang van een koloniaal project*, 157.

7 Cf. Ten Brummelhuis, *King of the waters. Homan van der Heide and the origin of modern irrigation in Siam*

8 Homan van der Heide, *Beschouwingen aangaande de volkswelvaart en het irrigatiewezen op Java in verband met de Solovalleiwerken*, 181.

9 Ott de Vries, 'De werkkring en taak van den ingenieur in Nederlandsch-Indië'.

10 Cf. Gordon, 'The collapse of Java's colonial sugar system and the breakdown of independent Indonesia's economy'.

11 In that group it was Widjojo Nitisastro who fulfilled the role of arch technocrat. See Anderson, *Language and power. Exploring political cultures in Indonesia*, 111-112.
12 In the years prior to 1965 the usual term for development was 'perkembangan' which literally means 'flowering' or 'blossoming' and which refers to a relatively organic process. See for an interesting explanation of the semantic differences between 'perkembangan' and 'pembangunan' (and the relevant political background) the article by Ariel Heryanto: 'The development of "development"'.
13 Anderson, *Language and power*, 111.

APPENDIX 3

1 Evidently water was drawn from a tapped natural spring.
2 Soerabaja has also processed drinking water from river water (Kali Mas) since 28-10-1922 by means of its pumping station, the sedimentation basin, the rapid filter and sterilisation (capacity 6000 m3 per 20 hour); mentioned in *De Ingenieur* [The Engineer] 1923.

APPENDIX 9

1 Emeritus Professor of irrigation, Delft University of Technology, the Netherlands. In this appendix the remarks made by professor Schoemaker were further completed and elucidated by Professor J.H. Kop C Eng FICE in order to make the text more publishable.
2 Begemann, 'Toepassing van de waarschijnlijkheidsleer op hydrologische waarnemingen'.
3 On the basis of sprinkling tests carried out on a scale-model. See also: Kraijenhoff van de Leur, 'Run-off models with linear elements'.
4 Supplementary remark made by Professor J.H. Kop: Present-day design formulas for unlined irrigation canals aim at conditions just below erosion-velocities (or tractive forces) and at constant or increasing sediment carrying capacities in the downstream direction of the flow.
5 Nugteren, 'On irrigation efficiencies'.
6 Clément, 'Calcul des débits dans les réseaux d'irrigation fonctionnant à la demande'.
7 Schoemaker, 'Wateraanvoer, verdeling en afvoer'.
8 In archaic Dutch: 'bericht en raad'.
9 *Mantri* is the Indonesian word for 'overseer'.
10 De Meyier, 'Gedenkboek, uitgegeven ter gelegenheid van het vijftigjarig bestaan van het Koninklijk Instituut van Ingenieurs, 1847-1897', 302.

Literature and sources

INTRODUCTION **COVETOUSNESS AND VOCATION**

M. Adas, *Machines as the measure of men. Science, technology, and ideologies of western dominance* (Ithaca/Londen 1989)

G. Basalla, 'The spread of Western science', in: *Science* 156 (1967) 611-622

H. Baudet and I.J. Brugmans (eds.), *Balans van beleid. Terugblik op de laatste halve eeuw van Nederlandsch-Indië* [Balance of policy. Reflections on the last half-century of the Dutch East Indies] (Assen 1961)

R.F. Beerling, *Heden en verleden. Denken over geschiedenis* [Present and past: Thinking about history] (Arnhem 1962)

J.H. Boeke, 'Van vier tot vierenveertig millioen zielen op Java' [From four to forty-four million souls in Java], in: W.H. van Helsdingen and H. Hoogenberk (eds.), *Daar wèrd wat groots verricht .... Nederlandsch-Indië in de XXste eeuw* [Great things were achieved there... the Dutch East Indies in the XX century] (Amsterdam 1941) 346-356

P. Boomgaard and A.J. Gooszen, *Population trends 1795-1942. Changing economy in Indonesia. A selection of statistical source material from the early 19th century up to 1940*, volume 11 (Amsterdam 1991)

D.H. Burger, *Sociologisch-economische geschiedenis van Indonesië* [Sociological-economic history of Indonesia], two parts (Amsterdam 1975)

W.E. Bijker, Th.P. Hughes and T. Pinch (eds.), *The social construction of technological systems. New directions in the sociology and history of technology* (Cambridge, Mass./London, England 1987)

J.N.F.M à Campo, *Koninklijke Paketvaart Maatschappij. Stoomvaart en staatsvorming in de Indonesische archipel 1888-1914* [The Royal Packet Shipping Company. Steam shipping and state formation in the Indonesian archipelago 1888-1914] (Hilversum 1992)

J.N.F.M. à Campo, *Engines of empire. Steamshipping and state formation in colonial Indonesia* (Hilversum 2005)

A.H.P. Clemens and J. TH. Lindblad (eds.), *Het belang van de Buitengewesten. Economische expansie en koloniale staatsvorming in de Buitengewesten van Nederlands-Indië* [The importance of the Outer Islands: Economic expansion and colonial state formation in the Outer Islands of the Dutch East Indies] (Amsterdam 1989)

J. Diamond, *Guns, Germs and Steel. The Fates of Human Societies* (New York/London 1997)

J. Diamond, Answer to W.H. McNeill, Review of J. Diamond, *Guns, Germs and Steel. The Fates of Human Societies* (New York/London 1997), *The New York Review of Books*, June 26, 1997

H.W. van den Doel, *De stille macht. Het Europese binnenlands bestuur op Java en Madoera, 1808-1942* [The quiet force. European inland administration in Java and Madura, 1808-1942] (Amsterdam 1994)

J.A.A. van Doorn, *The engineers and the colonial system: Technocratic tendencies in the Dutch East Indies* (Rotterdam 1982)

J.A.A. van Doorn, *De laatste eeuw van Indië. Ontwikkeling en ondergang van een koloniaal project* [The last century of the East Indies. The rise and fall of a colonial project] (Amsterdam 1994)

J.A.A. van Doorn, 'De eerste spoorweg op Java' [The first Javan railway], in: M.L. ten Horn-van Nispen, H.W. Lintsen and A.J. Veenendaal jr. (eds.), *Nederlandse ingenieurs en hun kunstwerken. Tweehonderd jaar civiele techniek* [Dutch engineers and their structural works. Two hundred years of civil engineering] (Zutphen 1994)

Encyclopaedia Britannica

M.W. Ertsen, *Prescribing perfection: Emergence of an engineering irrigation design approach in the Netherlands East Indies and its legacy 1830-1990* (Rotterdam 2005)

C. Fasseur, *Kultuurstelsel en koloniale baten. De Nederlandse exploitatie van Java 1840-1860* [Cultivation system and colonial profit. The Dutch exploitation of Java 1840-1860] (Leiden 1975)

C. Fasseur, *De weg naar het paradijs en andere Indische geschiedenissen* [The road to paradise and other East Indian histories] (Amsterdam 1995)
E.J. Fischer (ed.), *Geschiedenis van de techniek* [History of technology] (Den Haag 1980)
F.S. Gaastra, *De geschiedenis van de VOC* [The history of the VOC] (Zutphen 2002)
G.F.E. Gonggryp, *Geïllustreerde encyclopaedie van Nederlandsch-Indië* [Illustrated encyclopaedia of the Dutch East Indies] (Wijk en Aalburg 1991)
S. de Graaff and D.G. Stibbe (eds.), 'Junghuhn', *Encyclopaedie van Nederlandsch-Indië* [Dutch East Indian encyclopaedia], part 2 (The Hague/Leiden 1918, 2nd impression) 223-226
D.R. Headrick, *The tools of empire. Technology and European imperialism in the nineteenth century* (New York 1981)
D.R. Headrick, *The tentacles of progress. Technology transfer in the age of imperialism* (New York/Oxford 1988)
W.H. van Helsdingen and H. Hoogenberk (eds.), *Daar wèrd wat groots verricht .... Nederlandsch-Indië in de XXste eeuw* [Great things were achieved there... The Dutch East Indies in the XXth century] (Amsterdam 1941)
Th.P. Hughes, *Networks of power: Electrification in Western society, 1880-1930* (Baltimore 1983)
Th.P. Hughes, 'The evolution of large technological systems', in: W.E. Bijker, Th.P. Hughes and T. Pinch (eds.), *The social construction of technological systems. New directions in the sociology and history of technology* (Cambridge, Mass./London, England 1987) 51-82
De Ingenieur in Nederlands-Indië 23 (1935) VI
J.J.P. de Jong, *De waaier van het fortuin. Van handelscompagnie tot koloniaal imperium* [The fan of fortune: From trading company to colonial empire] (Den Haag 1998)
R.N.J. Kamerling (ed.), *Indonesië toen en nu* [Indonesia: Then and now] (Amsterdam 1980)
Koloniaal Verslag [Colonial Report], appendix to the *Handelingen der Staten-Generaal, Eerste en Tweede Kamer* [Proceedings of the States General, First and Second Chamber]
De koloniale roeping van Nederland [The colonial vocation of the Netherlands) (The Hague 1930)
M. Kuitenbrouwer, 'Het imperialisme-debat in de Nederlandse geschiedschrijving' [The imperial debate in Dutch historiography], in: *Bijdragen en mededelingen betreffende de geschiedenis der Nederlanden* 113 (1998) 1, 56-73
D.S. Landes, *The wealth and poverty of nations: Why some are so rich and some so poor* (New York, 1998)
D.S. Landes, *Arm en rijk. Waarom sommige landen erg rijk zijn en andere erg arm* [Dutch translation of 'The wealth and poverty of nations: Why some are so rich and some so poor'] (Utrecht 1998)
M. Leidelmeijer, *Van suikermolen tot grootbedrijf. Technische vernieuwing in de Java-suikerindustrie in de negentiende eeuw* [From sugar grinder to big company: technological innovations in the Javanese sugar industry during the nineteenth century] (Amsterdam 1997)
C. Lekkerkerker, *Land en volk van Java* [The country and population of Java] (Batavia 1938)
D. van Lente, H.W. Lintsen. M.S.C. Bakker, E. Homburg, J.W. Schot and G.P.J. Verbong, 'Techniek en modernisering' [Technology and modernisation], in: H.W. Lintsen (ed.), *Techniek in Nederland. De wording van een moderne samenleving 1800-1890*, deel I (Zutphen 1992) 19-36
H.W. Lintsen (ed.), *Techniek in Nederland. De wording van een moderne samenleving 1800-1890* [Technology in the Netherlands: The birth of a modern society 1800-1890], six volumes (Zutphen 1992-1995)
E. Locher-Scholten, *Ethiek in fragmenten; Vijf studies over koloniaal denken en doen van Nederlanders in de Indonesische archipel 1877-1942* [Ethics in fragments: Five studies on the colonial thinking and actions of the Dutch in the Indonesian archipelago 1877-1942] (Utrecht 1981)
E. Locher-Scholten, *Sumatraans sultanaat en koloniale staat. De relatie Djambi-Batavia (1830-1907) en het Nederlandse imperialisme* (Leiden 1994)
E. Locher-Scholten, *Sumatran sultanate and colonial state: Jambi and the rise of Dutch imperialism, 1830-1907* (Ithaca 2004)
C. Lorenz, *De constructie van het verleden. Een inleiding in de theorie van de geschiedenis* [The construction of the past: an introduction to the theory of history] (Meppel/Amsterdam 1990)
H. Maat, 'Techniek en het koloniale verleden' [Technology and the colonial past], in: J.W. Schot, H.W. Lintsen, A. Rip and A.A. Albert de la Bruhèze (eds.), *Techniek in Nederland in de twintigste eeuw* [Technology in the Netherlands in the Twentieth Century], volume 7 (Zutphen 2003) 175-195

W.H. McNeill, Review of J. Diamond's *Guns, germs and steel. The fates of human societies* (New York/London 1997), *The New York Review of Books*, May 15, 1997
R. Mrázek, *Engineers of happy land: Technology and nationalism in a colony* (Princeton, N.J. 2002)
A. Pacey, *Technology in world civilization. A thousand-year history* (Oxford 1990)
A. Pacey, *The maze of ingenuity. Ideas and idealism in the development of technology* (Cambridge, Massachusetts/London England 1992)
J. Paulus (ed.), 'Irrigatie' [Irrigation], *Encyclopaedie van Nederlandsch-Indië* [Dutch East Indian encyclopaedia], part 1 (The Hague/Leiden 1917, 2nd impression) 289-298
J. Paulus, S. de Graaff and D.G. Stibbe (eds.), *Encyclopaedie van Nederlandsch-Indië* [Dutch East Indian encyclopaedia], eight parts (The Hague/[Leiden] 1917 [part 1], 1918 [part 2], 1919 [part 3], 1921 [part 4], 1927 [part 5], 1932 [part 6], 1935 [part 7], 1939 [part 8], 2nd impression)
I.R. van de Poel and M.P.M Franssen (eds.), 'Understanding technical development: the concept of "technological regime"', special issue of *International Journal of Technology, Policy and Management* 2 (2002) 4
O. van den Muijzenberg and W. Wolters, *Conceptualizing development* (Dordrecht 1988)
PERPAMSI (Persatuan Perusahaan Air Minum Seluruh Indonesia), *Direktori 2000* (Jakarta 2000)
W. Ravesteijn, *De zegenrijke heeren der wateren. Irrigatie en staat op Java, 1832-1942* [The auspicious lords of the waters. Irrigation and state in Java, 1832- 1942] (Delft 1997)
W. Ravesteijn, 'Irrigation development in colonial Java: The history of the Solo Valley works from a regime perspective', in: *International Journal of Technology, Policy and Management* 2 (2002) 4, 361-386
W. Ravesteijn, 'Dutch engineering overseas: The creation of a modern irrigation system in colonial Java', in: *Knowledge, Technology & Policy* 14 (2002) 4, 126-144
W. Ravesteijn, L. Hermans and E. van der Vleuten (eds.), 'Water systems: Participation and globalization in water system building', special issue of *Knowledge, Technology & Policy* 14 (2002) 4
A. van Schaik, *Colonial control and peasant resources in Java. Agricultural involution reconsidered* (Amsterdam 1986)
J.W. Schot, H.W. Lintsen, A. Rip and A.A. Albert de la Bruhèze (eds.), *Techniek in Nederland in de twintigste eeuw* [Technology in the Netherlands in the Twentieth Century], seven volumes (Zutphen 1998-2003)
Staatsblad van Nederlandsch-Indië [Law Gazette for the Dutch East Indies], inclusive of *Bijblad* [Supplement].
J.H. Thal Larsen, *Grepen uit het verleden en het heden van het irrigatiewezen* [Examples of irrigation methods of the past and present] (Wageningen 1932)
P. van 't Veer, *De Atjeh-oorlog* [The Atjeh war] (Amsterdam 1980)
Verslag over de burgerlijke openbare werken in Nederlandsch-Indië [Civil public works in the Dutch-East Indies report] (Batavia/'s-Gravenhage 1893-1927).
G.J. Vink, 'De inheemsche landbouw' [The indigenous agriculture], in: W.H. van Helsdingen and H. Hoogenberk (eds.), *Daar wèrd wat groots verricht .... Nederlandsch-Indië in de XXste eeuw* [Great things were achieved there... the Dutch East Indies in the XX century] (Amsterdam 1941) 357-369
E. van der Vleuten, 'Twee decennia van onderzoek naar grote technische systemen: Thema's, afbakening, kritiek' [Two decades of research into large technological systems: Themes, demarcations, criticism], in: *NEHA-jaarboek voor de geschiedenis van bedrijf en techniek* 63 (2000) 328-364
H.C.P. de Vos, 'De strijd om en tegen het water' [The battle for and against water], in: W.H. van Helsdingen and H. Hoogenberk (eds.), *Daar wèrd wat groots verricht.... Nederlandsch-Indië in de XXste eeuw* [Great things were achieved there... the Dutch East Indies in the XX century] (Amsterdam 1941) 273-286
H.L. Wesseling, *Indië verloren, rampspoed geboren* [Goodbye East Indies, hello disaster] (Amsterdam 1988)
H.L. Wesseling, *Europa's koloniale eeuw* [Europe's colonial century] (Amsterdam 2003)
L. Winner, 'Upon opening the black box and finding it empty: Social constructivism and the philosophy of technology', in: *Science, Technology and Human Values* 18 (1993) 3, 362-378

CHAPTER 1 **TECHNOLOGY AND ADMINISTRATION**

A. Booth, 'The evolution of fiscal policy and the role of government in the colonial economy', in: A. Booth, W.J. O'Malley and A. Weidemann, *Indonesian economic history in the Dutch colonial era* (New Haven, Connecticut 1990)

M.J. van Bosse, 'J.A. de Gelder in het tijdperk van de bouw der haven van Tandjong Priok' [J.A. de Gelder at the time when the harbour of Tandjonk Priok was under construction], in: *De Ingenieur 27 (1912) 39, 786-789*

H.W. van den Doel, De stille macht. Het Europese binnenlands bestuur op Java en Madoera, 1808-1942 [The quiet force. European inland administration in Java and Madura, 1808-1942] (Amsterdam 1994)

J.A.A. van Doorn, *Indische lessen* [East Indian lessons] (Amsterdam 1995)

S. de Graaff and D.G. Stibbe (eds.),'Krakatau', *Encyclopaedie van Nederlandsch-Indië* [Encyclopedia of the Dutch East Indies], part 2 (The Hague/Leiden 1918-1939, 2e impression) 443

M. Kuitenbrouwer, 'Het imperialisme-debat in de Nederlandse geschiedschrijving' [The imperialist debate in Dutch historiography], in: *Bijdragen en mededelingen betreffende de geschiedenis der Nederlanden* 113 (1998) 1, 56-73.

D. Landes, The wealth and poverty of nations: Why some are so rich and some are so poor (New York, 1998)

J. Limburg et al., *De toekomst der academisch gegradueerden* [The future for academics] (Groningen/ Batavia 1936)

W.M.F. Mansvelt and P. Creutzberg (eds.), *Changing economy in Indonesia. A selection of statistical source material from the early 19th century up tot 1940*, four volumes (The Hague 1975-1978)

J.E. de Meyier, 'Waterstaat' [Public Works], in: D.G. Stibbe (ed.), *Encyclopaedie van Nederlandsch-Indië* [Encyclopaedia of the Dutch East Indies], part 4 (The Hague/Leiden 1921, 2e impression) 700-731

J. Paulus, S. de Graaff and D.G. Stibbe et al., (eds.), *Encyclopaedie van Nederlandsch-Indië* [Encyclopaedia of the Dutch East Indies], eight parts (The Hague/[Leiden] 1917 [part 1], 1918 [part 2], 1919 [part 3], 1921 [part 4], 1927 [part 5], 1932 [part 6], 1935 [part 7], 1939 [part 8], 2nd impression)

C.L.F. Post, *Over den waterstaat in Nederlandsch-Indië* [About public works in the Dutch East Indies] (Amsterdam 1879)

W. Ravesteijn, *De zegenrijke heeren der wateren. Irrigatie en staat op Java, 1832-1942* [The auspicious lords of the waters. Irrigation and state in Java, 1832-1942] (Delft 1997)

W. Ravesteijn, 'Irrigatie en koloniale staat op Java: De gevolgen van de hongersnoden in Demak' [Irrigation and the colonial state in Java: The consequences of the famine in Demak], in: *Jaarboek voor Ecologische Geschiedenis* 4 (2000) 87-112

R.A. van Sandick, 'Ter herinnering aan J.A. de Gelder' [In memory of J.A. de Gelder], in: *De Ingenieur* 27 (1912) 39, 785-786

Staatsblad van Nederlandsch-Indië [Dutch East Indian Bulletin of Acts, Orders and Decrees]

D.G. Stibbe and F.J.W.H. Sandbergen (eds.), 'Waterstaat', *Encyclopaedie van Nederlandsch-Indië* [Encyclopedia of the Dutch East Indies], part 8 (The Hague 1939, 2nd impression) 476-477

'Viering van de 10e dies natalis der TH Bandoeng' [Celebration of the 10th foundation day of the Bandoeng Technical College], in: *De Ingenieur in Nederlandsch-Indië* 2 (1935) I.115-I.123

CHAPTER 2 **THE ROAD TO A NEW EMPIRE**

Afdeeling Bruggen en Wegen van het Departement B.O.W., *Handleiding voor het traceeren van wegen* [Handbook for the tracing of roads] (Bandoeng 1928)

P. Bakker, *Bali in kleuren* (Joure/Utrecht s.a.)

P.J. Boreel, 'Mededeelingen betreffende de gewone wegen in Ned.-Indië en meer in het bijzonder omtrent den aanleg daarvan in de Buitenbezittingen' [Reports concerning ordinary roads in the Dutch East Indies and, more in particular, their construction in the Outer Regions], in: *De Ingenieur* 30 (1915) 808-817

H. van Breen, 'De Nederlandsch-Indische Wegenvereeniging' [The Dutch- East Indian Road Association], in: *De Ingenieur* 40 (1925) 649

J.C. van Citters, *Gedachten over de economische beteekenis van het wegenvraagstuk in Nederlandsch-Indië* [Thoughts about the economic significance of the roads issue in the Dutch East Indies] (Soerabaja 1929)

H. Cramer, *Het verkeerswezen in Zuid-Sumatra; de verbindingswegen* [Traffic in South-Sumatra; the connecting roads] (Weltevreden 1917)

H. Cramer, 'Het wegennet' [The roads network], in: *Gedenkboek voor Nederlandsch-Indië ter gelegenheid van het regeeringsjubileum van H.M. de Koningin 1898-1923* (Batavia 1923) 463-472

J.A.A. van Doorn, *Ontsporing van geweld. Over het Nederlands/Indisch/Indonesisch conflict* [Lapse of violence. About the Dutch/Indian/Indonesian conflict] (Rotterdam 1970)

J.A.A. van Doorn, *De laatste eeuw van Indië. Ontwikkelingen en ondergang van een koloniaal project* [The last century of the Dutch East Indies.Development and decline of a colonial project] (Amsterdam 1994)

F.G. Dumas en J.H. Blok, 'Aanleg en onderhoud van wegen in de buitengewesten. Prae-advies BOW' [Building and maintenance of roads in the outer regions. Preliminary recommendations for Civil Public Works], in: *Nederlandsch Indisch Wegen Congres Bandoeng. Prae-adviezen, prae-adviezen van het bestuur, verslagen* [The Dutch East Indian Road Congress in Bandoeng. Preliminary recommendations, prior recommendations brought out by the administration, reports] (Bandoeng 1924)

Gedenkboek voor Nederlandsch-Indië ter gelegenheid van het regeeringsjubileum van H.M. de Koningin 1898-1923 [Memorial book on the occasion of the government jubilee of Her Majesty the Queen 1898-1923 (Batavia/Weltevreden/Leiden 1923)

Grote Winkler Prins encyclopedie part 12 (1991)

J. Honing, *Wegenconstructies. Leerboek voor de Technische Scholen in Indonesië* [Road constructions. Study book for the Technical Schools in Indonesia] (Haarlem/Antwerpen/Djakarta 1950)

De Ingenieur in Nederlandsch-Indië 6 (1939) I.32

D. de Iongh Wzn., 'Regeling van het verkeer te land in Ned.-Indië' [Regulating the traffic on land in the Dutch East Indies], in: *De Ingenieur* 51 (1936) V.113-V.121

W.L.M.E. van Leeuwen, *Honderd jaar Nederland 1848-1948* [One hundred years of the Netherlands 1848-1948] (Hengelo 1948)

G.M. van de Linde, 'Wegenaanleg op Java' [Road Construction in Java], in: *Waterbouwkundig Tijdschrift OTAR* 10 (1925) 17-19 and 25-29

J.E. de Meyier, 'Wegen en bruggen in Nederlandsch-Indië' [Roads and bridges in the Dutch East Indies], in: *Gedenkboek Koninklijk Instituut van Ingenieurs 1847-1897* (Den Haag 1897) 301-303

Nederlandsch-Indisch Wegen Congres Bandoeng. Prae-adviezen, prae-adviezen van het bestuur, verslagen [The Dutch East Indies Roads Congress Bandoeng. Preliminary recommendations, preliminary administration recommendations, reports] (Bandoeng 1924)

Oosthoeks encyclopedie part 7 (1968)

C. Ortt, 'Verslag van het Ned.-Ind. Wegencongres 23-25 Juni 1924 te Bandoeng' [The Dutch East Indies Road Congress report 23-25 June 1924 in Bandoeng], in: *De Ingenieur* 39 (1924) 864-867, 892-895

C. Ortt, 'De ontwikkeling van de Ned.-Indische wegenbouw-techniek' [The development of Dutch East Indian road-building techniques]. Voordracht gehouden voor de Afdeeling Nederland der Vereeniging van Waterstaatsingenieurs in N.-I. op 26 November 1938 te 's-Gravenhage, in: *De Ingenieur* 53 (1938) A.523-A.525

C. Ortt, 'De wegen in Deli' [The Roads in Deli], in: *Wegen* 1939, 354-355

P.J. Ott de Vries, 'Het wegenvraagstuk in Nederlandsch-Indië' [The roads issue in the Dutch East Indies], in: *De Ingenieur* 35 (1920) 947-954

S.A. Reitsma, *De wegen in de Preanger* [The roads in the Preanger region] (Bandoeng 1912)

S.A. Reitsma, *De wegenkwestie op Java en Madoera* [The roads issue in Java and Madoera] (Bandoeng 1913)

R.A. van Sandick, 'Algemeen Ingenieurscongres te Batavia 8-15 Mei 1920' [The General Engineers' Congress in Batavia 8-15 May 1920], in: *De Ingenieur* 35 (1920)

F.W. Stapel, *Geschiedenis van Nederlandsch Indië* [History of the Dutch East Indies], part V (Amsterdam 1940)

D.G. Stibbe, J. Paulus and S. de Graaff (eds.), *Encyclopaedie van Nederlandsch-Indië* [Dutch East Indian encyclopaedia], eight parts (The Hague/Leiden 1917 [part 1], 1918 [part 2], 1919 [part 3], 1921 [part 4], 1927 [part 5], 1932 [part 6], 1935 [part 7], 1939 [part 8], 2nd impression)

D.G. Stibbe and F.J.W.H. Sandbergen (eds.), 'Wegverkeer. Wegverkeerswetgeving' [Road Traffic. Road Traffic Legislation], *Encyclopaedie van Nederlandsch-Indië* [Dutch East Indian encyclopedia], part 8 (The Hague 1939, 2nd impression) 477-493

D.G. Stibbe and J. Stroomberg (eds.), 'Materiaalonderzoek' [Materials research], *Encyclopaedie van Nederlandsch-Indië* [Dutch East Indian encyclopedia], part 6 (The Hague 1932, 2nd impression) 252-253

D.G. Stibbe (ed.), 'Wegen' [Roads], *Encyclopedie van Nederlandsch-Indië* [Dutch East Indian encyclopedia], part 4 (The Hague/Leiden 1921, 2nd impression) 739-750

P. van 't Veer, *Daendels Maarschalk van Holland* [Daendels. Marshall of Holland] (Zeist 1963)

De Waterstaats-ambtenaar in Nederlandsch Oost-Indië. Handleiding bevattende alle bepalingen en circulaires betreffende den Waterstaat in Nederlandsch Oost-Indië [The Public Works civil servant in the Dutch East Indies. Manual including all stipulations and circulars pertaining to Public Works in the Dutch East Indies] (Soerabaja 1905)

M. de Wolff, *Het wegenvraagstuk* [The roads issue] Bandoeng 1922)

Journals

De Ingenieur [The Engineer] 1888-1938

De Ingenieur in Nederlandsch-Indië [The Engineer in the Dutch East Indies] 1934-1939

De Waterstaats-Ingenieur [The Public Works Engineer] 1920-1933

Wegen [Roads, journal of the Vereeniging het Nederlandsch Wegen Congres, i.e. Association of the Dutch Roads Congress] 1925-1943

CHAPTER 3 **THE LOCOMOTIVE OF MODERNITY**

J.F.L. de Balbian Verster, *Gedenkschrift der Deli Maatschappij 1869-1919* [Memoirs of the Deli Company 1869-1919] (1919)

Biografisch woordenboek van Nederland [Biographical dictionary of the Netherlands], part 3 (The Hague 1989)

J.P. de Bordes, *De spoorweg Samarang-Vorstenlanden* [The Samarang Principalities railroad] (The Hague 1870)

J.P. de Bordes, Minutes of the meeting of 11th April 1871, appendix 21, in: *Tijdschrift van het Koninklijk Instituut van Ingenieurs* (1870-1871)

J. Breman, *Koelies, planters en koloniale politiek. Het arbeidsregime op de grootlandbouwondernemingen aan Sumatra's Oostkust in het begin van de twintigste eeuw* [Coolies, planters and colonial politics. The labour regime in the large agricultural enterprises on Sumatra's East coast at the beginning of the twentieth century] (Dordrecht 1987, 2nd impression)

R. Broersma, *Oostkust van Sumatra* [East coast of Sumatra], part 1, *De ontluiking van Deli* [The opening of Deli] (Batavia 1919)

R. Broersma, *Oostkust van Sumatra* [East coast of Sumatra], part 2, *De ontwikkeling van het gewest* [The development of the region] (Deventer 1922)

J. de Bruin, *Du Croo & Brauns locomotieven* [Du Croo & Brauns locomotives] (s.l., 1987)

J. Burman, *Early railways at the Cape* (Cape Town-Pretoria 1984)

J.N.F.M. à Campo, *Koninklijke Paketvaart Maatschappij. Stoomvaart en staatsvorming in de Indonesische archipel 1888-1914* [The Royal Packet Shipping Company. Steam shipping and state formation in the Indonesian archipelago 1888-1914] (Hilversum 1992)

G.P.J. Caspersz, 'De Atjehtram; haar geschiedenis, haar aandeel in de onderwerping en pacificatie van Atjeh; technische bijzonderheden' [The Atjeh Tram, its history and its part in the subjection and pacification of Atjeh; technical specifications], in: *De Ingenieur* 42 (1927) 865-879

G.P.J. Caspersz, 'De concurrentie van het motorvervoer aan de spoor- en tramwegen in Indië' [The competition between motor transport and the railways and tramways in the Indies], in: *De Ingenieur* 45 (1930) T.33-T.46

A.H.P. Clemens and J.Th. Lindblad, *Het belang van de buitengewesten. Economische expansie en koloniale staatsvorming in de buitengewesten van Nederlands-Indië 1870-1942* [The importance of the outer regions. Economic expansion and colonial state formation in the outer regions of the Dutch East Indies] (Amsterdam 1989)

J.A.A. Doorn, 'De eerste spoorweg op Java' [The first railway on Java], in: M.L. ten Horn-van Nispen, H.W. Lintsen en A.J. Veenendaal jr. (red.), *Nederlandse ingenieurs en hun kunstwerken. Tweehonderd jaar civiele techniek* [Dutch engineers and their structural works. Two hundred years of civil engineering] (Zutphen 1994)

A.E. Durrant, *The Mallet Locomotive* (Newton Abbot 1974)

A.E. Durrant, *Australian Steam* (Newton Abbot 1978)

R.A. Eekhout, *Aanleg van Staatsspoorwegen in Nederlandsch Borneo en Zuid-Sumatra* [The construction of State railways in Dutch Borneo and South Sumatra] (Leiden 1891)

J.Ph. Ermeling, 'De genie-uitrusting van de tweede expeditie tegen Atjeh, deel III, Het expeditionair tramway materieel' [The military equipment used during the second expedition against Atjeh, part III, The dispatched tramway rolling stock, in: *Tijdschrift van het Koninklijk Instituut van Ingenieurs. Afdeeling Nederlandsch-Indië* 1878-79, 23-24

A.J.H. van Ette, *Onze ministers sinds 1798* [Our ministers since 1798] (Alphen aan den Rijn 1948)

C. Fasseur, *Kultuurstelsel en koloniale baten. De Nederlandse exploitatie van Java 1840-1870* [The cultivation system and colonial profits. The Dutch exploitation of Java 1840-1870] (Leiden 1975)

C. Fasseur, 'De Nederlandse koloniën 1795-1914' [The Dutch colonies 1795-1914], in: *Algemene geschiedenis der Nederlanden* [General history of the Netherlands] (Weesp 1983) 11, 347-380

G. de Gelder, 'De electrificatie der spoorwegen in Nederlandsch-Indië' [The electrification of the railways in the Dutch East Indies], in: *De Ingenieur* 36 (1921) 642-644 en 38 (1923) 715-731

G. de Gelder, 'De electrische Staatsspoorwegen in Ned.-Indië' [The electric State Railways in the Dutch East Indies], in *De Ingenieur* 43 (1928) V.1-V.21

F.C. Gerretson, *Geschiedenis der `Koninklijke'* [The history of 'Royal Dutch'], four volumes (Haarlem 1932-1941)

W.T. Gevers Deynoot, *Herinneringen eener reis naar Nederlandsch-Indië in 1862* [Memories of a journey to the Dutch East Indies in 1862] (The Hague 1864)

Wat een goede couragie vermach [What a good courage can do] (Den Helder 1944)

G. Gonggrijp, *Schets eener economische geschiedenis van Nederlandsch-Indië* [Sketch of the economic history of the Dutch East Indies] (Haarlem 1938, 2nd impression)

G.F.E. Gonggryp, *Geïllustreerde encyclopaedie van Nederlandsch-Indië* [Illustrated encyclopedia of the Dutch East Indies] (Wijk en Aalburg 1991)

Het hellend vlak van Agudio en de stangenbanen. Rapport van den ingenieur J.L. Cluijsenaer over de waarde en bruikbaarheid dezer stelsels bij den aanleg van een spoorweg ter verbinding van de Ombiliën-kolenvelden op Sumatra met de Indische Zee [The inclined plane of Agudio and the rod tracks. Engineer J.L. Cluijsenaer's report on the value and usefulness of these systems in the construction of a railroad to connect the Ombilin coal fields on Sumatra with the Indian Ocean]. Departement van Koloniën (The Hague 1878)

W.H. van Helsdingen en H. Hoogenberk (eds.), *Daar wèrd wat groots verricht .... Nederlandsch-Indië in de XXste eeuw* [Great things were achieved there... the Dutch East Indies in the XX century] (Amsterdam 1941)

N.H. Henket, *De aanleg van spoorwegen op Midden-Java* [The Construction of railways in Central Java] (Amsterdam 1866)

J.D. van Herwerden, *De spoorweg-kwestie op Java* [The railway question in Java] (The Hague 1863)

J.D. van Herwerden, *Een spoorwegnet over Java* [A railway network through Java] (The Hague 1872)

H.M. Hirschfeld, *Het ontstaan van het moderne bankwezen in Nederland* [The rise of modern banking in the Netherlands] (Rotterdam 1922)

W.H.P.M. van Hooff, *De Nederlandse machinefabrieken 1825-1914* [The Dutch engineering works 1825-1914] (Amsterdam 1990)

H. Hovinga, *Dodenspoorweg door het oerwoud. Het vergeten drama van de Pakan-Baroespoorweg op Sumatra, aangelegd door krijgsgevangenen onder de Japanse bezetting* [The railway of the dead through the jungle. The forgotten tragedy of the Pakan-Baroe railway on Sumatra, constructed by prisoners of war during the Japanese occupation] (Franeker 1976)

H. Hughes, *Indian Locomotives*, Part 1, *Broad gauge 1851-1940* (Harrow 1990)

H. Hughes, *Indian Locomotives*, Part 2, *Metre gauge 1872-1940* (Harrow 1992)

'De Japanse bezetting' [The Japanese occupation], in: *Weerzien met Indië* [Return to the Indies], nr. 46 (Zwolle 1995)

J.H. Jonckers Nieboer, *Geschiedenis der Nederlandsche Spoorwegen* [History of the Dutch Railways] (Rotterdam 1938, 2nd impression)

J.J.P. de Jong, *De waaier van het fortuin. Van handelscompagnie tot koloniaal imperium. De Nederlanders in Azië en de Indonesische archipel 1595-1950* [The fan of fortune: From trading company to colonial empire] (Den Haag 1998)

J.J. Jonker, *Merchants, bankers, middlemen. The Amsterdam money market during the first half of the 19th century* (Amsterdam 1996)

Koloniaal Verslag (Colonial Report], Appendix to the *Handelingen der Staten-Generaal, Eerste en Tweede Kamer* [Proceedings of the States General, First and Second Chambers] (1896)

De koloniale roeping van Nederland [The colonial vocation of the Netherlands] (Den Haag 1930)

H.I. Keus and M.C. Brandes, 'Electrische locomotief voor de staatsspoorwegen in Ned.-Indië, gebouwd door Heemaf in samenwerking met Werkspoor' [Electric locomotives for the State Railways in the Dutch East Indies, built by Heemaf in cooperation with Werkspoor], in: *De Ingenieur* 40 (1925) 583-593.

J.A. Kool and N.H. Henket, *Onderzoek in hoeverre de smalle spoorwijdte van 1.00 à 1.10 meter voor de behoeften van het vervoer op Java toe te passen en uit een economisch oogpunt aan te bevelen zoude zijn* [Research as to how far the narrow gauge of 1.00 to 1.10 meter would be applicable to the needs of Javanese transport and recommendable from an economic point of view] (Rotterdam 1870)

J.Th. Lindblad (ed.), *New challenges in the modern economic history of Indonesia* (Leiden 1993)

W.M.F. Mansvelt, *Handelsstatistiek van Java 1823-1873* [Javanese trade statistics] (s.l. 1938)

H. Meijer, *De Deli Spoorweg Maatschappij* [The Deli Railway Company] (Zutphen 1987)

Nieuw Nederlandsch biografisch woordenboek [New Dutch biographical dictionary], part I (Leiden 1911), part II (Leiden 1912), part VI (Leiden 1924),

M. Middelberg en F.E. van Hennekeler, *Tramwegen op Billiton en Banka* [Tramways on Billiton and Banka] (Batavia 1917)

W.J.M. Nivel, *Verslag eener spoorwegverkenning in_Midden-Sumatra* [Report of railway exploration in Central Sumatra], two volumes (Weltevreden 1927)

J.J.G. Oegema, *De stoomtractie op Java en Sumatra* [Steam traction on Java and Sumatra] (Deventer 1982)

Rapport van den ingenieur J.L. Cluysenaer over den aanleg van spoorwegen in de Padangsche Bovenlanden ter verbinding van de Ombiliën-kolenvelden met de Indische Zee [Engineer Cluijsenaer's report on the construction of lines in the Padang Highlands in order to link up the coal mines of Ombilin with the Indian Ocean]. Departement van Koloniën (The Hague 1876)

Rapport van den ingenieur J.L. Cluysenaer over den aanleg van spoorwegen in de Padangsche Bovenlanden. [Engineer Cluijsenaer's report on the construction of lines in the Padang Highlands]. Departement van Koloniën (The Hague 1878)

Het rapport van den heer Stieltjes over verbeterde vervoermiddelen op Java, met kantteekeningen van een officier der genie van het Ned. Oost-Ind. Leger [The Stieltjes report on improved transport means in Java including comments made by one of the military officers in the Dutch East Indian Army] (Leiden/Batavia 1864)

S.A. Reitsma, *Gedenkboek der staatsspoor- en tramwegen in Nederlandsch-Indië, 1875-1925* [Commemorative book of the state railways and tramways in the Dutch East Indies, 1875-1925] (Weltevreden 1925)

S.A. Reitsma, *Korte geschiedenis der Nederlandsch-Indische spoor- en tramwegen* [A brief history of the Dutch East Indian railways and tramways] (Weltevreden 1928)

S.A. Reitsma, 'De spoor- en tramwegen in Nederlandsch-Indië' [The railways and tramways in the Dutch East Indies], in: *De Koloniale roeping van Nederland* (Den Haag 1930)

S.A. Reitsma, *Van Stockum's travellers' handbook for the Dutch East Indies* (The Hague 1930)

S.A. Reitsma, *De Staatsspoorweg ter Sumatra's Westkust (S.S.S.)* [The State Railway on Sumatra's West coast (S.S.S.)] (Den Haag 1943)
R.A. van Sandick, 'De Ombilienkolenvelden' [The Ombilin coal fields], in: *Vragen des Tijds* 1891, I, 245-78
J. Slim, 'De aanleg der nieuwe spoorverbinding Solo-Djocja' [The construction of the new Solo-Djocja rail connection] , in: *Spoor- en Tramwegen* 3 (1930)
Staatsblad (Bulletin of acts, orders and decrees)
F.W. Stapel, *Geschiedenis van Nederlandsch-Indië* [History of the Dutch East Indies] (Amsterdam 1930)
T.J. Stieltjes, *Eenige beschouwingen over spoorwegen op Java* [Various considerations on railroads in Java] (Rotterdam 1874)
T.J. Stieltjes, *Gegevens omtrent de zaak der spoorwegen op Java* [Details on the matter of the Javanese railways], six parts (The Hague 1863-1865)
T.J. Stieltjes, *Overzigt van hetgeen met de spoorwegen op Midden-Java is voorgevallen* [An overview of what happened to the railways in Central Java] (The Hague 1864, 2nd impression)
De stoomtramwegen op Java. Gedenkboek_samengesteld ter gelegenheid van het vijf en twintigjarig bestaan der Samarang-Joana Stoomtram-Maatschappij [The steamtramways of Java. Commemorative book on the occasion of the company's twenty-fifth anniversary] (The Hague 1907)
Tijdschrift van het Koninklijk Instituut van Ingenieurs. Afdeeling Nederlandsch-Indië [Journal of the Royal Institute of Engineers. The Dutch East Indies Division] (1876-1914)
A.J. Veenendaal jr., 'De kennisoverdracht op het gebied van de spoorwegtechniek in Nederland 1830-1870' [Knowledge transfer in the field of railway technology in the Netherlands], in: *Jaarboek voor de geschiedenis van bedrijf en techniek* 7 (Amsterdam 1990) 54-82
A.J. Veenendaal, jr., *Slow train to paradise. How Dutch investment helped build American railroads* (Stanford 1996)
A.J. Veenendaal jr., *De ijzeren weg in een land vol water. Beknopte geschiedenis van de spoorwegen in Nederland 1834-1958* [The iron road in a land full of water. Short history of the railways in the Netherlands] (Amsterdam 1998)
A.J. Veenendaal, jr., 'Techniek, economie of politiek? Het Nederlandse spoorwegelektrificatieplan van 1922' [Technology, economics or politics? The Dutch railway electricfication plan], in: *NEHA-Jaarboek 2000* (Amsterdam 2000) 251-72
Paul van 't Veer, *De Atjeh-oorlog* [The Atjeh war] (Amsterdam 1969)
L.M. Vilain, *Les locomotives articulées du système Mallet dans le monde* (Paris 1969)
Wat een goede_couragie vermach (Den Helder 1944)
E. de Waal, *Onze Indische financiën*, II [Our East Indian finances] (The Hague 1877)
J. Weisfelt, *De Deli Spoorweg Maatschappij als factor in de economische ontwikkeling van de Oostkust van Sumatra* [The Deli Railway Company as a factor in the development of the East coast of Sumatra] (Rotterdam 1972)
H. van der Wijck, *De Nederlandsche Oost-Indische bezittingen onder het bestuur van den kommissaris-generaal Du Bus de Gisegnies* [The Dutch East Indian possessions during the administration of the commissioner-general Du Bus de Gisegnies] (The Hague 1866)
J.J.A. Wijn (red.), *Tot in de verste uithoeken ... De cruciale rol van de Gouvernements Marine bij het vestigen van de Pax Neerlandica in de Indische archipel 1815-1962* [Into the farthest corners ... The crucial role of the Government's Navy in establishing Pax Neerlandica in the East Indian archipelago 1815-1962] (Amsterdam 1998).
J.W. IJzerman, *Dwars door Sumatra. Tocht van Padang naar Siak* [Straight through Sumatra. Journey from Padang to Siak] (Haarlem/Batavia 1895)

CHAPTER 4 **FREE EXTENDING OVER THE BANDJIR**

J. Alsdorf, 'Het verlengen van de brug over de Porrongrivier bij Tjepiples in de Staatsspoorweg Sidhoardjo-Soerakarta' [Extending the bridge over the Porrongrivier at Tjepiples along the Sidhoardjo-Soerakarta State Railway line], with plates, in: *Tijdschrift van het Koninklijk Instituut van Ingenieurs. Afdeeling Nederlandsch-Indië* 1891-1892, XXII-XXIV

Atlas van tropisch Nederland [Atlas of the tropical Netherlands] (Amsterdam 1990)
'Bamboezen brug over de Tjitandoei te Indihiang. Grens Preanger R.-Cheribon' [Bamboo bridge over the Tjitandoei at Indihiang. Border Preanger R.-Cheribon], with plates, Appendix B, in: *Tijdschrift van het Koninklijk Instituut van Ingenieurs. Afdeeling Nederlandsch-Indië* 1893-1894, V-VI
M. van Ballegoijen de Jong, *Spoorwegstations op Java* [Railway stations in Java] (Amsterdam 1993)
M. van Ballegoijen de Jong, *Stations en spoorbruggen op Sumatra 1876-1941* [Stations and railway bridges in Sumatra 1876-1941] (Amsterdam 2001)
'Berichten van allerlei aard, Bruggenbouw op Sumatra' [Various kinds of information, Bridges in Sumatra], in: *De Ingenieur* 53 (1938) A.167
F.K.J. Beukema toe Water, 'Vernieuwing der bruggen over de Serang-rivier in de lijn Samarang – Vorstenlanden' [Renewal of the bridges over the Serang river in the Samarang –Principalities, with illustrations, in: *De Ingenieur* 23 (1908) 772-776
N.J. Beversen, 'Brug over de rivier Ajer-Silau in den grooten verbindingsweg van Tandjong-Balei naar Deli (oostkust van Sumatra)' [Bridge over the river Ajer-Silau on the long connecting road between Tandjong-Balei and Deli (on the Sumatran east coast)], with two plates, in: *De Ingenieur* 12 (1897) 135-136
P. Boorsma, *Collectie Hartkamp* [Hartkamp Collection] (Amsterdam 1990)
P. Boorsma and J. Lucassen, *Gids voor de collecties van het Nederlandsch econ-hist archief te Amsterdam* [Guide to the Dutch economic-historical archive collections in Amsterdam] (Amsterdam 1992)
E. Breton de Nijs, *Tempo Doeloe, fotografische documenten uit het oude Indië 1870-1914* [Tempo Doloe, photographic documents from the East Indies in olden times 1870-1914] (Amsterdam 1961)
P.P. Bijlaard, 'Demontage van 3 bruggen van 60 m, vormende de oude S.S. brug over de Tji-Taroem bij Kedoenggedeh, en weder monteeren daarvan bij 3 afzonderlijke kunstwerken in de in aanleg zijnde lijn Koppo-Tjiwidei' [Disassembling the 3 60 m bridges that formed the old State Railway bridge over the Tji-Taroem at Kedoenggedeh, and assembling them again in the Koppo-Tjiwidei line under construction], with illustrations, in: *De Ingenieur* 39 (1924) 778-784
P.P. Bijlaard, 'Meer dan 50 meter vrije uitbouw van een vakwerkbrug van 60 meter in Zuid Sumatra' [More than 50 meters of free extension leading from a 60 meters truss bridge in South Sumatra], with illustrations, in: *De Ingenieur* 41 (1926) 691-694
P.P. Bijlaard, 'Montage van de brug over de Way-Oempoe, als laatste schakel in de spoorlijn Palembang-Telok Betong' [Assembling the bridge over the Way-Oempoe, the final link in the Palembang-Telok Betong line], with illustrations, in: *De Ingenieur* 42 (1927) 597-602
P.P. Bijlaard, 'Het verminderen van de blijvende doorbuiging van eenige statisch onbepaalde vakwerkbruggen in het lijngedeelte Padalarang-Poerwakarta' [Diminishing the permanent bending of some statically undetermined truss bridges in the Padalarang-Poerwakarta line section], with illustrations, in: *De Ingenieur 43 (1928) 69-73*
P.P. Bijlaard, 'De brug over de Kali Progo, ontworpen volgens een nieuw systeem' [The bridge over the Kali Progo, designed according to a new method], with illustrations, with errata, in: *De Ingenieur* 46 (1931) B.273-B.291
P.P. Bijlaard, *Vrije Uitbouw, uitgevoerde werken* [Free extension, the completed works] (Bandoeng 1933)
P.P. Bijlaard, *Factoren die het materiaalverbruik in draagconstructies beïnvloeden* [Factors influencing the use of material in supporting constructions] (Bandoeng 1934)
J.L. Cluysenaer, 'J.P. de Bordes', in: *De Ingenieur* 14 (1899) 61-65
CUR,'Betonplaten onder geconcentreerde belastingen' [Concrete plates subjected to concentrated stress], reports 16 A & B
P.A. Daum, *H. van Brakel Ing B.O.W* [H.van Brakel Public Works engineer] (Amsterdam 1976)
P. Dehing and C. Seegers, *Katalogus van gedenkboeken. De kollectie gedenkboeken van ondernemingen en organisaties in de Econ.- Hist. Bibliotheek* [Catalogue of memorial books. The collection of memorial books of companies and organisations in the Economic-Historical Library] (Amsterdam 1988)
Deutsches Normalprofilbuch für Walzeisen (Aken 1897)
W.J. van der Eb, 'Het Tjisomang-viaduct' [The Tjisomang viaduct], with illustrations, in: *De Ingenieur* 49 (1934) B.116-B.118.
W.J. van der Eb, 'Onderzoek van staalconstructies en vermoeiingsverschijnselen' [Research into steel

constructions and fatigue phenomena], in: *De Ingenieur* 64 (1949) O.51-O.57
B.M. Gratama, 'Het overtrekken der noordelijke overspanning der brug over de Kali Krassak in den stoomtramweg Djocja-Magelang op 30 Jan.1898' [Constructing the northern spanning over the Kali Krassak along the Djocja-Magelang steamtram line on 30 Jan. 1898], with two illustrations; including a report by T.J. Rosskopf, in: *De Ingenieur* 13 (1898) 305-307
B.M. Gratama, 'Mededeelingen omtrent de in aanleg zijnde lijn Goendih-Soerabaja der Nederlandsch-Indische Spoorweg-Maatschappij in het bijzonder met betrekking tot den metalen bovenbouw der bruggen' [Proceedings regarding the current construction of the Goendih-Soerabaja line by the DutchEast Indian Railway Company with particular attention to the metal superstructure of the bridges], with illustrations and plates,in: *De Ingenieur* 16 (1901) 66-76 and 90-95
B.M. Gratama, 'Mededeelingen omtrent den metalen bovenbouw der bruggen in de lijn Magelang-Willem I met zijtak Setjang-Perakan der Nederlandsch-Indische Spoorweg-Maatschappij' [Proceedings regarding the metal superstructure of the bridges in the Magelang-Willem I line with the Setjang-Perakan branch line of the Dutch East Indian Railway Company], with illustrations, in: *De Ingenieur* 24 (1909) 356-362
J.H.A. Haarman, 'Het ophangen van de spoorwegbruggen te Ngoedjang' [The suspending of railway bridges at Ngoedjang], in: *Tijdschrift van het Koninklijk Instituut van Ingenieurs. Afdeeling Nederlandsch-Indië* 1905-1907, 9-21
J.H.A. Haarman, 'Het ontwerp van den bovenbouw der overbrugging van de Serajoe-rivier in de aan te leggen lijn Cheribon-Kroja der Staatsspoorwegen op Java' [The design of the superstructure of the Serajoe river bridge that was to be constructed along the Cheribon-Kroja line], with illustrations, in: *De Ingenieur* 30 (1915) 641-655
J.H.A. Haarman, 'De algemeene voorschriften betreffende ijzeren bruggen en pijlers voor spoor- en tramwegen in Nederlandsch-Indië' [The general stipulations regarding iron bridges and pillars for railways and tramways], in: *De Ingenieur* 30 (1915) 108-111
J.H.A. Haarman, 'De groote bruggen in de spoorlijn Cheribon-Kroja der Staatsspoorwegen op Java'[Large bridges along the Cheribon-Kroja railway line belonging to the Javanese State Railway Company], in: *Indisch Tijdschrift voor Spoor- en Tramwegen* (1915) 207-212
J.H.A. Haarman, 'De montage van de brug over de Serajoe-rivier in de lijn Cheribon-Kroja der Staatsspoorwegen op Java' [The assembly of the bridge over the Serajoe river along the State Railway's Cheribon-Kroja railway line in Java], with illustrations, in: *De Ingenieur* 32 (1917) 57-62
J.H.A. Haarman, 'De berekening van ijzeren bruggen en de richting waarin die zich ontwikkelt' [The calculating of iron bridge dimensions and the direction in which they are developing], in: *De Ingenieur* 38 (1923) 695-698
E.C.U. Hartman, 'Mededeelingen omtrent de verzwaring van den bovenbouw der bruggen in de lijn Goendih-Soerabaja der Nederlandsch-Indische Spoorweg-Maatschappij' [Proceedings regarding the strengthening of the bridge superstructure along the Dutch East Indian Railway Company's Goendih-Soerabaja line], with illustrations, in: *De Ingenieur* 34 (1919) 148-161
B.G.H. van der Jagt, 'Brug over de kali Brantas bij de dessa Tegalsari' [Bridge over the kali Brantas at dessa Tegalsari], with illustrations, in: *De Ingenieur* 37 (1922) 464-466
Jules J.A. Jansen, *Bamboo in building structures* (Eindhoven 1981)
W. de Jongh, 'De overbrugging van het Tjisaät-ravijn in den Staatsspoorweg Tjitjalengka-Garoet' [The bridging of the Tjisaät ravine along the Tjitjalenka-Garoet State Railway line], in: *Tijdschrift van het Koninklijk Instituut van Ingenieurs. Afdeeling Nederlandsch-Indië* 1888-1889, 22-30
N.C. Kist, 'Een algemeene methode tot het schatten van het eigengewicht van hoofdliggers van vakwerkbruggen' [A general way of estimating the own weight of the main girders of trellis bridges], in: *De Ingenieur* 41 (1926) 60-63
N.C. Kist, 'Het eigen gewicht van Indische en Nederlandse bruggen' [The own weight of East Indian and Dutch bridges], with illustrations, in: *De Ingenieur* 43 (1928) B.197-B.200
N.C. Kist, 'Het meest aangewezen brugtype in bepaalde gevallen' [The most appropriate bridge types for specific cases], in: *De Ingenieur* 43 (1928) B.87
N.C. Kist, 'Het eigen gewicht van Nederlandsche en Indische bruggen' [The own weight of Dutch and East Indian bridges], in: *De Ingenieur* 52 (1937) B.107-B.112
J.A. van der Kloes and J.N. van Ruijven, *Het bouwen in overzeesche gewesten* [Building in overseas territories] (Leiden 1905)

'Mededeeling over het draagvermogen enz.' [Proceedings regarding carrying capacity etc.], Appendix D, in: *Tijdschrift van het Koninklijk Instituut van Ingenieurs. Afdeeling Nederlandsch-Indië* 1897-1898, XXV-XXVI

'Mededeelingen van verschillenden aard, Bescherming van houtwerken tegen witte mieren' [Information of various sorts, Protecting wooden structures against white ants], in: *Tijdschrift van het Koninklijk Instituut van Ingenieurs. Afdeeling Nederlandsch-Indië* 1880-1881, 36-38

J.E. de Meyier, *Tijdschrift van het Koninklijk Instituut van Ingenieurs. Afdeeling Nederlandsch-Indië* 1893-1894, appendix B.

R.R.L. de Muralt, 'Iets over de afvoerravijnen van den Smeroe en den bouw van de nieuwe brug over de rivier de Moedjoer' [Details about the conveyor ravines of the Smeroe and the construction of the new bridge over the Moedjoer river], met afbeeldingen, in: *De Ingenieur* 18 (1903) 311-320

J. Oosterhoff (ed.), *Bruggen in Nederland 1800-1940* [Bridges in the Netherlands], II (Utrecht 1998)

A. den Ouden, *Een hoekstaal van de maatschappij, constructie werkplaatsen in Nederland van 1840 tot heden (1994)* [A corner-steel of society, construction workshops in the Netherlands from 1840 until the present (1994)] (Stichting Nederlandse Staalbouw 1994)

Pioniers over Zee. De geschiedenis van het Bataljon Genietroepen van de Eerste Divisie (7 December/C-Divisie), 1945-1950 [Pioneers overseas. The history of the Engineering Batallion of the First Divison (s.l.e.a.)

W.J.G. Paardekooper, 'Houten bruggen in Zuidwest-Celebes' [Wooden bridges in Southwest Celebes], with illustrations, in: *De Ingenieur in Nederlandsch-Indië* 8 (1941) II.1-II.4

S.A. Reitsma, *Gedenkboek der staatsspoor- en tramwegen in Nederlandsch-Indië 1875-1925* [Commemorative book of the state railways and tramways in the Dutch East Indies 1875-1925] (Weltevreden 1925)

Ch.F. de Rochemont, 'Ontstaan en inrichting van de Atjeh-tram' [The rise and furnishing of the Atjeh tram], with illustrations, in: *De Ingenieur* 26 (1911) 117-128 (issue on the occasion of the 25 year jubilee 1886-1911)

R. Roosseno Soerjohadikoesoemo, *De Ingenieur in Nederlandsch-Indië* 6 (1939) II.174-II.176.

R. Roosseno Soerjohadikoesoemo, 'Het gewapend beton in den bruggenbouw in Indië' [Reinforced concrete in bridge building in the East Indies], in: *De Ingenieur in Nederlandsch-Indië* 7 (1940) I.143–I.156

J.J. Seegers, *Economisch-historische wegwijzer. Een gids voor het bronnen- en literatuuronderzoek van de Ned. Econ. geschiedenis* [Economic-historical guide. A guide for sources and literature research into Dutch Economic history] (Amsterdam 1990)

P.J. Seitert, *Handboek voor den technischen dienst van het Korps Genietroepen van het Nederlandsch-Indische Leger* [Handbook of the Engineering Corps of the Dutch East Indian Army, technical service] (Haarlem 1894)

W.K. Seyn, 'Technische werkzaamheden uitgevoerd door het Korps Genietroepen van de B-Divisie in West- en Midden-Java en na de politionele actie 1947' [Technical activities carried out by the Engineering Corps of B Division in West and Central Java after the Police Intervention of 1947], in: *De Ingenieur* 63 (1948) 340-347

J. Siegel, *Traditional bridges of Papua New Guinea* (Papua New Guinea 1982)

A. Snethlage, 'Brug over de Bekassierivier in den spoorweg Batavia-Kedong Gedeh' [Bridge over the Bekassie river in the Batavia-Kedong Gedeh railway line], with one plate, in: *De Ingenieur* 6 (1891) 151-152

J.C. Sprenger, *Historische bedrijfsarchieven. Bouwnijverheid en -installatiebedrijven. Een geschiedenis en bronnenoverzicht* [Historical company archives. Building and Installation companies. A history and sources overview] (Amsterdam 1993)

D.G. Stibbe and F.J.W.H. Sandbergen (eds.), *Encyclopaedie van Nederlandsch-Indië* [Dutch East Indian encyclopedia], part 8 (The Hague 1939) 477-493

J.P. Textot, 'Bij het overlijden van W.J. de Bordes c.i.' [In memory of W.J. de Bordes, civil engineer], with photo, in: *De Ingenieur* 21 (1906) 282-283

J.K.E. Triebart, 'Ontwerp van ene normaal-constructie voor stalen bruggen, van 12-24 meter spanning in Indië' [The designing of a normal construction for steel bridges in the East Indies with 12-24 meter spans], with one plate, in: *Tijdschrift van het Koninklijk Instituut van Ingenieurs. Afdeeling Nederlandsch-Indië* 1878-1879, 1-13

J. van Velzen, 'De ijzeren brug over de Kediri-rivier, ter hoofdplaats van de Residentie Kediri' [The iron bridge over the Kediri river at the central point in the Kediri Residence], met afbeeldingen, in: *Tijdschrift van het Koninklijk Instituut van Ingenieurs. Afdeeling Nederlandsch-Indië* 1877-1878, 65-72

H.J. Verdam, 'Bouw der bruggen over de Tjimanoek en de Tjiloetoeng in den groten weg van Cheribon naar Soemedang' [The construction of bridges over the Tjimanoek and the Tjiloetoeng on the main road from Cheribon to Soemedang], with illustrations, in: *Tijdschrift van het Koninklijk Instituut van Ingenieurs. Afdeeling Nederlandsch-Indië* 1896-1897, 1-8

J.N. Vermande, 'Vervanging van den bovenbouw der bruggen in de lijn Goendih-Soerabaja der Nederlandsch-Indische Spoorweg-Maatschappij' [Replacing the superstructure of the bridges in the Goendih-Soerabaja line of the Dutch East Indian Railway Company], with illustrations, in: *De Ingenieur* 36 (1921) 1040-1044

'Uit het verslag van de Burgerlijke Openbare Werken in Nederlandsch-Indië over 1912, II. Bruggen en wegen' [From the Public Works report of the Dutch East Indies for 1912. II Bridges and roads], in: *De Ingenieur 31 (1916) 648*

'Verslag van een excursie naar de in aanbouw zijnde S.S. Brug over de Kali Progo bij Sentolo, gehouden door Kring 3, Semarang, van het Kon. Instituut van Ingenieurs' [Report of an excursion to the State Railway bridge under construction over the Kali Progo at Sentolo, organized by Circle 3, Semarang, of the Royal Institute of Engineers], in: *De Ingenieur* 48 (1933) A.35

H.C.P. de Vos, 'De strijd om en in het water' [The battle for and against water], in: W.H. van Helsdingen en H. Hoogenberk (red.), *Daar wèrd wat groots verricht .... Nederlandsch-Indië in de XXste eeuw* [Great things were achieved there… The Dutch East Indies in the XX century] (Amsterdam 1941)

A.N.P. de Wit, 'Montage van een brug van 3 x 50 m over de Bogowontorivier in de S.S.-lijn Koetoardjo-Djokja op Java' [Assembly of a bridge of 3 x 50 m over the Bogowonto Rivier along the Koetoardjo-Djokja State Railway line in Java], with illustrations, in: *De Ingenieur* 43 (1928) B.160-B.162

L. Zwiers, *Bouwkundig woordenboek* [Structural engineering Dictionary], I (Amsterdam s.a.) 77

Archives

National Archive [Nationaal Archief] (NA), The Hague

-Inventory of the archives of the firm De Vries Robbé file reference: 3.21.02

-Inventory of the archives of the Koninklijk Instituut van Ingenieurs (KIVI) [Royal Institute of Engineers] 1847-1960, file reference: 2.19.047.01

-Inventory of the archives of Vereeniging van Burgerlijke Ingenieurs [Association of Civil Engineers] (1853-1900) and the Vereeniging van Delftsche Ingenieurs [Association of Delft Engineers] (1900-1960) (1853-1960), file reference 2.19.047.02

-Inventory of the archives of the editorial board of De Ingenieur [The engineer], file reference 2.19.047.03

-Inventory of the archives of the Ministerie van Koloniën [Ministery for the Colonies] 1900-1963, file reference 2.10.036.011

Inventory of the archives of the Ministerie van Koloniën [Ministery for the Colonies] 1900-1963, Archive 'Bannier'

-Inventory of the archives of the Ministerie van Koloniën [Ministery for the Colonies] 1900-1963, file reference 2.10.036.15, section 'Indisch'

-Inventory of the archives of the Ministerie van Koloniën [Ministery for the Colonies] 1900-1963, file reference 2.10.036.111, index/register on the newspapers of the Dutch East Indies and the Netherlands

-Inventory of the archives of the manuscript collection of the Koninklijk Instituut van Ingenieurs [Royal Institute of Engineers], file reference 2.22.02 (79 drawings regarding the construction of the bridge near Tjimanoek.1911)

Internationaal Instituut voor Sociale Geschiedenis [International Institute of Social History] (IISG, Eng. IISH)

Nederlandsch Economisch Historisch Archief te Amsterdam [Netherlands Institute for Economic and Business History] (NEHA)

Archive of the Hollandse Constructie Werkplaatsen te Leiden [Dutch Construction Yards in Leiden]

Koninklijk Instituut voor Taal- Land- en Volkenkunde [Royal Netherlands Institute of Southeast Asian and Caribbean Studies] (KITLV)
Koninklijk Instituut voor de Tropen [Royal Tropical Institute] (KIT)
Museum Bronbeek [Royal Home for Retired Servicemen. Museum relating to the colonial history of the Netherlands, particularly concerning military involvement in the former Dutch East Indies]
Geniemuseum (Van Brederodekazerne te Vught) [Museum on the history of the Dutch Royal Military Engineering Corps]

CHAPTER 5 **NODES IN THE MARITIME NETWORK**

A.P. van den Aardweg, N. Brugmans and N. Japikse, *Persoonlijkheden in het koninkrijk der Nederlanden in woord en beeld: Nederlanders en hun werk* [Personalities within the Dutch kingdom in word and image: Dutch people and their work] (Amsterdam 1938)
S. Abeyasekere, *Jakarta, a history* (Oxford 1987)
H.E. van Berckel, 'De zeehaven voor Batavia te Tandjong Priok' [The port of Batavia at Tandjong Priok], in: *Gedenkboek uitgegeven ter gelegenheid van het vijftigjarig bestaan van het Koninklijk Instituut van Ingenenieurs 1847-1897* [Commemorative book published on the occasion of the fifty-year jubilee of the Dutch Royal Institute of Engineers] (The Hague 1897) 305-307
N.P. van den Berg, *Munt-, crediet- en bankwezen, handel en scheepvaart in Nederlandsch-Indië: Historisch-statistische bijdragen* [Money, credit and bank affairs, trade and shipping in the Dutch East Indies: Historical-statistical contributions] (The Hague 1907)
J.H. Bird, *Seaports and seaport terminals* (London 1971)
M.G. de Boer, *Zeehaven en kolenstation Sabang 1899-1924* [Sabang sea port and coaling station 1899-1924] (Amsterdam 1924)
M.J. van Bosse, 'J.A. de Gelder in het tijdperk van den bouw der haven van Tandjong Priok. Een herinneringswoord' [J.A. de Gelder in the era when the harbour at Tandjong Priok was constructed. A commemorative word], in: *De Ingenieur* 27 (1912) 39, 786-789
'Brantas-delta-werken' [Brantas delta works], in: *Encyclopaedia van Nederlandsch-Indië* [Dutch East Indian encyclopedia], part 1 (The Hague/Leiden 1917, 2nd edition) 407-408
P. Bruining, 'De reede van Batavia' [The Batavia roadstead], in: *Verhandelingen en berigten betrekkelijk het zeewezen en de zeevaartkunde* 12 (1852) 201-217
P. Bruining, 'De reede van Batavia, II' [The Batavia roadstead, II], in: *Verhandelingen en berigten betrekkelijk het zeewezen en de zeevaartkunde* 13 (1853) 372-387
J.N.F.M. à Campo, *Koninklijke Paketvaart Maatschappij. Stoomvaart en staatsvorming in de Indonesische archipel 1888-1942* [The Royal Packet Shipping Company. Steam shipping and state formation in the Indonesian archipelago 1888-1914] (Hilversum 1992)
A. Claver, *Commerce and capital in colonial Java. Trade finance and commercial relations between the Europeans and Chinese, 1820s – 1942* (Amsterdam, 2006)
F. Colombijn, 'Uiteenlopende spoorrails. De verschillende ideeën over spoorwegaanleg en ontginning van het Umbilin-kolenveld in West Sumatra 1868-1891' [Divergent railway lines. The different views about railway construction and the opening up of the Ombilin coal mine in West Sumatra], in: *Bijdragen en Mededelingen betreffende de Geschiedenis der Nederlanden* 107 (1992) 437-458
W. Cool, *Technische lessen en vraagstukken op het gebied van de Indische havenbouw* [Technical lessons and questions in the area of East Indian harbour management] (Batavia 1916)
W. Cool, 'Belawan oceaanhaven' [Belawan ocean port], in: *De Waterstaats-Ingenieur* 6 (1918) 3, 101-131
W. Cool, *Nederlandsch-Indische havens* [Dutch East Indian harbours] (Batavia 1920)
H.W. Dick, 'Indonesian economic history inside out', in: *Review of Indonesian and Malaysian Affairs* 27 (1993) 1-12
H.W. Dick, 'The emergence of a national economy, 1808-1990's', in: J.T. Lindblad (ed.), *Historical foundations of a national economy in Indonesia, 1890s-1990s* (Amsterdam 1996) 21-51
H.W. Dick, *Soerabaja, city of work. A socio-economic history* (Athens 2002)
Division of Commerce of the Department of Agriculture, Industry and Commerce, *1930 Handbook of*

the Netherlands East-Indies (Buitenzorg 1930)
S. Dobbs, *An ecological history of the Singapore river. With particular reference to the lighterage industry* (Perth 1999)
J.A.A. van Doorn, *De laatste eeuw van Indië. Ontwikkeling en ondergang van een koloniaal project* [The last century of the East Indies. The rise and fall of a colonial project] (Amsterdam 1994)
Eene zeehaven voor Batavia, verzameling van de officiële bescheiden uitgegeven door de Kamer van Koophandel te Batavia [A sea port for Batavia, collection of official documents published by the Chamber of Commerce in Batavia]. (Batavia 1872)
Encyclopaedie van Nederlandsch-Indië [Dutch East Indian encyclopedia], eight parts (The Hague/Leiden 1917-1939, 2nd edition)
C. Fasseur, *Kultuurstelsel en koloniale baten. De Nederlandse exploitatie van Java 1840-1860* [Cultivation system and colonial profits. The Dutch exploitation of Java 1840-1860] (Leiden 1978, 2nd edition)
L.A. van Gasteren, H.J. Moeshart, H.C. Toussaint and P.A. de Wilde (eds.), *In een Japanse stroomversnelling: Berichten van Nederlandse watermannen – rijswerkers, ingenieurs, werkbazen – 1872-1903* [At a Japanese, accelerated pace: Reports from Dutch water men – osier workers, engineers, employers – 1872-1903] (Zutphen 2000)
J.A. de Gelder, 'Henri Emanuel van Berckel', in: *De Ingenieur* 17 (1902) 17, 297-301
F.M.M. de Goey, 'Water- en luchttransport in Nederland, 1814-1990' [Water and air transport in the Netherlands, 1814-1990], in: *Zee- binnen- en luchtvaart en hulpbedrijven van het vervoer. Een geschiedenis en bronnenoverzicht* (Amsterdam 1993) 5-59
Handleiding havenbeheer [Guide to harbour management] (Batavia 1919)
'Havenwerken', in: *Encyclopaedia van Nederlandsch-Indië*, Dutch East Indian encyclopedia, part 2 (The Hague/Leiden 1918, 2nd edition) 70-73
'De havenwerken in Nederlandsch-Indië' [The harbour works in the Dutch East Indies], in: *Tijdschrift van Nederlandsch-Indië* 23 (1861) 2, 208
J. Homan van der Heide, *Economische studiën en critieken met betrekking tot Java* [Economic studies and criticism relating to Java] (The Hague 1901)
W.G. Huff, *The economic growth of Singapore: Trade and development in the twentieth century* (Cambridge 1994)
J. Huijts, and S. Tils, 'De veranderende scheepvaart tussen 1870 en 1930 en de gevolgen voor de handel op Nederlands-Indië' [Changes in shipping between 1870 and 1930 and the consequences for trade in the Dutch East Indies], in: E.J. Fischer (ed.), *Katoen voor Indië: Sociale ondernemers op het spoor naar vooruitgang 1815-1940* [Cotton for the Indies: Social entrepreneurs on the road to progress 1815-1940] (Amsterdam 1994)
A.H. Kidwai, 'Conceptual and methodological issues: Ports, port cities and port-hinterlands', in: I. Banga (ed.), *Ports and their hinterlands in India, 1700-1950* (New Delhi 1992) 7-43
J. Kloppenburg, *De haven van Batavia* [The Batavia harbour] (Batavia 1866)
G.J. Knaap, *Transport 1819-1940. Changing Economy of Indonesia. A selection of statistical source material from the early 19th century up to 1940*, volume 9 (Amsterdam 1989)
G.J. Knaap, *Shallow water, rising tide. Shipping and trade in Java around 1775* (Leiden 1996)
J.A. Kok, *De scheepvaartbescherming in Nederland en Nederlandsch-Indië* [Shipping protection in the Netherlands and the Dutch East Indies] (Leiden 1931)
J. Kraus, 'Mededeelingen over eenige Nederlandsch-Indische havens. Voordracht gehouden in de Vergadering van het Koninklijk Instituut van Ingenieurs van 20 December 1910' [Information on some of the Dutch East Indian harbours. Presentation given during the meeting held by the Royal Institute of Engineers], copied from *De Ingenieur* 7 (1911) (The Hague)
J. Kraus and G.J. de Jongh, *Verslag over verbetering van de haventoestanden van Makassar* [Report on the improving of the harbour situation in Makassar] (Batavia 1910)
J. Kraus and G.J. de Jongh, *Verslag over verbetering van de havenstoestanden van Soerabaia* [Report on the improving of the harbour situation in Soerabaja] (Batavia 1910)
J. Kraus and G.J. de Jongh, *Verslag over verbetering van de havenstoestanden van Tandjong-Priok, Semarang en andere havens* [Report on the improving of the harbour situation in Tandjong Priok, Semarang and other harbours] (Batavia 1910)
W.F.J.M. Krul, 'Wouter Cool', in: *Jaarboek van de Maatschappij der Nederlandse Letterkunde te Leiden*

[Yearbook of the Dutch Literature Society in Leiden] (Leiden 1947-1949) 94-101

W.F. Lichtenauer, 'Jongh, Gerrit Johannes de (1845-1917)', *Biografisch woordenboek van Nederland* [Biographical dictionary for the Netherlands], part 1 (Den Haag 1979)

J.T. Lindblad, 'Between Singapore and Batavia: the Outer Islands in the Southeast Asian economy in the nineteenth century', in: C.A. Davids, W. Fritschy and L.A. van der Valk (eds.), *Kapitaal, ondernemerschap en beleid. Studies over economie en politiek in Nederland, Europa en Azië van 1500 tot heden. Afscheidsbundel voor prof. dr. P.W. Klein* [Capital, entrepreneurship and policy. Studies on economy and politics in the Netherlands, Europe and Asia from 1500 to the present. Valedictory volume for Professor P.W. Klein] (Amsterdam 1996) 529-548

J.T. Lindblad (ed.), *Historical foundations of a national economy in Indonesia, 1890s-1990s* (Amsterdam 1996)

Maandverslagen omtrent de werkzaamheden aan de bouw van de havenwerken van Batavia [Monthly reports about work on the construction of the harbour works in Batavia] (Batavia 1877-1885)

H. Meijer, *De stad en de haven. Stedebouw als culturele opgave in London, Barcelona, New York en Rotterdam: Veranderende relaties tussen stedelijke openbare ruimte en grootschalige infrastructuur* [The city and the port. Town planning as a cultural task in London, Barcelona, New York and Rotterdam: Changing relations between urban public space and large-scale infrastructure] (Rotterdam 1996)

T.J. Mesman and C.J. Vos, *The history of caisson construction within the Hollandsche Beton Groep nv from 1902 to 1977* (Rijswijk 1977)

J.E. de Meyier, *Aanleg van eene zeehaven te Tandjong Priok* [The construction of the seaport at Tandjong Priok] (The Hague 1893)

The ports of Tandjung Priok, Surabaya and Belawan. Port Survey Team of the UN Economic Commission for Asia (1967)

W. van Raders, 'Historische schets bij de aanbieding aan het Koninklijk Instituut van Ingenieurs van eenige bescheiden betreffende eene voorgestelde haven voor Batavia' [An historical overview on the occasion of the presentation of a number of plans concerning the proposed Batavia harbour], in: *Tijdschrift van het Koninklijk Instituut van Ingenieurs* 6 (1874-1875) 120-134

Rapport der kommissie ingesteld bij gouvernementsbesluit van den 9den Januarij 1873, no. 28 met het doel om de kwestie omtrent de geschikste plaats waar eene zeehaven voor Batavia kan worden daargesteld, te beoordelen [The report of the committee created by government decree on 9th January 1873, no. 28, for the purpose of finding a suitable location for the new Batavia harbour] (Batavia 1874)

W. Ravesteijn, *De zegenrijke heeren der wateren. Irrigatie en staat op Java, 1832-1942* [The auspicious lords of the waters. Irrigation and state in Java, 1832-1942] (Delft 1997)

S.A. Reitsma, *De verkeersbedrijven van den staat (spoorwegen, post, telegraaf en telefoondienst, havenwezen)* [Government communication companies (railways, postal, telegraph and telephone services, harbours)] (Weltevreden 1924)

S.T. Sulistiyono: *The Java Sea network: Patterns in the development of interregional shipping and trade in the process of national economic integration in Indonesia, 1870-1970s* (Leiden 2003)

G. Teitler, *Anatomie van de Indische defensie. Scenario's beleid, plannen 1892-1920* [Anatomy of the East Indian defence system. Policy scenarios, plans 1892-1920] (Leiden 1988)

The ports of Tandjung Priok, Surabaya and Belawan. Port survey team of the UN Economic Commission for Asia (1967)

J. Touwen, *Extremes in the archipelago. Trade and economic development in the Outer Islands of Indonesia, 1900-1942* (Leiden 1997)

J. Touwen, *Shipping and trade in the Java Sea region, 1870-1940. A collection of statistics of the major ports* (Leiden 2001)

E. de Waal, *Onze Indische financiën. Nieuwe reeks aanteekeningen* [Our East Indian finances. New series of memos] (The Hague 1879-1880)

'Werven en dokken' [Wharfs and docks], in: *Encyclopaedia van Nederlandsch-Indië* [Dutch East Indian encyclopedia], part 4 (The Hague/Leiden 2nd edition) 759-761

Archives

National Archive [Nationaal Archief] (NA), The Hague
-Collection Ministerie van Koloniеën [Ministery for the Colonies] (Miko) 1850-1900
-Collection Stoomvaart Maatschappij Nederland [Dutch Steam Shipping Company] (SMN)
-Collection Koninklijke Paketvaart Maatschappij [Royal Netherlands Packet Shipping Company]
-Collection Nederlandsch-Indische Spoorweg Maatschappij [Netherlands East Indian Railway Company]
-Collection Droogdok Maatschappij Tandjong Priok [Dry Dock Company Tandjong Priok]
-Collection N.P. van den Berg
-Collection Wouter Cool

Archive of the Koninklijk Huis [Dutch Royal House] (KHA), The Hague, Collection HRH Prins Hendrik (1820-1879)

National Archive of the Republic Indonesia, Jakarta, Indonesia, Collection Algemene Secretarie [General Secretary], Collection Binnenlandsch Bestuur [Inland Administration]

Arsip Departemen Pekerjaan Umum [Archive of the Department of Public Works], Citeureup, Indonesia, Collection Burgerlijke Openbare Werken [Civil Public Works], Afdeling H

CHAPTER 6 **LIVING WATER**

S.H.A. Begemann, 'Toegepaste waarschijnlijkheidsleer op hydrologische waarnemingen' [Applied probability theory in hydraulic observations], offprint from *De Waterstaats-Ingenieur* 19 (1931) 1-3

W.J. Betz, 'Capaciteitsvermindering van de toevoerleiding van het waterkrachtwerk Kratjak, ten gevolge van biologische aangroeiing' [Capacity reduction of the supply channel of the Kratjak hydropower plant due to biological growth], in: *De Ingenieur in Indonesië* 1 (1948) 1, I.27-I.28

W.J. van Blommestein, 'Een federaal welvaartsplan voor het westelijk gedeelte van Java' [A federal welfare plan for the western part of Java], in: *De Ingenieur in Indonesië* 1 (1949) 4, I.50-I.53, 5, I.61-I.82

W.J. van Blommestein, 'A development project for the islands of Java and Madura (Indonesia)' (Voorburg 1979, Jakarta 1981)

M.W. Ertsen, 'From trial and error to science. The development and application of hydrological knowledge on Java', in: *Colloquium History of hydrology (Dyon 2001)*

M.W. Ertsen, 'Irrigation traditions, roots of modern irrigation knowledge', in: International Journal of Technology, Policy and Management 2 (2002) 4, 387-406

M.W. Ertsen, 'Experimental learning in the Netherlands East Indies', in: Seminar 'Lessons from failures in irrigation, drainage, and flood control systems', 18th International Congress on Irrigation and Drainage (Montreal 2002) 15-25

M.W. Ertsen, 'Time and scale related perspectives on failures in water engineering', in: *Seminar 'Lessons from Failures in Irrigation, Drainage, and Flood Control Systems', 18th International Congress on Irrigation and Drainage* (Montreal 2002) 1-13

M.W. Ertsen, Water distribution on Java: A revision of a traditional perspective, in: *Indonesian Environmental History Newsletter* (1998) 11, 17-18

M.W. Ertsen, *Prescribing perfection: Emergence of an engineering irrigation design approach in the Netherlands East Indies and its legacy 1830-1990* (Rotterdam 2005)

G.S. Goemans, 'Een merkwaardig knikgeval' [An extraordinary case of buckling], in: *De Ingenieur in Indonesië* 6 (1954) 3

G.S. Goemans, 'Waterkrachtcentrale Sengguruh. De plaats van de centrale Sengguruh in het Oost-Java energieopwekkingsysteem' [The Sengguruh hydroelectric works. The part played by the Sengguruh power station within the East Javan energy generation system], in: *De Ingenieur in Indonesië* 6 (1954) 1, III.17-III.36

J.F. Graadt van Roggen, 'Plant- en waterregelingen in de provinciale waterstaatsafdeling "Pemali-Tjomal"' [Plant and water distribution rulings in the provincial water department of "Pemali-

Tjomal"], in: *De Ingenieur in Nederlandsch-Indië* 2 (1935) 4, VI.47-VI.56

A. Groothoff, 'Eenige mededeelingen over de waterkrachtindustrie in Scandinavië en over het waterkrachtvraagstuk in Nederlandsch-Indië' [Information on the waterpower industry in Scandinavia and waterpower issues in the Dutch East Indies], in: *De Ingenieur* 33 (1918) 2, 18-33 (including the reactions of C.W. Weijs and H.J.E. Wenkebach).

A. Groothoff, 'Verrichtingen en verwachtingen op waterkrachtgebied in Ned.-Indië' [Activities and expectations in the field of hydroelectric power in the Dutch East Indies], in: *De Ingenieur* 36 (1921) 48, 945-953

A. Groothoff, 'Een en ander over de ontwikkeling van waterkracht- en electriciteitsbedrijven voor de energievoorziening van belangrijke gebieden in Ned.-Indië' [Information on the development of waterpower and electricity plants for energy supplies for important areas in the Dutch East Indies], in: *De Ingenieur* 37 (1922) 39, 774-778

P. de Gruyter, 'Beschouwingen over aftapsluizen en meetinrichtingen voor bevloeiingswerken' [Reflections on tap sluices and measuring methods for irrigation works], in: *De Waterstaats-Ingenieur* 13 (1925) 2, 17-37, 13 (1925) 3, 53-74

P. de Gruyter, 'Een nieuwe aftap- tevens meetsluis en de resultaten van een proef met een dergelijk kunstwerk' [A new drawing-off and measuring weir and the results of a test with such a hydraulic structure], in: *De Waterstaats-Ingenieur* 14 (1926) 12, 391-408, 15 (1927) 1, 1-15, 15 (1927) 2, 25-34

C.F. van Haeften, 'Drukbuizen van gewapend beton' [Pressure pipes made of reinforced concrete], in: *De Waterstaats-Ingenieur* 10 (1922) 7, 238-241

P.L.E. Happé, 'Eenige beschouwingen over bevloeiingswerken op Java en Madoera' [Some reflections on irrigation works in Java and Madoera], in: *De Ingenieur in Nederlandsch-Indië* 6 (1939) 1, II.13-II.26, 2, II.33-II.42

F. van Iterson, *De Asahan-waterkracht en de Billiton maatschappij* [The Asahan waterpower plant and the Billiton company] (Batavia 1939)

'Irrigation management', Water users training project (Indonesia 1984)

F.A. Janssen van Raay, 'Eenige medeelingen over den bouw van grootere waterkrachtwerken door den Dienst voor Waterkracht en Electriciteit in Nederlandsch-Indië' [Information on the construction of larger waterpower plants built by the Service for Waterpower and Electricity in the Dutch East Indies], in: *De Ingenieur* 37 (1922) 35, 670-680

F.A. Janssen van Raay, 'Jaarverslag van den Dienst van WK en El. in NI over 1923' [Annual report of the Service for Waterpower and Electricity in the Dutch East Indies for 1923], in: *De Ingenieur* 40 (1925) 8, 159-160

F.H. van Kooten, *Eenige empirische methoden tot het berekenen van den maximum afvoer eener rivier uit de grootte van den regenval* [Various empirical methods for calculating the maximum discharge of a river according to the level of precipitation] (Amsterdam 1927)

C.R. Kras, 'Het berekenen van den maximum afvoer van stroomgebieden met een oppervlakte van 0-100 km^2' [The calculation of the maximum discharge of basins with a surface of 0-100 km^2], in: *De Ingenieur in Nederlandsch-Indië* 7 (1940) 4, II.46-II.49

A.G. Lamminga, 'Pekalen-irrigatiewerken in de afdeling Kraksaän der residentie Pasoeroean' [The Pekalen irrigation works in the Kraksaän division situated in the residence of Pasoeroean] in: *De Ingenieur* 20 (1905) 46, 758-766, 20 (1905) 47, 788-793

A.G. Lamminga, *Beschouwingen over den tegenwoordigen stand van het irrigatiewezen in Nederlandsch-Indië* [Reflections on the current state of irrigation in the Dutch East Indies] (The Hague 1910)

R. Lauterburg, 'Anleitung zur Berechnung der Quellen und Stromabflussmessungen aus der Regenmenge, Grösze und Beschaffenheit der Quellen- und Flussgebiete', in: *Allgemeine Bauzeitung* (1887)

J.N. van der Ley, 'Het nieuwe onderstation Badra van den Dienst voor Waterkracht en Electriciteit en de electriciteitsvoorziening van de Bandoengsche hoogvlakte' [The new Badra sub-station of the Waterpower and Electricity Service and the electricity supplies in the Bandoeng plain], in: *De Ingenieur in Nederlandsch-Indië* 1 (1934) 2, III.17- III.22

Th.D. van Maanen, *Irrigatie in Nederlandsch-Indië. Een handleiding bij het ontwerpen van irrigatiewerken ten dienste van studeerenden en practici* [Irrigation in the Dutch East Indies. An irrigation works design manual for students and practitioners] (Batavia 1931, second impression)

'Meetinrichtingen in de irrigatieafdeling Pemali-Tjomal' [Measuring devices in the Pemali-Tjomal irrigation division] (±1930), National Archive, Haringhuizen/Schoemaker Collection (list number 2.22.07), inv. no. 53

A.P. Melchior, 'De toepassing van de formules van Lauterburg voor de bepaling van den grootsten afvoer van de rivieren op Java' [Applying Lauterburg's formulae to determine maximum river discharges in Java], in: *Tijdschrift van het Koninklijk Instituut van Ingenieurs, afdeling Nederlands-Indië* 1895-1896, 16-58

W.A. van der Meulen, 'Irrigation in the Netherlands Indies', in: *Bulletin of the Colonial Institute of Amsterdam (1940) 3, 142-159*

J.E. de Meyier, *'Bevloeiingen'*, in: N.H. Henket, Ch.M. Schols and J.M. Telders (eds.), *Waterbouwkunde* [Hydraulic engineering], First part, section VII (The Hague 1891)

J.E. de Meyier, 'De legende der Solovallei-werken' [The legend of the Solo Valley works], in: *De Indische Gids* 32 (1910) I, 822-823

A.N. Nijman, 'Bepaling van de maximale afvoer van rivieren volgens Melchior' [Determining the maximum discharge of rivers according to Melchior], in: *De Waterstaats-Ingenieur* 21 (1933) 12, 325-327

W. Ravesteijn, *De zegenrijke heeren der wateren. Irrigatie en staat op Java, 1832-1942* [The auspicious lords of the waters: Irrigation and state in Java, 1832-1942] (Delft 1997)

W. Ravesteijn, 'Willem Johan van Blommestein', in: J. Bosmans et al. (eds.), *Biografisch woordenboek van Nederland* [Dutch biographical dictionary], part five (The Hague 2001) 43-45

W. Ravesteijn 'Controlling water, controlling people: Irrigation engineering and state formation in the Dutch East Indies', in: *Itinerario* 31 (2007) 1, 89-118

J.T. Rietveld, 'De Sampeanstuw honderd jaar' [One hundred years of the Sampean weir], in: *De Waterstaats-Ingenieur* 20 (1932) 9, 277-287

B.W.P. Roessel, 'Maximum afvoer en hydrologische empirie' [Maximum discharge and hydrological empirical data], in: *De Ingenieur in Nederlandsch-Indië* 7 (1940) 7, II.104-II.130

D.G. Romijn, 'Een regelbare meetoverlaat als tertiaire aftapsluis' [An adjustable measuring weir as a tertiary outlet], in: *De Waterstaats-Ingenieur* 20 (1932) 9, 287-292

R.A. van Sandick, 'De Dienst voor Waterkracht en Electriciteit' [The Waterpower and Electricity Service], in: *De Ingenieur* 42 (1927) 8, 120-127

H.M. Verweij, 'Het berekenen van den maximum-afvoer van stroomgebieden met een oppervlak van 0-100 km²' [The calculation of the maximum discharge of basins with a surface of 0-100 km²], in: *De Ingenieur in Nederlandsch-Indië* 6 (1939) 4, II.99-II.104

H. Vlugter, 'Honderd jaar irrigatie' [A hundred years of irrigation], in: *De Ingenieur in Indonesië* 1 (1949) 7, I.99-I.105

C.G.J. Vreedenburgh, 'Opmerkingen over het ontwerpen van een groot waterkrachtwerk in het bijzonder in Nederlandsch-Indië' [Comments on the design of large hydropower works, notably in the Dutch East Indies], in: *De Ingenieur* 45 (1930) 46, 297-305

J.J. de Vries, *Anderhalve eeuw hydrologisch onderzoek in Nederland* [One and a half centuries of hydrological research in the Netherlands] (Amsterdam 1982)

J.D. der Weduwen, 'Het berekenen van den maximum afvoer van stroomgebieden met een oppervlak van 0-100 km² [The calculation of the maximum discharge of basins with a surface of 0-100 km²], in: *De Ingenieur in Nederlands-Indië* 4 (1937) 10, II.135-II.156

Archive

National Archive, The Hague, Collection Haringhuizen/Schoemaker.

CHAPTER 7 **FOR KOTA AND KAMPONG**

H. Akihary, *Architectuur & stedebouw in Indonesië 1870/1970* [Architecture & town planning in Indonesia 1870/1970] (Zutphen 1990)

J.T. Bethe, *De hygiënische zijde van het volkshuisvestingsvraagstuk* [The hygenic aspect of public housing] (Semarang 1922).

'Bouwbeperkingencommissie' [Committee for building restrictions], in: *Locale Techniek* 1/2 (1932) 41
H. van Breen, *Overzicht der voorgestelde werken tot verbetering van den wateraf- en aanvoer ter hoofdplaats Batavia* [Overview of the works proposed for the improvement of the drainage and supply of water in the capital Batavia] (1913)
B. Brommer et al., *Semarang. Beeld van een stad* [Semarang. Image of a town] (Purmerend 1995)
H.J. Bussemaker, 'Hoe dient de toekomstige gemeentelijke organisatie te worden na de eventuele opheffing van het desa-verband binnen de gemeentegrens?' [What should the future organisation of the municipality be like after the abolishment of desa-coherence within the municipal boundaries], *Mededeeling Locale Belangen* 72 (1928) 1-36
Bijblad op het *Staatsblad van Nederlandsch-Indië* [Supplement to the Law Gazette for the Dutch East Indies] 6802 (1908)
Bijblad op het *Staatsblad van Nederlandsch-Indië* [Supplement to the Law Gazette for the Dutch East Indies] 11272 (1926)
Commissie voor de Kampongverbetering, *Eerste verslag van de kampongverbeteringscommissie, ingesteld bij Gouvernementsbesluit van 25 mei 1938, no.3* [First report of the Committee on kampong improvement, established by Government Decree on May 25, 1938, no. 3] (Batavia 1939)
J.A.A. van Doorn, *De laatste eeuw van Indië. Ontwikkeling en ondergang van een koloniaal project* [The last century of the East Indies. The rise and fall of a colonial project] (Amsterdam 1994)
P.J. Drooglever, 'Koloniaal beleid en Indische samenleving tot 1942' [Colonial policy and Indian society until 1942], in: P.J. Drooglever, *Indisch Intermezzo. Geschiedenis van de Nederlanders in Indonesië* [Indian interlude. History of the Dutch in Indonesia] (Amsterdam 1991) 19-32
'Eenige kwesties, welke zich na de instelling der gemeente over den afstand van domeingronden aan die gemeenschappen hebben voorgedaan' [Certain issues which arose after municipalities were established regarding the renouncing of national land to those same communities], *Mededeeling Locale Belangen* 66 (1927) 9
G. Flieringa, *De zorg voor de volkshuisvesting in de stadsgemeenten in Nederlands Oost-Indië in het bijzonder Semarang* [The concern for public housing in municipalities in the Netherlands East-Indies, particularly in Semarang] ('s-Gravenhage 1930)
F.C. Frumeau, 'De kosten der stadsvorming'[The cost of town planning], *Mededeeling Locale Belangen* 96 (1933) 13-21
F.C. Frumau, *De stedebouwkundige zijde van het volkshuisvestingsvraagstuk* [The town planning side of public housing] (Semarang 1922)
M.J. Granpré Molière, 'Indiese stedebouw door Ir. Th. Karsten' [Indian town planning by engineer T. Karsten], in: *Tijdschrift voor Volkshuisvesting* 9 (1922) 226-234
H. Heetjans, *De sociaal-politieke zijde van het volkshuisvestingsvraagstuk* [The social-political side of public housing] (Semarang 1922)
Ebenezer Howard, *Garden cities of to-morrow* (London 1902,)
D. de Iongh, 'Het woningvraagstuk' [The housing issue], *Mededeeling Locale Belangen* 26 (1918) 1-26
G. Jansen, 'De decentralisatie en de grond' [Decentralisation and the soil] in: F.W.M. Kerchman, *25 Jaren decentralisatie in Nederlands-Indië 1905-1930* [25 Years of decentralisation in the Dutch East Indies 1905-1930] (Weltevreden 1930) 148-158
Gerhard Jobst, *De stedebouwkundige zijde van het volkshuisvestingsvraagstuk* [The town planning side of public housing] (Semarang 1922)
J.J.P. de Jong, *De Waaier van het fortuin. Van handelscompagnie tot koloniaal imperium. De Nederlandsers in Azië en de Indonesische archipel 1595-1950* [The fan of fortune. From trading company to colonial empire. The Dutch in Asia and the Indonesian archipelago 1595-1950] (Den Haag 1998)
K., 'Indische stedenbouw' [Indian town planning], in: *P.E.B.* 42 (1922) 137-139
H.Th. Karsten, 'Indiese stedebouw' [Indian town planning], *Mededeeling Locale Belangen* 40 (1920)
H.Th. Karsten, *De woning. Technisch-architektonische zijde van het volkshuisvestingsvraagstuk* [The house. The technical-architectonic aspect of public housing] (Semarang 1922)
H.Th.Karsten, 'Volkshuisvesting I' [Public housing I], in: F.W.M. Kerchman, *25 Jaren decentralisatie in Nederlandsch-Indië 1905-1930* [25 Years of decentralisation in the Dutch East Indies 1905-1930] (Semarang/Weltevreden 1930) 159-161
F.W.M. Kerchman, *25 Jaren decentralisatie in Nederlands-Indië 1905-1930* [25 Years of decentralisation in the Dutch East Indies 1905-1930] (Semarang/Weltevreden 1930)

Coert Peter Krabbe, *Ambacht, Kunst, Wetenschap. Bevordering van de bouwkunst in Nederland (1775-1880)* [Craft, Art, Science. The promotion of architecture in the Netherlands (1775-1880] (Zwolle 1998)

J.J. van Lonkhuijzen, *De hygiënische zijde van het volkshuisvestingsvraagstuk* [The hygiene aspect of public housing] (Semarang 1922)

H.W. Nachenius, *Bijdrage tot de kennis van den stedenbouw* [Contribution to knowledge on town planning] (Haarlem 1880)

Nicole Niessen, *Municipal government in Indonesia. Policy, law, and the practice of decentralization and urban spatial planning* (Leiden 1999)

Thomas Nix, *Bijdrage tot de vormleer van de stedebouw in het bijzonder voor Indonesië* [Contribution to the morphology of town planning, particularly in Indonesia] Heemstede 1949)

A. Poldervaart, 'De kosten der stadsvorming' [The cost of town planning], *Mededeeling Locale Belangen* 96 (1933) 1-11

Regeeringsalmanak voor Nederlands-Indië en supplement [Government almanac and supplement for the Dutch East Indies] (Weltevreden 1931)

Regeeringsalmanak voor Nederlands-Indië en supplement [Government almanac and supplement for the Dutch East Indies] (Weltevreden 1932)

Regeeringsalmanak voor Nederlands-Indië en supplement [Government almanac for the Dutch East Indies and supplement] (Weltevreden 1935)

P. van Roosmalen, *De stedebouwkundige ontwikkeling van Bandoeng tussen 1906 en 1949* [The town planning development of Bandoeng between 1906 and 1949] (Amsterdam 1992)

J.J.G.E. Rückert, *Leidraad bij de uitvoering van kampongverbetering* [Guidelines for the implementation of kampong improvement] (1928)

S., 'Architectuur van openbare gebouwen' [Architecture of public buildings], in: *Indisch Bouwkundig Tijdschrift* 1 (1912) 2-3

S., 'Beschouwingen over bouwkunst I' [Contemplations on architecture I], in: *Indisch Bouwkundig Tijdschrift* 7 (1912) 118-120

S., 'Beschouwingen over bouwkunst II' [Contemplations on architecture II], in: *Indisch Bouwkundig Tijdschrift* 7 (1912) 150-151

'Samenstelling Planologische Studiegroep' [Composition of the Planological Study Group], *Locale Techniek/Indisch Bouwkundig Tijdschrift* 3 (1939) 84.

J.J. Schrieke, *Ontstaan en groei der stads- en landsgemeenten in Nederlands-Indië* [Origin and development of urban and rural councils in the Dutch East Indies] (Amsterdam 1918)

Camillo Sitte, *Der Städtebau nach seinen Künstlerischen Grundsätzen* [The artistic foundations of town planning] (Braunschweig 1983, herdruk)

R. Slamet, 'Hoe dient de toekomstige organisatie te worden na de eventuele opheffing van het desa-verband binnen de gemeentegrenzen?' [What should a future organisation of the municipality be like after the abolishment of desa-coherence within the municipal boundaries], *Mededeeling Locale Belangen* 73 (1929) 1-58

S. Snuyff, 'Over bouwkunst en bouwkunde in Indië' [On architecture and building in East India], in: *Indisch Bouwkundig Tijdschrift 8 (1908) 131*

M. Soesilo, 'De grondrechten en de gemeenschap' [Land rights and the community], in: *Kritiek en Opbouw* 8 (1939) 119-122

M. Soesilo, 'De grondrechten en de gemeenschap' [Land rights and the community], in: *Kritiek en Opbouw* 9 (1939) 132-134

M. Soesilo, 'De grondrechten en de gemeenschap' [Land rights and the community], in: *Kritiek en Opbouw* 10 (1939) 149-152

Staatsblad van Nederlandsch-Indië [Law Gazette for the Dutch East Indies] 168 (1948)

Staatsblad van Indonesië [Law Gazette of Indonesia] 250 (1948)

Staatsblad van Indonesië [Law Gazette of Indonesia] 331 (1948)

Staatsblad van Indonesië [Law Gazette of Indonesia] 241 (1949)

'Stadsvormings-commissie' [Town Planning Committee], in: *Indisch Bouwkundig Tijdschrift/Locale Techniek* 2 (1934) 36

Stadsvormingsordonnantie Stadsgemeenten Java [Town Planning Ordinance Municipalities Java] (Batavia 1938)

Th. Stevens, 'Indo-Europeanen in Nederlands-Indië. Sociale positie en welvaartsontwikkeling' [Indo-Europeans in the Dutch East Indies. Social positions and the development of affluence], in: P.J. Drooglever, *Indisch Intermezzo. Geschiedenis van de Nederlanders in Indonesië* [Indian interlude. History of the Dutch in Indonesia] (Amsterdam 1991) 33-45
H.F. Tillema, *Van wonen en bewonen. Van bouwen, huis en erf* [On living and inhabiting. On building, house and property] (Samarang 1913)
Toelichting op de Stadsvormingsordonnantie Stadsgemeenten Java [Elucidation on the Town Planning Ordinance Municipalities of Java] (Batavia 1938)
'Verslag van den Planologische Dag georganiseerd vanwege de Vereeniging voor Locale Belangen door de Planologische Studiegroep dier Vereeniging te Bandoeng op 1 en 2 juli 1939' [Report of the Planological Day organised in conjunction with the Local Interests Association by the Planological Study Group of the Bandoeng Society on 1st and 2nd July 1939], in: *Technische Mededeeling* 16 (1939) 2-47
Verslag Volkshuisvestingscongres 1922 [Report Public Housing Congres 1922] (Semarang 1922)
Verslag Volkshuisvestingscongres 1925 [Report Public Housing Congres 1925] (Semarang 1925)
'Het Volkshuisvestings-Congres' [The Public Housing Congres], in: *Indisch Bouwkundig Tijdschrift* 6 (1922) 108
C. van Vollenhoven, *De Indonesiër en zijn grond* [The Indonesian and his land] (Leiden 1932)
P.F. Woesthoff, *De Indische decentralisatie-wetgeving* [Indian decentralisation legislation] (Leiden 1915)

Archives

Arsip Nasional Republik Indonesia [National Archive of the Republic of Indonesia] (ANRI) Jakarta
-Inventary of the archive of the Algemene Secretarie en het Kabinet van de Gouverneur-Generaal [the General Secretary and the Cabinet of the Governor-General] 1944-1950, inv. nos. 924, 925
-Archive of the Departement van Binnenlands Bestuur [Department of Inland Administration], inv. nos. 1395, 1683, 1886
-Record of the Arsip Depo Bogor [Archive of the Depot in Bogor] (1891-1942) - BGS 6-3-1924, inv. no. 562
Nationaal Archief [National Archive] (NA), The Hague, Archives of the Ministerie van Koloniën [Ministery for the Colonies], file reference 2.10.036.04, inv. no. 1124
Nederlands Architectuur instituut [Netherlands Architecture Institute] (NAi), Rotterdam, Archive Thijsse (not indexed)
Gemeentearchief Amsterdam [Municipal Archive Amsterdam] (GAA), Amsterdam, Archive Publieke Werken [Public Works] file reference 580, inv. no. 5342.

CHAPTER 8 **WATER IN THE CITY**

Aardrijkskundig en statistisch woordenboek van Nederlandsch Indië [The Dutch East Indian geographical and statistical dictionary], three volumes (Amsterdam 1861-1869)
C. Baumgarten, 'Uit het Laboratorium voor Technische Hygiëne te Bandoeng. Over de gebruiksmogelijkheden van de bij rotting van afvalwaterslib vrijkomende gassen' [From the Laboratory for Technological Hygiene in Bandoeng. On the possible uses for gases released from waste water sludge during the rotting process], in: *De Ingenieur in Nederlandsch-Indië* 4 (1937) VI
W.J. van Blommestein, 'Een federaal welvaartsplan voor het westelijk gedeelte van Java' [A federal welfare plan for the western part of Java], in: *De Ingenieur in Indonesië* 1 (1948/1949) 4, I.50-I.53 and 5, I.61-I.82
P. Boomgaard and A.J. Gooszen, *Population trends 1795-1942. Changing economy in Indonesia. A selection of statistical source material from the early 19th century up to 1940*, volume 11 (Amsterdam 1991)
W.H. Brandenburg, 'De centrale drinkwatervoorziening van de gemeente Probolinggo' [Centralised

drinking water provisions for the town of Probolinggo], in: *De Waterstaats-Ingenieur in Nederlandsch-Indië* 15 (1927) 327-335
H. van Breen, 'Kleine werken ter verbetering van den gezondheidstoestand der hoofdplaats Batavia' [Small works in the interests of improving the state of people's health in the capital of Batavia], in: *De Waterstaats-Ingenieur* 7 (1919) 130 onwards
H. van Breen, 'Verbetering van den waterstaat van de hoofdplaats Batavia' [Improving water management in the capital of Batavia], in: *De Ingenieur* 38 (1923) 25, 483-490; 26, 515-527; 27, 537-549; 28, 562-566
C.J. de Bruijn, *Indische bouwhygiëne* [East Indian building hygienics] (Weltevreden 1927)
Division of Commerce of the Department of Agriculture, Industry and Commerce, *1930 Handbook of the Netherlands East-Indies* (Buitenzorg 1930)
H.W. van den Doel, *Afscheid van Indië; De val van het Nederlandse imperium in Azië* [Good-bye to the East Indies; The fall of the Dutch empire in Asia] (Amsterdam 2000)
Directie Handelsvereeniging 'Amsterdam', *Rede president-directeur op 31december 1928* [Speech made by the chairman of the board on the 31st of December 1928] (Amsterdam 1929)
J.A.A. van Doorn, *De laatste eeuw van Indië. Ontwikkeling en ondergang van een koloniaal project* [The Last century of the East Indies: The rise and fall of a colonial project] (Amsterdam 1996)
Ensiklopedi Indonesia (Jakarta 1980)
'De gezondmaking van Cheribon' [The sanitation of Cheribon], in: *De Waterstaats-Ingenieur* 6 (1918) 469
A. Goedhart, *Het wonder van Deli* [The miracle of Deli] (Alphen aan den Rijn 2002)
S.J. Gompert, 'Waterverbruik door minvermogende inlanders en vreemde oosterlingen bij Indische drinkwatervoorzieningen' [Water consumption for less wealthy natives and foreign easterners within the East Indian drinking water provision system], in: *De Waterstaats-Ingenieur* 4 (1916) 11-17
G.F.E. Gonggryp, *Geïllustreerde encyclopaedie van Nederlandsch-Indië* [Illustrated encyclopedia of the Dutch East Indies] (Wijk en Aalburg 1991)
W.J. van Gorkom, 'Ongezond Batavia' [Unhealthy Batavia], in: *Tijdschrift van het Koninklijk Instituut van Ingenieurs. Afdeeling Nederlandsch-Indië* 1913, 1
J.A. Groen, *Een cent per emmer. Het Amsterdamse drinkwater door de eeuwen heen* [One cent per bucket. Amsterdam drinking water through the ages] (Amsterdam 1990)
W.H. van Helsdingen and H. Hoogenberk (red.), *Daar wèrd wat groots verricht .... Nederlandsch-Indië in de XXste eeuw* [Great things were achieved there... The Dutch East Indies in the XX century] (Amsterdam 1941)
J. Huisman, 'Epidemiologie' [Epidemiology] (Delft)
A.F. Kamp, *De Technische Hogeschool te Delft, 1905-1955* [The Technical College in Delft] ('s-Gravenhage 1955)
J.A. van der Kloes and J.N. van Ruijven, *Het bouwen in overzeesche gewesten, Tweede afdeeling, Onderwerpen uit de Weg- en Waterbouwkunde* [The building in overseas territories, Second division, Subjects from civil engineering] (Leiden 1906)
G.J. Knaap, P.J. Drooglever, Th. Stevens, E. Touwen-Bouwsma, J.J.P. de Jong, R. Voorneman and J.M. van der Hoeven, *Indisch intermezzo; Geschiedenis van de Nederlanders in Indonesië* [East Indian intermezzo; History of the Dutch in Indonesia] (Amsterdam 1994)
Koloniaal Verslag [Colonial Report], appendix to the *Handelingen der Staten-Generaal, Eerste en Tweede Kamer* [Proceedings of the States General, First and Second Chamber]
D.A. Koster, 'Drinkwatervoorzieningen in tropische landen, in het bijzonder in Nederlandsch-Indië' [Drinking water provision in tropical countries and in particular in the Dutch East Indies] in: *De Ingenieur* 16 (1901) 10, 157-168
C.A.E. van Leeuwen, 'Beginselen in acht te nemen bij het ontwerpen van rioleerings- en afwateringswerken in groote gemeenten' [Principles to be observed when designing sewerage and drainage works for large municipalities], in: *De Waterstaats-Ingenieur* 7 (1919) 12, 571 onwards
C.A.E. van Leeuwen, 'Het rioleeringsvraagstuk in Nederlandsch-Indië' [The sewage question in the Dutch East Indies], in: *De Waterstaats-Ingenieur 8 (1920) 5, 196-212*
C. Lekkerkerker, *Land en volk van Java* [The land and people of Java] (Batavia 1938)
Margono Djojohadikusomo, *Herinneringen aan 3 tijdperken* [Memories of 3 eras] (Amsterdam 1970)

A. Massink, 'De drinkwatervoorziening van Bandoeng' [The drinking water provisions for Bandoeng], in: *De Ingenieur* 49 (1934) B.240-B.241

NEDECO, 'Masterplan for drainage and flood control of Jakarta' (Jakarta/The Hague 1973)

PERPAMSI (Persatuan Perusahaan Air Minum Seluruh Indonesia), 'Direktori 2000' (Jakarta 2000)

H.G. Nieuwenhuis, 'Stadshygiëne door bodem-assaineering' [City hygiene resulting from soil sanitation], in: *De Waterstaats-Ingenieur* 11 (1923) 181-195

Petra University Surabaya, 'The heritage and impact of Dutch architecture and civil engineering in Surabaya and Malang' (Surabaya/Delft 2003)

J.J.G.E. Rückert, 'Klein assaineeringswerk' [Small water management Works], in: *De Waterstaats-Ingenieur* 7 (1919) 378 onwards

R.A. van Sandick, 'De drinkwatervoorziening van Soerabaja' [The drinking water provision for Soerabaja], in: *De Ingenieur* 18 (1903) 48, 825-830

N.D.R. Schaafsma, 'Septic tanks en zinkputten voor de faecaal afvoer in de tropen' Septic tanks and cesspits for faeces conveyance in the tropics], in: *I.B.T. Locale Techniek* III 1934

L.H. Slinkers, 'Het verslag over de burgerlijke openbare werken in Ned.-Indië over het jaar 1902' [The report on civil public works in the Dutch East Indies in the year 1902], in: *De Ingenieur* 19 (1904) 45, 803-807

L.H. Slinkers, 'Uit het verslag over de burgerlijke openbare werken in Ned.-Indië over het jaar 1908' [From the report on civil public works in the Dutch East Indies in the year 1908], in: *De Ingenieur* 26 (1911) 31, 748-750

L.J.A. Tollens, *Warnasarie: Indisch jaarboekje voor 1858* [Warnasarie: East Indian name book] (Batavia 1858)

P.J.L. van Wiechen, 'De drinkwatervoorziening van Malang' [The drinking water provisions for Malang], in: *De Waterstaats-Ingenieur* 5 (1917) 408 onwards

G.W.F. de Vos, *Indische bouw-hygiene* [East Indian building hygienics] (Batavia 1891)

T.J. Willer, *Volkstelling in Nederlandsch Indië* [Census in the Netherlands East Indies] ('s Gravenhage 1861)

C.W. Wormser, F.H. van Naerssen, J.A.M. Bruineman, W.J.A. Kernkamp, K.J. Brouwer, L. van Rijckevorsel, J. de Jong, F.A. Wagner, F.H. van Naerssen, L.J.J. Caron, C.J.J. van Hall, F.P.C.S. van der Ploeg and W.L. Utermark, *Wat Indië ontving en schonk* [What the East Indies received and gave] (Amsterdam 1946)

Journals

De Ingenieur [The Engineer] 1886-2002

De Waterstaats-Ingenieur [The Public Works Engineer] 1913-1926

De Waterstaats-Ingenieur in Nederlandsch-Indië [The Public Works Engineer in the Dutch East Indies] 1927-1933

De Ingenieur in Nederlandsch-Indië [The Engineer in the Dutch East Indies] 1934-1941

De Ingenieur in Indonesië [The Engineer in Indonesia] 1948/1949-1955

H_2O, Tijdschrift voor watervoorziening en waterbeheer [H_2O, Journal for water supply and water management] 1969-2003

Tijdschrift van het Koninklijk Instituut van Ingenieurs. Afdeeling Nederlandsch-Indië [Journal of the Royal Institute of Engineers. The Dutch East Indies division]

CHAPTER 9 **FROM TECHNICAL ASSISTANCE TO GOTONG ROYONG**

P. Ankum, 'Management of irrigation rehabilitation projects. An experience from Indonesia', in: *Asian regional symposium on the modernisation and rehabilitation of irrigation and drainage schemes, the Development Academy of The Philippines 13-15 February 1989* (Wallingford 1988)

S. Arlosoroff, *Community water supply. The handpump option* (Washington 1987)

A. van Balen, *A survey of the drinking water and sanitation policy in the Dutch East Indies 1872-1942* (The Hague 1987)

E. van Beek, 'Planning of integrated water resources development. Development of a national strategy – BTA 155' (Delft 1990)
E. van Beek, 'Cisedane – Cimanuk integrated water resources development – BTA 155' (Delft 1990)
J.W.J. Beek 'Indische spoorwegen na de Japanse bezetting' [East Indian railways after Japanese occupation], in: *Voordrachten voor het Koninklijk Instituut van Ingenieurs* 2 (1950) 573-590
F.M.C. Berkhout, 'Het waterstaatswerk in Indonesië na de oorlog' [Public works activities in Indonesia after the war], in: *Voordrachten voor het Koninklijk Instituut van Ingenieurs* 2 (1950) 317-337
J. Boes, W. Ravesteijn and W. Riedijk (eds), *Appropriate technology in industrialized countries* (Delft 1989)
W.J. van Blommestein, 'Een federaal welvaartsplan voor het westelijk gedeelte van Java' [A federal welfare plan for the western part of Java], in: *De Ingenieur in Indonesië* 1 (1948/1949) 4, I.50-I.53 and 5, I.61-I.82
A. Deeleman and J.H. Kop, 'Water resources development/Water resources management Indonesia' (The Hague 1990/1991)
'Djakarta flood control, Preliminary survey and recommendations' (Djakarta 1970)
Internationale Samenwerking, Monthly magazine on international cooperation published by the Dutch Ministry of Foreign Affairs, The Hague, November 2003
G. de Jager, 'Wadaslintang irrigation and drainage scheme, Central Java, Indonesia' (Jakarta 1989)
'Jakarta drainage and flood control', in: *De Raadgevend-Ingenieur* February 1977
J.M. Kalbermatten, *A planners's guide* (Washington 1980)
J.H. Kop, W.H. van den Toorn, L.J. van Veen and T. van der Zee, 'Irrigation sector mission 1984, main report and annexes' (Jakarta/Wageningen 1984/1985)
B.H.A. van Kreel, 'De Deli Spoorweg Maatschappij' [The Deli Railway Company], in: *Voordrachten voor het Koninklijk Instituut van Ingenieurs* 2 (1950) 591-602
Mconsult, 'Twins, godfathers and sandwiches, evaluation report on the twinning arrangements between Dutch and Indonesian water supply companies' (Utrecht/Woerden 1991)
Ministerie van Buitenlandse Zaken, *Projektlijst bilaterale projekten in het kader van de Nederlandse financiële en technische samenwerking met ontwikkelingslanden* [Project list relating to bilateral projects undertaken in conjunction with Dutch financial and technological cooperation with developing countries] (The Hague 1965 -1986).
NEDECO, 'Masterplan for drainage and flood control of Jakarta' (Jakarta/The Hague 1973)
PERPAMSI (Persatuan Perusahaan Air Minum Seluruh Indonesia), 'Direktori 2000' (Jakarta 2000)
PERPAMSI/ITB, 'Enquête 2001-2003' [Survey 2001-2003] (Bandung 2003)
Petra University Surabaya, 'The heritage and impact of Dutch architecture and civil engineering in Surabaya and Malang' (Surabaya/Delft 2003)
'Pluit Polder works Jakarta Indonesia', in: *Land + Water International* 47 (1982)
Proceedings symposium lowland development in Indonesia-Jakarta 1986 (Wageningen 1987)
C.L.P.M. Pompe, W.R. van Kerkvoorden, 'Ferrocement watertowers in Indonesia', in: *Journal of Ferrocement* (1985) 1
W. Ravesteijn, *De zegenrijke heeren der wateren. Irrigatie en staat op Java, 1832-1942* [The auspicious lords of the waters, Irrigation and state in Java, 1832-1942] (Delft 1997)
W. Ravesteijn, 'Een ingenieur met visie. Prof. dr. ir. Willem Johan van Blommestein (1905-1985)' [An engineer with vision: Professor Willem Johan van Blommestein (1905-1985)], in: *Tijdschrift voor Waterstaatsgeschiedenis* 11 (2002) 1: 6-11
Water supply and sanitation decade (New York 1980)
'Reconnaissance survey Djratunseluna area' (1968)
Resolution 2626 (24th October 1970) of the General Assembly of the United Nations
W. Riedijk, *Technology for liberation* (Delft 1986)
W. Riedijk (ed.), *Appropriate technology for developing countries* (Delft 1982)
Ministerie van Buitenlandse Zaken, *De geschiedenis van 50 jaar Nederlandse Ontwikkelingssamenwerking* [The history of 50 years of Dutch development cooperation] (The Hague 1999)
A. Schrevel, *Access to water* (The Hague 1993)
A.R. Soehoed, 'Asahan, Impian yang menjadi kenyataan' (Jakarta 1983)
R. Tirtotjondro, 'Siphon bertekanan rendah Sunter-Jakarta, Laporan penyelidikan hidrolis dengan model, No. 403' (Bandung 1977)

M.N. Vivekananthan, 'Rehabilitation of irrigation systems in East Java, Indonesia', in: *13th International congress on irrigation and drainage, International Commission on Irrigation and Drainage* (Rabat 1987) 147-173
Waterbedrijf Europoort, 'Tien jaar Twinning' [Ten years of twinning] (Rotterdam/Bandung 1997
Water supply and sanitation decade (New York 1980)

Journals

De Ingenieur in Indonesië [The engineer in Indonesia] 1948/1949 t/m 1955
H2O, Tijdschrift voor watervoorziening en waterbeheer [H_2O, Journal for water supply and water management] 1969 t/m 2000

CHAPTER 10 **INDIGENOUS OR INTERNATIONAL**

W.H.G. Armytage, *A social history of engineering* (London 1976)
H. Baudet, *De lange weg naar de TU Delft* [The long road to Delft University of Technology].Part I. *De Delftse ingenieursschool en haar voorgeschiedenis.* The Delft engineering school and its past history] (The Hague 1992)
K. van Berkel, *In het voetspoor van Stevin. Geschiedenis van de natuurwetenschap in Nederland 1580-1940* [In the footsteps of Stevin. The history of natural sciences in the Netherlands 1580-1940] (Amsterdam 1985)
W. Boiten, 'De Hobrad stuw: Regelen en meten van debieten' [The Hobrad weir: The regulating and measuring of flow rates], in: *PT/Civiele techniek* 42 (1987) 1, 51-57
A.R.H. Brouwer, *Waterkracht perspectieven* [Waterpower perspectives] (Delft 1955)
R. Brouwer, 'Design and application of an automatic check gate for tertiary turnouts'. in: *13th International congress on irrigation and drainage, International Commission on Irrigation and Drainage* (Rabat 1987) 671-683
H.C.F. ten Brummelhuis, *De waterkoning. J. Homan van der Heide, staatsvorming en de oorsprong van moderne irrigatie in Siam, 1902-1990* [The water king. J. Homan van der Heide, state formation and the origin of modern irrigation in Siam, 1902-1990] (Amsterdam 1995)
C.M. Burt, R. Angold, M. Lehmkuhl and S. Styles, 'Flap gate design for automatic upstream canal water level control', in: *Journal of Irrigation and Drainage Engineering* 127 (2001) 2, 84-91
J.M. Dirkzwager, *Water – van natuurgebeuren tot dienstbaarheid* [Water – from natural phenomenon to useful resource] (The Hague 1977)
C. Disco, *Made in Delft: Professional engineering in the Netherlands, 1880 – 1940* (Amsterdam 1990)
J.A.A. van Doorn, *De laatste eeuw van Indië. Ontwikkeling en ondergang van een koloniaal project* [The last century of the East Indies. The rise and fall of a colonial project] (Amsterdam 1994)
J.J.L. Duyvendak, *De Hangende Drievoet, indrukken bij een weerzien met China* [The Hanging Tripod, impressions upon revisiting China] (Arnhem 1936)
M.W. Ertsen, 'Irrigation traditions, roots of modern irrigation knowledge', in: *International Journal of Technology, Policy and Management* 2 (2002) 4, 387-406
M.W. Ertsen, *Prescribing perfection: Emergence of an engineering irrigation design approach in the Netherlands East Indies and its legacy 1830-1990* (Rotterdam 2005)
L.A. van Gasteren, H.J. Moeshart, H.C. Toussaint and P.A. de Wilde (eds.), *In een Japanse stroomversnelling: Berichten van Nederlandse watermannen – rijswerkers, ingenieurs, werkbazen – 1872-1903* [At a Japanese, accelerated pace: Reports from Dutch water men – osier workers, engineers, employers – 1872-1903] (Zutphen 2000)
M. Groen, *Het wetenschappelijk onderwijs in Nederland van 1815 tot 1980. Een onderwijskundig overzicht* [Academic education in the Netherlands between 1815 and 1980: An educational overview]. Part 2. *Wis- en natuurkunde, letteren, technische wetenschappen, landbouwwetenschappen* [Mathematics, physics, language and literature, engineering sciences] (1988)
J. Haringhuizen, *De Indische waterbouwkunde* [East Indian hydraulic engineering] (Wageningen 1919)
J. Homan van der Heide, 'Irrigation and drainage in the lower Menam Valley' (1903)

S. Hoogewerff, C.W. Weys and R.A. Sandick, *Het leerplan der op te richten Nederlandsch-Indische Technische Hoogeschool. Verslag der Sub-Commissie, ingesteld door de Commissie van advies voor onderwijs van het Koninklijk Instituut voor Hooger Technisch Onderwijs in Nederlandsch-Indië* [The curriculum for the new Dutch-East Indian Technical College. The report of the Sub-Committee, established by the Advisory Committee for education for the Royal Institute of Technical Education in the Dutch-East Indies] (The Hague 1918)

M.L. ten Horn-van Nispen, H.W. Lintsen and A.J. Veenendaal (Jr.), *Wonderen der techniek. Nederlandse ingenieurs en hun kunstwerken. 200 Jaar civiele techniek* [The wonders of technology. Dutch engineers and their structural works: 200 Years of civil engineering] (Zutphen 1994)

L. Horst, *Irrigation water division technology in Indonesia. A case of ambivalent development* (Wageningen 1996)

Irrigation design standards (Jakarta 1986)

A.W.A. Jacometti, *Handleiding bij het onderwijs in de waterbouwkunde aan de technische scholen in Nederlandsch-Indië* [Manual for hydraulic engineering education at technical schools in the Dutch East Indies] (Weltevreden 1924)

J. de Jong, *De waaier van fortuin. Van handelscompagnie tot koloniaal imperium* [The fan of fortune: From trading company to colonial empire] (The Hague 2000)

A.F. Kamp (ed.), *De Technische Hogeschool te Delft 1905-1955* [The Technical College in Delft 1905-1955] (The Hague 1955)

J. Klopper, *Het leerplan der op te richten Nederlandsch-Indische Technische Hoogeschool. Opmerkingen naar aanleiding van het ontwerp der Sub-Commissie, ingesteld door de Commissie van advies voor onderwijs van het Koninklijk Instituut voor Hooger Technisch Onderwijs in Nederlandsch-Indië* [The curriculum for the new Dutch-East Indian Technical College. Remarks following the Sub-Committee's plans, introduced by the educational Advisory Committee for the Royal Institute of Technical Education in the Dutch-East Indies] (The Hague 1918)

M. Leidelmeijer, *Van suikermolen tot grootbedrijf* [From sugar grinder to big business] (Amsterdam 1997)

J.L.W. van Leur and R.P.J. Ammerlaan, *De Indische instelling te Delft. Méér dan een opleiding tot bestuursambtenaar* [The East Indian institute in Delft: More than an education as an administrator] (Delft 1989)

Th.D. van Maanen, *Irrigatie in Nederlandsch-Indië* [Irrigation in the Dutch East Indies] (Weltevreden 1924; Batavia 1931, 2nd edition)

J.P. Mazure, 'De verjonging van een oude techniek' [The rejuvenation of an old technique], in: A.F. Kamp (ed.), *De Technische Hogeschool te Delft 1905-1955* [The Technical College in Delft 1905-1955] (The Hague 1955) 193-212

Th. Metz, *Java-Sumatra-Bali; Über Kolonialpolitik im tropischen Holland* (Leipzig 1932)

'Ontwerp meetstuwen in het infiltratie-gebied' [Designing measuring weirs for the infiltration area] 1: 'Duinwaterwinplaats Vogelenzang: Verslag onderzoek' [Dune water-collection area Vogelenzang: Research report] (Delft 1976)

Nippon Koei, 'Way Sekampung irrigation project. Design report on Bekri and West Rumbia irrigation systems', Volume 1, 'Main report' (1996)

H. Plusquellec, C. Burt and H.W. Wolter, *Modern water control in irrigation: Concepts, issues, and applications* (Washington 1994)

J.A. Ringers, *Een eeuw Nederlandse waterbouw* [A century of Dutch hydraulic engineering] (Amsterdam 1947)

H. Schippers, *Van tusschenlieden tot ingenieurs. De geschiedenis van het hoger technisch onderwijs in Nederland* [From middlemen to engineers. The history of higher technical education in the Netherlands] (Hilversum 1989)

H. Stevens, 'De Nederlandse ingenieur' [The Dutch engineer], in: L.A. van Gasteren, H.J. Moeshart, H.C. Toussaint and P.A. de Wilde (eds.), *In een Japanse stroomversnelling: Berichten van Nederlandse watermannen – rijswerkers, ingenieurs, werkbazen – 1872-1903* [At a Japanese accelerated pace: Reports from Dutch water men – rice workers, engineers, employers 1872-1903] (Zutphen 2000)

W.K. Storey, *Science and power in colonial Mauritius* (New York 1998)

G.P. van de Ven, *Leefbaar laagland. Geschiedenis van de waterbeheersing en landaanwinning in Nederland* [Liveable lowlands: The history of water management and land reclamation in the Netherlands] (Utrecht 1993)

H.C.P. de Vos, 'De strijd om en tegen het water' [The battle for and against water], in: W.H. van Helsdingen and H. Hoogenberk (eds.), *Daar wèrd wat groots verricht.... Nederlandsch-Indië in de XXste eeuw* [Something great was achieved there ... the Dutch East Indies in the 20th century] (Amsterdam 1941) 273-286

Waterloopkundig Laboratorium, 'Meetstuwen Linge-project: Bepaling van de afvoerbetrekkingen voor twee types meetoverlaten in de Linge' [Measuring weirs in the Linge project: Determining the water conveyance rates for two types of metre overflows in the Linge] (Delft 1981)

CONCLUDING CONSIDERATIONS **WHERE THE HIGH MOUNTAINS RISE...**

M. Adas, *Machines as the measure of men. Science, technology, and ideologies of western dominance* (Ithaca/London 1989)

F. Bacon, *The New Atlantis* (1626/1995); http://oregonstate.edu/instruct/phl302/texts/bacon/atlantis.htmlG. Basalla, 'The spread of Western science', in: *Science* 156 (1967) 611-22

G. Basalla, *The evolution of technology* (Cambridge 1987)

J. Boes, W. Ravesteijn and W. Riedijk (eds), *Appropriate technology in industrialized countries* (Delft 1989)

R. Brouwer, *Irrigatie* [Irrigation] (Delft 1991)

H.C.F. ten Brummelhuis, *De Waterkoning. J. Homan van der Heide, staatsvorming en de oorsprong van moderne irrigatie in Siam, 1902-1990* [The Water King. J. Homan van der Heide, state formation and the origin of modern irrigation in Siam, 1902-1990] (Amsterdam 1995)

J.A.A. van Doorn, *De laatste eeuw van Indië. Ontwikkeling en ondergang van een koloniaal project* [The final East Indian century. The development and decline of a colonial project] (Amsterdam 1994)

N. Elias, *Het civilisatieproces. Sociogenetische en psychogenetische onderzoekingen* [The civilisation process. Socio-genetic and psycho-genetic investigations] (Utrecht 1983)

G. Gonggrijp, *Schets ener economische geschiedenis van Indonesië* [(A sketch of the economic history of Indonesia] (Haarlem 1957)

J. Goudsblom, *Het regime van de tijd* [The regime of time] (Amsterdam 1997)

J. Goudsblom, *Stof waar honger uit ontstond. Over evolutie en sociale processen* [What created hunger. On evolution and socially processes] (Amsterdam 2001)

V.L. Grottanelli, *Het leven der volken* [The life of peoples], ten parts (Baarn 1970)

D. R. Headrick, *The tools of empire. Technology and European imperialism in the nineteenth century* (New York 1981)

D. R. Headrick, *The tentacles of progress. Technology transfer in the age of imperialism* (New York/Oxford 1988)

M. Hollick, 'The appropriate technology movement and its literature: A retrospective', in: *Technology in Society* 4 (1982) 213-229

Th.P. Hughes, 'The evolution of large technological systems', in: W.E. Bijker, Th.P. Hughes and T.J. Pinch (eds.), *The social construction of technological systems. New directions in the sociology and history of technology* (Cambridge, Mass./London, England 1987) 51-82

Irrigation design standards (Jakarta 1986)

J. Jongejans, *Land en volk van Atjeh vroeger en nu* [The land and people of Atjeh of the past and today] (Baarn 1939)

A.D.A. de Kat Angelino, *Staatkundig beleid en bestuurszorg in Nederlandsch-Indië* [Political policy and administrative matters in the Dutch East Indies] (The Hague 1929)

F.M. Keesing, *Inleiding tot de culturele antropologie* [Introduction to cultural anthropology] (Utrecht/Antwerp 1965)

O. Kroesen and W. Ravesteijn, 'De toekomst als opdracht. Utopie, revolutie en techniek in Europa' [The future as an assignment. Utopia, revolution and technology in Europe], in: J. van Burg, P.A. van Gennip and E. Korthals Altes (eds.), *Europa: Balans en richting* [Europe: Balance and direction] (Tielt 2003) 137-156

J.Th. Lindblad (ed.), *Historical foundations of a national economy in Indonesia, 1890s-1990s* (Amsterdam 1996)

H.W. Lintsen, 'Wees utopisch!' [Be utopian!], in: *De Ingenieur* 21 (1999) 26-33
H.W. Lintsen (ed.), *Techniek in Nederland. De wording van een moderne samenleving 1800-1890* [Technology in the Netherlands. The creation of a modern society 1800-1890], six parts (Zutphen 1992-1995)
T.J. Mesman and Ch.J. Vos, The history of caisson construction within the Hollandsche Beton Groep nv from 1902 to 1977 (Rijswijk 1977)
W. Ravesteijn, 'Dutch engineering overseas: The creation of a modern irrigation system in colonial Java', in: *Knowledge, Technology & Policy* 14 (2002) 4, 126-144
W. Ravesteijn, 'Between globalization and localization. The case of Dutch civil engineering in Indonesia, 1800-1950', in: *Comparative Technology Transfer and Society* 5 (2007) 1, 32-64
W. Riedijk, *Technology for liberation* (Delft 1986)
W. Riedijk (ed.), *Appropriate technology for developing countries* (Delft 1982);
J. W. Schot, H. W. Lintsen, A. Rip and A.A .Albert de la Bruhèze (eds.), *Techniek in Nederland in de twintigste eeuw* [Technology in the Netherlands in the twentieth century], seven parts (Zutphen 1998-2003)
F. Spier, *The stucture of big history. From the big bang until today* (Amsterdam 1995).
W. K. Storey, 'Colonial technology: Science and the transfer of innovation to Australia', in: *Social Epistomology* 12 (1998) 135-141
H.J. Störig, *Geschiedenis van de filosofie* [History of philosophy], part 1 (Utrecht/Antwerp 1974)
E. van der Vleuten, 'Twee decennia van onderzoek naar grote technische systemen: Thema's, afbakening en kritiek' [Two decades of research into large technological systems: Themes, demarcations and criticism], in: *NEHA-jaarboek voor economische, bedrijfs- en techniekgeschiedenis* 63 (2000) 328-364
B. Witjes, F. Hüsken and J. Banning, *Indonesië* [Indonesia] (Utrecht/Zutphen 1990)

AFTERWORD

M.M. Adas, *Machines as the measure of men. Science, technology, and ideologies of western dominance* (Ithaca/London 1989)
B.R.O'G. Anderson, *Language and power. Exploring political cultures in Indonesia* (Ithaca/London 1990)
H. ten Brummelhuis, *King of the waters. Homan van der Heide and the origin of modern irrigation in Siam* (Leiden 2005)
Ch.G. C. Cramer, *De ingenieur in Ned.-Indië op technisch en sociaal gebied* [The engineer in the Dutch East Indies in technological and social fields] (Amsterdam 1914)
H.W. van den Doel, *Het rijk van Insulinde. Opkomst en ondergang van een Nederlandse kolonie* [The empire of Insulinde. The rise and fall of a Dutch colony] (Amsterdam 1996)
J.A.A. van Doorn, *De laatste eeuw van Indië. Ontwikkeling en ondergang van een koloniaal project* [The last century of the East Indies. The rise and fall of a colonial project] (Amsterdam 1994)
C. Geertz, *Agricultural involution. The processes of ecological change in Indonesia* (Berkeley 1963)
A. Gordon, 'The collapse of Java's colonial sugar system and the breakdown of independent Indonesia's economy', in: F. van Anrooij et al. (eds.), *Between people and statistics. Essays in modern Indonesian history presented to P. Creutzberg* (The Hague 1979) 251-265
W.H. van Helsdingen and H. Hoogenberk (eds), *Daar wèrd wat groots verricht .... Nederlandsch-Indië in de XXste eeuw* [Great things were achieved there...The Dutch East Indies in the 20th century] (Amsterdam 1941)
W.H. van Helsdingen and H. Hoogenberk (eds), *Mission interrupted. The Dutch in the East Indies and their work in the XXth century* [abridged English edition by J.J.L. Duyvendak] (Amsterdam 1945)
A. Heryanto, 'The development of "development"', in: *Indonesia* 46 (1988) 1-24
J. Homan van der Heide, *Beschouwingen aangaande de volkswelvaart en het irrigatiewezen op Java in verband met de Solovalleiwerken* [Considerations concerning public health and irrigation affairs in Java in connection with the Solo Valley Works] (Batavia/The Hague 1899)
H.H. van Kol, 'Ontwerp-program voor de Nederlandsche koloniale politiek' [A draft program for Dutch colonial policy], in: *De Nieuwe Tijd*, 1 April 1901

R. Mrázek, *Engineers of happy land: Technology and nationalism in a colony* (Princeton, N.J. 2002)
P.J. Ott de Vries, 'De werkkring en taak van den ingenieur in Nederlandsch-Indië' [The position and task of the engineer in the Dutch East Indies], in: *De Ingenieur* 37 (1922) 37, 736-744

APPENDIX 9

S.H.A. Begemann, 'Toepassing van de waarschijnlijkheidsleer op hydrologische waarnemingen' [Application of the theory of probability to hydrologic data], in: *De Waterstaats- Ingenieur* 1931: 2 onwards, 55 onwards and 103 onwards
R. Clément, 'Calcul des débits dans les réseaux d'irrigation fonctionnant à la demande', La Houille Blanche, no. 5, 1966
Grondslagen voor de samenstelling van begrootingen ten dienste van het Departement der Burgerlijke Openbare Werken in Nederlandsch-Indië [Fundamentals for drafting estimates for the use of the Department of Civil Public Works] (Batavia 1885)
P. Honig and F. Verdoorn, 'Science and scientists in the Dutch East Indies', Board for the Dutch East Indies, Surinam and Curaçao (New York City, 1945)
D.A. Kraijenhoff van de Leur, 'Run-off models with linear elements', in: *Recent trends in hydrograph synthesis*, Proceedings of technical meeting 21 of the Committee for Hydrological Research TNO (The Hague 1966)
J.E. de Meyier, 'Gedenkboek, uitgegeven ter gelegenheid van het vijftigjarig bestaan van het Koninklijk Instituut van Ingenieurs, 1847-1897' [Commemorative book published on the occasion of the fiftieth anniversary of the Royal Institute of Engineers, 1847-1897]
J. Nugteren, 'On irrigation efficiencies', Publication 19 of the International Institute for Land Reclamation and Improvement (ILRI) (Wageningen 1974)
H.J. Schoemaker, 'Wateraanvoer, verdeling en afvoer' [Lecture notes on water supply, distribution and discharge], Delft University of Technology (1980)
A.F. Wehlburg, The diaries of an engineer (on irrigation and transmigration), Library of the Royal Tropical Institute (Amsterdam).

Authors and editors

MICHEL BAKKER, M.A. (1958) studied architectural history and classical archaeology at Leiden University where he specialised in early Christian architecture and the structural ground plans of Pompeii. Between 1989 and 1997 he was employed by the Municipal Monument Conservation Bureau in Amsterdam. He has a research and advisory consultancy for architectural history, structural history and archaeology: Bureau Bakker in Heemstede. Michel is also affiliated to the national Cultural Heritage Section and to the International Council of Monuments and Sites (ICOMOS). Since 1998 and in conjunction with monument conservation he has carried out research in Aruba, the Dutch Antilles and Surinam. Within the Dutch Bridges Association he is also editor of the *Bruggen* [Bridges] periodical. He has published a number of works including 'Many waters to bridge' (in: *The Low Countries*), 'Bruggen van hout' [Wooden bridges] (in: *Bruggen in Nederland* [Bridges in the Netherlands]) and *De Mirakelbrug in Amsterdam* [The Miracle Bridge in Amsterdam].

Dr. MAURITS W. ERTSEN (1968) graduated in 1993 from the Agricultural University of Wageningen in the field of tropical agricultural technology. He gained his doctorate in 2005 with the thesis entitled *Prescribing Perfection: Emergence of an Engineering Irrigation Design Approach in the Netherlands East Indies and its Legacy 1830-1990*. He is currently employed as a lecturer at the Faculty of Civil Engineering (departments of Civil Business Administration and Land and Water Management). He has published various papers and articles on the scientific aspects of Indonesia's irrigation history.

DR. MARIE-LOUISE TEN HORN-VAN NISPEN (1944) read sociology and social history at the University of Tilburg where she gained her doctorate in 1971 on the basis of research into *Jan B.M. van Besouw, een sociaal geïnspireerd ondernemer rond 1900* [Jan B.M. van Besouw, a socially inspired entrepreneur of around 1900]. After doing a considerable amount of freelance research she became affiliated to Delft University of Technology in1989 where she has since lectured within the History of Technology Department and, in recent times, within the Technology Dynamics and Sustainable Development Department in the Technology, Policy and Management Faculty. Her specialist field is the history of public works and traffic infrastructure. She has contributed to publications such as *Leefbaar laagland* [Liveable Netherlands], *Nederlandse ingenieurs en hun kunstwerken* [Dutch engineers and their structural works], *Geschiedenis van Noord-Brabant* [The history of North-Brabant], *Twee eeuwen Rijkswaterstaat* [Two centuries of the Dutch Public Works agency] and part 1 of the *Geschiedenis van de techniek in Nederland in de 20ste eeuw* [The history of technology in the Netherlands in the 20th century]. She is currently working on the biographies of various public works engineers. Since 1998 she has been on the assessment commission for the Registration of Rail-Linked Historic Objects (Mobile Collection of the Netherlands, Stabien). In 2001 she became secretary of the Public Works History Society.

FRIDA DE JONG, M.A. (1951) graduated from Utrecht University in the fields of social and economic history. After having completed various research projects she was appointed senior lecturer within the History of Technology Department at Delft University of Technology. She has published papers and articles on such subjects as: the education of engineers, industrial research, women at DUT and student life.

PROFESSOR JOHANNES H. KOP, MSc CEng FICE (born in 1930 in Djatiroto, East Java) was educated in Sumatra and the Netherlands. During the Second World War he was detained in concentration camps. He graduated in the field of civil engineering in 1957 and his subjects were irrigation and hydropower. He worked as an irrigation engineer on the Ganges-Kobadak project in Bangladesh under the supervision of Professor W.J. van Blommestein (the 'father' of the 1948 West Java welfare plan) and he was head of the Architecture, Construction, Hydraulic Engineering and Sanitary Engineering department at the Dutch consultancy firm Grontmij NV during the time of the creation and partial execution of the NEDECO Master Plan for the Drainage and Flood Control of Jakarta. From

1980 till 1985 he was head of the Planning Bureau for the Association of Water Companies in the Netherlands (VEWIN). In 1984 he led a delegation to Indonesia in conjunction with the International Directorate for Technological Aid of the Ministry of Foreign Affairs which was created for the purposes of selecting irrigation areas that were to be renovated and newly developed. From 1985 until the time of his emeritus in 1994 he was professor of sanitary engineering – particularly in relation to public drinking water supply matters – at Delft University of Technology. He has published numerous lecture notes, articles and papers in the fields of sanitary engineering, environmental engineering, irrigation, water management and urban drainage and flood control.

DR. WIM RAVESTEIJN (1954) is a senior lecturer affiliated to the Technology Dynamics and Sustainable Development Department at Delft University of Technology (within the Technology, Policy and Management Faculty). He conducts research and lectures in the fields of technology history and technology dynamics as well as on the sociology of labour, organisation and economic order. He graduated from the University of Amsterdam in the fields of anthropology and non-western sociology. His specialist research field is water management from an historical and international perspective. He has published works on: appropriate technology, organisation culture, the educating of engineers, the history of television, the development of modern irrigation and other public works in the Dutch East Indies/Indonesia, the history of Europe and Dutch versus European water management. He gained his doctorate in 1997 with a publication entitled *De zegenrijke heeren der wateren. Irrigatie en staat op Java, 1832-1942* [The auspicious lords of the waters. Irrigation and state in Java, 1832-1942], an English translation of which is forthcoming. Together with other academics he edited a special on water systems for the American journal *Knowledge, Technology and Policy* in 2002.

PAULINE K.M. VAN ROOSMALEN, M.A. (1964) graduated from Amsterdam's Free University (the VU) in the fields of art history and archaeology where, in 1992, she rounded off her master's degree by specialising in architectural history. Since January 2000 she has been employed by Delft University of Technology as an assistant researcher within the Faculty of Architecture where she is currently working on her doctoral dissertation which focuses on *Urban Development in the Dutch East Indies (1905-1950)*. She has published various articles and papers on architecture and urban development, including in Indonesia, and has had articles placed in *ArchiNed*.

DR. AUGUSTUS J. VEENENDAAL JR. (1940) read history at Utrecht University. In 1976 he gained his doctorate from Nijmegen's Radboud University. Until recently he was a senior researcher at the Institute for Social History in The Hague and also a part-time company historian for the Dutch Railway Company (NS). He has written a number of books and some twenty articles on the history of the railways in the Netherlands, the USA and South Africa.

ARJAN J.A. VEERING, M.A. (1969) studied social history at the Erasmus University in Rotterdam within the Faculty of History and Arts. He is currently working on a doctoral thesis entitled *Tanjung Priok, History of a Java Sea Port* and in that connection he was affiliated to the Free University in Amsterdam. He is also a freelance writer, his specialist fields being the environment and water management.

A list of the illustration sources

INTRODUCTION

14 KITLV/Royal Netherlands Institute of Southeast Asian and Caribbean Studies, Leiden [5857]
18 Royal Tropical Institute, Amsterdam (KIT) coll. no. 60016856
22 H.M. de Vries, *The Importance of Java seen from the air* (Batavia 1928) 20
26 *De Ingenieur in Nederlandsch-Indië* 1 (1934) 5, V
28 *10 Vervlogen jaren. Uitgegeven ter gelegenheid van het tienjarig bestaan van de Koninklijke Nederlandsch-Indische Luchtvaartmaatschappij. 1 November 1928-1938* [10 Bygone years. Published on the occasion of the tenth anniversary of the Royal Dutch East Indian Air Force. 1st November 1928-1938] (Batavia 1938)
31 Photo taken by W. Ravesteijn
33 KIT coll. no. 10007726
35 *De Waterstaats-Ingenieur* 8 (1920) 10, plate 1
36 Photo taken by W. Ravesteijn
42 Photo taken by J.H. Kop
44-45 Map drawn by D. v/d Spek

CHAPTER 1

46 KITLV 32349
48 left *Tijdschrift van het Koninklijk Instituut van Ingenieurs* (*Verhandelingen*) 1886/87
48 right *De Ingenieur* 27 (1912) 39, 786
49 left *De Ingenieur* 33 (1918) 9, 153
49 right *De Ingenieur 36* (1921) 20, 372
54 Photo taken by M. Ertsen
55 Photo taken by M. Ertsen

CHAPTER 2

68 KITLV 7375
70 Iconographic Bureau, The Hague
72 National Archive (NA). Bik Coll., N 302. additional no. 2.21.024, inv. no. 28
76 KITLV 7129
77 *De Ingenieur* 30 (1915) 40, 815
81 *De Ingenieur* 46 (1931) 40, A.413
87 KITLV 26335
90-91 Map drawn by D. v/d Spek

CHAPTER 3

92 Collection of the Dutch Association for Rail and Tramway Enthusiasts (NVBS)
106 NVBS collection
107 Picture postcard, NVBS collection
108 Picture postcard, NVBS collection

110 Map drawn by D. v/d Spek
115 Map drawn by D. v/d Spek
120 Photo taken by L.J. Biezeveld, NVBS collection
123 Picture postcard, NVBS collection
127 Picture postcard, NVBS collection
129 Map drawn by D. v/d Spek
130 Picture postcard, NVBS collection
132 Picture postcard, NVBS collection
134-135 Map drawn by D. v/d Spek

CHAPTER 4

136 KIT coll. no. 60011310
138 P.P. Bijlaard, *Vrije uitbouw, uitgevoerde werken* [Free extension, the completed works] (Bandoeng 1933) 9
143 A. Kingma collection
147 H. van Maarschalkerwaart collection
148 P.P. Bijlaard, *Vrije uitbouw, uitgevoerde werken* [Free extension, the completed works] (Bandoeng 1933) 6
149 above S.A. Reitsma, *Gedenkboek der Staatsspoor- en Tramwegen in Nederlandsch-Indië, 1875-1925* [Commemorative book of the State Railways and Tramways in the Dutch East Indies 1875-1925] (Weltevreden 1925) 123
149 below J. van Velzen, 'De ijzeren brug over de Kediri-rivier, ter hoofdplaats van de Residentie Kediri' [The iron bridge over the Kediri river at the central point in the Kediri Residence], in: *Tijdschrift van het Koninklijk Instituut van Ingenieurs. Afdeeling Nederlandsch-Indië* (1877-1878) plate VIII
150 KITLV 27713
151 W. de Jongh, 'De overbrugging van het Tjisaät-ravijn in den Staatsspoorweg Tjitjalenka-Garoet' [The bridging of the Tjisaät ravine along the Tjitjalenka-Garoet State Railway line], in *Tijdschrift van het Koninklijk Instituut van Ingenieurs. Afdeeling Nederlandsch-Indië* (1888-1889) page 1
153 *De Ingenieur* 18 (1903) 19, 314
156 left *De Ingenieur* 18 (1903) 19, 319
156 right *De Ingenieur* 18 (1903) 19, 319
157 above *De Ingenieur* 18 (1903) 19, 320
157 below *De Ingenieur* 32 (1917) 4, 61
158 M. van Ballegoijen de Jong collection
161 *De Ingenieur* 49 (1934) 32, B. 118
162 *De Ingenieur* 49 (1934) 32, B. 117
164 above left and below P.P. Bijlaard, *Vrije uitbouw, uitgevoerde werken* [Free extension, the completed works] (Bandoeng 1933) 24
164 above right P.P. Bijlaard, *Vrije uitbouw, uitgevoerde werken* [Free extension, the completed works] (Bandoeng 1933) 47
170 left *De Ingenieur in Nederlandsch-Indië* 7 (1940) 8, I.151
170 right *De Ingenieur in Nederlandsch-Indië* 7 (1940) 8, I.150
171 *De Ingenieur in Nederlandsch-Indië* 7 (1940) 8, I.152
174 KITLV 334979
176 The Military Museum in Vught
177 *Tijdschrift van het van het Koninklijk Instituut van Ingenieurs. Afdeeling Nederlandsch-Indië* (1893-1894) Appendix B, V-VI
178 above *De Ingenieur* 12 (1897) 11, 126
178 below *De Ingenieur* 12 (1897) 11, 122
179 *De Ingenieur* 12 (1897) 11, 122
180 *De Ingenieur in Nederlandsch-Indië* 8 (1941) 1, II, plate I

181 above *De Ingenieur in Nederlandsch-Indië* 8 (1941) 1, II, plate I
181 below *De Ingenieur in Nederlandsch-Indië* 6 (1939) II.176
183 KITLV [
188 *Pioniers over Zee. De geschiedenis van het Bataljon Genietroepen van de Eerste Divisie (7 December/C-Divisie), 1945-1950* [Pioneers overseas. The history of the Engineering Batallion of the First Divison (s.l.e.a.) 283
190 Departemen Pekerjaan Umum Direktorat Jenderal Bina Marga

CHAPTER 5

192 The World Museum in Rotterdam
196 Map drawn by D. v/d Spek
205 Map drawn by D. v/d Spek
210 KITLV 3287
217 KITLV 38679
218 W. Cool, *Nederlands Indische havens* [Dutch East Indian harbours] (Batavia 1920)
219 W. Cool, *Nederlands Indische havens* [Dutch East Indian harbours] (Batavia 1920)
223 W. Cool, *Nederlands Indische Havens* [Dutch East Indian harbours] (1920)
224 W. Cool, *Nederlands Indische havens* [Dutch East Indian harbours] (Batavia 1920)
225 Annual report on the Tandjongpriok harbour 1918 (Weltevreden 1919)
229 The Maritime Museum in Rotterdam
231 W. Cool, *Nederlands Indische havens* [Dutch East Indian harbours] (Batavia 1920)

CHAPTER 6

238 Photo taken by W. Ravesteijn
249 Photos taken by W. Ravesteijn
250 Photo taken by W. Ravesteijn
255 *De Ingenieur in Nederlandsch-Indië* 7 (1941) 6, II plate II
256 D.G. Romijn, 'Een regelbare meetoverlaat als tertiaire aftapsluis' [An adjustable measuring weir as a tertiary outlet], in: *De Waterstaats-Ingenieur* 20 (1932) 9, 287-292
263 *De Ingenieur in Nederlandsch-Indië* 1 (1934) 2, III, plate I
264 above *De Ingenieur in Nederlandsch-Indië* 7 (1940) 9, II.145
264 below *De Ingenieur in Nederlandsch-Indië* 7 (1940) 9, 11, plate VI
266 Photo taken by W. Ravesteijn
270-271 Map drawn by D. v/d Spek

CHAPTER 7

272 KIT, Tillema Collection, inv. no. 10017598
280 H.F. Tillema, *Van wonen en bewonen. Van bouwen, huis en erf* [On living and inhabiting. On building, house and property] (Samarang 1913) plate VI
281 left KIT, Tillema Collection, inv. no. 10017351
281 right KIT, Tillema Collection, inv. no. 1001477
282 NA, Ministry for the Colonies, Archive 2.10.036.04, inv. no. 1124
283 Dutch Architectural Institute, De Bazel Archive, inv. no. 1158, negative 18498
284 above left KIT
284 above right KIT
284 below KIT
291 above KIT
291 below KIT, Collection of Java Maps, inv. no. 3844
292 above KITLV

292 below left KITLV
292 below right KITLV
294 above left ANRI, Archive of the Arsip Depo Bogor Register (1891-1942) – BGS 6-3-1924, inv. no. 562
294 above right ANRI, Archive of the Arsip Depo Bogor Register (1891-1942) – BGS 6-3-1924, inv. no. 562
294 below left ANRI, Archive of the Arsip Depo Bogor Register (1891-1942) – BGS 6-3-1924, inv. no. 562
294 below right ANRI, Archive of the Arsip Depo Bogor Register (1891-1942) – BGS 6-3-1924, inv. no. 562
296 KIT
299 ANRI, Archive of the Inland Administration Department, inv. no. 1686
302 ANRI, Inventory of the archive of the General Secretary and the Cabinet of the Governor-General 1944-1950, inv. no. 928

CHAPTER 8

308 Photo taken by J.H. Kop
315 left J.A. van der Kloes and J.N. van Ruijven, *Het bouwen in overzeesche gewesten, Tweede Afdeeling, Onderwerpen uit de Weg- en Waterbouwkunde* [The building in overseas territories, Second division, Subjects from civil engineering] (Leiden 1906) 715
315 right *Gedenkboek Koninklijk Instituut van Ingenieurs 1847-1897* [Commemorative book of the Royal Institute of Engineers 1847-1897] (The Hague 1897), plate LXXX
318 *De Waterstaats-Ingenieur* 5 (1917) 8, plate II
322 left *De Ingenieur in Nederlandsch-Indië* 1 (1934) 3, VI
322 right *De Ingenieur* (1934) 34, B. 136
323 *De Ingenieur* (1934) 34, B. 140
330 *De Ingenieur* 38 (1923) 25, 484
332 *De Ingenieur in Nederlandsch-Indië* 3 (1936) 5, VI
336 *De Ingenieur* 38 (1923) 27, 544
338 *De Ingenieur* 18 (1903) 48, 827
340 G.F.E. Gonggryp, *Geïllustreerde encyclopaedie van Nederlandsch-Indië* [Illustrated encyclopaedia of the Dutch East Indies] (Leiden 1934)
341 *De Ingenieur* (1904) 45, 806
342 left Bos atlas (Groningen/Batavia 1947) 34C
342 right *De Ingenieur* 38 (1923) 25, 489
343 Photo taken by J.H. Kop

CHAPTER 9

344 Photo taken by J.H. Kop
350 Photo taken by J.H. Kop
352 Photo taken by J.H. Kop
353 Photo taken by J.H. Kop
354 TU Delft
359 above Photo taken by J.H. Kop
359 below Photo taken by J.H. Kop
361 Photo taken by J.H. Kop
363 above Photo taken by A.D. Maier (engineer)
363 below Photo taken by A.D. Maier (engineer)
369 left Photo taken by J. de Lange (engineer)
369 right Photo taken by J.H. Kop
371 IWACO photo

372 IWACO, West Java Rural Water Supply Project (July 1985)
373 Photo taken by J.H. Kop
375 Study report on East Jakarta Flood Control Project (Dec. 1985) 150
378 left *Land + Water International* 1982, 47, 21
378 right Photo taken by J.H. Kop
379 Photo taken by J.H. Kop

CHAPTER 10

380 Photo taken by M. Ertsen
382 Photo taken by M. Ertsen
384 Photo taken by M. Ertsen
388 *De Ingenieur* 35 (1920) 38, 693
389 *De Ingenieur* 35 (1920) 38, 692
397 Photo taken by R. Brouwer
399 Photo taken by M. Ertsen

CONCLUSION

402 KIT coll. no. 10007500
405 Photo taken by W. Ravesteijn
407 Photo taken by M. Ertsen
409 T.J. Mesman and C.J. Vos, The history of caisson construction within the Hollandsche Beton Groep nv from 1902 to 1977 (Rijswijk 1977) 43
411 KIT coll. no. 10007320
413 KIT coll. no. 10007684
416 *De Ingenieur* 18 (1903) 19, 318
423 *De Ingenieur in Nederlandsch-Indië* 7 (1940) 9, II, plate VI
424 Photo taken by W. Ravesteijn
425 Photo taken by W. Ravesteijn

AN INDONESIAN VIEW

426 Photo taken by J.H. Kop
428, 432, 433, 434, 435, 438, 439, 442, 443, 446 and 450 Photos and maps from the Petra Christian University of Surabaya research team led by Harry Patmadjaja, MT.

AFTERWORD

454 Photo taken by J.H. Kop

Index of persons

Glossary of words

Malay / Bahasa Indonesia – English

air (aer) belanda	'Dutch water', soda water
air bersih	clean water
air minoem/air minum	drinking water
adat	traditional rights
aloon aloon/alun alun	square
atap	roof, leaf of the Nipa palm
bajam	type of wood
bamboe/bambu	bamboo (panicum montanum)
bandeng	vegetation-eating fish
bandjir/banjir	river flood, torrent, inundation
bangkirai	type of wood
banting	sailing boat (two-masted, from Atjeh)
bersiap	be prepared, to spring into action to prepare oneself (synonymous with the bloody uprising of the Indonesian resistance in 1945)
boeijanger	labourer without a family
brondjong/bronjong	gabion
dessa/desa	village, also a village-like district in or near a city
djalan/jalan	road, street
djati/jati	type of wood, teak
djoeragan/juragan	ship's captain, batik company owner
gantang	capacity/measure of volume of 3.125 kg
goenoeng/gunung	mountain
gotong	carry/bear
gotong royong	solidarity, mutual aid
grobak	two-wheeled freight cart
ibukota	capital
idjoek/ijuk	fibre of the Aren palm, the fibrous fabric, fibre cloth
kabupaten	regency, administrative unit within a province (district)
kali	river
kampong/kampung	village, also a village-like district near or in a city
kecamatan	sub-district
kepala timah	tin head, type of fish
kerdja/kerja	work
kerjasamaan	cooperation
kota	city
kotamadya	provincial capital
kraton/keraton	palace of a Javanese ruler
kroëng	river
kroman	proa, prahn, pirigue

ladang	non-irrigated field
lahar	lava, stream of volcanic mud
loerah/lurah	village chief
maidan	square
mandi	bathe, to take a bath
mandibak	water tank in the bathroom
mantri	overseer
margaraden	town council (Sumatra)
moesim/musim	season, monsoon
musim penghujan	rainy season, wet monsoon (in Java the west monsoon)
musim kering	dry period, dry monsoon (in Java the east monsoon)
musholla	prayer house
oeloe oeloe/ulu ulu	water master
Orde Baru	the New Order, the New Era (Soeharto)
Orde Lama	the Old Order, the Old Era (Soekarno)
pasang	flood
pasanggrahan	lodging house
pasar	market
pekarangan	property, non-irrigated land
pembangunan	development, building
perahu	proa, prahn, pirigue
proyek pasang surut	tidal-reliant irrigation-cum-drainage project
saluran	ditch, canal
sampan	proa, Chinese jonk
sassak/sasak	interweaving
sawa/sawah	irrigated field, paddy field
selamatan	religious meal
slokan/selokan	ditch, trench
soengei (soengai)/sungai	river
soeroet/surut	ebb
soebak/subak	Balinese water board
tali	string (tali cloth = cloth made of woven idjoek)
tegalan	dry field, non-irrigated land
tempo doeloe/ tempo dulu (dahulu)	in (good) olden times
tji/ci	river
tjikar/cikar	ox-cart, ox-drawn wagon
tjoenia/cunia	proa, prahn, pirigue
wadoek/waduk	stomach, water reservoir